Nanotechnology in Dermatology

Adnan Nasir • Adam Friedman
Steven Wang
Editors

Nanotechnology in Dermatology

Springer

Editors
Adnan Nasir, MD, PhD
Assistant Professor Dermatology
UNC School of Medicine
3100 Thurston Bowles Bldg
Chapel Hill, NC 27554

Steven Wang, MD
Dermatology Service
Memorial Sloan-Kettering
Cancer Center
160 E 53rd Street
New York, NY 10022

Adam Friedman, MD, FAAD
Assistant Professor of Dermatology
Assistant Professor of Physiology
and Biophysics
Director of Dermatologic Research
Associate Residency Program Director
Albert Einstein College of Medicine
111E 210th Street
Bronx, NY 10467

ISBN 978-1-4614-5033-7 ISBN 978-1-4614-5034-4 (eBook)
DOI 10.1007/978-1-4614-5034-4
Springer New York Heidelberg Dordrecht London

Library of Congress Control Number: 2012954290

Printed on acid-free paper

Springer is part of Springer Science+Business Media (www.springer.com)

Preface

Dermatology is at the leading edge of discoveries related to nanotechnology. The skin is the first point of contact of nanomaterials, and many of the greatest advances of nanotechnology over the past decade have taken place in dermatology. Nanotechnology represents a convergence of many scientific disciplines, and in nanodermatology these include biology, chemistry, physics, engineering, information science, pharmacology, and manufacturing. In dermatology, nanotechnology discoveries, tools, and techniques include consumer products, novel drug delivery methods, and diagnostic devices. The outcome of these advances is already being seen in more precise and targeted methods of skin health maintenance, skin disease prevention, earlier and more accurate skin disease diagnosis, and more effective disease management. These advances have crossed an inflection point, and are expected to grow exponentially over the coming decade. The successful implementation and integration of nanotechnological innovations has challenged regulatory and public policy officials trying to keep abreast of a rapidly moving target.

Our aim in this book is to bring together the most current understanding of nanotechnology in dermatology in a format that is clear and succinct and one that encompasses the many facets of nanodermatology in a manner which is useful to clinicians, basic scientists, biologists, physiologists, pathologists, industry experts, public safety and regulatory policy experts. Ideally, the broad approach taken in this text will enable stakeholders from a variety of overlapping fields to have the necessary fundamentals to delve into their particular area of interest in greater depth while gaining a solid foundation in related subjects outside their area of expertise. We hope that this approach will serve as a springboard for readers to further their work, whether it is grasping the fundamentals of nanotechnology in dermatology, advancing their own research, or broadening their horizons through collaborations with experts in related fields. The ultimate goal, of course, in addition to the advancement of science, is to improve the quality of life of consumers and patients.

Chapel Hill, NC, USA — Adnan Nasir
Bronx, NY, USA — Adam Friedman
New York, NY, USA — Steven Wang

Introduction

Nanotechnology is the study of materials in the size range of 100 nm or less. It has made great strides in materials science, electronics, engineering, consumer goods, and medicine. Nanotechnology traces its origins to physics in the 1950s. It capitalizes on the unique properties of matter on this scale. For example, materials such as glass, which shatter in bulk form, are flexible in the nanometer scale. Nano counterparts of graphite and table salt—which are brittle insulators—become malleable and taffy-like conductors. It is the rational design and precise synthesis of this new class of matter which defines nanotechnology and which offers so much promise for medicine and dermatology. The confluence of the various disciplines involved has reached an inflection point of growth in clinical applications.

In addition to its composition, physical and chemical properties, the shape of matter determines its behavior in the nanometer size range. Crystals of titanium dioxide are less reactive in one configuration (rutile) than another (anastase). Antigens may be more or less immunogenic and RNA aptamers may have differing binding affinities for ligands based upon their 3-dimensional conformation. The surface to volume ratio of nanomaterials increases as particle size shrinks, and this increase is exponential. If the surface groups of nanomaterials are reactive with their environment, for example, by perforating cell walls of bacteria or by sequestering contact allergens such as nickel, the reactivity per particle increases with shrinking size.

The nanoscale is an ideal scale for the exploration of interface between technology and biology. Many of the processes central to life occur on the nanoscale, which is larger than an atom and smaller than a microbe. While naturally occurring nanomaterials—volcanic ash, viruses, tattoo pigments—have been in existence for millennia, some of the most significant materials have been discovered quite recently, and the preponderance of advances have been made in dermatology.

Among the largest patent holders of nanotechnology are pharmaceutical companies and cosmetics companies. Sunscreens incorporating nanomaterials are in widespread use. Emollients containing biocompatible lipids have been formulated in nanoparticles for the management of atopic dermatitis. Antioxidants such as ascorbic acid have been stabilized in nanoparticles to retain and maintain their activity in topical applications. Volatile antimicrobial compounds such as nitric oxide are being stabilized for drug delivery in the treatment of acne, wound infection, and for wound healing. Topical steroids in nanostructured lipid carriers can localize in the epidermis or passively migrate to sites of inflammation, to maximize their efficacy and minimize their side effects. Nanoparticles targeted with high affinity ligands and receptors have been used to deliver drugs and gene inhibitors for the treatment of inflammatory skin disease, genodermatoses, and malignancies such as melanoma. Micro- and nano-diameter needles are being used to deliver drugs and vaccines to precise depths in the skin for optimal effect or to sample the skin in a minimally invasive and painless fashion. Biosensors using carbon nanotubes, quantum dots, nanocantilevers, and gold nanoparticles are being used singly and in arrays for the detection of skin infection, and for mapping genomes and proteomes from infinitesimal quantities of substrate with high sensitivity and specificity. Combination agents, called theranostic agents, are being investigated to simultaneously diagnose and treat skin disease. For example, biologics can be used as nanoparticles alone, or in conjunction with inorganic imaging molecules to visualize, target, and treat skin disease in one step.

Despite the tremendous growth of nanotechnology in medicine, training among dermatologists in nanoscience is fundamentally lacking. Clinicians faced with patient questions about nanotechnology, whether regarding consumer products, such as sunscreens, or novel diagnostic or therapeutic modalities, lack adequate or accurate information to properly address their concerns. As the pace of innovation accelerates, more clinicians and scientists with a knowledge of nanodermatology will be called upon to provide counsel in the intellectual property, regulatory policy, and public health dimensions of nanotechnology. A deeper collaboration between bench and bedside will be needed to point the way toward fruitful avenues of research and development. A number of formal and informal surveys have revealed significant gaps in education on nanotechnology even among major dermatology teaching programs. Despite dermatology's central role in nanoscience, medical specialties such as oncology, cardiology, and radiology have surpassed dermatology in implementing nanotechnology teaching and research in their training programs.

Changes in healthcare are isolating dermatology from its peers in the basic and clinical sciences. Dermatology training programs are becoming delocalized from hospital campuses in favor of community-based clinics. Dermatologists are attending fewer hospital-based patients and conferences. If this trend continues, dermatology will lag in its own "stovepipe" of knowledge and lore, without the benefit of—or perhaps through the redundant duplication of efforts among—neighboring disciplines. This geographic isolation is in danger of stifling interdisciplinary interaction and collaboration, both of which are acutely needed in advancing medicine, particularly nanodermatology. In numerous surveys, dermatology faculty and trainees nearly unanimously agree on the importance of nanotechnology for dermatology, and on the need for further education, research, and funding in nanotechnology.

Nanotechnology offers tremendous opportunities for dermatology. Only by incorporating nanotechnology into training curricula and fostering teaching, research, and collaboration using the latest advances in nanotechnology for patient care will dermatology training programs prepare for the future of nanomedicine. This book aims to be a vade mecum for both novices and experts. The convergence of diverse scientific disciplines has led to a golden era of nanotechnology, and the judicious incorporation of these advances in dermatology will lead to a golden era of nanodermatology.

Acknowledgements

This book is dedicated to our families, Angela and Sofia; Oliver and Sarah; and Judy. We thank them for being pillars of patience, strength, and selflessness throughout the course of this project. Maria Smilios has been a tireless and dedicated shepherd of this project with an infectious wit and a sharp eye for detail. Rebekah Amos has been a Delphic champion of nanodermatology, keenly divining its importance at around the same time as the editors.

Contents

Contributors

Zahra Afraz, MD, PhD Department of Dermatology and Allergy, Charité-University Medicine Berlin, Berlin, Germany

Karin Blecher, MD Department of Dermatology, Montifiore Medical Center, Bronx, NY, USA

Andrew R. Binder, PhD Department of Communication, North Carolina State University, Raleigh, NC, USA

Behazine Combadière Laboratory of Immunity and Infection, Institut National de Santé et de Recherche Médicale, INSERM UMR-S 945 and Université Pierre et Marie Curie (UPMC—Paris 6), Paris, France

David L. Chen, B.Sc Department of Dermatology, Northwestern University, Chicago, IL, USA

Lucy L. Chen, BA Dermatology Service, Memorial Sloan-Kettering Cancer Center, New York, NY, USA

Zoe Diana Draelos, MD Consulting Professor of Dermatology, Department of Dermatology, Duke University School of Medicine, High Point, NC, USA

Lindsey A. Hunter-Ellul, MD Department of Internal Medicine, University of Texas/Medicine Branch, Galveston, TX, USA

Dirk M. Elston, MD Ackerman Academy of Dermatopathology, New York, NY, USA

Luis Joaquim Pina da Fonseca, PhD BioEngineering Department, Instituto Superior Técnico—Technical University of Lisbon, Lisbon, Portugal

Adam Friedman, MD, FAAD Department of Medicine, Division of Dermatology, Albert Einstein College of Medicine, Bronx, NY, USA

Anthony Gaspari, MD Department of Dermatology, University of Maryland Medical Center, Baltimore, MD, USA

Julian P. Hewitt, BA JPH SunCare Technologies Ltd, Durham, England, UK

Peter D. Kaplan, PhD Montclair, NJ, USA

Julie K. Kim, MD Department of Dermatology, Duke University Medical Center, Durham, NC, USA

Sancy A. Leachman, MD, PhD Department of Dermatology, University of Utah Hospital, Salt Lake City, UT, USA

Puiyan Lee, PhD Department of Surgery, The University of Hong Kong, Hong Kong, SAR, China

Xuelai Liu, MD Department of Surgery, The University of Hong Kong, Hong Kong, SAR, China

Alison C. Lowe, MS University of Texas Medical Branch School of Medicine, Galveston, TX, USA

Howard I. Maibach, MD Department of Dermatology, University of California, San Francisco, CA, USA

Maria da Luz Martins, MD Instituto de Higiene e Medicina Tropical, Laboratório de Micologia, Lisbon, Portugal

Jeffrey K. Mills, Ph.D., Esq. Fanelli Haag PLLC, Washington, DC, USA

Philip J. Moos, PhD Department of Pharmacology and Toxicology, University of Utah, L.S. Skaggs Pharmacy, Salt Lake City, UT, USA

Adnan Nasir, MD, PhD Department of Dermatology, UNC School of Medicine, Chapel Hill, NC, USA

Kelly C. Nelson, MD Department of Dermatology, Duke University Medical Center, Durham, NC, USA

Mai A. Ngo, PhD Department of Pesticide Regulation, California Environmental Protection Agency, Sacramento, CA, USA

Michael O'Malley Center for Health and the Environment, University of California, Davis, CA, USA

Amy S. Paller, MD Department of Dermatology, Northwestern University, Chicago, IL, USA

Fiorenza Rancan, PhD Department of Dermatology and Allergy, Charité-University Medicine Berlin, Berlin, Germany

Nancy A. Monteiro-Riviere, PhD Department of Anatomy and Physiology—CVM, KSU, Manhattan, KS, USA

Angèle Soria, MD Laboratory of Immunity and Infection, Institut National de Santé et de Recherche Médicale, INSERM UMR-S 945 and Université Pierre et Marie Curie (UPMC—Paris 6), Paris, France

Doug Schoon, MS Chemistry Schoon Scientific + Regulatory Consulting, LLC, Dana Point, CA, USA

Fernando Sérgio Rodrigues Ribeiro Teles, PhD Laboratory of Mycology/ Microbiology Unit and Center for Malaria and Tropical Diseases (CMDT), Instituto de Higiene e Medicina Tropical (IHMT), Lisbon, Portugal

Ian Tooley, PhD Physical Chemistry Croda Europe Ltd, Sun Care and Biotechnology, Cheshire, UK

Ulrike Blume-Peytavi, MD, PhD Department of Dermatology and Allergy, Charité-University Medicine Berlin, Berlin, Germany

John M. Veranth, PhD Department of Pharmacology and Toxicology, University of Utah, L.S. Skaggs Pharmacy, Salt Lake City, UT, USA

Maria Raquel Vieira, MD Serviço de Dermatologia, Hospital Curry Cabral, Lisbon, Portugal

Annika Vogt, MD, PhD Department of Dermatology and Allergy, Charité-University Medicine Berlin, Berlin, Germany

Roger Ronghua Wang, PhD Nano-Tex Inc, Shatin, Hong Kong

Steven Wang, MD Dermatology Service, Memorial Sloan-Kettering Cancer Center, New York, NY, USA

Michael G. Wilkerson, MD Department of Dermatology, University of Texas/Medicine Branch, Galveston, TX, USA

Kenneth K.Y. Wong, M.D, PhD Department of Surgery, The University of Hong Kong, Hong Kong, SAR, China

Dan Zheng Department of Chemistry, Northwestern University, Chicago, Evanston, IL, USA

Formulating with Nanotechnology in Skin Care Opportunities and Challenges

1

Julian P. Hewitt

1.1 Introduction

Nanoparticles and other nanotechnologies are finding application in many areas of skin care. In most cases, the nano-ingredients are present as functional ingredients, conferring a particular benefit to the skin or, in the case of inorganic sunscreens, protecting the skin against external damaging agents such as UV light. Such ingredients often require specific formulating techniques to ensure that they remain stable and effective until, and after, they are delivered to the skin. However, there are also cases in which nanotechnologies can assist the formulator to achieve the desired properties of their formulation, to protect functional ingredients, or deliver them to the correct site in the skin.

This chapter, therefore, deals with two distinct aspects of formulating with nanotechnologies in skin care:

- Opportunities: nanotechnologies that can assist the formulator to solve formulating problems
- Challenges: formulating approaches to incorporating nanoparticles in a finished formulation, where these are desired for a particular functional benefit

J.P. Hewitt, B.A. (✉)
JPH SunCare Technologies Ltd.,
7 Patterdale Close, Durham DH1 2UF, England, UK
e-mail: julian@jphsuncare.com

1.2 The Challenges of Cosmetic Formulation

Modern skin care emulsions are complex systems that present a number of different technical challenges, all of which the formulator must address in order to create an effective and appealing product.

1.2.1 Stability

The majority of skin care products are formulated as emulsions, which are thermodynamically unstable systems, and so achieving emulsion stability is always a challenge for the cosmetic formulator. Most products are either oil-in-water (O/W) or water-in-oil (W/O) emulsions; O/W systems tend to be the most common and most popular, for several reasons:

- Skin feel: with water as the external phase of the emulsion, there is an immediate feeling of hydration, and O/W emulsions feel less oily and less greasy than W/O systems.
- Cost: water typically accounts for more than 70% of a cosmetic O/W emulsion, and production processes are usually less costly than for W/O systems.
- Ease of manufacture: production processes tend to be shorter and use less energy.
- Flexibility and ease of formulation: there is a vast range of emulsifiers, emulsion stabilisers and rheological additives available for O/W emulsions, giving the formulator plenty of

A. Nasir et al. (eds.), *Nanotechnology in Dermatology*,
DOI 10.1007/978-1-4614-5034-4_1, © Springer Science+Business Media, LLC 2013

options for achieving a stable, effective product with the desired skin feel.

However, W/O emulsions also have advantages:

- Moisturisation: W/O emulsions give a more occlusive film on skin and are therefore more effective at preventing transepidermal water loss. Also, as the water in a W/O emulsion is contained within the internal phase, it is held in close proximity to the skin, giving a longer-lasting moisturising effect than is achieved with an O/W emulsion.
- Efficacy: in certain applications, for example in sun care, W/O emulsions show greater efficacy for a given concentration of active ingredient.
- Water resistance: with oil as the external phase, W/O emulsions are inherently more water-resistant than O/W, which is an advantage in, for example, make-up and sun care applications.

Achieving a stable emulsion is a greater challenge in W/O systems than in O/W, because the internal phase content is typically much higher in a W/O emulsion. This makes for a more "crowded" system, with the emulsion droplets more numerous and closer together, so the droplets are more likely to flocculate and/or coalesce. Stabilisation systems in W/O emulsions therefore need to be very effective in protecting the droplet interface and preventing coalescence.

1.2.2 Aesthetic Properties

The importance of the appearance, odour and skin feel of a skin care product cannot be overstated; no matter how effective a product is, if it is not pleasant to use then consumers will not buy it. Unfortunately, in many cases, ingredients and formulation systems that are beneficial for either stability or efficacy have an adverse effect on the aesthetic properties of an emulsion. For example, thickeners are usually needed to aid stability, to provide the desired viscosity for the intended application and to optimise product efficacy. But many hydrocolloids used in O/W emulsions can make the product feel sticky, while other thickening ingredients such as waxes can make the product feel "heavy" and difficult to apply. Also, in W/O emulsions, to meet the stability challenge as outlined above can require relatively high contents of emulsifier, which can make the product feel greasy on skin.

1.2.3 Efficacy

The product must also do the job it is intended for. The ideal moisturiser, for example, should give an immediate sensation of hydration or moisturisation when it is first applied, and then maintain a prolonged moisturising effect. O/W emulsions achieve the first requirement but are not ideal for long-lived moisturisation; W/O emulsions are better for this but are less effective in terms of immediate hydration.

If a product contains active ingredients for specific effects, e.g. anti-ageing, these must be delivered at the right time and to the correct site in the skin in order to be effective. The science of skin delivery is a complex field and many different aspects of the formulation need to be considered [1]. In addition, some active ingredients are incompatible with other necessary formulation components, or can be degraded by external influences such as UV light, so an additional challenge for the formulator is to protect such ingredients in the product itself, so that the actives can be delivered intact when the product is applied.

1.3 Nanotechnologies as Formulating Aids

Nanocarrier systems, such as nanoemulsions or nanoliposomes, are finding increasing application as a means of incorporating, stabilising or delivering active ingredients in personal care or pharmaceutical preparations. Such systems are sometimes termed "labile nanoparticles" as they are formed in situ and are designed to eventually break down (and thus release the encapsulated active), either during or after application of the product. These nanocarrier systems are covered elsewhere in this book and so will not be discussed in this chapter, which rather will focus on the formulation benefits of "persistent" nanoparticles, i.e. materials that are incorporated in a particulate form and remain in this form throughout the formulation and application of the product.

1.3.1 Emulsion Stabilisation

A method of stabilising emulsions that has long been known but is rarely discussed (at least in personal care and pharmaceutical applications) is the concept of Pickering emulsions. In a Pickering emulsion, the emulsion droplets are protected from coalescence by adsorption of particles at the interface between the external and internal phases. Such adsorption occurs when there is a great difference between the radius of the emulsion droplets and the radius of the stabilising particles; since nanoparticles have a much smaller size than even the smallest droplets of a conventional emulsion, they are an ideal candidate to stabilise emulsions in this way.

Dahms [2] showed how this concept can be advantageously applied in cosmetic water-in-oil emulsions. Such emulsions have advantages over the more common oil-in-water emulsions in certain applications, but W/O systems present a greater challenge than O/W emulsions when it comes to achieving stability. High concentrations of emulsifier and/or rheological additives are often required, and this can have an adverse effect on skin feel, making the emulsions feel "heavy" or greasy in application. Dahms showed that incorporation of an oil-based dispersion of inorganic sunscreen into a W/O system has a stabilising effect on the emulsion, due to a Pickering-type mechanism, with particles of the inorganic sunscreen being adsorbed at the interface and providing a physical barrier to droplet coalescence. Using this approach, he was able to formulate stable W/O sunscreen systems with less than 1% emulsifier, thus enabling an elegant skin feel to be achieved.

Nanoparticles can also have a stabilising effect in O/W emulsion systems, via their use in conjunction with lamellar liquid crystal structures. Liquid crystals are present in many cosmetic emulsions, particularly O/W systems [3–6]. These liquid crystals can form various structural phases; of most interest is the lamellar phase, in which layers of water are trapped between layers of lipid emulsifier. These lamellae, in turn, can take different forms, such as oleosomes, in which the lamellae form concentric spheres around the oil droplets of an emulsion. Another form is lamellar gel networks, also known as hydrosomes, in which low-HLB lipid emulsifiers are used to form a liquid crystalline network throughout an O/W emulsion, thereby stabilising it.

It is in these hydrosome systems that nanoparticles can provide an additional stabilising effect. Studies using freeze-fracture electron microscopy have shown that, with a suitable dispersing agent, hydrophilic TiO_2 nanoparticles tend to locate themselves within the lamellar gel network structure [7]. The result is a kind of "symbiotic" relationship; the nanoparticles act as a "swelling agent", increasing the thickness of the water layers in the lamellar structure and thus making it more stable [2]. In turn, the lamellar structure helps to maintain the dispersion of the nanoparticles, and provides a vehicle by which the hydrophilic particles may be effectively dispersed in the oil film that is left on skin after an O/W emulsion has been applied and the bulk water has evaporated [8].

1.3.2 Rheology Modification

The rheological behaviour of skin care emulsions is of vital importance. It influences consumer perception of the product, skin feel and also can affect efficacy. Many different ingredients are used to confer the desired rheological properties on the finished formulation. Nanoparticles can play a role here also.

Fine particle silica, usually referred to as fumed silica or colloidal silica, is used as a rheological additive in a variety of personal care products. It can be used to gel cosmetic oils, giving products with a pseudoplastic flow behaviour. Hydrophilic grades are particularly effective in this application; hydrophobic grades do not thicken as much but may be better suited to some types of oils and for use in emulsions. For example hydrophobic silica grades are often used to thicken W/O emulsions. Unlike many other rheological additives, for example waxes, fumed silica can be incorporated cold, allowing for energy and cost savings by facilitating cold production of the emulsion. Another advantage of formulations formulated with fumed silica is their low sensitivity to temperature, electrolytes and pH. In addition, fumed silica counteracts any oily or greasy skin feel in a formulation. With suitable surface treatments, fumed silica can also help to stabilise emulsions via the Pickering effect as described earlier, i.e. by adsorbing at the droplet interface. In sun care emulsions, besides improving storage and temperature stability, hydrophobic fumed silica can improve water resistance in particular—a necessary feature of modern sunscreens. In addition, hydrophobic grades can produce a more homogeneous distribution of physical and chemical UV filters, raising the sun protection factor.

There are applications in colour cosmetics also. In lipsticks, silica improves the homogeneous distribution of the pigments. Once applied, the silica prevents the pigments from migrating or "bleeding" into the fine lines of the lips. In nail polishes, silica improves the distribution of pigments and prevents them from settling. Also, the rheological behaviour can be adjusted as required, thus improving the application properties. Products such as face powder, rouge, eye shadow, mascara and cosmetic pencils can be optimised when silica products are used. As an effective free flow agent, silica improves the flow of powders thereby improving production productivity. They also enable the use of high pigment contents in cosmetic products by acting to prevent re-agglomeration of the pigments.

1.3.3 Delivery and Efficacy

Encapsulation technologies have been, and continue to be, investigated for many applications in both the personal care and health care fields, as well as in a number of other industries. The thinking behind the use of encapsulation is usually based on one or more of the following:

- Controlled release of actives over an extended period of time
- Enhancing delivery of actives to a particular site
- Protection of sensitive actives from interaction with other components of a formulation

Encapsulation usually involves surrounding a core of the active with a shell (typically composed of a polymer) or a fatty wall (in the case of liposomes). However, in some cases there is a more intimate mixture between the active and the carrier.

Two similar technologies which have been the subject of considerable research in recent years are solid lipid nanoparticles (SLN) and nanostructured lipid carriers (NLC). These systems involve the incorporation of an active into a lipid matrix. The difference between them is that SLN are based on a pure solid lipid, whereas NLC systems use a blend of solid and liquid lipids. The delivery form of the system is typically an aqueous dispersion. SLN have been used as a carrier system for tocopheryl acetate (vitamin E acetate) [9] and chemical sunscreens [10], providing protection of the active against chemical degradation and increasing efficacy. Increased moisturisation, via an occlusive effect preventing trans-epidermal water loss (TEWL), has also been reported for SLN systems [9, 10]. Although such systems have been claimed to give improved UV absorbance, they are not yet in use in commercial sunscreen systems. The most likely reasons for this are that the manufacturing processes involved are quite complex, involving high temperatures and high pressure homogenisation, and the resultant loadings of active UV filter in the final dispersions tend to be quite low. However Xia et al. [11] reported NLC systems in which up to 70% chemical sunscreen could be incorporated. The use of TiO_2, rather than organic UV filters, in SLN systems has also been reported [12], but this does not appear to be an encapsulation of TiO_2. The particle sizes reported for the SLN in this case were in the same range as the typical size for the TiO_2 particles; it seems unlikely that the TiO_2 is within the lipid nanoparticles. The authors suggested that TiO_2 "molecules" might be absorbed on the surface of the lipid nanoparticles, but given the insolubility of TiO_2, and the size of the TiO_2 particles, this too seems unlikely. The authors reported that their SLN systems containing TiO_2 showed enhanced UV absorption, but this is more likely to be due to light scattering from the lipid nanoparticles themselves. Dahms [13] referred to this as a "pigment effect" of SLN systems, and described a similar SPF-boosting effect from the use of membrane-structured solid nanoparticles (MSSN). These are similar in concept to SLN systems, and their main purpose, as with SLN, is encapsulation of actives. The claimed advantage

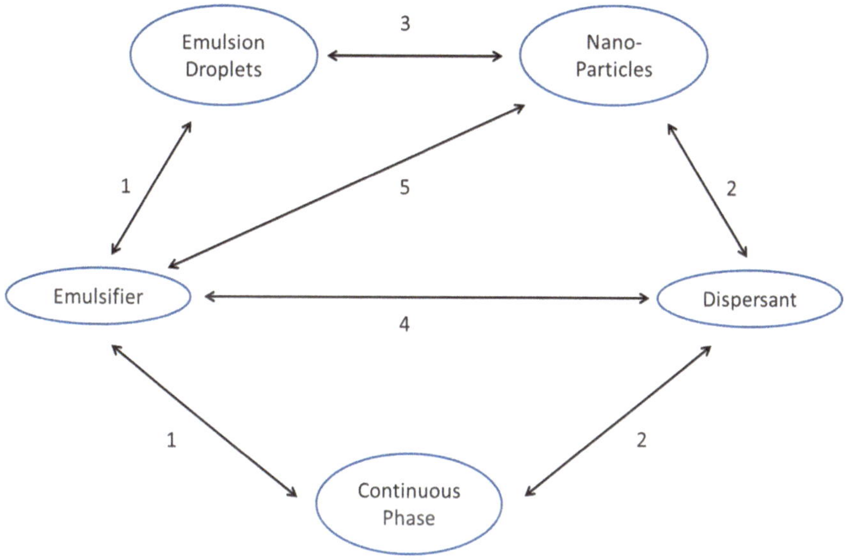

Fig. 1.1 Interactions in a suspoemulsion system. (1) emulsification; (2) dispersion; (3) flocculation; (4) desorption; (5) networking

of MSSN is that each particle consists of a continuous liquid crystalline membrane, which facilitates the homogeneous incorporation of amphiphilic molecules such as retinol or tocopherol. Dahms reported an SPF boosting effect when unloaded MSSN were incorporated into a sunscreen formulation containing ethylhexyl salicylate. This improvement in UV efficacy, however, was attributed not to encapsulation of the UV filter but rather to scattering of light, thus increasing the path length through the sunscreen film.

1.4 Formulating with Nanoparticle Active Ingredients

Since most skin care products are formulated as emulsions, the principles of incorporating nanoparticles into such products can be conveniently discussed with reference to the scheme shown in Fig. 1.1, as described by Anderson et al. [14]. This scheme shows the interactions in a "suspoemulsion" system, i.e. an emulsion in which solid particles are suspended or dispersed. The term 'suspoemulsion' was originally coined to describe agricultural emulsions containing suspended particles of solid pesticide. The term can equally well be applied to a cosmetic emulsion containing a nanoparticle active. Figure 1.1 shows the interactions present in such a system; some of these are desirable or "positive" interactions, some are undesirable or "negative" interactions, while one type can have either a positive or negative effect depending on the particular system. The following sections will discuss each of these interactions in turn.

1.4.1 Positive Interactions

1.4.1.1 Emulsification

This is the action of the emulsifier to stabilise the emulsion droplets within the external phase of the emulsion. Emulsifying agents are typically surfactants, with a polar hydrophilic "head" group and a non-polar lipophilic "tail". The relative sizes of the hydrophilic and lipophilic parts determine whether the emulsifier preferentially forms oil-in-water or water-in-oil emulsions. More hydrophilic molecules tend to be O/W emulsifiers, while the more lipophilic types have a greater tendency to stabilise W/O systems. Often, particularly in O/W emulsions, more than one emulsifier is used. By using two different molecules with different size heads and tails, more efficient packing at the interface can be achieved, making the droplets more stable. Most emulsifiers that are used in modern personal care products are non-ionic, and make use of steric repulsion to prevent droplets from flocculating together or coalescing. However, ionic emulsifiers are used in some cases; here, electrostatic repulsion also plays a role in stabilising the droplets.

1.4.1.2 Dispersion

This is the action of the dispersing agent to stabilise the particles within the continuous phase. In this context, "continuous phase" means the phase in which the particles are dispersed, which could be either the internal or external phase of the emulsion—or both. For example, inorganic sunscreens are available with either hydrophilic or hydrophobic surface treatments, or as either water-based or oil-based dispersions.

Where nanoparticle actives are used in cosmetic formulations, it is usually important that they remain reasonably well dispersed in the final formulation in order to perform their function. Some degree of aggregation and/or agglomeration is usually inevitable, and in fact is often assumed in the design of the actives; however excessive aggregation or agglomeration is likely to have an adverse effect on efficacy, so the dispersion process is critically important in using these materials.

A dispersant or dispersing agent is usually necessary to ensure efficient dispersion. Note that the term "dispersant" is frequently misused, particularly in literature from ingredient suppliers, who often tend to describe certain cosmetic oils, esters etc. as good "dispersants" for particulate ingredients, when what they actually mean is that these oils or esters are good carrier fluids for particular types of pigments or particles. A dispersant is an ingredient that is added to assist in the homogeneous dispersion of the particles within the carrier fluid. Dispersants are almost always surfactants. In choosing a suitable dispersant, the formulator needs to consider the nature of the surface of the particles, and also the carrier fluid. The ideal dispersant molecule should consist of one part that binds strongly to the particle surface, and another part that is well solvated by the carrier fluid. The amount of dispersant required depends on how well the dispersant binds to the particle surface, and on the surface area of the particles. In the case of nanoparticles, the specific surface area is considerably larger than, for example, that of pigments; as a result, the concentration of dispersant needed, relative to the weight of the particles, may be an order of magnitude higher than is the case for pigment particles. Optimising the type and concentration of dispersant is, in itself, a considerable technical challenge, and as a result many suppliers of nanoparticle actives seek to make life easier for their customers by supplying the actives in a pre-dispersed form.

Even so, the formulator must still consider the compatibility of the dispersion system with other components added to the formulation. For example, a nanoparticle dispersion in a polar ester, designed with a dispersant appropriate for polar oils, can be destabilised if combined with an excess of non-polar or silicone oils. In the case of aqueous dispersions, if ionic dispersing agents are used, the formulator must be mindful of electrolyte content and pH of the system.

In the case of inorganic particles dispersed in water, particularly important is the iso-electric point, or point of net zero charge. All inorganic particles carry charges on their surface. When such particles are dispersed in water, the surface charges can play a major role in the interparticle forces. Both positive and negative charges are present; at high pH there is a net negative charge, while at low pH the net charge is positive. In between, there is a certain pH at which the positive and negative charges exactly balance each other and there is no net charge; this is the iso-electric point. The iso-electric point is characteristic of the surface of the particles, and therefore depends on the coatings applied and also on the dispersing agent used.

In formulation, the iso-electric point must be avoided, since the lack of electrostatic repulsion at this point means that the particles can agglomerate. These agglomerates are difficult to break up again once formed. Therefore control of emulsion pH is an important aspect of formulating with hydrophilic particles in the aqueous phase.

1.4.2 Negative Interactions

1.4.2.1 Flocculation

If the particles flocculate around the emulsion droplets, this can encourage aggregation of the particles. As discussed earlier, in some cases, the tendency of small particles to flocculate around larger droplets can have a positive effect on emulsion stability (Pickering effect), but such aggregation is likely to adversely affect the efficacy of nanoparticle actives. In order to avoid this, the formulator needs to ensure that the dispersion effect described above is optimised as much as possible.

1.4.2.2 Desorption

Interactions between the emulsifier and the dispersant can result in either, or both, becoming desorbed from the interfaces that they are intended to stabilise. This phenomenon is sometimes known as competitive adsorption. It can be substantially avoided by choosing dispersing agents that have a low affinity for oil/water interfaces. Ideally, polymer-stabilised systems should be used, with the concentration of dispersant optimised such that there is no free surfactant that could affect the emulsifier system [2].

1.4.3 Other Interactions

1.4.3.1 Networking

As indicated earlier, many cosmetic emulsions contain liquid crystalline structures, and nanoparticles can be incorporated into these structures. Whether this type of interaction is positive or negative depends on the type of structure formed.

In W/O emulsions, an excess of emulsifier can lead to the formation of localised liquid crystalline aggregates. These should be avoided in any case, whether particles are present or not, as they can lead to shear-thickening or rheopectic behaviour (in which the emulsion increases in viscosity as shear is applied). This is undesirable in skin care emulsions as it makes the product very difficult to spread on the skin. If nanoparticle actives are associated with such structures, this has an adverse effect on the homogeneous dispersion of the particles.

In O/W systems, as described earlier, the association of hydrophilic nanoparticles with lamellar liquid crystalline structures can be beneficial, as it helps to strengthen the lamellar structures and increase stability. Also, in the case of delocalised networks, such liquid crystal structures can help to maintain a homogeneous dispersion of the particles throughout the emulsion.

1.5 Conclusions

The skin care formulator must address a number of challenges in order to create a stable, effective and appealing product. Some nanotechnologies can help to address these issues, improving emulsion stabilisation, rheological properties or delivery of active ingredients. Where the active ingredients themselves are nanoparticles, this can create additional challenges that the formulator must address, in particular to ensure a homogeneous dispersion of the particles in the product in order to maintain efficacy.

References

1. Wiechers JW. Optimizing skin delivery of active ingredients from formulations—from theory to practice (Chap. 20). In: Rosen MR, editor. Delivery system handbook for personal care and cosmetic products, technology, applications, and formulations. Norwich: William Andrew; 2005. p. 409–36.
2. Dahms GH. Formulating with a physical sun block. Cosm Toil. 1994;107(10):87–92.
3. Junginger HE. Crystalline gel structures in O/W creams. Skin Care Forum Nr.5 Dusseldorf: Henkel KGaA, Feb 1993.
4. Dahms GH. Properties of O/W emulsions with anisotropic lamellar phases. Cosm Toil. 1986;101(11):113–5.
5. Dahms GH. Optimised formulations for skin care products 1990. SÖFW J. 1990;10:388–92.
6. Loll P. Liquid crystals in cosmetic emulsions. Cosmetics & Toiletries Manufacture Worldwide: Aston; 1994. p. 108–20.
7. Catlow B. Formulating with ultrafine TiO_2. SÖFW J. 1993;119:497–500.
8. Hewitt JP. SPF modulation: optimizing the efficacy of sunscreens (Chap. 19). In: Shaath NA, editor. Sunscreens: regulation and commercial development.

3rd ed. Boca Raton: Taylor & Francis; 2005. p. 385–412.
9. Wissing SA, Muller RH. A novel sunscreen system based on tocopherol acetate incorporated into solid lipid nanoparticles. Int J Cosm Sci. 2001;23(4): 233–43.
10. Wissing SA, Muller RH. Cosmetic applications for solid lipid nanoparticles (SLN). Int J Pharm. 2003;254(1):65–8.
11. Xia Q, Saupe A, Muller RH, et al. Nanostructured lipid carriers as novel carrier for sunscreen formulations. Int J Cosm Sci. 2007;29(6):473–82.
12. Cengiz E, Wissing SA, Muller RH, et al. Sunblocking efficiency of various TiO_2-loaded solid lipid nanoparticle formulations. Int J Cosm Sci. 2006;28(5):371–8.
13. Dahms GH. Membrane structured solid nanoparticles—a novel nanotechnology for delivery of cosmetic active ingredients. IFSCC Magazine. 2008;8(3):193–8.
14. Anderson MW, Hewitt JP, Spruce SR. Broad spectrum physical sunscreens: titanium dioxide and zinc oxide (Chap. 18). In: Lowe NJ, Shaath NA, Pathak MA, editors. Sunscreens: development, evaluation, and regulatory aspects. 2nd ed. New York: Marcel Dekker; 1997. p. 353–97.

2 Nanotechnology in Photoprotection

Lucy L. Chen, Ian Tooley, and Steve Q. Wang

2.1 Introduction

Excessive exposure to ultraviolet (UV) radiation induces a wide range of adverse effects such as sunburn, photoaging, photoimmunosuppression, and photocarcinogenesis. Use of sunscreen is an important practice by the public to protect against excessive UV exposure and reduce UV damages. In general, inorganic-based sunscreen composed of mineral UV filters, such as titanium dioxide (TiO_2) and zinc oxide (ZnO), work by reflecting and scattering UV radiation. These agents are regarded as safe and effective. Compared to organic UV filters, such as avobenzone and oxybenzone, inorganic filters are less irritating on individuals with sensitive skin and chronic skin disorders. For these reasons, TiO_2 and ZnO have been widely recommended as the safest UV filters in sunscreen products. Despite these benefits, older sunscreens containing these ingredients were limited in popularity by their poor cosmetic appearance. Due to the broad particle size distribution and poor dispersive qualities of the TiO_2 and ZnO particles, these sunscreens left a white or opaque film, as well as grainy-residue on the skin. The diminished aesthetics of these sunscreens hindered wide acceptance by the public.

This problem was met with a solution in nanotechnology. Nanotechnology involves the design, production, and application of materials in the size range of 1–100 nm. As existing materials are reduced to this size, a new set of physical, chemical, mechanical, and electrical properties are revealed. Application of this technology has aided in the development and advancement of new tools in numerous fields. Today, nanomaterials are found in electronics, paints, foods, cosmetics, and coatings and are increasingly applied to the medical field in diagnosis and drug delivery. The advent of nanotechnology also brought about great concern for the potential risks and toxicity of these foreign materials. Concerns surrounding nanotechnology in cosmetics and sunscreens predominantly surround penetration into human skin and possible systemic exposure from topical application.

2.2 History of Nanosized TiO_2 and ZnO

The broad distribution of particle size in older-generation sunscreens caused excessive whitening of the skin, resulting in consumer reluctance to use products containing TiO_2 and ZnO. The solution arose from reducing their size to nanoparticles. The average size of these minerals was <100 nm,

L.L. Chen, B.A. • S.Q. Wang, M.D. (✉)
Dermatology Service, Memorial Sloan-Kettering Cancer Center, 160 E 53rd Street, New York, NY 10022, USA
e-mail: wangs@mskcc.org

I. Tooley, Ph.D.
Croda Europe Ltd, Sun Care and Biotechnology, Foundry Lane, Ditton, Widnes, Cheshire WA8 8UB, UK

A. Nasir et al. (eds.), *Nanotechnology in Dermatology*,
DOI 10.1007/978-1-4614-5034-4_2, © Springer Science+Business Media, LLC 2013

had superior UV protection, and had improved cosmetic appearance. By the 1980s, patents were filed and commercial sunscreens containing TiO_2 nanoparticles were introduced on a large scale by 1990. Nanosized ZnO was used in the later part of the decade [1]. Since that time, nanosized TiO_2 and ZnO ingredients have been approved and deemed safe for use by various consortiums and organizations including the US Food and Drug Administration (FDA). These materials now represent one of the largest applications of oxide nanoparticles with estimations that 70 % of titanium sunscreens and 30 % of zinc sunscreens are formulated with nanoingredients [2].

The current sunscreens on the market are more aesthetically acceptable and superior to older-generation sunscreens due to nanoparticle technology. In terms of cosmetic elegance, the smaller particle sizes minimize visible light scattering so the resulting topical application appears "transparent." However, it is a misconception that the nano- TiO_2 and ZnO particles are truly transparent. In reality, they appear transparent at or below concentration thresholds due to increased light transmittance in the visible light range. Not only must particle size remain small, but also the particle size distribution must be narrow to avoid any whitening effect. Mineral particles of TiO_2 reflect and scatter UV light most efficiently at a size of 30–100 nm; whereas ZnO has an optimal size of 60–100 nm particles [3, 4]. Another benefit of nanoingredients is the greater ease of the product to blend into the skin, due to the small particulate size. The reduction in particle size scatters and reflects UV more efficiently, improving human skin protection against UV-induced damage. For all of these reasons, nanometal oxides have been included into sunscreen products.

2.3 Three States of Nano TiO_2 and ZnO

Nanosized TiO_2 and ZnO can exist in three different states, encountered during the manufacturing process: primary nanoparticles (5–20 nm), aggregates (30–150 nm), and agglomerates (1–100 μm) (Fig. 2.1). The first stage in manufacturing nanoparticles is creation of the primary particles. The strong crystal attractions force primary particles to cluster together, forming chemically bound aggregates. These aggregates represent the smallest units that actually occur in a final sunscreen formulation because the forces required to break apart aggregates are far greater than those encountered during production of sunscreen formulations or application onto the skin. Aggregates may also form loosely bound agglomerates with sizes greater than 1 μm (1000 nm). This is the typical size of nano- TiO_2 and ZnO powder and results from the drying and heat treatment processes of manufacturing. The sunscreen's vehicle formulation can also cause formation of large agglomerates

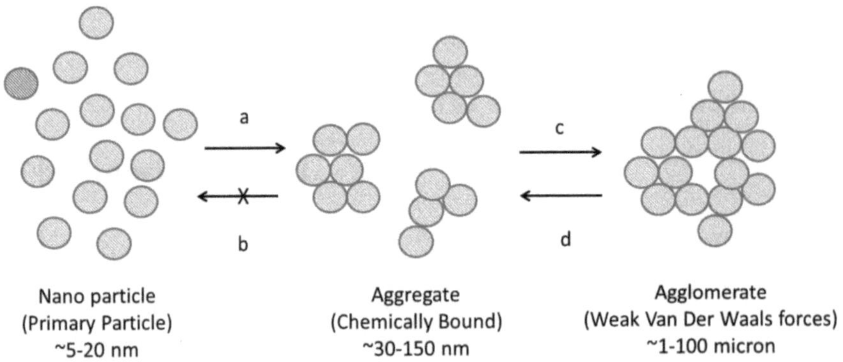

Fig. 2.1 Formation of aggregates and agglomerates from nanoparticle building blocks

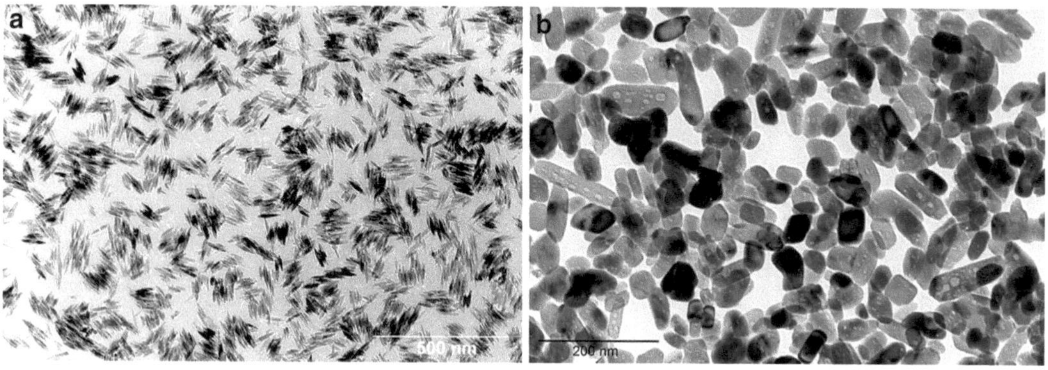

Fig. 2.2 (**a**) 15 nm TiO_2 and (**b**) 35 nm TiO_2 primary particles and aggregates measured by TEM

which do not provide effective UV attenuation. Special inert coating materials (commonly silica or dimethicone) can be applied to the surface of mineral particles to improve dispersion in sunscreen products and resist the formation of large particle agglomerations [5]. Nanoparticle-containing sunscreen emulsions can be visualized under transmission electron microscopy (TEM) (Fig. 2.2a, b). Typically, individual TiO_2 and ZnO nanoparticles cannot be detected, but present mainly in clusters (usually aggregates of 30–150 nm in size). On the market, sunscreen grade nanosized TiO_2 ranges from an ultra fine size of 30 nm to micro-sized aggregates. On the other hand, ZnO particles are used in the size range between 60 and 200 nm and the aggregate form is the smallest size that will occur in a sunscreen formulation.

2.4 UV Attenuation Based on Particle Size

Nanosized ultrafine inorganic sunscreens have the advantage of having both UV-reflecting and UV-absorbing properties. UV attenuation is the sum of scattering and absorption and depends on several factors including (1) dispersion of the inorganic particles (2) particle size and (3) refractive index [1]. With decreasing particle size, UV protection shifts towards protection to shorter UV wavelengths. In other words, smaller particles attenuate shorter wavelengths of UV radiation. By TEM studies of UV attenuation and particle size, it is also known that nanoparticles reflect and absorb UV light most efficiently at the aggregate size (Fig. 2.3). At the average aggregate size of approximately 100 nm (red curve), TiO_2 particles offer effective UVA and UVB coverage. However, significant scattering is noted in the visible region. In comparison, TiO_2 particles with an average aggregate size of approximately 50 nm (green curve) have less visible light scattering and offer higher UVB and lower UVA protection. TiO_2 particles with an average aggregate size of 20 nm (blue curve) offer significantly lower protection from UVA and UVB radiation compared to 50- and 100-nm aggregates. The sunscreen with a smaller particle size will be a more transparent product; however, its UV efficacy will require additional UVA filters to achieve broad spectrum protection. Thereby, the best sunscreen would be one containing a mixture of TiO_2 nanoparticle aggregate sizes ranging between 50 and 120 nm [3]. ZnO has an optimal size between 60 and 100 nm particles. To achieve the optimal size and desired UV absorption profile, nanometal oxides are usually stabilized with dispersing agents to maintain stability and prevent the reformation of agglomerates. Therefore, sunscreen formulators must find an optimum particle size in order to achieve the desired UV absorption profile while delivering a pleasing aesthetic formula.

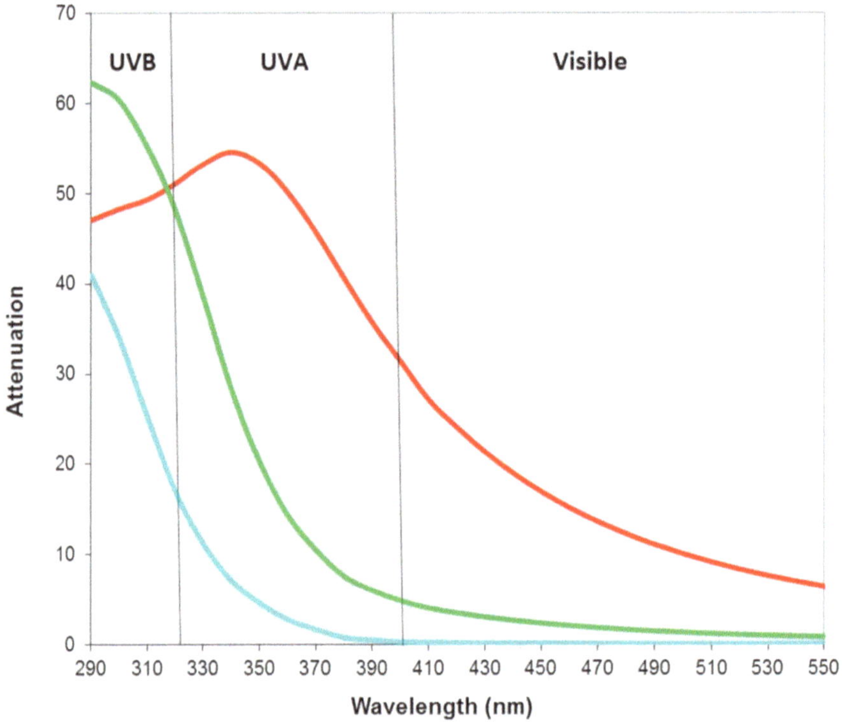

Fig. 2.3 UV attenuation vs. wavelength for spherical TiO_2 of varying particle size. *Blue line*, 20 nm; *green line*, 50 nm; and *red line*,100 nm

2.5 Safety Concerns for Nano TiO^2 and ZnO

With the advent of TiO_2 and ZnO-based sunscreens, the safety concerns are centered around several issues: the endogenous toxicity of nanoparticles, the potential toxicity on the epidermis, and the percutaneous penetration of nanomaterials through skin. In their small size, nanoparticles are able to form protein complexes, evade immunologic defense mechanisms, and most importantly, induce free radical formation [6]. Because of the small size of nanoparticles and their increased surface reactivity, there is a greater potential to cause harm than larger particles. For example, the large surface interface can result in nanoparticles reacting with biologic proteins. When nanoparticles bind to proteins, they form complexes which can result in protein surface modifications and downstream signaling pathways and metabolic routes which are different from the original protein. In addition, nanoparticles can evade the human defense system by escaping phagocytosis, either due to their small size or by carrying unrecognizable surface protein signals. Nanoingredients may also be intentionally engineered to avoid clearance in order to achieve therapeutic effects resulting in a prolonged half-life, ability to penetrate the blood brain barrier, and preference to collect in specific organs.

Besides the endogenous toxicities of nanoparticles, there is concern that TiO_2 and ZnO in topical sunscreens generate free radicals during exposure to UV radiation. When exposed to UV radiation, TiO_2 and ZnO generate electrons in the local media which in turn can induce the formation of peroxides, free radicals, and other reactive oxygen species (ROS). In turn, these ROS can cause damage to proteins, lipids, and DNA. This conglomeration of damage can alter the genetic code, as well as irreversibly injure cells and tissue.

The potential phototoxicity of TiO_2 nanoparticles was first demonstrated as DNA damage from hydroxyl radicals (·OH) after exposure to UV irradiation [7–10]. In the presence of free radical quenchers such as dimethyl sulfoxide and mannitol, the DNA damage could be suppressed. The target damage included oxidative biomarkers such as 8-oxoguanosine [11], DNA strand breakage [7, 10], and structural chromosomal aberrations [12]. TiO_2 and ZnO in combination had similar genotoxic effects [10]. Further *in vitro* studies demonstrated that the DNA damage resulting from TiO_2 and ZnO agents are cytotoxic in the presence of cell cultures [11–13]. By using the technique of electron spin resonance spin trapping to catch these short-lived, volatile molecules, other ROS have been proposed to be generated in the process: hydrogen peroxide (H_2O_2) [14, 15], superoxide anion (O_2^-) [16], singlet oxygen ($^1O_2$) [16, 17], and carboxyl radical anion ($\cdot CO_2^-$) [18]. The size and crystal form of the metal oxide also affects toxicity: the anatase form of TiO_2 is more effective at generating ·(middot) OH radicals than the rutile form [9]. These studies suggest that metal oxides are phototoxic by way of generating free radical species upon UV irradiation. In order to determine whether the phototoxicity was truly due to the presence of metal oxides alone, Dufour et al. [12] studied mammalian cells to examine the ability of ZnO to damage DNA in the presence of UV light when compared to its activity in the dark. The authors measured the effects of ZnO on Chinese hamster ovarian cells under three conditions: in the dark, pre-irradiation followed by ZnO treatment (PI) and simultaneous irradiation with ZnO treatment (SI). Interestingly, the nature, incidence, and severity of chromosomal aberrations (CA) in the PI and SI group were nearly identical. Therefore, the timing of exposure of the test agent and cells to UV radiation appeared to have had no effect on the frequency of CA induction. This is in contrast to true photoclastogenic agents, in which clastogenic potency would be significantly higher under simultaneous irradiation conditions. Therefore, the authors concluded that UV radiation mediated enhanced susceptibility of mammalian cells to ZnO, but ZnO itself was non-photoclastogenic.

There are several arguments that support the safety profile of nanoparticles. In the previous studies, the behavior of nanoparticles has only been elucidated *in vitro*, not in a true biological setting. In fact, the skin has in place an elaborate antioxidant mechanism, composed of enzymatic and nonenzymatic molecules, to quench ROS. Therefore, ROS generated by TIO_2 and ZnO during UV exposure can be neutralized by the body's natural defense mechanism. Besides endogenous protection, there are numerous synthetic techniques which have been developed to reduce these potential risks and decrease the reactivity of these nanoparticles. Pflucker et al. [19] ; Livraghi S, et al. [20] tested a carbon coating on TiO_2 powder and demonstrated significant reductions in the formation of superoxide anion ($\cdot O_2^-$) and hydroxyl radicals (·OH), even after UV exposure. Pan et al. [21] grafted an organic coating which could prevent nanoparticle adherence to the cell membranes of human dermal fibroblasts, and effectively reduce the cytotoxicity of ROS. Other techniques such as manganese doping and hydrophobic polymers have been proposed to prevent the contact of the TiO_2 surface with oxygen and water to inhibit the formation of radical species [22, 23]. The final argument in support of nanotechnology in sunscreens is the overall safety record of TiO_2 and ZnO. Both metal oxides have been used in mineral sunscreens and other topical products for years, with no studies demonstrating adverse effects of potentially harmful free radicals. TiO_2 can be found in toothpaste, lotion, skimmed milk, and cottage cheese, and ZnO is a major component in many baby powders, antidandruff shampoos as well as barrier creams.

Another component of the safety of nanoparticles is predicated on the dermal penetration of these materials and whether these small-sized minerals can penetrate the human skin. Are there new risks of cosmetic formulations containing nanosized particles when compared to traditional ingredients? The obvious concern is that the smaller particle size can enhance skin penetration, pass the skin barrier, and increase the risk for systemic toxicity and exposure. For most substances, the stratum corneum (SC) is the rate-limiting barrier against the percutaneous penetration of topically applied substances. This non-viable outermost layer of the epidermis is approximately 15–20 μm thick in humans. Nanoparticles must pass the SC to reach the living skin in order to exert its toxicities. Most studies done under in vitro or in vivo conditions must be interpreted cautiously for several reasons. Animal skin samples have differing permeability for certain materials. In general, the order from most permeable to least permeable is rabbit, rodent, pig, and lastly, human skin. In addition, penetration of substrates may be favored if skin occlusion methods (which increases swelling of the corneocytes) were used in the studies. Finally, destruction methods such as tape stripping of the skin may artificially capture particles in the deep hair follicles or skin furrows to incorrectly conclude that a substance has penetrated the epidermis. In light of these study limitations, the consistent findings from these studies suggest that TiO_2 and ZnO nanoparticles remain on the surface of the SC and are unable to reach the living skin layers (Tables 2.1, 2.2, 2.3). Most studies have reported that nanoparticles only penetrate into hair follicle openings and skin furrows, with negligible materials found below the SC. In 2000, the EU Scientific Committee on Cosmetics and Non-Food Products (SCCNFP) summarized a series of ten studies all investigating the percutaneous penetration of nanosized TiO_2 pigments [24]. Nanoparticles were detected using a myriad of imaging techniques including electron and light microscopy, X-ray fluorescence, particle-induced X-ray emission (PIXE), and electron emission spectrophotometry. All studies concluded that nanosized TiO_2 particles remain on the outer layers of the SC and do not penetrate into the living skin. Similarly, ZnO nanoparticles showed negligible penetration beyond the SC in animal and human skin studies. These skin penetration studies consistently summarize that neither nanoparticle can penetrate beyond the SC.

A number of factors may explain the poor penetration through the stratum corneum of intact and healthy human skin. From the field of transdermal drug delivery, it is known that the

Table 2.1 TiO2 skin penetration studies (Adapted with permission from Newman et al. [6])

Study	Material	Particle size	Skin model/design	Results
Tan et al. [22]	TiO_2 (no coating specified)	Not specified	Human skin, in vitro	No significant penetration into skin
Lademann et al. [32]	TiO_2 (Al_2O_3, stearic acid coated)	150–170 nm	Human skin biopsy	Penetration into upper layers of stratum corneum; ~1 % of particles in ostium of follicle
European Union SCCNFP Opinion [24]	TiO_2 (anatase and rutile forms, various coatings)	14–200 μm	Pig skin, in vitro Human skin, tape stripping, or biopsy	No penetration beyond the stratum corneum in any study
Pflucker et al. [19]	TiO_2 (SiO_2, Al_2O_3, $Al_2O_3+SiO_2$ coated)	10–100 nm	Human skin biopsy	Penetration into upper layers of the stratum corneum
Schulz et al. [33]	TiO_2 ($SiO_2 \pm Al_2O_3$ coated	10–100 nm	Human skin biopsy	Penetration into the upper layers of the stratum corneum
Gottbrath and Muller-Goymann [34]	TiO_2-containing sunscreen (no coating specified)	Not specified	Human skin, tape stripping	Particles into upper layers of stratum corneum
Menzel et al. [13]	TiO_2 (various forms, no coating specified)	45–150 nm	Pig skin, in vitro	Particles in stratum corneum; minimal penetration into stratum granulosum
Popov et al. [3]	TiO_2 (rutile form)	100 nm	Human skin, tape stripping	No penetration beyond the stratum corneum
Mavon et al. [23]	TiO_2 (SiO_2 coated)	20 nm	Human skin, tape stripping and in vitro	Penetration in upper layers of stratum corneum

Table 2.2 ZnO skin penetration studies (Adapted with permission from Newman et al. [6])

Pirot et al, 1996 [35]	ZnO (no coating specified)	Not specified	Human skin, in vitro	0.36 % penetration in 72 h
European Union SCCNFP Opinion [36]	ZnO	Not specified	Pig skin, in vitro Human nonpsoriatic and psoriatic skin	No increase in plasma zinc levels; in vitro, penetration <1 % of dose; most ZnO recovered from stratum corneum
Cross et al. [4]	ZnO (siliconate coated)	15–30 nm	Human skin, in vitro	<3 % of applied Zn recovered in stratum corneum; penetration into upper layers of the stratum corneum
Zvyagin et al. [26]	ZnO (uncoated)	26–30 nm	Human skin, in vivo and in vitro	No penetration beyond stratum corneum, accumulation in skin folds and hair follicles

Table 2.3 Combination TiO2 and ZnO skin penetration studies (Adapted with permission from Newman et al. [6])

Lansdown and Taylor [37]	TiO_2, ZnO (no coating specified)	<2–20 μm	Rabbit skin, in vivo	Penetration into stratum corneum and outer hair follicle
Dussert et al. [38]	TiO_2, ZnO (no coating specified)	TiO_2: 50–100 nm ZnO: 20–200 nm	Human skin, in vitro	Penetration into upper layers of stratum corneum
Gontier et al. [39]	TiO_2, ZnO (Al_2O_3)	Not specified	Human, pig, mouse skin, in vitro	TiO_2 found in intercellular space between corneocytes of upper layers of stratum corneum
Gamer et al. [40]	TiO_2 (SiO_2, dimethicone coated) ZnO (uncoated)	TiO_2: 30–60 nm ZnO: <160 nm	Pig skin, in vitro	Penetration into upper layers of stratum corneum; 0.8–1.4 % of applied dose recovered
Filipe et al. [41]	TiO_2 ($SiO_2 \pm Al_2O_3$ -coated ZnO	TiO_2: 20 nm ZnO: 20–60 nm	Human skin (intact, compromised, and psoriatic), tape stripping	No penetration beyond stratum corneum

upper limit of molecular size of a drug capable of skin penetration is 500 Da (2.5 nm) [10, 25]. Common dermatological drugs such as corticosteroids and antifungals all fall below this size and can penetrate the stratum corneum. As mentioned, nanosized TiO_2 and ZnO exist as aggregates and agglomerates in sunscreen products, their final size often exceeding 100 nm, 40 times larger than the upper limit. Besides molecular size, the drug substance must meet the criteria for polarity, concentration, and melting temperature. Nanoparticles are also considered insoluble substances, lacking a diffusion driving force to promote penetration into the skin. Besides the physiochemical properties of nanoparticles, the local skin environment may be preventative of skin penetration. The stratum corneum is constantly turning over and this constant shedding process prevents long-term accumulation and penetration of nanoparticles into the viable components of the skin tissue. Finally, there is a potential penetration pathway of nanomaterials through the follicles, first proposed by Lademann et al. [26]. Hair follicles, with its tight network of capillaries, are an important target for transdermal drug delivery and potential nanoparticle uptake. TiO_2 nanoparticles were found collecting in the "follicular sink" (the hair follicle openings and superficial portion of the follicles) at concentrations two orders of magnitude smaller than the upper part of the stratum corneum. However, it was later determined that the penetration process depends on a certain phase of the hair growth cycle; nanoparticles are only found when hair growth and sebum production are active. Therefore, despite the discovery of nanomaterials in the follicular orifices, the potential for them entering the living skin tissue is negligible, mainly because growing hair shafts push these materials to the surface of the skin [27, 28].

It is hypothesized that the percutaneous absorption of topically applied substances will be altered if the skin barrier is compromised or disrupted. In general, it is unknown what degree of compromised skin integrity will significantly increase skin penetration of small-sized particles. In hyperkeratotic skin conditions such as psoriasis vulgaris, a thicker epidermis actually enhances the skin's barrier function, which reduces penetration of topical drugs. Similarly, it has been found that there is identical or lower percutaneous absorption on intact compared to inflamed skin (from UVB-induced sunburn) following application of a [14C]methylprednisolone aceponate-containing lotion. The percutaneous absorption increased after skin tape stripping (removal of the stratum corneum) [29]. Skin conditions which disrupt the skin barrier such as eczema have increased penetration of topically applied substances [30]. In a recent study of four sunscreen formulations containing nanosized TiO_2 and ZnO (primary particle size 10–50 and 140 nm, respectively), UVB irradiation slightly enhanced the SC penetration of both types of nanoparticles in pig skin (TiO_2 deeper than ZnO). However, there was no evidence of systemic absorption [31]. Though a large amount of studies has rendered nanoparticle safety in cosmetic products in normal skin, their safety should be further assessed taking into account abnormal skin conditions and the possible impact of mechanical effects on skin penetration.

2.6 Conclusion

Sunscreen will continue to be an important component of photoprotection. The advancement in nanotechnology has allowed nanosized TiO_2 and ZnO ingredients to be amply incorporated into more effective and cosmetically acceptable sunscreen products. In developing modern inorganic sunscreens based on nanoparticles, formulators need to understand the UV-visible properties of TiO_2 and ZnO nanomaterials in relation to their particle size. A systemic and methodical approach is required to maintain the optimal size of these nanoparticles in the final sunscreen formulation. The safety debate is ongoing, despite overall consistent findings from laboratory studies demonstrating no penetration of nanoparticles into the living layers of the skin. Despite their inert behavior, nanoparticles could still pose a risk on compromised, sun damaged skin or from inhalation (such as from spray sunscreens). Such questions remain to be answered.

References

1. Mitchnick MA, Fairhurst D, Pinnell SR. Microfine zinc oxide (Z-cote) as a photostable UVA/UVB sunblock agent. J Am Acad Dermatol. 1999;40(1): 85–90.
2. Therapeutic Goods Administration. *A Review of the scientific literature of the safety of nanoparticulate titanium dioxide or zinc oxide in sunscreens*. Australian Government Department of Health and Ageing, 2006.
3. Popov AP, et al. Effect of size of TiO_2 nanoparticles embedded into stratum corneum on ultraviolet-A and ultraviolet-B sun-blocking properties of the skin. J Biomed Opt. 2005;10(6):064037.
4. Cross SE, et al. Human skin penetration of sunscreen nanoparticles: in-vitro assessment of a novel micronized zinc oxide formulation. Skin Pharmacol Physiol. 2007;20(3):148–54.
5. Nohynek GJ, et al. Grey goo on the skin? Nanotechnology, cosmetic and sunscreen safety. Crit Rev Toxicol. 2007;37(3):251–77.
6. Newman MD, Stotland M, Ellis JI. The safety of nanosized particles in titanium dioxide- and zinc oxide-based sunscreens. J Am Acad Dermatol. 2009; 61(4):685–92.
7. Dunford R, et al. Chemical oxidation and DNA damage catalysed by inorganic sunscreen ingredients. FEBS Lett. 1997;418(1–2):87–90.

8. Serpone N, Salinaro A, Emeline A. Deleterious effects of sunscreen titanium dioxide nanoparticles on DNA: efforts to limit DNA damage by particle surface modification. Proc SPIE. 2001;4258:86–98.
9. Uchino T, et al. Quantitative determination of OH radical generation and its cytotoxicity induced by TiO_2–UVA treatment. Toxicol In Vitro. 2002;16(5):629–35.
10. Bos JD, Meinardi MM. The 500 Dalton rule for the skin penetration of chemical compounds and drugs. Exp Dermatol. 2000;9(3):165–9.
11. Wamer WG, Yin JJ, Wei RR. Oxidative damage to nucleic acids photosensitized by titanium dioxide. Free Radic Biol Med. 1997;23(6):851–8.
12. Dufour EK, et al. Clastogenicity, photo-clastogenicity or pseudo-photo-clastogenicity: Genotoxic effects of zinc oxide in the dark, in pre-irradiated or simultaneously irradiated Chinese hamster ovary cells. Mutat Res. 2006;607(2):215–24.
13. Menzel F, et al. Investigations of percutaneous uptake of ultrafine TiO_2 particles at the high energy ion nanoprobe LIPSION. Nucl Instrum Methods Phys Res, Sect B. 2004;219–220(1–4):82–6.
14. Cai R, et al. Induction of cytotoxicity by photoexcited TiO_2 particles. Cancer Res. 1992;52(8):2346–8.
15. Hirakawa K, et al. Photo-irradiated Titanium Dioxide Catalyzes Site Specific DNA Damage via Generation of Hydrogen Peroxide. Free Radic Res. 2004;38(5): 439–47.
16. Konaka R, et al. Irradiation of titanium dioxide generates both singlet oxygen and superoxide anion. Free Radic Biol Med. 1999;27(3–4):294–300.
17. Brezova V, et al. Reactive oxygen species produced upon photoexcitation of sunscreens containing titanium dioxide (an EPR study). J Photochem Photobiol B. 2005;79(2):121–34.
18. Dodd NJ, Jha AN. Titanium dioxide induced cell damage: a proposed role of the carboxyl radical. Mutat Res. 2009;660(1–2):79–82.
19. Pflücker F, et al. The human stratum corneum layer: An effective barrier against dermal uptake of different forms of topically applied micronised titanium dioxide. Skin Pharmacol Appl Skin Physiol. 2001;14 Suppl 1:92–7.
20. Livraghi S, et al. Decreasing the oxidative potential of TiO(2) nanoparticles through modification of the surface with carbon: a new strategy for the production of safe UV filters. Chem Commun(Camb), 2011. 46(44):8478–80.
21. Pan Z, et al. Adverse Effects of Titanium Dioxide Nanoparticles on Human Dermal Fibroblasts and How to Protect Cells. Small. 2009;5(4):511–20.
22. Tan MH, et al. A pilot study on the percutaneous absorption of microfine titanium dioxide from sunscreens. Australas J Dermatol. 1996;37(4):185–7.
23. Mavon A, et al. In vitro percutaneous absorption and in vivo stratum corneum distribution of an organic and a mineral sunscreen. Skin Pharmacol Physiol. 2007;20(1):10–20.
24. European Union's Scientific Committee on Cosmetic Products and Non-Food Products. Opinion of the scientific committee on cosmetic products and non-food products intended for consumers concerning zinc oxide; 2000 (cited 2012 December 20). http://ec.europa.eu/health/ph_risk/committees/04_sccp/04_sccp_en.htm.
25. Subedi RK, et al. Recent advances in transdermal drug delivery. Arch Pharm Res. 2010;33(3):339–51.
26. Zvyagin AV, et al. Imaging of zinc oxide nanoparticle penetration in human skin in vitro and in vivo. J Biomed Opt. 2008;13(6):064031.
27. Lademann J, et al. Investigation of follicular penetration of topically applied substances. Skin Pharmacol Appl Skin Physiol. 2001;14 Suppl 1:17–22.
28. Lademann J, et al. Hair follicles—an efficient storage and penetration pathway for topically applied substances. Summary of recent results obtained at the Center of Experimental and Applied Cutaneous Physiology, Charite-Universitatsmedizin Berlin, Germany. Skin Pharmacol Physiol. 2008;21(3):150–5.
29. Gunther C, et al. Percutaneous absorption of methylprednisolone aceponate following topical application of Advantan lotion on intact, inflamed and stripped skin of male volunteers. Skin Pharmacol Appl Skin Physiol. 1998;11(1):35–42.
30. Korting HC, et al. Liposome encapsulation improves efficacy of betamethasone dipropionate in atopic eczema but not in psoriasis vulgaris. Eur J Clin Pharmacol. 1990;39(4):349–51.
31. Monteiro-Riviere NA, et al. Safety evaluation of sunscreen formulations containing titanium dioxide and zinc oxide nanoparticles in UVB sunburned skin: an in vitro and in vivo study. Toxicol Sci. 2011;123(1):264–80.
32. Lademann J, et al. Penetration of titanium dioxide microparticles in a sunscreen formulation into the horny layer and the follicular orifice. Skin Pharmacol Appl Skin Physiol. 1999;12(5):247–56.
33. Schulz J, et al. Distribution of sunscreens on skin. Adv Drug Deliv Rev. 2002;54 Suppl 1:S157–63.
34. Gottbrath S, Mueller-Goymann CC. Penetration and visualization of titanium dioxide microparticles in human stratum corneum—effect of different formulations on the penetration of titanium dioxide. SOFW J. 2003;129(3):11–7.
35. Pirot F, et al. In vitro study of percutaneous absorption, cutaneous bioavailability and bioequivalence of zinc and copper from five topical formulations. Skin Pharmacol. 1996;9(4):259–69.
36. European Union's Scientific Committee on Cosmetic Products and Non-Food Products. Opinion of the scientific committee on cosmetic products and non-food products intended for consumers concerning zinc oxide; 2003 (cited 2011 December). Available from: http://ec.europa.eu/health/ph_risk/committees/04_sccp/04_sccp_en.htm.
37. Lansdown ABG, Taylor A. Zinc and titanium oxides: Promising UV-absorbers but what influence do they have on the intact skin? Int J Cosmet Sci. 1997;19(4):167–72.
38. Dussert AS, Gooris E, Hemmerle J. Characterization of the mineral content of a physical sunscreen emulsion and its distribution onto human stratum corneum. Int J Cosmet Sci. 1997;19(3):119–29.

39. Gontier E, et al. Is there penetration of titania nanoparticles in sunscreens through skin? A comparative electron and ion microscopy study. Nanotoxicology. 2008;2(4):218–31.
40. Gamer AO, Leibold E, van Ravenzwaaay B. The in vitro absorption of microfine zinc oxide and titanium dioxide through porcine skin. Toxicol In Vitro. 2006;20(3):301–7.
40. Filipe P, et al. Stratum corneum is an effective barrier to TiO_2 and ZnO nanoparticle percutaneous absorption. Skin Pharmacol Physiol. 2009;22(5):266–75.

Nanotechnology in Nail Care

3

Doug Schoon

3.1 Nail Anatomy and Structure

The nails of the fingers and toes can be divided into several major parts that function together to create the nail unit, as shown in Fig. 3.1.

3.1.1 The Nail Plate and Distal Phalange

The nail plate is a hardened, elongated, and curved structure that covers and protects the dorsal aspect of the distal phalanges, which is important since they are smaller and more delicate than the proximal or intermediate phalanges. Ossification of these bones begins during the eighth week of fetal life and continues development for another 18–20 years before reaching full maturity [1]. The nail plate helps protect against distal phalangeal fractures, which are most commonly caused when the fingertip is crushed by a perpendicular force.

As with all types of hair, claw, horn, or quill, the chief structural protein in the nail plate is alpha keratin. Nail keratin is a very tough and flexible material that is well suited to withstand the rigors and stresses to which it is often subjected. Ancient people relied on their natural nails as tools, as well as for protection, so their nails needed to be tough and durable. Almost two-thirds of the keratin found inside nail cells exists as coiled strands of amino acids stacked neatly into tight bundles called fibrils. The fibrils are embedded in semi-solid keratin with a gel-like consistency and encased within a cell membrane. The fibrils stacks are arranged end to end and lay perpendicular to the direction of nail growth. This arrangement helps prevent propagating cracks from traversing the length of the nail plate and helps ensure crack growth is predominately across the width of the nail plate. Figure 3.2 demonstrates several potential crack failure modes for the natural nail plate.

3.1.2 Nail Matrix, Lunula, and Plate Growth

The most important component of the nail unit is the matrix, located several millimeters above the dorsal midpoint of the distal phalange. Nail cells are produced from specialized epithelia tissue found in the matrix region, which when surgically exposed appears as a small, glistening opaque, bluish-white, almost rectangular patch of tissue. An excellent histological description of this region is provided by Hashimoto et al. [2]. The keratin cells produced here become the building blocks of the nail plate, stacking 100–150 nail cells deep to create a thin, yet tough and durable structure. The overwhelming majority of nail cells are produced in the matrix area. However, nuclear DNA labeling in the nail bed basal cells suggests that a

D. Schoon, Ms (✉)
Schoon Scientific and Regulatory Consulting, LLC, 33935 Crystal Lantern St., Dana Point, CA 92629, USA
e-mail: dschoon@schoonscientific.com

A. Nasir et al. (eds.), *Nanotechnology in Dermatology*,
DOI 10.1007/978-1-4614-5034-4_3, © Springer Science+Business Media, LLC 2013

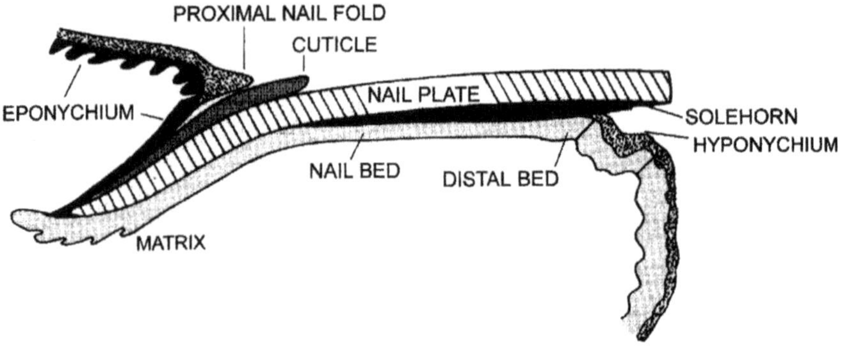

Fig. 3.1 Cross-section of the nail unit. (Courtesy Nail Structure and Product Chemistry, Second Edition, D. Schoon)

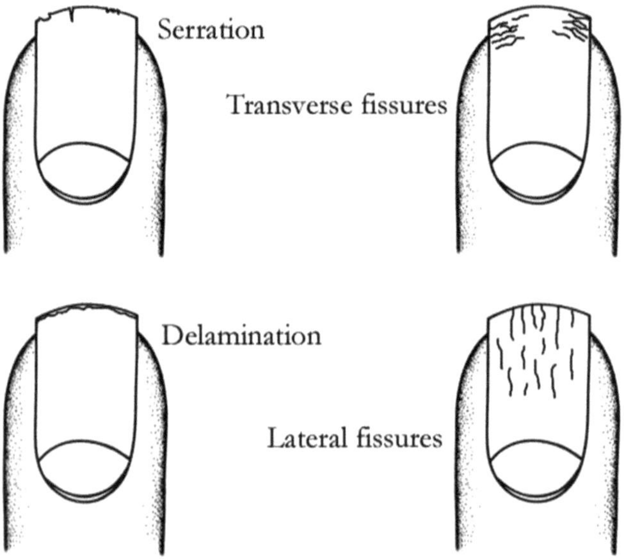

Fig. 3.2 Common nail plate crack failure modes. (Courtesy Nail Structure and Product Chemistry, Second Edition, D. Schoon)

germinative cell population is situated immediately distal to the lunula. A very detailed description of the nuclear DNA labeling studies and a discussion of nail growth kinetics are provided by Zaias [3]. The size and shape of the matrix area determines plate thickness, width, and curvature. For instance, a matrix with greater width produces a wider plate, while those with a matrix of greater length produce a thicker nail plate.

The lunula is the most distal portion of the growth matrix area, usually visible at the base of the nail plate and colloquially referred to as the nail "moon." This could in part account for the observation that the nail plate above the lunula is slightly thinner and easier to puncture than the distal plate, especially with sharp implements, so manicuring this area must be performed with care. Interestingly, the thumb with the largest lunula is most often on the dominant hand. The curvature of the matrix area correlates to the curvature of the plate, while the shape of the matrix's distal edge determines the shape of the distal nail plate, often referred to as the "free edge" in lay text. A recent study by Murdan quantified the transverse curvature dimensions of adult nail by gender, age, hand dominance, and hand size [4]. Animals with

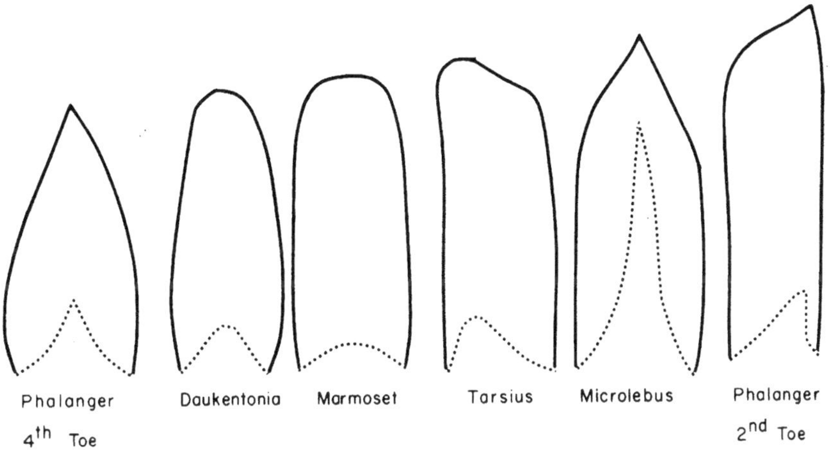

Fig. 3.3 The correlation of the claw-like nail of primates with the shape of the lunula

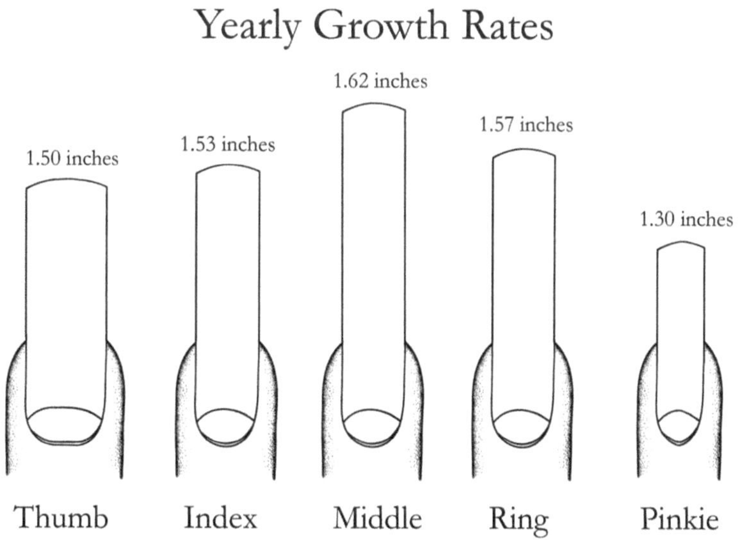

Fig. 3.4 The yearly nail plate growth rate for each finger

different shaped lunula have claws that also match as shown in Fig. 3.3 [5].

When the keratin nail cells are formed in the matrix, they are pushed upward and outward by the rhythmic flow of newly developed nail cells. The newer cells push older nail cells along a predetermined pathway until the new growth emerges from under the proximal nail fold of the eponychium on its journey toward the distal plate. When these cells leave the matrix they are plump and whitish in appearance, but quickly flatten and become highly translucent as they undergo normal cell differentiation [6]. On the average, the normal thumbnail will grow about 0.1 in. (0.25 cm)/month or 1.5 in. (3.8 cm)/per year, therefore requiring 5 or 6 months to completely replace the entire nail plate. Toenails grow much slower than fingernails and may take a year or longer to replace. As a rule, the longer the finger the faster the nail plate will grow as shown in Fig. 3.4 [7].

Nail plates grow about 20 % faster in the summer; men's nails grow faster than women, especially on the dominant hand. Nail plates grow fastest during pregnancy (Fig 3.5), which is more likely due to hormonal changes than prenatal vitamins, as commonly believed in some places.

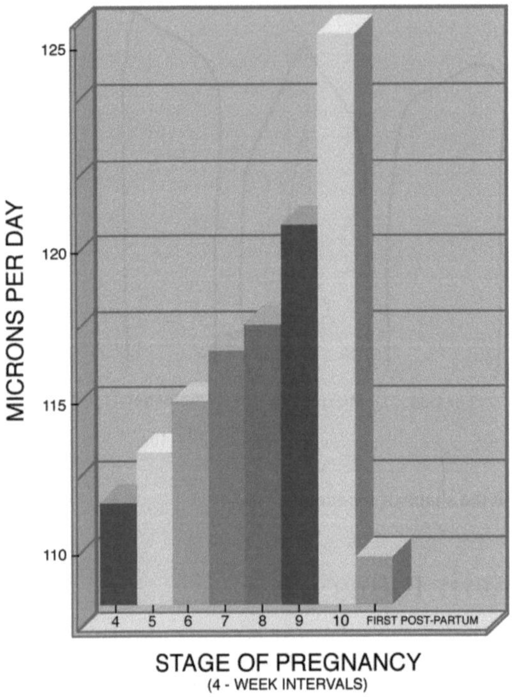

Fig. 3.5 Nail growth during the last two trimesters of pregnancy

Table 3.1 Ranges of results of elemental analysis of healthy nail plates reported in various studies

Element	mg/g
Aluminum	0.00000000045
Boron	0.007–0.006
Calcium	0.67–0.806
	3.64 ± 1.77
Copper	0.29–0.89
Gold	0.00044 ± 0.0006
Iron	0.29–0.64
Lead	0.0097–0.24
Magnesium	0.1–0.1
	3.45 ± 1.5
Phosphorus	0.00008–0.00027
Silicone	0.17–0.54
Silver	0.000000048
Sodium	2.4 ± 1.8
	3.4 ± 1.4
Strontium	0.00000016
Sulfur	36.6 ± 2.1
Titanium	0.000016
Zinc	0.106–0.154

Nail plate growth increases by about 3.5 % between the fourth and eighth month of pregnancy and between the ninth month until just after delivery, nail growth rate increases by 20 %. Age also affects the growth rate. Growth peaks between the ages of 10 and 14 years and slowly declines after age 20. Nail biting, accidental damage or loss causes nail plates to grow faster and while other factors slow plate growth, e.g., immobilization or paralysis, poor circulation, malnutrition, lactation, serious infections, psoriasis, and certain medications [8].

Various analysis of nail clippings have found a wide range of elements including iron, aluminum, copper, silver, gold, titanium, phosphorus, zinc, sodium, and calcium [9]. Each of these is found in extremely low concentrations. Sodium, magnesium, and calcium have been reported over a much wider range than others, probably due to their abundance in the environment. Much of the calcium associated with nails is likely isolated to the surface and a result of hand washing in hard water (Table 3.1).

3.1.3 Nail Plate Ultrastructure

Cross-sections of the plate demonstrate that various regions of the matrix produce visibly different types of nail plate as shown in Fig. 3.6. Three distinct layers are seen via scanning electron microscopy (SEM). The dorsal layer which produces a harden surface is derived from the proximal matrix, while the more fibrous region in the center of the plate is produced by the intermediate matrix and a thin layer seen on the ventral plate is derived from the lunula and the adjacent nail bed.

Sulfur cross-links are a special type of covalent bond created between two cysteine amino acids found on separate protein chains. These two sulfur-containing amino acids can link these chains together like rungs on a ladder, converting a collection of single chains into an ultra-tough, net-like structure. Cross-linking also creates surfaces that are more resistant to stains or the damaging effect of solvents, cleaners, etc. The high level of cysteine cross-linking in keratin is the primary source of hardness, strength, and rigidity. Individual nail cells are locked into a gigantic, interconnecting network chiefly by desmosomes, ring-like

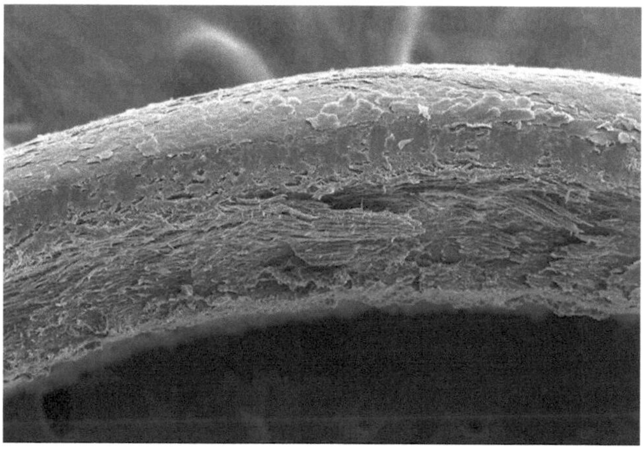

Fig. 3.6 Three distinct layers in nail plate cross-sections are seen via SEM

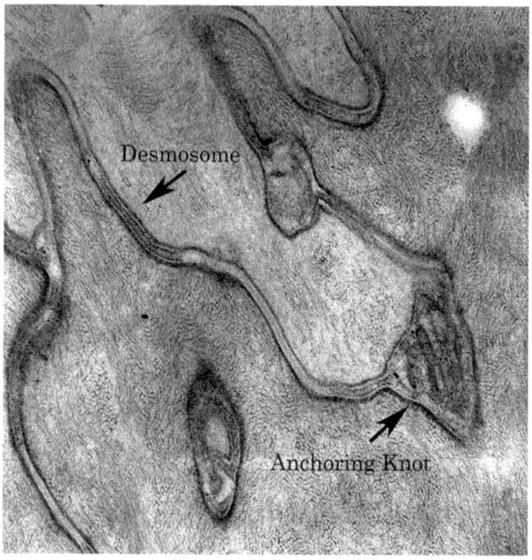

Fig. 3.7 Ventral part of the nail plate showing desmosomes and anchoring knot between two corneocytes

junctions which act like a tiny rivet between two nail cells; each cell contains several of these permanent connections [10]. Desmosomes are a unique type of cell junction. The circular junctions are partially hollow, but not empty. Desmosomes are filled with keratin fibrils, which further reinforce the connection between each cell, making the nail plate a continuous, inter-connected mass of keratin fibrils as shown in Fig. 3.7. The cell wall can become contorted and folded into anchoring knots which also play an important role and help bind cells together.

Table 3.2 Amino acid and sulfur percent content of human nails, hair, and stratum corneum

Amino acid	Nail	Hair	Stratum corneum
Lysine	3.1	2.5	4.2
Histidine	1.0	0.9	1.5
Arginine	6.4	6.5	3.8
Aspartic acid	7.0	5.4	7.9
Threonine	6.1	7.6	3.0
Serine	11.3	12.2	13.6
Glutamic acid	13.9	12.2	12.6
Proline	5.9	8.4	3.0
Glycine	7.9	5.8	24.5
Alanine	5.5	4.3	4.4
Valine	4.2	5.5	3.0
Methionine	0.7	0.5	1.1
Isoleucine	2.7	2.3	2.7
Leucine	8.3	6.1	6.9
Tyrosine	3.2	2.2	3.4
Phenylalanine	2.5	1.7	3.2
Half-cystine	10.6	15.9	1.2
Sulfur	3.2	4.5	1.4

Many natural oils and proteins are found in the nail plate. Many studies indicate the main oils found in the nail plate are cholesterol and squalene and in greater abundance than found in human sebum. Recent proteomic analysis of the nail and hair has provided interesting insights about which proteins are typically found in these structures [11, 12]. An amino acid analysis and sulfur content of nail plates compared with human hair and the stratum corneum is found in Table 3.2 [13].

3.1.4 Nail Folds and Related Structures

The skin bordering the proximal nail plate is called the eponychium. It does not end at the nail plate, but instead folds back underneath to create a tight seal which prevents pathogens or contaminants from gaining access to the matrix area. The eponychium covers approximately 20 % of the nail plate and serves to help protect and cushion the matrix. This tissue is often mistakenly confused with the cuticle. The cuticle is actually dead tissue that derives from the ventral eponychium to adhere and bind tightly to the growing nail plate causing a thin layer of tissue to detach "hitch a ride" on the nail plate as shown in Fig. 3.8. The cuticle is a vital part of the seal that protects the matrix from pathogenic invasion, which explains why this area should be treated with care when manicuring.

On each side of the plate are the lateral nail folds which create tight seals along the sidewalls. The underside to the nail plate is covered with a tissue called bed epithelium which adheres tightly to the bottom of the nail plate. This tissue allows the nail plate to slide along the nail bed as it grows. The nail bed has many ridges and groove-like channels and the softer bed epithelium flows into these channels which help to guild the plate's growth [14]. The bed epithelium remains attached and when found on the underside of the nail plate beyond the distal bed, this thin layer of cloudy, yellowish tissue is referred to as the "solehorn." As the nail plate grows past the distal nail bed, the bed epithelium is temporarily forced together into a small zone located just before the point where the nail plate grows beyond the tip of the finger nail. This zone stretches the width of the nail plate to form the onychodermal band, a grayish, glassy-looking narrow zone visible through the nail plate. This zone coincides with another seal which protects the distal nail bed from infection, the hyponychium. This tissue forms a watertight seal that prevents pathogenic invasion of the nail bed.

3.2 Properties of the Nail

Nail protein chains linked together by sulfur-to-sulfur cross-linkages across two cysteine amino acids to create chemical bonds which are difficult to break. Nail plates and hair are among the most highly cross-linked tissues in the body, which is one reason for their great strength, toughness, and durability. Cross-linking also helps to regulate porosity to both hydrophobic and hydrophilic substances. Dense cross-linking increases the nail's barrier resistance by reducing absorption rates of most substances, other than water, into the nail plate, included drugs and cosmetic ingredients. The combination of crystalline keratin fibrils arranged in an orderly fashion and embedded in non-crystalline keratin gel-like fractions provides the healthy nail plate with much of its physical

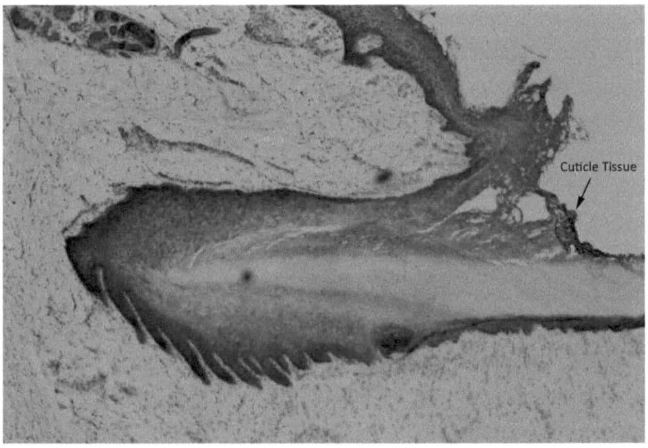

Fig. 3.8 The cuticle tissue derives from the ventral side of the eponychium

strength, hardness, flexibility, toughness, and wear resistance. It is important to understand what these terms mean, in order to understand the physical properties of the natural nail plate.

- Strength is the ability to resist cracking or breaking when a load (stress) is applied. This property is often confused with durability or toughness, which is described below.
- Flexibility is a material's ability to bend to absorb an applied force or impacts that may normally lead to fractures. Normally, nail plates are highly flexible and will usually bend before reaching the breaking point. Repeated or long-term exposure to harsh cleaners and solvents can also make nail plates brittle and less flexible. Environmental effects such as heat, cold, sunlight, and moisture can affect flexibility.
- Toughness demonstrates the ability of a material to yield and deform under an applied force. The durability of the natural nail is largely due to its toughness, which in turn can be attributed to the synergistic balance between the nail plates inherent strength and flexibility. If either became out of balance in the nail plate, toughness will be diminished.
- Brittleness is the property associated with a nail plate that lacks toughness, usually due to a lacking of sufficient flexibility.
- Hardness is the measure of a surfaces resistance to scratching or denting when an external force is applied. Nail plates that are softer than normal are more prone to scratching, peeling, or pitting. Harder than normal nail plates are more susceptible to fracturing, chipping, or splitting. For instance, over use of nail hardening products may cause overly soft nail plates to become excessively hard and develop reduced flexibility, causing the natural nail to become more brittle. Hardness can be measured in many ways, but the most common method utilizes the Mohs scale (ratings range from 1 to 10), which is useful to determine relative scratch or abrasion resistance. Hard surfaces are found on materials that are either weak or strong, so clearly hardness should not be confused with strength or durability.
- Wear resistance is another other property that affects durability. Wear resistance is the ability to resist abrasion or rubbing. The nail plate has a lower hardness (Mohs 2.5) than the materials coating the surfaces of nail files. These coatings are usually either white crystalline silicon carbide (Mohs hardness 9), or a less aggressive black crystalline aluminum oxide coating (Mohs hardness 7.5).

A more detailed discussion concerning the bulk and surface properties of materials is available [15]. Normal nail plates on average contain about 18 % water by weight and plate flexibility is highest when it is completely saturated with water, about 25 % by weight [16]. The diffusivity of water into the plate at 32 °C was determined from cadaver finger nails having water contents ranging from 0.536 g H_2O/g dry nail to 0.035 g H_2O/g dry nail that increase with increasing RH to achieve a more than 400-fold increase at 100 % RH, which was consistent with measurements of other keratinized tissues including wool, horn, and the corneocyte phase of stratum corneum [17]. The combination of the concentration-dependent water diffusivity with the natural water activity gradient in nail in vivo leads to the prediction of a nonlinear steady-state water concentration profile in human nail in vivo which, in turn, has implications for drug and cosmetic ingredient delivery. Even so, too much moisture can weaken the plate and lead to damage. Nails that are constantly exposed to water can become soft and weak. But, too little moisture can leave the plate dry and brittle, since water is a chief plasticizer for the plate. Typically the natural oil content of the plate (lipids such as squalene and cholesterol) has a lower concentration than water in the plate, 5 % compared to 18 % moisture, respectively. Even so, both play a significant role in nail plate flexibility and toughness.

3.3 Consumer Nail Care Needs and Challenges

Dehydration of the nail plate will lower flexibility and reduce toughness. Many solvents and surfactants can remove plasticizing moisture and oil. Excessive exposure to dish washing detergent and common solvents, e.g., acetone, ethyl acetate, and methyl ethyl ketone, can contribute to nail plate brittleness. Damage becomes more likely when solvent exposure is excessive. Unlike the dry surface appearance caused by infrequent solvent exposure, e.g., nail polish remover, which quickly corrects itself as the nail plate seeks to maintain the established equilibrium gradient that normally exists throughout. Oils and moisture are constantly migrating from outside environments, nail bed, and surrounding skin. Surfactants found in detergents, cleaners, and soaps solubilize and remove oils from nails and surrounding skin. Excessive hand washing and over use of hand sanitizers are often the culprits. Nail cells created near the back of the matrix comprise the topside of the nail plate and also must travel further to reach the distal nail plate, making nail cells on top of the nail plate almost 2 months older that those on the bottom of the plate. In combination with additional opportunity for environmental exposure, it is not surprising that nail plates tend to peel at the dorsal surface of the nail plate, rather than the ventral surface. Nail hardening products often contain approximately 0.6–1.2 % methylene glycol (1–2 % formalin), which is the protein cross-linking agent in formalin and not "formaldehyde" as has been incorrectly taught for over 100 years [18]. Overly flexible nail plates are problematic for those who desire long nail plates, since excessive flexing can create stress fractures in the lateral side walls of the plate. Regular application of nail hardener stiffens the plate, but over use creates excessive cross-linking, reduction of flexibility, increases stiffness, and reduced durability. Consumers misinterpret this course of events to assume they should increase usage of nail hardeners, which perpetuates the problem. This misuse of nail hardeners may lead to localized irritation, increased potential for allergic dermatitis, onycholysis, subungual hyperkeratosis, and even subungual hemorrhages have been reported [19]. Such products should be used only for thin and/or overly flexible nail plates and should be discontinued when the desired hardness is reached and before the onset of nail brittleness.

Other common problems include sagittal ridging of the nail plate which is normally seen with advancing age; however transverse ridging or transverse leukonychia may result from overly aggressive use of manicure implements in the lunula area or improper use of high pH cuticle removers (pH 12–14). Nail discoloration can result from a number of reasons. Improper sugar metabolism in diabetics can increase the concentration of yellow glycosylation end products in the plate [20]. Yellows, deeper shades of red and brown and green nail polish often contain soluble dyes that can pool just below the upper surface to cause mild staining, depending on nail conditions and porosity [21]. Other nail polish dyes such as D & C Red no. 6, 7, and 34; FD & C yellow no. 5 and their corresponding lake pigments have been implicated. The lake pigments may become destabilized in some solvent blends and allow fugitive dye to bleed into the nail plate. Surface granulation has been occasionally observed in some individuals who apply fresh coats of enamel on top of old, worn enamel, for several weeks in succession which can result in superficial friability [22]. However, it is a common practice among nail polish wearers to pick or pry polish from the nail plate and the destructive forces created by scraping or prying implement (or teeth) can damage the nail plate. For example, small pieces of UV cured colored nail coatings residue can remain on the nail plate after improper removal procedures. These remaining areas are tightly adhered to the nail plate so forcible prying or scraping will result in similar surface disruptions as shown in Fig. 3.9. This damage was created by a wooden pusher used to force a small piece of UV cured nail color from a nail. This area is slightly smaller than the diameter of a human hair.

When occurring in clusters these disruptions can appear as superficial, diffuse white to off-white areas that are typically 0.01–0.1 in (0.25–2.5 cm). Onycholysis can result from overly zealous manicuring or injury, e.g., improper use of abrasive files or excessive elongation of the distal nail plate beyond the free edge with an artificial nail can increase the potential of an accident, leading to this type of physical trauma.

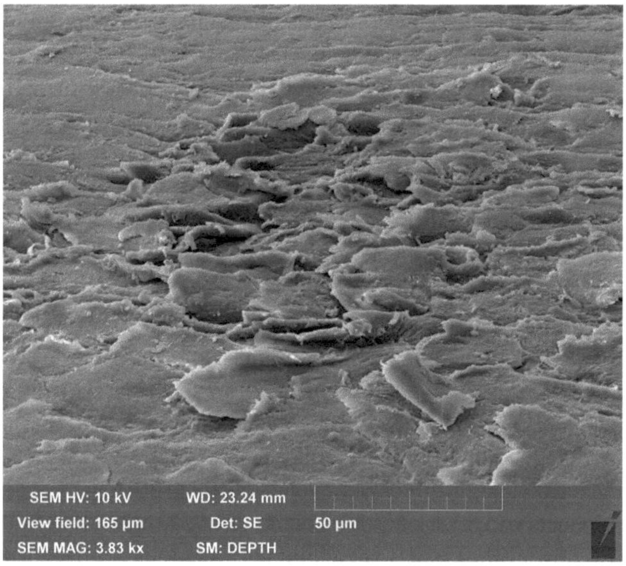

Fig. 3.9 Surface damage to the dorsal nail caused by improper removal of a UV-curing colored nail coating

3.4 Use and Future Nanotechnology Applications for Nail Care

Presently, the main uses of nanotechnology for nail care are mainly restricted to manicure products claiming to use delivery systems based on nanoencapsulation technologies such as nanoliposomes; vascular structures within aqueous core surrounded by a hydrophobic lipid bilayer usually created by extrusion of phospholipids which are "generally regarded as safe" materials (GRAS). Liposomes can vary in size from 15 nm to several microns, but nanostructures are typically 50–100 nm. Liposomes are typically too brittle, therefore a new type called a transfersome (200–300 nm) has been developed which is claimed to be more flexible and have improved efficiency [23]. These are self-assembling lipid droplets with elastic bilayers and improved penetration through the stratum corneum which may be better suited to overcome the highly efficient barrier properties of the nail plate. Nanoliposomes are also useful for enhanced delivery to the eponychium and nail folds. These could be used to deliver biomolecules, antioxidants such as tocopherol or lycopene, or plasticizing agents such as jojoba oil or vasodilators, e.g., caffeine.

Niosomes are nonionic surfactant base vesicles that are otherwise similar to phospholipid-type liposomes and are also formed via self-assembly in aqueous solutions [24]. These are claimed to be more suitable for delivery of poorly absorbed ingredients, but relative to liposomes, are suspected to have an increased potential for irritating because they do not use GRAS surfactants. Due to low permeability of the nail plate, these may be more useful tools for delivering useful ingredients into the natural nail, e.g., moisturizers, plasticizers, antioxidants, etc.

Solid lipid nanoparticles (SLN) are nanometer size particles within a lipid matrix that is normally solid at body temperature [25]. They are used when the encapsulated ingredient must be protected from degradation over prolonged periods, e.g., UV absorbers [26]. Nanostructured lipid carriers (NLC) are similar to SLN except their matrix structure is distorted, creating spaces to accommodate guest ingredients, giving them a high loading capacity and therefore may be a better choice for nail-related applications. Both SLN and NLC have properties more suited to these types of applications than nanoliposomes. A useful application of these technologies could be to deliver lubricating molecules which could increase wear resistance and while regulating water diffusion rates in the nail plate, which would level and control dimensional stability of the nail plate. When the nail plate is soaked in water it will significantly swell only to shrink again as excess moisture is lost through evaporation. This can lead to internal stress damage in the plate which could later serve as starter seeds for crack development, especially for brittle nails.

Nanotechnology has been used in manicure products as carriers for enzymes, zinc oxide, and anti-inflammatory ingredients, but these uses have been limited to very few product offerings. The Woodrow Wilson Project on Emerging Nanotechnologies, a web-based nanotechnology tracking report, presently reports 424 uses of nanotechnology in cosmetics and personal care products of which only four claims to be useful for finger or toe nails. The claim purposes stated by manufacturers are for improved hydration and/or elasticity, UV protection, and more efficient penetration of useful ingredients.

Although not a cosmetic issue, other than their effect on nail appearance, the medical condition onychomycosis (tinea unguium) is a fungal infection of the nail plate that affects about 8 % of the adult population. The certain members of the fungus genus Trichophyton are common organisms implicated in fungal infections on the finger and toe nail. Delivery of active ingredients to the infection site is of key importance. To maximize drug efficacy it is important to maintain the appropriate concentrations of the active ingredient at the target location for specific duration, while minimizing unwanted side effects. Lotions and creams have proven ineffective on nail plates because of their highly resistant natural barrier properties which significantly hinder appreciable penetration of most substances. To improve penetration of active agents, nanocarriers in various forms have been developed to include nanoemulsions, nanoliposomes, and quantum dots. Quantum dots are extremely small nanoparticles, typically 10 nm or less, and they present several unique advantages. Quantum dots can be magnetized and guided to specific places in the body by utilizing an appropriate electromagnetic field. Quantum dots can also be made to carry two different drugs which can be released at different times depending on the strength of a remotely generated electromagnetic pulse. Although use of nanotechnology in delivery of effective antifungal therapies is in its infancy, this is likely to be a valuable use of this technology. This type of technology could also be useful for other targeted natural nail treatments and therapies. Certain types and sizes of quantum dots are also known for their ability to emit visible light of various wavelengths. These may perhaps be used in the future to produce a nail polish that may be reversibly switched between a range of colors and/or patterns.

3.4.1 Dendrimers and Unsymmetrical Hyper-branching Polymers

Dendrimers are unimolecular, micellular structures around 20 nm in size having well-defined, systematically and symmetrically branched structures with chemically functional groups position at their periphery that are tailored to be either chemically reactive or inert. Dendrimers have the advantages of covalently bonding to the nail plate. These 3D structures are capable of holding guest molecules, however of likely greater importance are the peripheral surface chemical functionalities which create a type of nanoscaffold or array [27]. With sufficient density these highly functionalized materials have the ability to significantly modify the surface properties of the nail plate. Multifunctional peripheral groups are also possible. Utilizing a wide range of chemical functional groups it may be possible to specific tailoring of the surface properties of the natural nail plate and to enhance or diminish existing characteristics to improve its overall properties, e.g., stain resistance, improved adhesion or reactivity with other surface coatings, moisture/diffusion regulation, gloss, hardness, UV protection, penetration enhancement of useful substances such as antioxidants, crack/damage repair, etc. If the appropriate chromophores could be attached to a dendrimer, it may be possible to control the coloration and/or appearance of the nail plate. In future applications these may be used to create interchangeable patterns or designs on the plate.

Hyper-branched polymers are similar to dendrimers, but with one notable difference, these are highly unsymmetrical with disorganized structures. These are usually made in a single polymerization step, while dendrimer synthesis typically involves many multiple steps. Hyperbranched polymers are therefore easier to make and probably more cost-effective for cosmetic applications. Patents have already been issued for use of these types of materials to use in cosmetic surface treatments, e.g., hair, skin, and nail care [28–30].

3.4.2 Artificial Nails

Much of the discussion has focused on the use of nanotechnology in natural nail applications; however, there is also potential applicability for improvements to artificial nail coatings. Improving critical properties of artificial nail coatings is important to their usefulness and future success. Presently artificial nails suffer from several disadvantages, adhesion loss, potential for microbial colonization at the natural nail/artificial nail interface as well as cracking, chipping, scratching, dulling, discoloration from staining or UV exposure, etc. Many of these problems may be solved by use of the appropriate nanomaterials. Artificially coating the nail with self-repair agents would lead to targeted nail repair, and could improve nail integrity, color stability, and clarity if the agents were encapsulated in nanoparticles capable of release responsive to nail damage. Also an important breakthrough would be obtained by gaining better control over the adhesion of artificial nail coatings. If adhesion could be easily reversed to facilitate simple and quick removal, this could help minimize the potential for damage caused by improper or overly aggressive removal techniques or procedures.

3.5 Additional Research Is Needed

Other than the need for safety substantiation, as discussed elsewhere in this book, there exists a need for additional research to gain a better understanding of interaction of nanomaterials with the nail. Nanometrology uses advanced techniques to make scientific measurements at the nanoscale. Atomic force microscopy (AFM) or other types of scanning probe microscopy can be used for measuring surface area and topography, surface friction, and scratch resistance of the natural nail plate and can be used to monitor other surface changes induced by nanomaterials, such as self-assembly of monolayers or covalent attachment of dendrimers. For example, AFM has been previously used to study stress-induced fatigue in hair [31]. Nanomeasurement of mechanical properties, such as the stress–strain curve of an isolated collagen fibril under constant and cyclic loading conditions, has been previously reported [32]. Tip-enhanced Raman spectroscopy is a unique application for scanning probe microscopy that combines chemical and structural characterization capability (NPL) [33] This is done by combining Raman spectroscopy chemical characterization abilities with the spatial resolution of atomic force microscopy to achieve chemical characterization at resolutions of less than 30 nm. Several types of nanoindentation measuring devices have been developed and are also useful for measuring surface properties such as hardness and scratch resistance. Indentation time can also be used to measure important properties such as toughness, yield resistance, and elastic flow or to monitor nanoscale strains and develop stress maps.

Improved understanding of nanomaterial deposition and their effects on the properties of both natural and artificial nail coatings is needed. Obtaining this information is likely to require the use of advanced techniques to measure the usefulness of nanomaterials for nail cosmetic applications. This type of instrumentation and expertise is not a normal competency for most cosmetic companies, so it likely to take time before nanomaterials begin to provide true benefits for nails; however, it is expected that drug-deliver application for treating onychomycosis will continue and improve since these are being driven by much larger and more technologically sophisticated pharmaceutical companies. For these reasons and others, joint collaborations between the pharmaceutical and cosmetic companies would have a great potential to produce exciting innovations in this category.

References

1. Gray H. Anatomy of the human body, 20th ed. Lea & Febiger, 1918; Bartleby.com, 2000. www.bartleby.com/107/. Accessed Jan 2012.
2. Hashimoto K, Gross BG, Nelson R, et al. Ultrastructure of the skin of human embryos. The formation of the nail on 16–18 week old embryos. J Invest Dermatol. 1966;47(3):205–17.
3. Zaias N. The nail in health and disease. 2nd ed. Stamford: Appleton & Lange; 1990. p. 33–44.
4. Murdan S. Transverse fingernail curvature in adults: a quantitative evaluation and the influence of gender, age, and hand size and dominance. Int J Cosmet Sci. 2011;33(6):509–13.
5. LeGros C. The problems of the claw in Primates. Proc Zool Sovc. 1936;1:1.
6. Kitahara T, Ogawa H. Coexpression of keratins characteristic of skin and hair differentiation in nail cells. J Invest Dermatol. 1993;100(2):171–5.
7. Schoon D. Nail structure and product chemistry. 2nd ed. Mason: Cengage Learning; 2005. p. 20.
8. Hewitt D, Hillman R. Relation between rate of nail growth in pregnant women and estimated previous general growth rate. Am J Clin Cutr. 1966;19:436.
9. Zaias N. The nail in health and disease. 2nd ed. Stamford: Appleton & Lange; 1990. p. 33–44.
10. Delva E, Tucker D, Kowalcykz A. The Desmosome. Cold Spring Harb Perspect Biol. 2009;1:a002543.
11. Rice R, Xia Y, et al. Proteomic analysis of human nail plate. J Proteome Res. 2010;9(12):6752–8.
12. Rice R. Proteomic analysis of hair shaft and nail plate. J Cosmet Sci. 2011;62(2):229–36.
13. Gupchup G, et al. Structure and permeability of the human nail. J Cosmet Sci. 1999;50:363–85.
14. Sher R, Daniel C. Nails: therapy, diagnosis, surgery. 2nd ed. Philadelphia: W.B. Saunders; 1997. p. 29.
15. Ratner B, Hoffman A, et al. Biomaterial science—an introduction to materials in medicine, Part 1. Boston: Academic; 1996. p. 11–35.
16. Schoon D. Nail structure and product chemistry. 2nd ed. Mason: Cengage Learning; 2005. p. 35.
17. Gunt H, Miller M, Kasting G. Water diffusivity in human nail plate. J Pharm Sci. 2007;96(21):3353–62.
18. Bedino JH. The Expanding Encyclopedia of Mortuary Practices. Champion Company; Springfield, OH: 2001. No. 649, p. 2619–20.
19. Lazar P. Reactions to nail hardeners. Arch of Dermatology. 1966;94:446–8.
20. Oimomi M, Maeda Y, et al. Glycosylation levels of nail proteins in diabetic patients with retinopathy and neuropathy. Kobe J Med Sci. 1985;31:183–8.
21. Calnan C. Reactions to artificial colouring materials. J Soc Cosmet Chem. 1967;18:215–23.
22. Baran R. Cosmetics and the fingernails; old and new facts. Society of Cosmetic Scientists symposium on recent advances in skin biology in relation to cosmetics, Nov, 1985, Bristol. Paper 14, p. 11–12.
23. Cevc G. Transfersomes, liposomes and other lipid suspension on the skin: permeation enhancement, physical penetration, and transdermal drug delivery. Crit Rev Ther Drug Career Syst. 1996;13: 257–388.
24. Uchegbua I, Vyas S. Non-ionic surfactant base vesicles (niosomes) in drug delivery. Int J Pharm. 1998;172(1–2):33–70.
25. Muller R, Radtke M, Wissing S. Solid lipid nanoparticles (SLN) and nano structured lipid carriers (NLC) in cosmetic and dermatological preparations. Adv Drug Deliv Rev. 2002;54 Suppl 1:S131–55.
26. Song C, Liu S. a new healthy sunscreen system for human: Solid lipid nanoparticles as carrier for three, four, five-trimethoxybenzoylchitin and the improvement by adding vitamin E. Int J Biol Macromol. 2005;36:116–9.
27. Svenson S, Tomalia D. Dentrimers in biomedical applications-reflections on the field. Adv Drug Deliv Rev. 2005;57:2106–29.
28. Maignan J, Genard S. Use of hyperbranched polymers and dendrimers comprising a particular group as film-forming agents, film-forming compositions comprising same and use particularly in cosmetics and pharmaceuticals, US Pat. 6432423. Paris: L'Oreal; 2002.
29. Tournilhac F, Simon P. Cosmetic or dermatological topical compositions comprising dendritic polyesters, US Pat. 6287552. Paris: L'Oreal; 2001.
30. Franzke M, Steinbrecht K, et al. Cosmetic compositions for hair treatment containing dendrimers or dendrimer conjugates, US Pat. 6068835. Darmstadt: Wella; 2000.
31. Seeshadri I, Bhushan B. Effect of ethnic city and treatments on in situ tensile response and morphological changes of human hair characterized by atomic force microscopy. Acta Mater. 2008;56:3585–97.
32. Eppel S, Smith B, et al. Nano measurements with micro-devices: mechanical properties of hydrated collagen fibrils. J R Soc Interface. 2005;3(6):117–21.
33. NPL http://www.npl.co.uk/.

Nanoscience on Shampoo and Conditioner

4

Peter D. Kaplan

4.1 Introduction

The nanoscience of shampoo and conditioners touches on our everyday experience and bridges the traditional field of colloid science to emerging topics in nanoscience and biomaterials. This chapter, written for dermatologists, aims to demystify the hair care process focusing on the interactions of water, shampoo, and conditioners with hair as a biomaterial. The practical goal, to provide language that can be used when discussing hair care products, is met by reading through the discussion of the physical interactions between proteins, water, surfactants, oils, polymers, and hair. We conclude by reviewing several nanotechnology-based approaches found in hair care today as well as some emerging approaches we expect to see in future products.

4.2 Structure of the Hair Fiber

Known for its strength, durability, flexibility, and relative ease of acquisition, hair has long been a preferred subject in the study of biomaterial structure. Especially notable was the use of hair and wool in the early study of protein structure by X-ray scattering [1, 2]. In the scientific community, protein structure sometimes overshadows structure at fiber level by the microscopy community. Knowledge of fiber structure, however, is critical to understanding care and conditioning technologies.

The fiber's layered structure also lends itself to microscopic study. As we look with greater and greater resolution, each magnification step reveals something new. By eye, on the millimeter scale, we see a fine fiber notable for its color and curl. Human terminal fiber widths range from just over 1/20th to just over 1/10th of a millimeter. The most striking differences between fibers are related to curl. Fibers naturally curl with a radius of curvature of anywhere from a few millimeters to tens of centimeters. Zoom in a bit and view the fiber in cross section and you will generally observe an elliptical cross section, especially in the shorter and kinkier hair often found in Africans. The cross section of the long straight hair typically found in Asians tends to have a rounder cross section.

With micron resolution, we find three distinct zones in the cross section. The outermost cuticle consists of thin hard scales surrounding the cortex, which is a mix of highly organized hydrophobic columns embedded in a largely disordered "matrix." These cortical columns are made principally of keratin, which forms intermediate filaments with their own hierarchical structure [3, 4]. At the center of the fiber we often find a medulla. The medulla is the one part of the fiber that retains the shape of the progenitor keratinocytes and does not show the effects of collapse

P.D. Kaplan, Ph.D. (✉)
220 Park Street, Montclair, NJ 07042, USA
e-mail: peter.kaplan@gmail.com

A. Nasir et al. (eds.), *Nanotechnology in Dermatology*,
DOI 10.1007/978-1-4614-5034-4_4, © Springer Science+Business Media, LLC 2013

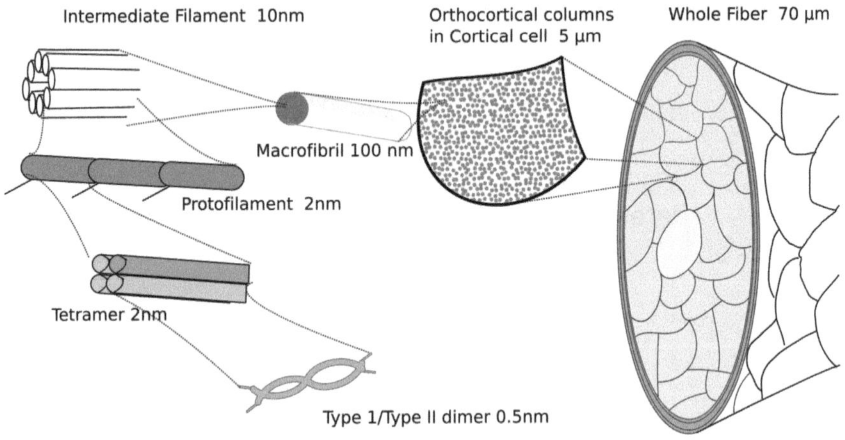

Fig. 4.1 Hierarchical structure of the hair fiber. Key concepts: structural keratin is both hydrophobic and highly structured in bundles in the cortex that are embedded in hydrophilic proteins

and hardening associated with the apoptotic degranulation that occurs during hair growth (Fig. 4.1).

At nanometer scale using electron microscopy, the detailed structure of cuticular cells and cortical columns can be probed. At chemical scales we have long studied fibers through elemental and amino acid analysis [5] but in this era of proteomics, there is an opportunity to acquire new information about protein modification and protein reactivity as a result of fiber damage [6, 7].

4.2.1 Cuticle

The cuticle includes the physical surfaces that cover the fiber. These surfaces are critically important in hair care. And a simple analysis of dimensions shows a tremendous amount of surface present on each styled head of hair. Using round numbers, a human head with about 300,000 15 cm long fibers will have a fiber surface area of 100,000 cm^2 or a fiber/scalp surface ratio of 700:1. By way of contrast, a human arm, covered with a lower density of fine vellus fibers, has a fiber/skin surface area ratio closer to one to one.

The cuticle consists of layers of thin, pancake-shaped cells. At a pitch of roughly 10 µm, each 0.3 µm thick cuticle cell has a hard, dense outer layer over a soft easily swollen base. On emergence from the scalp, there are typically about ten layers of cuticle cells, laid out like crackers tilted with a 10-µm pitch. After hundreds of washes, years of brushing, and other sources of fiber aging, several inches from the scalp, we may find only a few layers of cuticle, typically with rough looking, apparently cracked cell boundaries (Fig. 4.2).

Cuticle structure explains a great deal about how hair care products work. The hard, brittle hydrophobic epicuticle on the outer surface does not take up water. It bends but does not stretch; it is essentially impervious to almost everything that goes in or out of the fiber including water. The fiber would be rigid and brittle if this hard barrier was not divided into pieces, individual cuticles, each of which sits above a flexible endocuticular layer. Stated simply, the fiber might not be able to swell in water without the pathway through the endocuticule to the cortex. The epicuticle is only a few nanometers thick. It is generally considered to be covered with a monolayer of lipid (18-MEA) that may play an important role in reducing fiber friction and surely plays a huge role in interaction with product. Under the epicuticle is roughly 100 nm of exocuticle, which is also highly cross-linked, rigid, and dense with cross-linked keratin. The porous endocuticle is far less cross-linked and composed primarily of globular proteins that plasticize easily in water.

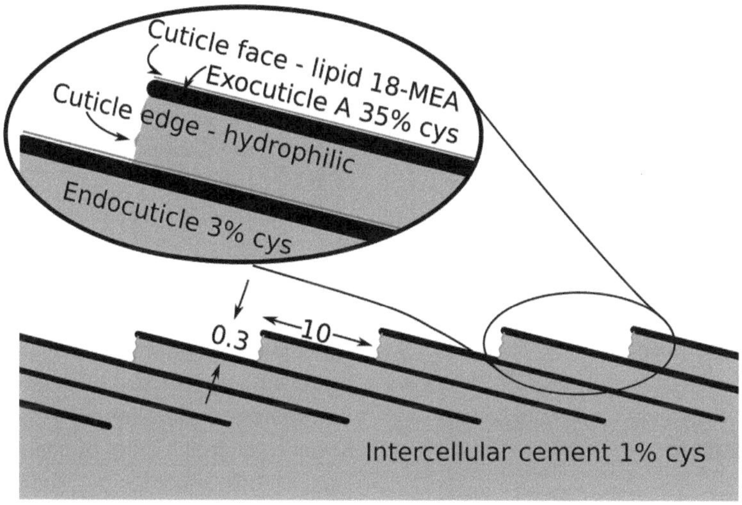

Fig. 4.2 Structure of cuticle demonstrating levels of cross-linking and role of cuticle edge and endocuticle in delivery into the fiber

We have discussed the structure of cuticle cells in some detail. To understand the role of the cuticle in the hair fiber, we need to zoom out and see how it is connected to the rest of the fiber (Fig. 4.2). The leading edge of the cuticle is often cracked, exposing the endocuticle, which reaches down into fiber. We should consider the cuticle edges as part of the hair surface. In fact, the simplest functional way to think about the cuticle is that it is a surface with two zones—the hydrophobic cuticle face and the hydrophilic and permeable cuticle edge. Hair care products deliver beneficial agents through the cuticle edge while caring for and lubricating the cuticle surface.

4.2.2 Cortex

In many semi-technical discussions the cortex becomes the synonym for the fiber and the keratin in the cortex is the only source of fiber strength and resiliency. This view encourages the use of keratin as a prominent hair care ingredient. This, too simple, explanation leaves out the fact that hair swells in water and neglects the possibility that other proteins and peptides may be as good or better than keratin at stabilizing fiber structure.

Like myosin and microtubulin, keratin forms carefully assembled intermediate filaments that provide mechanical strength and chemical connectivity (Fig. 4.1). Hair keratins are largely bundled into cortical columns. These are found in two orientations one runs parallel and the other orthogonal to the fiber axis. These well-structured columns are responsible for the most easily analyzed part of the X-ray scattering data from the fibers. When the fiber is stretched, the patterns tighten as expected. But when swollen with water, the patterns remain largely unchanged (at room temperature and zero strain) which is evidence of the bundles' hydrophobic nature [8, 9]. On the other hand, the fibers do take on 25 % of their weight in water plumping their diameter by about 5 % while oddly increasing their length barely at all [10]. What is the material that swells? Primarily the globular proteins found in the matrix between the cortical columns.

The proteins of the matrix include the keratin-associated proteins (KAP) which is a larger, more diverse family of proteins spread more widely through the genome than keratin[11]. Work on specific amino acids damaged in hair coloring, ultraviolet and mechanical damage has a rich and detailed history while research on the specific proteins and domains is in its early days. The location of all the keratins in the hair have been imaged and mapped over the last decade [12–14]. While it is not always clear what to do with this information, the biggest change in our understanding of hair structure in the last few decades is knowledge of exactly which keratins and KAPs are where in the fiber. This provides considerable detail to the more generic description given earlier in this chapter and in other reviews that focus on disulfide content and general descriptions of structured vs. open domains within the fiber [15].

4.3 Products

4.3.1 Protein Swelling and Surfactants

Proteins swell in water; as they swell, different residues are exposed creating the opportunity to bind ingredients, in particular surfactants. Protein denaturing gels use salt, heat, and sodium dodecyl sulfate (SDS) to unfold essentially any protein into a surfactant-coated linear structure. And so it is no surprise that shampooing is a treatment with salt, heat, and SDS. This mix not only cleans the fiber surface, it allows for the small rearrangements needed to bind beneficial agents such as conditioning polymers and peptides. Each surfactant system has its own protein interactions. In the skin care literature there is considerable attention paid to surfactant protein interactions in the pursuit of mildness [16–20]. Zein solubility is one common measure of irritation potential in skin. While there is little indication, however, that mildness to hair can or should be measured the same by Zein solubility, it still makes sense to consider the swelling that results from protein surfactant interactions when considering approaches to structured interactions with the mostly proteinaceous hair fiber.

4.3.2 Shampoo

Shampoo at its simplest is a surfactant solution made thick by choosing a surfactant mix that forms long thin worm-like micelles. These solutions foam easily and are effective cleansers. The simplest shampoos are generally made using sodium lauryl sulfate (SLS) as a primary surfactant and betaine as a secondary surfactant. With salt and appropriate pH adjustment this is a good cleaning system.

Shampoo as marketed today is far more complex. Surfactant choices have changed over the years and now most products favor a mix of ethoxylated surfactants for improved mildness. Other choices of surfactants include using sugar-based surfactants and substituting ammonium for the sodium counterion. Far more interesting, however, is the ubiquity of conditioning ingredients in modern shampoos. Originally marketed as two-in-one (shampoo plus conditioner) products, today the vast majority of marketed shampoos have significant conditioning benefits including better alignment, lubricity, and shine [21]. Many conditioning shampoos include a hydrophobic conditioning agent, often silicone oil, in an aqueous system designed to remove oil. The tricks used to deposit oil are a real example of nanotechnology in consumer products.

The most discussed approach to a conditioning shampoo is the use of dilution deposition polymers. These are polymers with the unusual behavior that they precipitate when diluted. Dilution-deposition has been well studied [18, 22] and is used to deliver conditioning actives, most commonly silicone oils, in shampoos and body washes [22–24].

4.3.3 Conditioners

Conditioners are typically made by mixing at least two lipids one of which has a positively charged head group that will bind to the cuticle surface, especially to the disorganized cuticle edge and to damaged sites in the 18-MEA layer and making the surface smoother and more uniform. The most common combination is cetearyl alcohol and the positively charged cetrimonium chloride. Together, after appropriate processing these molecules with alkyl chains of 16–18 carbons form a lameller liquid-crystalline phase.

The mode of action for conditioners is simpler than for conditioning shampoos. Conditioners stick and hopefully do not rinse away too quickly [25]. Binding studies of conditioners and shampoos typically show that a large fraction of the bound material adheres to the cuticle edge—with much of the penetration of large molecules being found within the first fraction of a micron into the cuticle [26]. Since this is also the location that water enters and leaves the fiber, the cuticle edge is a key location for conditioning treatments.

Commercial conditioners typically contain multiple mesophases. Often there is an emulsion of active ingredients and a liquid-crystalline phase with the more basic conditioning agents. The function of the emulsified ingredients varies between conditioners but they typically deliver lubricity, shine, and any actives intended to penetrate into the fiber.

4.4 Nanoscience of Hair Care

4.4.1 Drexler's Nanotech Vision

The nanotechnology movement was strongly influenced by K. Eric Drexler's analysis of science behind potentially very small machines published in the early 1990 [27]. This book taught profoundly simple approaches to analyzing the possible and summarized some of the key phenomena that occur with objects bigger than a nanometer and much smaller than a micron. Drexler's vision included self-assembling nanorobots that could be used, ribosome like, to create new nano-bots. The vision was also neutral with respect to how exactly this would be accomplished; it is clearly that biological nanotechnology such as self-replicating molecular systems lay some of the groundwork for human engineered nanotech [28]. Chemical scale engineering relies on phenomena such as self-assembly and the phase behavior of surfactants, polymers, lipids, and other ingredients that are commonly used in personal care products. In the case of hair care, the fact that the fiber itself is highly structured on the nanometer scale makes hair a natural substrate for treatment by nanotechnology. To discuss nanotechnology in hair care, we will begin with the preexisting nanotechnology of basic hair cleansing and continue with specific examples of current and future nano-approaches to conditioning and coating hair fibers.

4.4.2 Review: The Nano-Story of Washing Hair

During the days and nights before shampooing, the enormous surface area of a hairstyle is exposed to the environment, which includes not only particles of dirt, smog, and smoke but also the fluctuating humidity that causes the fibers to swell, shrink, and bend. In addition, sebaceous oil wicks from scalp along the fibers forming a thin coating that can actually hold particles and also creates the weak adhesion between fibers, gradually causing hair to look heavy and greasy on the head.

The wash begins with lots of water. Salts from the fiber begin to migrate out [29]. Hydrophobic domains and oily spots on the surfaces are wet by the force of fingers massaging the hair. The water swells the fibers, stretching oily droplets and pulling at debris lying against cuticle edges.

Shampoo is then diluted and rubbed creating a suspension of tiny oil particles, cationic polymers, and lots of spherical micelles [22]. These various materials are rubbed against the hair, destabilizing and suspending oily deposits and particulate dirt. At the same time some of the product ingredients begin adhere to the hair, mostly through ionic bonding. The surfactants coating the fiber form a highly lubricious hexatic phase which makes the hair manageable during shampooing. Even as this phase rinses away, conditioning agents continue to provide some lubrication during rinse as the original oils and dirt particles are carried away.

Next, conditioner is rubbed into the style, mechanically presenting a variety of surface-active agents [30] to the cuticle face and edge. These complex phases lubricate the surface and gradually allow binding, conditioning agents to stick to damaged surfaces. Binding is especially driven by quaternary ammonium groups that create a replacement oily layer on the cuticle face. During conditioning, small molecules including amino acids, small peptides, and antioxidants have the opportunity to slowly diffuse into and through the endocuticle. Some of these molecules even reach the cortex of the fiber. This diffusion process is partly reversed during the rinse, which continues long enough for the hair to feel uncoated. Hopefully the user stops rinsing before all the conditioning agents have diffused back out of the fiber.

Following rinse, the fiber dries. During drying, water leaves previously hydrated residues exposed resulting in little change to the cortical columns but enormous changes in the rigidity of the matrix and hence at least doubling the fiber's elastic modulus [7, 25].

4.4.3 Dilution Deposition

Most materials, when added to water, are at first a cloudy suspension and then gradually clear as we add enough water and wait enough time for them to fully dissolve. Many clear shampoos, interestingly, go through this process in reverse. They become cloudy, due to the formation of a precipitate, when diluted. Schematically, we understand this as follows. The shampoo contains a solution of charged polymers that are associated with counterions from salts and surfactants. These polymer–surfactant complexes are electrically neutral and fully dissolved. In solution they behave, in a statistical mechanical sense, as a single object whose entropy has contributions from both its position and its configuration. On dilution, there is the opportunity for surfactants and counterions to dissociate gaining their own positional entropy. Due to the shear number of countercharges, their separate entropy eventually overcomes other energies causing the suspended polymer to become insoluble, falling out of solution as a suspended coacervate. The scaling argument in the footnote is an example of the kind of scaling and thermodynamic argument often used in nano-engineering.[1] There are many additional details and complications to this simple explanation of dilution deposition, for example something must join the coacervate to balance its charge. These complications are at the heart of the underlying physical chemistry and substantially all the published literature focuses on polymer–surfactant interactions rather than the overall thermodynamics of precipitation. It is the complications that comprise the art and science of formulating shampoos. We note here that coacervates can carry or associate with a wide variety of benefit agents including silicones, polymers, emulsified droplets of beneficial oils, or even solid particles.

4.4.4 Amino Acids–Peptides–Proteins

Drexler described future nano-bots circulating in our bodies seeking out and repairing damage. Since our hair is regularly damaged by environment and chemical treatments, there is a strong desire to deliver nano-bots or at least their equivalent into the fiber to repair it. As the damaged material has a significant amount of structural keratin, it is generally accepted that keratin itself is a good candidate for a repair material. With a molecular weight in excess of 50,000 Da, however, intact monomeric keratin will not penetrate into the hair fiber. Seeking alternatives, it is common to find fragments of keratin or other proteins in hair care products. The composition of these nano-ingredients includes both uniquely synthesized and random peptides: protein hydrolysates

[1] A polymer chain in solution has $S = Sp + Sc = (1/2)kT \ln V + (3/2)k(R/b)2/N$ where k is Boltman's constant, T the temperature in Kelvin, R the polymer end to end length, N the length of the polymer measured in units of the persistence length b. On dilution the polymer can give up Nc counterions increasing the positional entropy by $NcSp$ and decreasing the configurational entropy of the polymer, which collapses into a non-flexible aggregate by Sc. The overly simplified result is that the cloud point is found when $\ln(V/V0) > 3(R/b)2(Nc/N)$. Since $R/b – N1/2$ and $Nc – N$, the dilution factor for any particular polymer/counter ion system ends up depending on the polymer molecular weight $V/V0 – \exp(MW)$.

as small as single amino acids; and amino acids and their mixtures. There is evidence that many of these materials can penetrate into the fiber and there is also evidence that once penetrated, these materials can modify the mechanical properties of the fiber [31–34].

4.4.5 Fluoropolymers

Looking for materials with stronger ability to repel water, there has been interest in adapting Teflon-like fluoropolymers to hair care [35, 36]. To this end there are several patents and some products that claim to use these polymers to reduce frizz and improve manageability. The literature on the subject is primarily either marketing material or published in patents and is therefore difficult to review in this chapter.

4.4.6 Surface Binding

The cuticle surface is well known to change from hydrophobic to hydrophilic after the use of bleach, relaxer, permanent wave, and other chemical treatments. It is generally accepted that this is a result of the loss of molecules of 18-MEA from the epicuticle's surface lipid layer. Chemically this is generally a result of the cleavage of a thioester bond which leaves a negatively charged –SH4 group exposed. Cationic quaternary ammonia groups are attracted to the ionic sulfates. In addition to Alkyl quats (behenyl quat, etc.) conditioners are often formulated with ingredients from the family of polyquats. Each polyquat has at least one and often several cationic ammonia groups. The polyquats are named in the order that they were registered as cosmetic ingredients. Commonly used examples are PQ-10 (a functionalized cellulose), PQ-7 (which includes polyacrylamide), and PQ-22 (which includes polyacrylic acid). These provide different degrees of binding and substantivity. Polyquat binding is not specific to the surface of the fiber.

The concept of specific binding, the generation of molecules that will be attracted to hair and not to other biological surfaces, is intriguing and could be accomplished by any one of the many routes used to develop targeted therapies in pharmaceuticals including specific molecular design, screening libraries of compounds (existing compounds and randomly generated compounds), and directed evolution [37, 38].

4.5 Conclusions and Future Directions

We have reviewed the nanoscale structure of the hair fiber and its interaction with shampoos and conditioners. This has helped the reader build an understanding of the mode of action of today's hair care at the level of nanoscience (also known as colloid and surface science). As our understanding and technology improves, this nanoscale understanding of biomaterials and treatment systems will continue to create new technologies and modes of action.

There are many stories of how advanced products will protect hair fibers. Most of those stories can be understood within the framework of hair care presented in this chapter. Strengthening actives, for example, are generally understood to either improve the fiber's lubricity or transport through the cuticle to the cortex where they bind providing added strength. Materials that claim to seal the fiber may enter the cuticle face and bind to the endocuticle clogging the channels used for transport.

Some directions of ongoing technical interest in the field include specific binding, the use of naturals at a colloidal level rather than as molecular extracts, and the use of molecules that become chemically active during heat treatments.

Hair fibers are organized biomaterials. They are a textbook substrate for self-assembling nanoparticles and smart nanomaterials to usher in the next generation of hair care products.

References

1. Astbury WT, Woods HJ. X-ray studies of the structure of hair, wool, and related fibers. II.—The molecular structure and elastic properties of hair keratin. Philos Trans R Soc London A. 1933;232:333–94.

2. Pauling L, Corey RB. The Structure of Hair, Muscle, and Related Proteins. Proc Natl Acad Sci U S A. 1951;31(5):261–71.
3. Goldman RD, Grin B, Mendez MG, Kuczmarski ER. Intermediate filaments: versatile building blocks of cell structure. Curr Opin Cell Biol. 2008;20(1): 28–34.
4. Dufresne ER, Noh H, Saranathan V, Mochrie SGJ, Cao H, Prum RO. Self-assembly of amorphous biophotonic nanostructures by phase separation. Soft Matter. 2009;5(9):1792–5.
5. Kaplan PD, Polefka T, Grove G, Daly S, Jumbelic L, Harper D, et al. Gray hair: clinical investigation of changes in hair fibers with loss of pigmentation in a photoprotected population. Int J Cosmet Sci. 2011;33:171–82.
6. Rogers MA, Langbein L, Winter H, Beckmann I, Praetzel S, Schweizer J. Hair keratin associated proteins: Characterization of a second high sulfur KAP gene domain on human chromosome 21. J Invest Dermatol. 2004;122(1):147–58.
7. Dyer JM, Plowman JE, Krsinic GL, Deb-Choudhury S, Koehn H, Millington KR, et al. Proteomic evaluation and location of UVB-induced photo-oxidation in wool. Photochem Photobiol B. 2010;98(2):118–27.
8. Strelkov SV, Kreplak L, Herrmann H, Aebi U. Intermediate filament protein structure determination. Methods Cell Biol. 2004;78:25–43.
9. Kreplak L, Franbourg A, Briki F, Leroy F, Dalle D, Doucet J. A new deformation model of hard alpha-keratin fibers at the nanometer scale: Implications for hard alpha-keratin intermediate filament mechanical properties. Biophys J. 2002;82(4):2265–74.
10. Robbins CR. The chemical and physical behavior of human hair. New York: Van Nostrand Reinhold; 1979.
11. Schweizer J, Bowden PE, Coulombe PA, Langbein L, Lane EB, Magin TM, et al. New consensus nomenclature for mammalian keratins. J Cell Biol. 2006;174(2):169–74.
12. Langbein L, Schweizer J. Keratins of the human hair follicle. Int Rev Cytol. 2005;243:1–78.
13. Langbein L, Rogers MA, Winter H, Praetzel S, Schweizer J. The catalog of human hair keratins—II. Expression of the six type II members in the hair follicle and the combined catalog of human type I and II keratins. J Biol Chem. 2001;276(37):35123–32.
14. Langbein L, Rogers MA, Winter H, Praetzel S, Beckhaus U, Rackwitz HR, et al. The catalog of human hair keratins I. Expression of the nine type I members in the hair follicle. J Biol Chem. 1999; 274(28):19874–84.
15. Popescu C, Hocker H. Hair—the most sophisticated biological composite material. Chem Soc Rev. 2007;36(8):1282–91.
16. Ananthapadmanabhan KR, Yang L, Vincent C, Lips A. Recent advances in surfactant interactions with proteins and stratum corneum. J Cosmet Sci. 2009;60(3):377.
17. Yang L, Vincent C, Ananthapadmanabhan KP, Lips A. Enhancing mildness of syndet cleansing bars. J Am Acad Dermatol. 2005;52(3):P37.
18. Schnittger S, Sinha M, Mehta SC, Misra M, Mukherjee S, O'Lenick AJ, et al. The materials science of cosmetics. MRS Bull. 2007;32:760–9.
19. Deo N, Jockusch S, Turro NJ, Somasundaran P. Surfactant interactions with zein protein. Langmuir. 2003;19(12):5083–8.
20. Ananthapadmanabhan KP, Huang L, Lips A, Somasundaran P, Deo N. Mechanism of surfactant induced dissolution of zein protein. Abstr Pap Am Chem Soc. 2002;224:081-COLL.
21. Kaplan PD, Yang K, Park K, Ramaprasad R. Shine and Color: Interplay of Angle-Dependent Optical Measurements and Visual Attributes with Cosmetic Treatment. J Soc Cosmet Chem. 2009;60:111–23.
22. Goddard ED. Polymer/surfactant interaction: Interfacial aspects. J Colloid Interface Sci. 2002;256(1):228–35.
23. Ananthapadmanabhan KP, Goddard ED, Chandar P. A study of the solution, interfacial and wetting properties of silicone surfactants. Colloids Surfaces. 1990;44:281–97.
24. Li W, Jordan SLP, Zhang XD, Amos J, Davis C. Syrnergistic effects of high molecular weight polyethylene oxide (PEO) and cationic cellulosic polymers on conditioning properties of hair care products. J Cosmet Sci. 2004;55(1):123–4.
25. Lee YJ, Rice RH, Lee YM. Proteome analysis of human hair shaft - From protein identification to post-translational modification. Mol Cell Proteomics. 2006;5(5):789–800.
26. Staudgel JA, Bunasky K, Gamsky CJ, Wagner MS, Stump KJ, Baker JM, et al. Use of quaternized cassia galactomannan for hair conditioning. J Cosmet Sci. 2007;58(6):637–50.
27. Drexler KE. Nanosystems: molecular machinery, manufacturing, and computation. New York: Wiley; 1992.
28. Browne WR, Feringa BL. Making molecular machines work. Nat Nanotechnol. 2006;1(1):25–35.
29. Evans AO, Marsh JM, Wickett RR. The structural implications of water hardness metal uptake by human hair. Int J Cosmet Sci. 2011;33(5):477–82.
30. Mancini RA, Hunt MC, Kropf DH. Reflectance at 610 nanometers estimates oxymyoglobin content on the surface of ground beef. Meat Sci. 2003;64(2): 157–62.
31. Gummer CL. Elucidating penetration pathways into the hair fiber using novel microscopic techniques. J Cosmet Sci. 2001;52(5):265–80.
32. Keis K, Huemmer CL, Kamath YK. Effect of oil films on moisture vapor absorption on human hair. J Cosmet Sci. 2007;58(2):135–45.

33. Fregonesi A, Scanavez C, Santos L, de Oliveira A, Roesler R, Escudeiro C, et al. Brazilian oils and butters: The effect of different fatty acid chain composition on human hair physiochemical properties. J Cosmet Sci. 2009;60(2):273–80.
34. Wortmann FJ, Zahn H. THE STRESS/STRAIN CURVE OF ALPHA-KERATIN FIBERS AND THE STRUCTURE OF THE INTERMEDIATE FILAMENT. Tex Res J. 1994;64(12):737–43.
35. Anderson, Griffith D, Puerta, Thomas D, Akcasu, Scott B et al., inventors. In situ polymerization for hair treatment. Cambridge: Living Proof Inc.; 2011.
36. Anderson DG, Nashat A, DeRosa MJ, Puerta DT, McLaughlin RP, Akcasu BS et al., inventors. Hair care compositions and methods of treating hair using same. Cambridge: Living Proof Inc.; 2011.
37. Tao HY, Cornish VW. Milestones in directed enzyme evolution. Curr Opin Chem Biol. 2002;6(6): 858–64.
38. Szostak JW, Bartel DP, Luisi PL. Synthesizing life. Nature. 2001;409(6818):387–90.

Nanotechnology-Based Fabrics

5

Roger Ronghua Wang and Adnan Nasir

5.1 Introduction to Nanotechnology in Textiles

There are no uniform definitions of the term nanotechnology in textiles. Nanotechnology has been defined as systematically arranged functional structures which consist of particles with size-dependent properties by the Hohenstein international textile research center—an internationally recognized textile research and service center based in Boenigheim, Germany [1]. The nanotechnology therefore should only be perceptible to the wearer by means of a demonstrably improved function, and should only have a negligible effect on the textile properties. Any applications existing in theoretical form or as prototypes in textiles would be grounded in the area of nanoscience and nanotechnology defined by the National Nanotechnology Initiative (NNI) [2]. The NNI serves as the central point of communication, cooperation, and collaboration for all Federal agencies engaged in nanotechnology research, bringing together the expertise needed to advance this broad and complex field. Under its definition, nanotechnology is science, engineering, and technology conducted at the nanoscale, from about 1 to 100 nm. Based on the above standards, Nano-Tex [3] has defined that nanotechnology in textiles must meet criteria of (1) technology development at the nanoscaled size (1–100 nm), (2) the ability to control or manipulate the ordered assembly of components at the nanometer scale, and (3) the size and smart assembly of the nanoscale architecture creates demonstrable novel properties with negligible adverse effect on the textile properties.

Nanotechnology-based enhancements for fiber/yarns/fabrics and garments have been one of the most important components of textile processing for end-users. The textile industry has taken an innovative stance to develop high value-added products in order to maintain and further enhance its competitiveness in the global marketplace. Increasing demands for functionality and intelligence of textiles pose great challenges in product development for both civilian and military fields. For civilian uses, the functionality includes comfort, healthcare, protection against water, oil, soil, malodor, sunlight, and so on. For military uses, it is important for the textiles to have proactive ability, to give protection against chemical and biological hazards, and to in the example of combat wear combine bulletproof armor, with lightweight, temperature regulation, and insect repellence, etc. The ability to transfer nature's wisdom into fabrics is one of the key issues of wearable nanobionics.

R.R. Wang, Ph.D., (✉)
Nano-Tex Inc., Room 1-7, 10/F, Wah Yiu Industrial Center, 30-32 Au Pui Wan Street, Shatin, Hong Kong

MaNano-Tex Inc., Room 1-7, 10/F, Wah Yiu Industrial Center, 30-32 Au Pui Wan Street, Shatin, Hong Kong
e-mail: wang.roger@nano-tex.com

A. Nasir, M.D. Ph.D.,
Department of Dermatology, UNC Chapel Hill, Chapel Hill, NC 67552, USA
e-mail: anasir@live.com

A. Nasir et al. (eds.), *Nanotechnology in Dermatology*,
DOI 10.1007/978-1-4614-5034-4_5, © Springer Science+Business Media, LLC 2013

To date, there are some proactive participants on nanotechnology in textiles in academia. The Institute for Soldier Nanotechnologies (ISN) of MIT [4] was founded in March 2002 with a US $50 million 5-year contract (2002–2007, phase I) from U.S. Army Research Office (ARO) and consequently the second 5-year contract (2007–2012, Phase II). ISN's goal is to dramatically improve the survivability of the soldier by working at and extending the frontiers of nanotechnology through fundamental research and transitioning with US Army and industry partners. This mission includes not only decreasing the weight that soldiers carry but also improving blast and ballistic protection, creating new methods of detecting and detoxifying chemical and biological threats, and providing physiological monitoring and automated medical intervention. The ultimate goal is to help the Army create an integrated system of nanotechnologies for soldier protection. The Hong Kong Research Institute of Textiles & Apparel (HKRITA) [5] was founded in April 2006 with a US $34.5 million (HK $269.1 million) 5-year contract (2006–2011, Phase I) from Innovation & Technology Commission (ITC), HKSAR and a consequent 3-year contract (2011–2014, Phase II) with a US $38.3 million (HK $298.8 million). HKRITA is claimed to be a leading center of excellence in research, development, and technology transfer in fashion and textile technologies. Its strategic research areas focus on (1) new materials and textiles and apparel products, (2) advanced textiles and clothing production technologies, (3) innovative design and evaluation technologies, and (4) enhanced industrial systems and infrastructure. National Textile Center (NTC)-University Research Consortium [6] was found in 1991. NTC is a federally funded research consortium of eight universities: Auburn University, Clemson University, Cornell University, Georgia Institute of Technology, North Carolina State University, UC-Davis, UMass-Dartmouth, and Philadelphia University. The institutions share human resources, equipment, and facilities. The final amount granted in 2008 was US $4.7 million (senate request: US $13 million). Its strategic research areas includes (1) design, development and measurement of natural and synthetic polymeric materials and fibers, including polymer mixtures and additives, (2) fabrication, processing and manufacture of fibrous structures and fabricated products, (3) chemical applications to, and modifications of, fibers and fiber substrates, including dyeing, finishing and waste reduction, and (4) management of product design, sourcing, production, distribution and consumption systems.

In the textile industrial sector, Nano-Tex [3] was the first company to bring nanotechnology to the textile industry. It is headquartered in Oakland, California with operations and technical services in North America, South America, mainland China, Hong Kong, Taiwan, South Korea, India, Japan, and Europe. It develops and markets a family of nanotechnology-based textile enhancements that dramatically improve the performance of everyday fabrics. Its diversified products portfolio includes Resists Spills (liquid repellency), Coolest Comfort (moisture wicking for synthetics and wrinkle-free cotton), Resists Spills & Releases Stains, Resists Static (first permanent anti-static treatment for synthetics), Repels and Releases Stains (market's best-performing stain release capabilities with repellency), Aquapel (fluorine free water repellency), Dry-Inside (one-way moisture management), Neutralizer (anti-odorant), Fortify DP Wrinkle-free, etc. It is claimed that common spills that Resists Spills has been shown to protect against can be Coca Cola™, drinking water, milk-whole, orange juice, cranberry juice, apple juice, champagne, cream (half & half), saline solution, generic Cola, ginger ale, coffee, sunflower oil, coffee with milk, Worcestershire sauce, Kool-aid™, grape juice, chocolate syrup, beet juice, red wine, soy sauce, soy milk, chocolate milk, tomato juice, soybean oil, peanut oil, hot sauce, corn oil, salad dressing, and olive oil, etc. Today more than 80 textile mills worldwide use Nano-Tex enhancements in products sold by more than 250 leading apparel and interior furnishings brands. The company's patented technology delivers superior durability and performance without compromising the look, feel or comfort of the fabric, a unique benefit which has driven brand and consumer adoption of Nano-Tex treatment in everyday wear, activewear, outwear, uniforms and interior furnishings.

5.2 Benefits of Nanotechnology in Textiles

The textile industry probably is the first manufacturing industry to come up with nanotechnology-based enhancements. The role of nanotechnology in the textile industry has accelerated with a positive momentum during the past decades. It has served as a catalyst for growth by helping retailers bring differentiation and competitiveness to the selling floor, offering consumers new reasons to buy and driving repeat purchases. Nanotechnology has versatile benefits in textiles with water/oil/soil stains repellency (casual wear, suits, shirts, undergarments, outdoor, home textiles, upholstery), waterproofing (capability to submerge under pressure), moisture retention (wet wipes, facial mask), fire/flame resistance (furnishings), mosquito repellence (military uniform), wrinkle-resistance (shirting, pants), one-way moisture management (sportswear), anti-mold (outdoor patio), antibacterial (medical cases, uniforms), deodorant (carpets, tees), anti-static (synthetics), ultraviolet (UV) protection (shirts, undershirts), anti-reflective coating/moth eye structure, iridescent colors/photonic crystals, insulation, lightweight structure, stain elimination (upholstery), shape memory, anti-fading (outdoor synthetics), scratch resistance (high density polyethylene, lacquers), anti-fog, anti-fingerprint, anti-graffiti, sealing/dappling, diffusion barriers (packaging foils), anti-friction (handbags, footwear), tensile/tearing/impact strength retention (outdoor synthetics), compression (for athletes), and soil release, self-cleaning (carpets, furnishings, upholstery, outdoor patio, tents), etc. Some have been already available in the markets. Some are still on the stage of very early prototypes.

The benefits are encouraged to be assessed by the internationally recognized standard testing method such as Oeko-Tex standard 100 [7], the American Association of Textile Chemists and Colorists (AATCC) Technical Manual [8], ASTM international [9], Guo Biao (GB, Chinese Standards Service Network) [10], Japanese Industrial Standards (JIS) [11], British Standards Institution (BSI) [12], International Organization for Standardization (ISO) [13]. The Oeko-Tex Standard 100 [7] is a globally uniform testing and certification system for textile raw materials, intermediate and end products at all stages of production. AATCC [8] is the world's leading not-for-profit association serving textile professionals. It provides test method development, quality control materials, and professional networking for thousands of members in 60 countries throughout the world. ASTM International [9], formerly known as the ASTM, is a globally recognized leader in the development and delivery of international voluntary consensus standards. Today, some 12,000 ASTM standards are used around the world to improve product quality, enhance safety, facilitate market access and trade, and build consumer confidence. Japanese Industrial Standard Committee (JISC) [11] consists of many national committees and plays central role in standardization activities in Japan. The task of JISC is establishment and maintenance of JIS, administration of accreditation and certification, participation and contribution in international standardization activities, and development of measurement standards and technical infrastructure for standardization. BSI British Standards [12] is the UK's national standards organization that produces standards and information products that promote and share best practice. The BSI was founded in 1901 and subsequently received a Royal Charter which meant that it became the recognized national body responsible for the development and agreement of standards. This was done in response to the demand from buyers that products would meet defined specifications and performance measures. ISO [13] is a non-governmental organization, the world's largest developer and publisher of International Standards with a Central Secretariat in Geneva, Switzerland.

5.3 Examples of Nanotechnology-Based Fabrics in the Marketplace

5.3.1 Antimicrobial

Antimicrobial fibers/yarns/fabrics/garments and textiles are fiber-based substrates to which antimicrobial agents have been applied at the surface, or incorporated into the fibers, rendering a product that kills or inhibits the growth of microorganisms [14]. Antimicrobial treatment of textiles has three key benefits [15], i.e., odor control, providing an improved sense of freshness, protection against discoloration and degradation of fabrics, and improving general cleanliness. The below testing method for bacteria and mildew/fungi are commonly used within the textile industry sector.

Antimicrobial textile tests for bacteria

- AATCC TM 100-2004 [16]: Antibacterial finishes on textile materials: assessment of (quantitative).
- AATCC TM 147-2011 [17]: Antibacterial activity assessment of textile materials: parallel streak method (qualitative).
- AATCC TM 174-2011 [18]: Antimicrobial activity assessment of new carpets (qualitative and quantitative).
- EPA Carpet Sanitization Test Method [19]: Current Environmental Protection Agency (EPA) method for carpet sanitizers.
- JIS L 1902-2008 [20]: Japanese test for antimicrobial activity and efficacy of textile products (quantitative).
- ISO 20743-2007 [21]: Textiles—determination of antibacterial activity of antibacterial finished products (quantitative).
- ASTM E2149-2010 Shake flask method [22]: Standard test method for determining the antimicrobial activity of immobilized antimicrobial agents under dynamic contact conditions (quantitative screening test).

Antimicrobial textile tests for mildew and fungi

- AATCC TM 100-2004 [16]: AATCC TM 100 method modified for fungi (quantitative).
- AATCC TM 174-2011 [18]: Antimicrobial activity assessment of new carpets (quantitative).
- AATCC TM 30-2004 [23]: Antifungal activity, assessment on textile materials: mildew and rot resistance of textile materials (qualitative).

Gao [24] reviewed the requirements for antimicrobial finishing, qualitative and quantitative evaluations of antimicrobial efficacy, the application methods of antimicrobial agents and some of the most recent developments in antimicrobial treatments of textiles using various active agents such as metal and metal salts (e.g., silver, copper, zinc, cobalt), quaternary ammonium salts (e.g., the BIOGUARD® produced by AEGIS Environments with active agent of 3-trimethoxysilylpropyldimethyloctadecyl ammonium chloride), polyhexamethylene biguanide (PHMB), triclosan (2,4,4′-trichloro-2′-hydroxydiphenyl ether), chitosan (the deacetylated derivative of chitin), dyes (e.g., metallic dyestuff), regenerable *N*-halamine compounds and peroxyacids.

Antimicrobial could be combined with other function/performance/protection enhancements while did not compromise the look, hand feel, durability, and comfort of the fabric. The typical market needs focus on the combo among antimicrobial + stain resistant (repellency/proof) + UV protection (for outdoor tents, yacht covers, golf cases, patio fabrics, etc.), antimicrobial + stain resistant (repellency/proof) + deodorant (for carpets, indoor furnishings, upholstery, etc.), antimicrobial + water repellency (for school uniforms, casualwear, denims and medical cases, etc.), antimicrobial + deodorant + stain elimination (for office cubic, outdoor patio, etc.), water repellency + water proof + one-way moisture management (for footwear, outwear), antimicrobial + moisture retention (for nonwoven products wet wipers, face masks, tampons and diapers, etc.), antimicrobial + water repellency + deodorant + anti-static (for synthetics) and antimicrobial + one-way moisture management (e.g., Dry-Inside, Dry Fit) + UV protection (for sportswear), low friction fabrics (for atopic dermatitis, for example Isoda et al. [25]). Currently, more and more tailor-made developments for customized clients appreciate the synergy effect of combination enhancements.

Therefore, the following test methods should be taken into account for parallel assessment and quality control/assurance.

- AATCC TM 22-2010 [26]: Water repellency spray test
- AATCC TM 130-2010 [27]: Soil release oily stain release method
- AATCC TM 118-2007 (revised) [28]: Oil Repellency: hydrocarbon resistance test
- AATCC 193-2007 [29]: Aqueous liquid repellency: water/alcohol solution resistance test
- AS/NZS 4399-1996 [30]: Sun protective clothing—evaluation and classification
- AATCC 16-2004 [31]: Colorfastness to light

A number of laboratories have intensively investigated and reviewed some other topics on actively moving polymers in textile applications [32], polymer nanocomposites for multifunctional finishing of textiles [33], designing the waterproof breathable fabrics [34, 35], modern applications of nanotechnology in textiles [36], fire-protective and flame-retardant coatings [37], flammability of natural fiber-reinforced composites and strategies for fire retardancy [38], surgical gown [39], structures of novel antibacterial agents for textiles [40], insect repellant fabrics having nanocapsules with insecticide [41] and the promising stain eliminations, etc. [42–61].

5.4 Examples of Nanotechnology-Based Fabrics with Clinical Applications

5.4.1 Compression Garments

Well-designed compression garments can be used for the enhancement of athletic performance and reduction of injury. The significant benefit of compression garments is that it increases venous return, helps guard exercise induced muscle damage (EIMD) and fatigue, accelerates recovery processes, reduces lactic acid build up during prolonged effort, remains lightweight and comfortable, increases strength and power, improves endurance, increases muscle oxygenation, improves thermoregulation, reduces in-flight ankle edema and wicks sweat away from the body to prevent chafing and rashes. Common brands are Adidas, Compressport, Easton, McDavid, Nike, Skins, Under Armour and Zensah, etc. The material (normally polyurethane spandex-type and Lycra spandex) and functional/geometric properties (one-way moisture management) of fabric for compression sportswear are vital in achieving compression effects. Lin et al. [62] evaluated and optimized the performance of fabric using the design of experiment (DOE) methods, the response surface method (RSM), and the finite element (FE) model. In order to enhance sports performance through gradient compression garment design, McLaren [63] developed a wearable wireless monitoring device which included six 18 mm diameter low profile pressure sensors that were placed on the body at various positions, e.g., on the calf, thigh, and buttocks regions, to sense garment pressure due to the dynamic motion in the range of 5–50 mmHg. It was claimed to be acceptable for further field studies in terms of portability, dynamic response, accuracy, precision, resolution, and general utility.

5.4.2 Fabrics with Photonic Textiles

Electronics firms, fashion houses, and medical-equipment suppliers are all busy developing textiles that blend photonics technology with materials science [64]. The idea of trying to integrate wearable electronics and photonics has been more than a decade [65, 66]. Customized and wearable health monitoring devices, incorporated into textiles and garments, are expected to become medical attendants in the future for continuous and autonomous monitoring of vital physiological indicators of professionals, patients, and elderly persons [67–69]. Objects such as dancing clothing, towels, upholstery, flags, and drapes would seem unlikely places on which to place intelligent and interactive systems. However efforts have been made from both industry sector e.g., Philips and France Telecom and academia e.g., X. M. Tao's group from the Hong Kong Polytechnic University have developed their own technology and demonstrated clothes that boast dynamic displays that the user can control at will. Lumalive led by the Philips Research Photonics Textiles group have been demonstrably striking due to the use of the surface fabrics to diffuse the light from the LED array. Other products from the counterparts like Lumitex's Bili-Blanket and LightMats. Biomedical sensors, integrated into textiles would enable monitoring of many vitally important physiological parameters during our daily life. Rothmaier et al. [70] demonstrated the design and performance of a textile-based pulse oximeter, operating on the forefinger tip in transmission mode. In their experiments thy used a two-wavelength approach (690 and 830 nm) for pulse wave acquisition and arterial oxygen saturation calculation. The results show that textile-based oximetry is feasible with sufficient data quality and its potential as a wearable health monitoring device is promising. Fink and colleagues have developed fabrics impregnated with nano-optical fibers capable of longitudinally transmitting light incident on the fabric. With computer processing, images exposed on woven fabric can be reconstructed, resulting in a wearable camera [71]. This type of device may have utility in monitoring skin lesions such as nevi and dermatidies. Fabrics can also be designed to transmit laser light [72]. If done in a pattern to match skin lesions, a phototherapeutic garment could be manufactured to treat complex patterns of skin disease, such as guttate psoriasis.

5.5 Examples of Smart Clothing for the Future

Smart and intelligent clothing is made from fabrics that are wireless and washable that integrate computing fibers and materials into the integrity of the fabrics. The pioneers include Georgia Tech (e.g., GTWM), MIT, University of California at Davis and the Hong Kong Polytechnic University (e.g., RadianFab) that conduct research in the area of intelligent textiles and apparel. Georgia Tech developed a Wearable Motherboard (GTWM), which was initially intended for use in combat conditions. It is currently being manufactured for commercial use under the name Smart Shirt. Sensatex/Lifelink is manufacturing the Smart Shirt. The company plans to develop relationships with firefighter groups, doctors, and others in order to create wearable motherboards that meet their different needs. The commercial applications for the Smart Shirt are: medical monitoring, disease monitoring, infant monitoring, obstetrics monitoring, clinical trials monitoring, athletics, biofeedback, and military uses.

5.6 Safety Concerns of Nanotechnology Fabrics

Textiles are a relatively small sector for nanotechnology. More than 250 European and US fashion houses offer garments with nanotechnology enhancements which are supplied by the Oakland-based nanotechnology fabric company Nano-Tex. Issues of chemical safety and environmental responsibility are the subject of increasing public interest. The EPA and counterparts are studying many compounds to determine the safe levels of these materials in the environment to protect human health. In recent years about 200 compounds, both synthetic and derivative have been found in the blood stream of the general population, according to U.S. Centers for Disease Control data. Scientists and watchdogs however are uncertain about the health implications of these compounds, many of which circulate in trace amounts. Levels considered safe have yet to be explored. In the textile industry sector, several factors have been taken into account for safety. One of the most important standards is Oeko-Tex standard 100. The standard applies to the analysis of the fully processed and finished textile fabric or garments, in all cases the finished textiles must be tested by a certified test institute to determine if it complies with the Oeko-Tex standard 100. Nanoparticles are given special consideration due to the widespread application of nanoscaled TiO_2, ZnO, silver, etc. Schilling et al. [73] conducted a human safety review of nanoscaled titanium dioxide and zinc oxide. Their conclusion is based on the current weight of evidence of all available data, the risk for humans from the use of nanostructured titanium dioxide (TiO_2) or zinc oxide (ZnO) currently used in cosmetic preparations or sunscreens is considered negligible. Other concerns about the textile auxiliary chemicals include perfluorooctane sulfonate (PFOS) and perflurorooctanoic acid (PFOA). Trace amounts of PFOA/PFOS should be controlled in some repellency products as an unintended byproduct of the manufacturing process. Typical levels of PFOA in the repellency products are favorable when they are below 1 ppm or undetectable. Regarding skin safety, Nano-Tex has claimed in the following statement. "Fabrics treated with Nano-Tex technologies have been tested in both animals and humans for their potential to cause skin irritation and skin sensitization. In all of the studies, the treated fabrics were considered to be non-irritating and non-sensitizing. Clinical testing was performed under occluded patch conditions and the results were reviewed by a board-certified dermatologist. Under these stringent conditions, fabrics treated with Nano-Tex technologies were non-irritating and non-allergenic."

5.7 Regulatory Concerns of Nanotechnology Fabrics

Currently, there is no regulation controlling the release of nanomaterials manufactured in the U.S. into the environment. Existing regulations on the textile industry include standards, such as requirements of flame retardancy, sun protection, water repellency spray test, soil release, oily stain release, oil repellency, hydrocarbon resistance, aqueous liquid repellency, water/alcohol solution resistance, wrinkle recovery of fabrics, appearance, colorfastness to light and antimicrobial properties. These standards will likely also apply to nanotextile products.

5.8 Summary

Nanotechnology is making great strides in the consumer and defense textile industry, from smart fabrics, to stain, bacteria, and friction resistant fabrics, to photonic textiles capable of diagnosing and treating skin disease. While many products are already available on the market, the industry is still in its infancy and poised to move forward at an accelerating rate as materials science, biology, and information technology converge. Currently available data do not indicate safety issues with nanotextiles.

References

1. http://www.hohenstein.com/en/home/home.xhtml
2. http://www.nano.gov
3. http://www.nanotex.com/index.html
4. http://web.mit.edu/isn/index.html
5. http://www.hkrita.com/html
6. http://www.ntcresearch.org
7. http://www.oeko-tex.com
8. http://www.aatcc.org
9. http://www.astm.org
10. Chinese Standards. http://www.cssn.net.cn
11. Japanese Industrial Standard. http://www.jisc.go.jp/eng/jis-act/index.html
12. British Standards Institution. http://www.bsigroup.com
13. http://www.iso.org
14. Tanner BD. Antimicrobial fabrics-issues and opportunities in the era of antibiotic resistance. AATCC Rev. 2009;9(11):30–3.
15. Swofford HW. Antibacterial efficacy of polymers containing triclosan and other antimicrobial additives. Am J Infect Control. 2001;29(6):428–30.
16. AATCC TM 100-2004. Antibacterial finishes on textile materials: assessment of. AATCC Technical Manual, vol. 85; 2010, p. 142–4.
17. AATCC TM 147-2011. Antibacterial activity assessment of textile materials: parallel streak method. AATCC Technical Manual, vol. 85; 2010, p. 251–2.
18. AATCC TM 174-2011. Antimicrobial activity assessment of new carpets. AATCC Technical Manual, vol. 85, p. 314–6; 2011.
19. EPA Carpet Sanitization Test Method: current EPA method for carpet sanitizers. U.S. Environmental Protection Agency.
20. JIS L 1902-2008 (JAFET/JSA). Testing for antibacterial activity and efficacy on textile products, Japanese Industrial Standard; 2008.
21. ISO 20743-2007. Textiles—determination of antibacterial activity of antibacterial finished products. 1st ed; 2007.
22. ASTM E2149-2010. Standard test method for determining the antimicrobial activity of immobilized antimicrobial agents under dynamic contact conditions. ASTM International; 2010.
23. AATCC TM 30-2010. Antifungal activity, assessment on textile materials: mildew and rot resistance of textile materials. AATCC Technical Manual, vol. 85, p. 76–9; 2010.

24. Gao Y, Cranston R. Recent advances in antimicrobial treatments of textiles. Text Res J. 2008;78(1):60–72.
25. Isoda K, Takagi Y, Kitahara T, Sano Y, Sugano I, Umeda-Togami K, et al. Treatment of cloth with a fabric softener ameliorates skin dryness. J Dermatol. 2011;38(7):685–92.
26. AATCC TM 22-2010. Water repellency spray test. AATCC Technical Manual, vol. 85, p. 65–7; 2010.
27. AATCC TM 130-2010. Soil release oily stain release method. AATCC Technical Manual, vol. 85, p. 215–7; 2010.
28. AATCC TM 118-2007 (revised). Oil repellency: hydrocarbon resistance test. AATCC Technical Manual, vol. 85, p. 187–9; 2007.
29. AATCC 193-2007. Aqueous liquid repellency: water/alcohol solution resistance test. AATCC Technical Manual, vol. 85, p. 376–8; 2007.
30. AS/NZS 4399-1996. Sun protective clothing—evaluation and classification.
31. AATCC 16-2004. Colorfastness to light. AATCC Technical Manual, vol. 85, p. 23–34; 2004.
32. Hu JL, Chen SJ. A review of actively moving polymers in textile applications. J Mater Chem. 2010;20(17):3346–55.
33. Gowri S, Almeida L, Amorim T, Carneiro N, Souto AP, Esteves MF. Polymer nanocomposites for multifunctional finishing of textiles—a review. Text Res J. 2010;80(13):1290–306.
34. Mukhopadhyay A, Midha VK. A review on designing the waterproof breathable fabrics part I: fundamental principles and designing aspects of breathable fabrics. J Ind Text. 2008;37(3):225–38.
35. Mukhopadhyay A, Midha VK. A review on designing the waterproof breathable fabrics part II: construction and suitability of breathable fabrics for different uses. J Ind Text. 2008;38(1):17–25.
36. Sawhney APS, Condon B, Singh KV, Pang SS, Li G, Hui D. Modern applications of nanotechnology in textiles. Text Res J. 2008;78:731–9.
37. Weil ED. Fire-protective and flame-retardant coatings—a state-of-the-art review. J Fire Sci, online. doi: 10.1177/0734904110395469; 2011.
38. Chapple S, Anandjiwala R. Flammability of natural fiber-reinforced composites and strategies for fire retardancy: a review. J Thermoplast Compos Mater. 2010;23:871–93.
39. Behera BK, Arora H. Surgical gown: a critical review. J Ind Text. 2009;38(3):205–31.
40. Simoncic B, Tomsic B. Structures of novel antimicrobial agents for textiles—a review. Text Res J. 2010;80(16):1721–37.
41. Xin JH, Fei B. Insect repellant fabrics having nanocapsules with insecticide. United States Application US2009/0010977 A1; 2009.
42. Wang RH, Wang XW, Xin JH. Advanced visible-light-driven self-cleaning cotton by Au/TiO2/SiO2 photocatalysts. ACS Appl Mater Interfaces. 2010;2(1):82–5.
43. Wang RH, Xin JH. Advanced visible light self-cleaning treatment for textile materials. In: The 86th textile institute world conference 2, conference proceedings, p. 89–96; 2008.
44. Wang RH, Hu ZG, Liu YY, Lu HF, Fei B, Chan WL, et al. Self-assembled gold nanoshells on biodegradable chitosan fibers. Biomacromolecules. 2006;7(10):2719–20.
45. Wang RH, Xin JH, Tao XM. The UV-blocking property of dumbbell-shaped ZnO crystallites on cotton fabrics. Inorg Chem. 2005;44(11):3926–30.
46. Wang RH, Xin JH, Tao XM, Daoud WA. ZnO nanorods grown on cotton fabrics at low temperature. Chem Phys Lett. 2004;398(1–3):250–5.
47. Wang RH, Xin JH, Yang Y, Liu HF, Xu LM, Hu JH. The characteristics and photocatalytic activities of silver doped ZnO nanocrystallites. Appl Surf Sci. 2004;227(1–4):312–7.
48. Fei B, Yang ZY, Yang HR, Hu ZG, Wang RH, Xin JH. Schizophrenic copolymer from natural biopolymer by facile grafting. Polymer. 2010;51(4):890–6.
49. Wang RH, Xin JH, Tao XM. Multi-functional protective nano-finishing of textile materials 1: self-cleaning surfaces. In: 230th American Chemical Society annual meeting. Inorganic, Washington, DC, 28 Aug to 1 Sept, p. 273; 2005.
50. Fei B, Qian BT, Yang ZY, Wang RH, Liu WC, Mak CL, et al. Coating carbon nanotubes by spontaneous oxidative polymerization of dopamine. Carbon. 2008;46(13):1795–7.
51. Lu HF, Fei B, Xin JH, Wang RH, Li L, Guan WC. Synthesis and lubricating performance of a carbon nanotube seeded miniemulsion. Carbon. 2007;45(5):936–42.
52. Liu YY, Tang J, Wang RH, Lu HF, Li L, Kong YY, et al. Artificial lotus leaf structures from assembling carbon nanotubes and their applications in hydrophobic textiles. J Mater Chem. 2007;17(11):1071–8.
53. Daoud WA, Leung SK, Tung WS, Xin JH, Cheuk K, Qi KH. Self-cleaning keratins. Chem Mater. 2008;20(4):1242–4.
54. Qi KH, Chen XQ, Liu YY, Xin JH, Mak CL, Daoud WA. Facile preparation of anatase/SiO2 spherical nanocomposites and their application in self-cleaning textiles. J Mater Chem. 2007;17(33):3504–8.
55. Qi KH, Daoud WA, Xin JH, Mak CL, Tang W, Cheung WP. Self-cleaning cotton. J Mater Chem. 2006; 16(47):4567–74.
56. Daoud WA, Xin JH. Nucleation and growth of anatase crystallites on cotton fabrics at low temperatures. J Am Ceram Soc. 2004;87(5):953–5.
57. Liu YY, Chen XQ, Xin JH. Can superhydrophobic surfaces repel hot water? J Mater Chem. 2009; 19:5602–11.
58. Liu YY, Wang XW, Qi KH, Xin JH. Functionalization of cotton with carbon nanotubes. J Mater Chem. 2008;18:3454–60.
59. Fei B, Yang ZY, Xin JH. Growing nanoballoons and nanotubes of pure polymer from a microcapsule. Macromol Rapid Commun. 2008;29:1882–6.
60. Liu YY, Chen XQ, Xin JH. Super-hydrophobic surfaces from a simple coating method: a bionic nanoengineering approach. Nanotechnology. 2006;17(13):3259–63.
61. Lu HF, Fei B, Xin JH, Wang RH, Li L. Fabrication of UV-blocking nanohybrid coating via miniemulsion polymerization. J Colloid Interface Sci. 2006; 300(1):111–6.

62. Lin YL, Choi KF, Zhang M, Li Y, Luximon A, Yao L, et al. optimized design of compression sportswear fabric using numerical simulation and the response surface method. Text Res J. 2012;82(2):108–16.
63. McLaren J. Preliminary development of a wearable device for dynamic pressure measurement in garments. Procedia Eng. 2010;2(2):3041–6.
64. Graham-Rowe D. Photonic fabrics take shape. Nat Photonics. 2007;1:6–7.
65. Tao X. Smart fibres, fabrics and clothing. Boca Raton, FL: CRC Press; 2001.
66. Tao X. Wearable electronics and photonics. Boca Raton, FL: CRC Press; 2005.
67. Zheng JW, Zhang ZB, Wu TH, Zhang Y. A wearable mobihealth care system supporting realtime diagnosis and alarm. Med Biol Eng Comput. 2007; 45:877–85.
68. Scanaill CN, Carew S, Barralon P, Noury N, Lyons D, Lyons GM. A review of approaches to mobility telemonitoring of the elderly in their living environment. Annu Rev Biomed Eng. 2006;34:547–63.
69. Post ER, Orth M, Gershenfeld PR, Russo N. E-broidery-design and fabrication of textile-based computing. IBM Syst J. 2000;39:840–60.
70. Rothmaier M, Selm B, Spichtig S, Haensse D, Wolf M. Photonic textiles for pulse oximetry. Opt Express. 2008;16(17):12973–86.
71. Sorin F, Lestoquoy G, Danto S, Joannopoulos JD, Fink Y. Resolving optical illumination distributions along an axially symmetric photodetecting fiber. Opt Express. 2010;18(23):24264–75.
72. Shapira O, Kuriki K, Orf ND, Abouraddy AF, Benoit G, Viens JF, et al. Surface-emitting fiber lasers. Opt Express. 2006;14(9):3929–35.
73. Schilling K, Bradford B, Castelli D, Dufour E, Nash JF, Pape W, et al. Human safety review of "nano" titanium dioxide and zinc oxide. Photochem Photobiol Sci. 2010;9:495–509.

Skin Penetration of Engineered Nanomaterials

6

Nancy A. Monteiro-Riviere

Over the past decade, the field of nanoscience has grown tremendously because nanomaterials have widespread applications in material science, engineering, and medicine. Although this work has received a great deal of attention, many challenges must be overcome before nanotechnology can be routinely applied in nanomedicines or in consumer products. Nanomaterials are substances having a physicochemical structure on a scale greater than atomic/molecular dimensions but less than 100 nm, and exhibit physical, chemical, and/or biological characteristics associated with its nanostructure. It is these unique characteristics that made them central components in an array of emerging technologies. Many new companies have emerged to commercialize these products. However, before many of these nanomaterials can be used in biological systems, their toxicology must first be evaluated under realistic environmental, occupational, and medicinal exposure conditions. Engineered nanoparticles (NP) are used in many household products, personal care products including sunscreens and cosmetics, or as drug delivery devices and as contrast imaging agents. The focus of this chapter is to depict how skin can serve as a potential route of exposure to several types of nanomaterials, and will discuss how size, shape, charge, surface properties, and vehicles can be important determinants on the penetration through the rate-limiting lipid barrier of the stratum corneum.

The transdermal flux of NP has obvious implications to toxicology, therapeutics, and drug delivery, but there is also a need to define the depth and mechanism of NP penetration into the epidermis and dermis after passage through the rate-limiting barrier of the stratum corneum. Penetration into skin without further systemic absorption could modulate immunological function and local toxicity. Consumers or workers often have compromised skin with an altered penetration barrier from conditions such as atopic dermatitis, psoriasis or eczema on their hands or other parts of their body which could influence the rate of penetration and even allow absorption of nanomaterials through skin. Several studies have evaluated the hypothesis that NP may become lodged in the lipid matrix of the skin [1–3] promoting their dermal residence and potential adverse effects. It will be shown how NP size, shape, charge, surface properties, and vehicle as well as animal species will modulate penetration and must be considered when designing studies to assess whether or not NP penetration occurs. There is limited information on the in vivo dermal penetration of NP in humans; however, there are several published studies in various in vitro model systems and on excised animal and human skins that shed light on this issue.

N.A. Monteiro-Riviere, Ph.D. (✉)
Department of Anatomy and Physiology, CVM,
28 Coles Hall, Manhattan, KS 66506-5604, USA
e-mail: monteiro@ksu.edu

A. Nasir et al. (eds.), *Nanotechnology in Dermatology*,
DOI 10.1007/978-1-4614-5034-4_6, © Springer Science+Business Media, LLC 2013

6.1 Titanium Dioxide and Zinc Oxide NP Penetration

The use of sunscreens is highly recommended to prevent skin cancer, sunburn, photoaging, and skin wrinkles. Some small molecule-based sunscreens may irritate the skin and others are very thick and appear white, making them aesthetically unacceptable to the consumer. Therefore, the use of nanoscale titanium dioxide (TiO_2) and micronized zinc oxide (ZnO) NP less than 100 nm have been incorporated into sunscreens to provide a broad spectrum physical sun blocker in a transparent form more acceptable to consumers. TiO_2 and ZnO production for use in skincare products have been reported to be 1,000 tons [4]. The Australian Therapeutic Goods Administration reports that 70 % of all sunscreens contain TiO_2 and 30 % contain ZnO NP [5]. The penetration of TiO_2 and ZnO emulsions on excised human skin tended to remain on the surface of the stratum corneum [6] but when 80–200 nm and up to 1 μm particles were applied to in vitro porcine skin showed no penetration [7]. Micronized TiO_2 applied to the skin of older patients for 9–31 days detected insignificant levels of Ti in the dermis [8]. A small amount of TiO_2 were detected within the follicular openings but most remained on the surface [9]. Others have reported that ZnO and TiO_2 NP do not permeate the skin [10, 11]. Exposure of 20 nm TiO_2 in a water/oil (w/o) emulsion for 5 h to in vitro and in vivo human skin showed no penetration into the viable epidermal layers [12]. Additional studies using different size micronized TiO_2 in oil/water (o/w) emulsions applied non-occlusive to human forearms for 6 h showed the emulsion remained on the surface of the stratum corneum and neither particle shape, formulation, nor exposure had a significant impact on penetration [11]. TiO_2 in o/w emulsions applied to in vitro cultures for 24 h and to human forearms for 6 h showed penetration was greater in vitro than in vivo, but tape stripping indicated NP remained within the stratum corneum [13]. Transparent ZnO NP of 26–30 nm in oil/water formulations were topically applied to in vitro human skin in static cells for 24 h and found ZnO NP on the surface of the stratum corneum [14]. TiO_2 in suncreens were applied to minipigs four times per day for 5 days per week for 4 weeks had high levels of Ti in the epidermis by transmission electron microscopy (TEM)/Energy-dispersive X-ray spectroscopy (EDX), in the stratum corneum layers and near hair follicle openings, with only a few particles of TiO_2 in the dermis which were regarded as contamination [15].

The European NANODERM project provided data from in vivo and in vitro experiments with TiO_2-based sunscreens on human and porcine skin and with human foreskin transplanted to immunodeficient mice. Some investigators detected a small amount of TiO_2 in the deep viable epidermal layers as analyzed by induced X-ray emission (PIXE) and scanning transmission ion microscopy (STIM) which tracked Ti particles to the stratum granulosum layer of the epidermis [16]. Studies conducted for 8–24 h on in vivo porcine skin suggested hair follicles were not important for penetration but detected four different formulations of TiO_2 on the stratum corneum and in the stratum granulosum with a high energy ion probe [17]. No TiO_2 penetration was detected through intact skin of SCID models [18]. This European project confirmed the safety of the sunscreens formulation containing TiO_2 NP, and that there was no reported evidence of NP penetration [19]. Previous reports by the European Union (EU) Scientific Committee on Cosmetics and Non-Food Products [20] found no evidence of TiO_2 NP penetration either in vitro and in vivo skin. In 2007, the EU Scientific Committee on Consumer Products stated that a safety assessment of nanosized TiO_2 in cosmetics should consider abnormal skin conditions and the impact of mechanical effects on skin penetration [21]. This will be discussed below.

The above discussion suggests that there are many conflicting reports as to the penetration of TiO_2. Many of these studies cannot be compared for the variables defining each study are so diverse. There are species specific anatomical differences in skin, hair follicle density, regional differences in absorption, vehicle differences, and duration of most of these experiments are very short. For a more complete discussion on the aspects of skin penetration with NP see the review by Monteiro-Riviere et al. [1], Monteiro-Riviere and Baroli [22], and Monteiro-Riviere and Riviere [23]. However, it appears that penetration is minimal and primarily limited to the upper layers of the stratum corneum with no significant transdermal flux detected.

6.2 Silver (Ag) NP Penetration

The almost ubiquitous occurrence of Ag NP has brought them to the attention of regulatory agencies in multiple jurisdictions. The Environmental Protection Agency (EPA) regulates all self-identified antimicrobials under the Federal Insecticide, Fungicidal & Rodenticide Act (FIFRA). Ag is such a substance. Currently, regulation of Ag NP is challenged by insufficient characterization and standardized methodology due to the various methods for Ag synthesis. The European Union (EU) has established the project "Engineered Nanoparticles: Review of Health and Environmental Safety (ENRHES)" to assess hazard, exposure and risk information for nanomaterials, including Ag. The goals of both of these agencies is to consider the various sources, pathways of exposure, and impacts on health and the environment to generate information and data that is suitable for risk assessment.

Ag in bulk form has been used for thousands of years in jewelry, currency, dental alloys, photography, medicines, and as an antimicrobial agent. Due to Ag enhanced antibacterial activity at the nanoscale level, it is used in the professional and personal healthcare fields through the incorporation of Ag NP into many products; including surgical and food handling tools, clothing, cosmetics, contact lens cases, catheters, and even children's toys. This is far from a modern phenomenon. The Phoenicians placed silver coins into large water jugs as a preservative. Solutions of Ag nitrate formed the mainstay of antimicrobial therapy before penicillin was developed to prevent bacterial growth in open wounds and burns. The use of Ag sulfadiazine creams are used in burn wards to treat a patient's denuded skin to prevent antibacterial infections. Ag NP can function as effective microbicidal agents against a broad spectrum of both Gram-positive and Gram-negative bacteria, as well as against many antibiotic-resistant bacterial strains and a variety of fungi. The Woodrow Wilson International Center for Scholars database on Nanotechnology states that Ag NP are the most used material in all nanotechnology consumer products [24]. The concentration of Ag NP within dietary supplements to cosmetics is in the range of <6 to 10,000 ppm.

Ag NP exhibit unique physicochemical properties that differ from, and provide distinct advantages over, those of their bulk metallic constituent. Ag in its bulk form is inert and exerts no biocidal action, but ionizes in the presence of water or tissue fluids to release antimicrobial Ag ions. The augmented properties of Ag NP are likely due to the increased surface area to volume ratio at the nanoscale. The generation of Ag NP may be achieved by several different methods including spark discharging, electrochemical reduction, chemical reduction, solution irradiation, laser ablation, and cryochemical synthesis [25]. Depending on the synthesis method, Ag NP may be of different shapes including spheres, rods, cubes, hollow tubes, multifacets, and films.

Unlike other heavy metals, the speciation of Ag does not play a large role in modulating toxicity; the toxicity of Ag^{+3} or Ag^{+2} is unlikely due to the transient nature of these forms of Ag. The rate of Ag ion production, as well as the distribution of Ag in tissue, may differ between Ag-NP and other sources of Ag ions due to the ability of Ag NP to release ions locally into specific tissues, cell membranes, or cells, thereby increasing the toxic potential. Ag ions released from wound dressings can be triggered by wound fluids and tissue exudates. The amount of Ag in dressings range from <0.1 mg/cm^2 to more than 1 mg/cm^2, with the amount of Ag ions released ranging from <1 to >70 ppm. There have been reports of Ag toxicity from such uses showing elevated serum and liver enzymes as well as argyria [26].

Ag NP absorption through intact skin is low (<1 ppm) since much of the free ion is precipitated as Ag sulfide in the superficial layers of the stratum corneum. Studies in our laboratory involved daily repetitive in vivo skin exposure for 14 days with 25 and 35 nm carbon-coated AgNP, as well as freshly synthesized or thoroughly washed 20, 50, and 80 nm Ag NP topically applied to the backs of weanling pigs [27]. Ag NP penetration occurred only to the superficial layers of the stratum corneum as assessed by TEM and confirmed by EDX. Skin penetration of 25 nm Ag NP was demonstrated in intact and damaged in vitro human skin in static diffusion cells [28] but with low flux compared to the absorption rates for metal powders [29]. TEM confirmed the presence of AgNP into the stratum corneum and the outermost surface of the epidermis but not into the dermis. These results suggest that a small fraction of the NP were dissolved and diffused through the skin layers as elemental Ag.

Our laboratory has also investigated the cytotoxicity and inflammatory potential of Ag NP in human epidermal keratinocytes (HEK) following 24 h exposure to 25 and 35 nm carbon-coated Ag NP, as well as freshly synthesized or thoroughly washed 20, 50, and 80 nm Ag NP at concentrations ranging from 0.54 ppb to 1.7 ppm [27]. Cytotoxicity was only observed with freshly synthesized Ag NP. The same Ag NP of different sizes, surface conditions, and synthesis methods were evaluated in *Escherichia coli*, Ag-resistant *Escherichia coli*, *Staphylococcus aureus*, methicillin-resistant *Staphylococcus aureus* (MRSA), and *Salmonella* sp. Freshly synthesized Ag NP were toxic to all bacterial strains between 3.0 and 8.0 ppm, the washed Ag NP and the carbon-coated Ag NP were toxic to all bacterial strains between 64.0 and 1024.0 ppm [30]. HEK and fibroblasts treated with Ag NP obtained from commercially Ag-based wound dressings depicted toxicity at 15 μg/mL [31]. Ag NP from 7 to 20 nm were studied in human skin carcinoma and fibrosarcoma cells depicted apoptosis at 0.78 μg/mL and 1.56 μg/mL respectively [32]. All of these studies confirm the potential toxicity of Ag NP to various tissues if concentrations due to NP localization and release are sufficient.

6.3 Gold (Au) NP Penetration

The skin penetration of three sizes of Au NP (15, 102, and 198 nm) were studied in in vitro rat skin diffusion cells and showed a size-dependent permeation. TEM showed small NP within the deep layers of skin but larger particles were located in the epidermis and dermis [33]. This was expected because the larger particles were not of true nanosize (<100 nm). Other authors have confirmed that 12.6 nm Au NP can penetrate intact and damaged human skin. This study showed an increase in penetration associated with an increase in Au NP concentration, forming a gradient of Au NP from the superficial to deep layers of the skin. Damaged skin showed a significant increase in penetration within the skin layers but not within the receptor compartment [34], again minimizing potential for transdermal transport with resulting systemic effects.

6.4 Quantum Dot (QD) NP Penetration

QD are a diverse class of engineered nanostructures that are highly variable in chemical composition, size, and shape. QD are a noncarbon-based semiconductor nanocrystal NP that emits a strong fluorescence signal and have potential for use as drug delivery, diagnostic, and imaging agents in biomedicine, and as semiconductors in the electronics industry. The QD center is comprised of a heterogeneous colloidal core of inorganic atoms (e.g. CdS, CdSe, CdTe) that is surrounded by a shell or "cap" of materials such as ZnS or CdS which can reduce leaching of core metals [35]. Frequently, several coatings may be applied to the core/shell complex to customize QD for different applications, such as increased solubility in biological media or conjugation to macromolecules for drug delivery and diagnostics [36]. Hydrophilic coating can increase biological compatibility by increasing QD solubility in water and cell medium, as well as facilitating QD conjugation to additional macromolecules, ligands, and antibodies. Customized QD are now readily available to researchers from a number of commercial sources.

Skin penetration of QD565 and QD655 with diverse physicochemical properties was studied in porcine flow-through diffusion cells at an occupationally relevant dose over the duration of a work

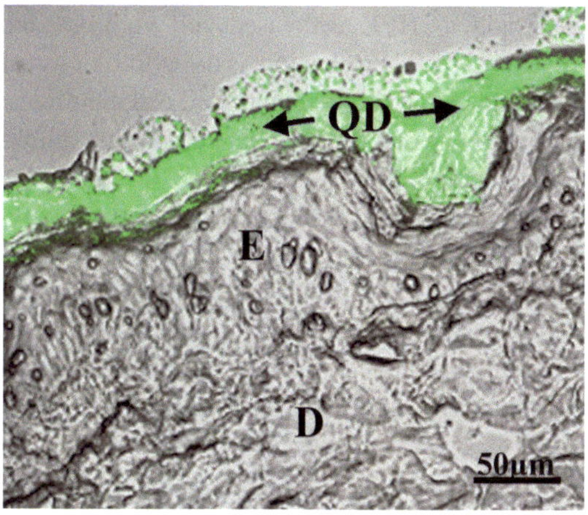

Fig. 6.1 Laser confocal micrograph of human skin topically dosed with QD565-COOH in flow-through diffusion cells after 8 h. QD depicted by *arrows* remained on the surface of the epidermis. Epidermis (E), dermis (D), and quantum dots (QD)

day exposure. QD565 with polyethylene glycol (PEG), PEG-NH_2, or COOH coatings showed penetration into the stratum corneum and localization within the epidermal and dermal layers by 8 h. PEG and PEG-NH_2 coated QD655 were localized within the epidermal layers by 8 h. The penetration of QD655-COOH was evident only at 24 h in the epidermal layers [37]. Skin absorption of QD 565 with three surface modifications was evaluated in human skin with in vitro flow-through and static diffusion cells at various pHs. Laser scanning confocal microscopy detected penetration into intact viable epidermis of skin for QD-PEG at pH 8.3, but not at pH 7.0 [38]. QD did not penetrate beyond the stratum corneum of intact skin (Fig. 6.1), even after tape stripping 30 times to remove most of the stratum corneum layers, QD were not found in the viable epidermal layers [39]. QD621 topically applied to dermatomed porcine skin in flow-through diffusion cells were located primarily in the normal stratum corneum, although a small amount of fluorescence was detected in the upper epidermal layers. Occasionally, QD621 were seen in the outer root sheath of the hair follicle [3]. By TEM, the QD were limited to the upper intercellular spaces of the stratum corneum of porcine skin. These studies show similar behavior of QD penetration into porcine and human skin, with a great dependence upon formulation factors (e.g., applied pH) and integrity of the stratum corneum. These factors are further discussed below.

The question often arises about what happens to QD if there were absorbed into the systemic circulation. This was addressed in our laboratory using ex vivo perfused skin. Neutral and charged QD621 were infused into the arterial vascular supply of the isolated perfused skin flap model to mimic systemic distribution after exposure by any route. The biodistribution of QD621 showed that QD621 can migrate out of the capillaries into the surrounding tissue [40]. Similarly, QD621 that were intradermally injected into SKH-1 hairless mice migrated from the injection site to regional lymph nodes through the lymphatic duct system and then to the liver and other organs [41].

Despite this widespread availability, much is still not known about QD cellular uptake or whether QD can cause effects on cell cytotoxicity or cause proinflammatory cytokine release. Commercially available QD with two core/shell sizes and shapes and three different surface coatings were assessed for cell toxicity potential in HEK [42]. The spherical QD565 (4.6 nm diameter) and ellipsoid QD655 (12 nm by 6 nm) have a surface coating of PEG (neutral charge), carboxylic acid (COOH, negative charge) or polyethylene glycol-amine (PEG-NH_2, positive charge). HEK were exposed to 20, 2, or 0.2 nM concentrations of QD for 24 or 48 h. Confocal and TEM imaging localized QD in the cytoplasm of cells by 24 h. Cytotoxicity was observed for QD565 and QD655 coated with COOH or PEG-NH_2 at 20 nM by 48 h, with little cytotoxicity observed for PEG-coated QD. In addition, only QD-COOH significantly increased the release of IL-1β, IL-6, and IL-8. These data indicate that the QD surface coating is a primary determinant of cytotoxicity in HEK. QD621, nail-shaped NP consisting of a cadmium/selenide core with a cadmium sulfide shell and coated with PEG, were also found to localize in cytoplasmic vacuoles of HEK. Cell viability was significantly less than controls from a QD concentration of 1.25–10 nM, while IL-8 and IL-6 were significantly greater than the controls from 2.5 to 10 nM [3]. It is critical to determine the nontoxic concentration of QD in vitro to minimize the cytotoxicity when QD are used in bioimaging, especially for organelle tracking in live cells.

While HEK internalize QD with surface coatings of PEG, PEG-NH_2, and COOH in variable amounts, the mechanisms of uptake have yet to be determined. HEK were treated with QD655 to study the effects of QD surface coating, temperature, culture medium supplements, and endocytic inhibitors on the uptake of QD into the cell. QD interactions with HEK were found to be coating dependent, with QD-COOH internalization greater than QD-PEG-NH_2 which in turn was greater than QD-PEG. Low temperature (4 °C) and nonsupplemented culture medium decreased QD uptake. Genistein and nocodazole decreased HEK interactions, depending on the coating; cytochalasin D potentiated these interactions over 200 % for the three coated QD. Overall, this indicates that regulation of QD uptake by cell cycle or endocytic mechanisms other than macropinocytosis, with QD surface groups playing a role [43]. The mechanism of QD cellular uptake was studied with 24 interfering agents conducted in our laboratory with carboxylic acid coated QD that showed lipid rafts not clathrin- or caveolae-mediated pathways. These studies showed that QD endocytic pathways are primarily regulated by the G protein coupled receptor-associated pathway and low density lipoprotein receptor/scavenger receptor, while other endocytic interfering agents may play a role but with less of an inhibitory effect. These studies also suggest that surface coating, size, and charge of QD NP are important determinants on how NP uptake occurs in mammalian cells for cancer diagnosis and treatment, and drug delivery [44].

The affect of mechanical strain on HEK uptake was investigated with QD. PEG-coated QD 655 were introduced in the HEK culture at a concentration of 3 nM, and a 10 % average strain was applied to the cells for 4 h. Twenty-four hours following treatment, QD uptake into the strained HEK was significantly greater than in nonstrained HEK, with cell viability significantly less in the strained compared to the nonstrained. These studies concluded that in the presence of cyclic strain, QD655 concentrations that were otherwise nontoxic to HEK decreased cell viability and increased IL-8 production. The cytotoxicity of QD may be the result of increased cell membrane permeability due to the tensile strain that allows QD to enter cell [45].

Studies with QD allowing investigators to modulate shape, size, and surface properties clearly show the importance of these factors on NP deposition and resulting cellular uptake and toxicity. Although it is evident that in normal skin minimal QD systemic absorption may occur, there is clear evidence in humans and animal studies that skin penetration into the viable epidermis can occur and epidermal cells are capable of QD cellular uptake.

6.5 Alterations to the Barrier and Effect on Nanoparticle Skin Penetration

Many factors can influence the rate of penetration of chemicals or NP through skin. The studies discussed thus far have primarily looked at intact and healthy human and animal skin. Alterations that influence the rate of NP and extent of NP uptake in skin could include the barrier integrity, contaminated surface, anatomical structure, skin diseases such as allergic and contact dermatitis, eczema and psoriasis [23, 46, 47]. The stratum corneum lipid barrier may be altered by mechanical action, ultraviolet radiation, chemical agents, or after occupational or accidental spillage of chemicals, or topical creams such as sunscreens, cosmetics, insect repellents, or drugs. It has been well documented that solvents may cause a range of structural alterations to the intercellular lipid bilayers of the stratum corneum by increasing the fluidization, and delipidization of the lipid, thereby changing the polarity, and phase separation causing vesicles within the lipid layers. Structural alterations to desmosomes may create fissures within the lipid bilayers and damage the fibrillar keratin of the stratum corneum, thereby creating vacuoles. Many of these solvent interactions will influence the rate and extent of chemical and small molecule absorption [47–51]. Little is known about their effects on NP.

6.5.1 Solvent Effects

Our laboratory studied the effect of solvents on pristine fullerenes (C_{60}) dermal penetration from toluene, cyclohexane, chloroform, and mineral oil vehicles in weanling pigs. The solvent and fullerene mixtures were topically applied to the skin for 4 days to simulate occupational exposures. Skin sites were tape-stripped 26 times for quantitative analytical determination by simplified liquid-liquid extraction followed by HPLC analysis. This study showed that pristine fullerenes penetrated deeper into the stratum corneum layers when dosed in toluene, cyclohexane, or chloroform, but more C_{60} was detected in the stratum corneum layers when dosed in chloroform. C_{60} dosed in mineral oil was not detected in the skin [52]. In vitro flow-through diffusion cells studies showed similar results but no fullerenes were detected in the receptor fluid, suggesting an absence of transdermal flux.

Terpenes are highly lipophilic aliphatic compounds found in many essential oils. Although they are generally regarded as safe, workers found terpenes capable of altering the stratum corneum lipid organization from orthorhombic to hexagonal configuration [51–53]. We evaluated whether terpene exposure would modulate topical NP dermal penetration by pretreating human skin for 1 h with terpenes (menthol, eucalyptol, limonene), or ethanol vehicle. In vitro flow-through diffusion cell studies were conducted for 8 h and the skin dosed with two different sizes of QD. Laser scanning confocal microscopy on the human skin showed no QD penetration into the viable layers, but detected QD in layers of the upper stratum corneum. TEM showed QD penetration slightly deeper into the stratum corneum in terpene pretreated skin, yet QD were not detected in the perfusate. These studies concluded that terpenes caused only a slight enhancement of QD penetration into the stratum corneum layers but no transdermal flux occurred across the skin [54].

6.5.2 Mechanical Action

In occupational settings, repetitive motions including constant hand movements and wrist flexion could accentuate biomechanical forces that may alter the structure of skin potentially resulting in an increase in the permeability of the skin in stressed areas to topically applied agents. Mechanical stressors could alter both the pathway and mechanism of how NP traverse the skin. The effects of flexing for 30 min on normal skin with FITC-conjugated dextran beads having a diameter of 0.5 μm showed penetration through the stratum corneum layer and into the viable epidermis [55]. A study employed topically applying very small fullerene amino acid-derivatized peptide nanoparticles of 3.5 nm to porcine skin in flow-through diffusion cells flexed at a 45° angle at a frequency of 20 flexes/min for 60 or 90 min. After 8 and 24 h, there was an increase in penetration through the epidermis and dermis in the flexed skin, compared to non-flexed skin where penetration was limited to the upper epidermal layers. TEM revealed NP localization within the intercellular spaces of the stratum granulosum, suggesting that penetration of the derivatized fullerene occurs through intercellular lipid moieties [1, 56]. The effects of applied mechanical strain on QD cellular uptake showed that strain may increase both uptake of QD into cells and result in cytokine production. This study showed that application of physiological load conditions increases the permeability of the cell membrane allowing for increase QD uptake [45]. In contrast to these studies, QD applied to rat skin flexed for 60 min showed that larger NP QD655 COOH (18 nm) and QD565 COOH (14 nm) did not penetrate by 8 and 24 h [2].

An experimental approach to assess the effect of barrier perturbation on NP penetration is to use tape stripping or mechanical abrasion to abrogate the stratum corneum barrier. Penetration of carboxylic acid coated QD of two sizes was assessed in intact, tape-stripped, and abraded rat skins in flow-through diffusion cells for 8 and 24 h. QD remained on the surface of the stratum corneum of intact skin (Fig. 6.1). Tape-stripped skin also showed QD on the surface of the viable epidermis, while in abraded skin QD penetrated into the viable dermis at both 8 and 24 h [2]. Similarly, laser confocal microscopy of topically applied QD565 COOH in 8 h flow-through diffusion cells with human skin tape-stripped 30 times to remove the stratum corneum, showed QD localized to the surface of the viable epidermal skin [39]. These studies confirm observations that removal of the stratum corneum barrier opens a window to the viable epidermis for NP exposure.

6.5.3 UVB-Damaged Skin

By far one of the most important and widespread mechanisms for damaging the epidermal barrier is exposure to solar ultraviolet B (UVB) radiation resulting in sun burned skin. Does such common damage alter the penetration of nanomaterial components of sunscreen? Studies with sunscreens containing zinc isotopes (^{68}Zn) topically applied in humans for 5 days have been conducted on sunburned UVB-damaged skin and showed higher levels of zinc in the blood and urine of individuals receiving the nano-sunscreen. However, whether ^{68}Zn was absorbed as ZnO particles or soluble Zn is unknown [57]. Recently, our laboratory conducted in vitro and in vivo studies with TiO_2 and ZnO NP to UVB-damaged porcine skin with four different sunscreen formulations. Flow-through diffusion cells were treated 24 h with four sunscreen formulations: 10 % coated-TiO_2 o/w; 10 % coated TiO_2 in w/o; 5 % coated ZnO in o/w; and 5 % uncoated ZnO in o/w. TiO_2 primary particle size was 10×50 nm with mean agglomerates of 200 nm; ZnO mean was 140 nm. Skin was evaluated by LM, SEM, TEM, and time-of-flight-secondary-ion-mass-spectrometry (TOF-SIMS). UVB-exposed skin showed TiO_2 NP down to 17 layers into the stratum corneum compared to normal skin, while ZnO remained on the surface. TOF-SIMS analysis showed TiO_2 and ZnO penetration. The perfusate analyzed by TEM/EDX or inductively-coupled-plasma-mass-spectrometry (ICP-MS) detected no Ti or Zn, indicating minimal transdermal absorption. The in vivo porcine skin was dosed for 24 h occluded with the same formulations at 48 h. The TiO_2 in o/w formulation penetrated 13 layers deep into UVB-damaged stratum corneum (Fig. 6.2), while only 7 layers deep in non UVB treated skin; TiO_2 in w/o penetrated deeper in the UVB-damaged stratum corneum. Coated and uncoated Zn NP in o/w were localized to the upper 1–2 superficial stratum corneum in UVB-damaged (Fig. 6.3) and normal skin. Only by using more sophisticated techniques like TOF-SIMS could Ti be seen within the epidermis and superficial dermis, with Zn limited to the stratum corneum and upper epidermis. In summary, UVB-damaged skin showed a slightly enhanced penetration of TiO_2 or ZnO in sunscreen formulations but no transdermal absorption was detected [58]. If TiO_2 NP accumulate in skin, they could potentially decrease the threshold for allergies or cause an allergic reaction.

Fig. 6.2 Transmission electron micrograph of in vivo porcine skin showing the localization of TiO_2 NP within the stratum corneum intercellular lipid space after UVB exposure treated with a sunscreen formulation for 48 h

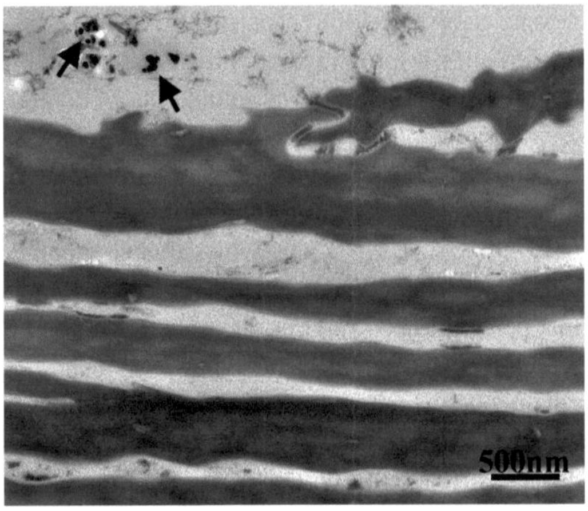

Fig. 6.3 Transmission electron micrograph of in vivo porcine skin showing the localization of ZnO NP on the surface of the stratum corneum layer of skin after UVB exposure to a sunscreen formulation for 48 h

6.6 Concluding Discussions

The above studies reviewing skin penetration of various types of topical NP illustrate both the breath of studies conducted, but also some common findings that begin to define properties and experimental parameters that must be considered when studying NP dermal penetration. Penetration of NP into the skin is particle dependent and sensitive to the vehicles used to administer them. In almost all cases, transdermal flux has not been detected yet various degrees of stratum corneum and epidermal penetration are common. This is particularly important for immunological endpoints since large amounts of penetrated NP are not necessary to illicit a biologically and clinically relevant response. Abrogation of the stratum corneum barrier by abrasion or tape stripping promotes NP penetration into deeper layers, as does mechanical deformation. In contrast, damage to the skin by UVB solar radiation does not significantly increase penetration of topical TiO_2 or ZnO sunscreens. Finally, vehicles can modulate NP penetration. Uptake into epidermal cells has also been demonstrated for most types of NP, with some clearly exhibiting cytotoxicity or inflammatory mediator release. The QD studies clearly showed specific pathways involved and responses that were coating dependent.

There is no doubt that dermal exposure to NP in occupational, cosmetic and dermatological preparations will continue to occur. It is crucial that experiments be properly conducted and endpoints defined that actually measure intact NP. Studies of chronic exposure and evaluation for toxicity are lacking making definitive statements of safety impossible to make.

References

1. Monteiro-Riviere NA, Inman AO, Ryman-Rasmussen JP. Dermal effects of nanomaterials. In: Monteiro-Riviere NA, Tran CL, editors. Nanotoxicology: characterization, dosing, and health effects. New York, NY: Informa Healthcare; 2007. p. 317–38.
2. Zhang LW, Monteiro-Riviere NA. Assessment of quantum dot penetration into intact, tape stripped, abraded and flexed rat skin. Skin Pharmacol Physiol. 2008;21:166–80.
3. Zhang LW, Yu WW, Colvin VL, Monteiro-Riviere NA. Biological interactions of quantum dot nanoparticles in skin and in human epidermal keratinocytes. Toxicol Appl Pharmacol. 2008;228:200–11.
4. Borm PJA, Robbins D, Haubold S, Kuhlbusch T, Fissan H, Donaldson K, et al. The potential risks of nanomaterials: a review conducted out for ECETOC. Part Fibre Toxicol. 2006;3:11.
5. Faunce T, Murray K, Nasu H, Bowman D. Sunscreen safety: the precautionary principle, the Australian Therapeutic Goods Administration and nanoparticles in sunscreens. Nanoethics. 2008;2:231–40.
6. Dussert AS, Gooris E. Characterization of the mineral content of a physical sunscreen emulsion and its distribution onto human stratum corneum. Int J Cosmet Sci. 1997;19:119–29.
7. Gamer AO, Leibold E, van Ravenzwaay B. The in vitro absorption of microfine zinc oxide and titanium dioxide through porcine skin. Toxicol In Vitro. 2006;20:301–7.
8. Tan MH, Commens CA, Burnett L, Snitch PJ. A pilot study on the percutaneous absorption of microfine titanium dioxide from sunscreens. Aust J Dermatol. 1996;37:185–7.
9. Lademann J, Weigmann HJ, Rickmeyer C, Barthelmes H, Schaefer H, Mueller G, et al. Penetration of titanium dioxide in sunscreen formulation into the horny layer and the follicular orifice. Skin Pharmacol Appl Skin Physiol. 1999;12:247–56.
10. Pflücker F, Wendel V, Hohenberg H, Gärtner E, Will T, Pfeiffer S, et al. The human stratum corneum layer: an effective barrier against dermal uptake of different forms of topically applied micronized titanium dioxide. Skin Pharmacol Appl Skin Physiol. 2001;14(S1): 92–7.
11. Schulz J, Hohenberg H, Pflücker F, Gartner E, Will T, PfeiVer S, et al. Distribution of sunscreens on skin. Adv Drug Deliv Rev. 2002;54(S1):S157–63.
12. Mavon A, Miquel C, Lejeune O, Payre B, Moretto P. In vitro percutaneous absorption and in vivo stratum corneum distribution of an organic and mineral sunscreen. Skin Pharmacol Physiol. 2007;20:10–20.
13. Bennat C, Müller-Goymann CC. Skin penetration and stabilization of formulations containing microfine titanium dioxide as physical UV filter. Int J Cosmet Sci. 2000;22:271–83.
14. Cross SE, Innes B, Roberts M, Tsuzuki T, Robertson TA, McCormick P. Human skin penetration of sunscreen nanoparticles: in-vitro assessment of a novel micronized zinc oxide formulation. Skin Pharmacol Physiol. 2007;20:148–54.
15. Sadrieh N, Wokovich AM, Gopee NV, Zheng J, Haines D, Parmiter D, et al. Lack of significant dermal penetration of titanium dioxide from sunscreen formulations containing nano- and submicron-size TiO2 particles. Toxicol Sci. 2010;115:156–66.
16. Kertész Z, Szikszai Z, Gontier E, Moretto P, Surlève-Bazeille JE, Kiss B, et al. Nuclear microprobe study of TiO2-penetration in the epidermis of human skin xenografts. Nucl Instrum Methods Phys Res, Sect B. 2005;231:280–5.

17. Menzel F, Reinert T, Vogt J, Butz T. Investigations of percutaneous uptake of ultrafine TiO2 particles at the high energy ion nanoprobe LIPSION. Nucl Instrum Methods Phys Res, Sect B. 2004;219–220:82–6.
18. Kiss B, Biró T, Czifra G, Tóth BI, Kertész Z, Szikszai Z, et al. Investigation of micronized titanium dioxide penetration in human skin xenografts and its effect on cellular functions of human skin-derived cells. Exp Dermatol. 2008;17(8):659–67.
19. NANODERM. Quality of skin as a barrier to ultrafine particles. Final Report. 2007 (Project Number: QLK4-CT-2002-02678). http://www.uni-leipzig.de/~nanoderm/. Accessed 16 April 2012.
20. SCCNFP. Opinion of the scientific committee on cosmetic products and non-food products intended for consumers concerning titanium dioxide. European Commission, Brussels, Belgium. Colipa No S 75; 2000.
21. SCCP. Preliminary opinion on safety of nanomaterials in cosmetic products. European Commission, Brussels, Belgium, 2007. http://ec.europa.eu/health/archive/ph_risk/committees/04_sccp/docs/sccp_o_123.pdf. Accessed 16 April 2012.
22. Monteiro-Riviere NA, Baroli B. Nanomaterial penetration. In: Monteiro-Riviere NA, editor. Toxicology of the skin-target organ series. New York, NY: Informa Healthcare; 2010. p. 333–46.
23. Monteiro-Riviere NA, Riviere JE. Interaction of nanomaterials with skin: aspects of absorption and biodistribution. Nanotoxicology. 2009;3(3):188–93.
24. Woodrow Wilson International Center for Scholars. A nanotechnology consumer products inventory. 2011. http://www.nanotechproject.org/consumerproucts. Accessed 16 April 2012.
25. Chen X, Schluesener HJ. Nanosilver: a nanoproduct in medical application. Toxicol Lett. 2008;176:1–12.
26. Wijnhoven SWP, Peijnenburg WJGM, Herberts CA, Hagens WI, Oomen AG, Heugens EHW, et al. Nanosilver: a review of available data and knowledge gaps in human and environmental risk assessment. Nanotoxicology. 2009;3(2):109–38.
27. Samberg ME, Oldenburg SJ, Monteiro-Riviere NA. Evaluation of silver nanoparticle toxicity in skin in vivo and keratinocytes in vitro. Environ Health Perspect. 2010;118(3):407–13.
28. Larese FF, D'Agostin F, Bovenzi M, Crosera M, Adami G, Romano C, et al. Human skin penetration of silver nanoparticles through intact and damaged skin. Toxicology. 2009;255:33–7.
29. Larese FF, Adami G, Venier M, Coceani N, Bussani R, Massiccio M, et al. In vitro percutaneous absorption of cobalt. Int Arch Occup Environ Health. 2004;77:85–9.
30. Samberg ME, Orndorff PE, Monteiro-Riviere NA. Antibacterial efficacy of silver nanoparticles of different sizes, surface conditions and synthesis methods. Nanotoxicology. 2011;5(2):244–53.
31. Burd A, Kwok CH, Hung SC, Chan HS, Gu H, Lam WK, et al. A comparative study of the cytotoxicity of silver-based dressings in monolayer cell, tissue explant, and animal models. Wound Repair Regen. 2007;15:94–104.
32. Arora S, Jain J, Rajwade JM, Paknikar KM. Cellular responses induced by silver nanoparticles: in vitro studies. Toxicol Lett. 2008;179:93–100.
33. Sonavane G, Tomoda K, Sano A, Ohshima H, Terada H, Makino K. In vitro permeation of gold nanoparticles through rat skin and rat intestine: effect of particle size. Colloids Surf B Biointerfaces. 2008;65:1–10.
34. Larese FF, Crosera M, Adami G, Bovenzi M, Rossi F, Maina G. Human skin penetration of gold nanoparticles through intact and damaged skin. Nanotoxicology. 2011;5:493–501.
35. Derfus AM, Chan WCW, Bhatia S. Probing the cytotoxicity of semiconductor nanocrystals. Nano Lett. 2004;4:11–8.
36. Michalet X, Pinaud FF, Bentolila LA, Tsay JM, Doose S, Li JJ, et al. Quantum dots for live cells, in vivo imaging, and diagnostics. Science. 2005;307:538–44.
37. Ryman-Rasmussen J, Riviere JE, Monteiro-Riviere NA. Penetration of intact skin by quantum dots with diverse physicochemical properties. Toxicol Sci. 2006;91:159–65.
38. Prow TW, Monteiro-Riviere NA, Inman AO, Grice JE, Chen X, Zhao X, et al. Quantum dot penetration into viable human skin. Nanotoxicology. 2012;6:173–85.
39. Monteiro-Riviere NA, Inman AO. Evaluation of quantum dot nanoparticle penetration in human skin. Toxicologist. 2008;102 Suppl 1:211.
40. Lee HA, Imran M, Monteiro-Riviere NA, Colvin VL, Yu WW, Riviere JE. Biodistribution of quantum dot nanoparticles in perfused skin: evidence of coating dependency and periodicity in arterial extraction. Nano Lett. 2007;9:2865–70.
41. Gopee NV, Roberts DW, Webb P, Cozart CR, Siitonen PH, Warbritton AR, et al. Migration of intradermally injected quantum dots to sentinel organs in mice. Toxicol Sci. 2007;98:249–57.
42. Ryman-Rasmussen JP, Riviere JE, Monteiro-Riviere NA. Surface coatings determine cytotoxicity and irritation potential of quantum dot nanoparticles in epidermal keratinocytes. J Invest Dermatol. 2007;127:143–53.
43. Ryman-Rasmussen JP, Riviere JE, Monteiro-Riviere NA. Variables influencing inter-actions of untargeted quantum dot nanoparticles with skin cells and identification of biochemical modulators. Nano Lett. 2007;7:1344–8.
44. Zhang LW, Monteiro-Riviere NA. Mechanisms of quantum dot nanoparticle cellular uptake. Toxicol Sci. 2009;111:138–55.
45. Rouse JG, Haslauer CM, Loboa EG, Monteiro-Riviere NA. Cyclic tensile strain increases interactions between human epidermal keratinocytes and quantum dot nanoparticles. Toxicol In Vitro. 2008;22:491–7.
46. Monteiro-Riviere NA. Anatomical factors that affect barrier function. In: Zhai H, Wilhelm KP, Maibach

HI, editors. Dermatotoxicology (Chap. 4). 7th ed. New York, NY: CRC Press; 2008. p. 39–50.

47. Riviere JE, Monteiro-Riviere NA. Toxicokinetics: dermal exposure and absorption of chemicals and nanomaterials. In: McQueen C, editor. Comprehensive toxicology. 2nd ed. New York, NY: Elsevier; 2010. p. 111–22.
48. Monteiro-Riviere NA, Inman AO, Mak V, Wertz P, Riviere JE. Effect of selective lipid extraction from different body regions on epidermal barrier function. Pharm Res. 2001;18:992–8.
49. Riviere JE, Brooks JD. Predicting skin permeability from complex chemical mixtures. Toxicol Appl Pharmacol. 2005;208:99–110.
50. Riviere JE, Brooks JD. Predicting skin permeability from complicated chemical mixtures: dependency of quantitative structure permeability relationships (QPSR) on biology of skin models used. Toxicol Sci. 2011;119:224–32.
51. Roberts MS, Gierden A, Riviere JE, Monteiro-Riviere NA. Solvents and vehicle effects on the skin. In: Roberts MS, Walters KA, editors. Dermal absorption and toxicity assessment. 2nd ed. New York, NY: Informa Healthcare; 2008. p. 433–47.
52. Xia XR, Monteiro-Riviere NA, Riviere JE. Skin penetration and kinetics of pristine fullerenes (C60) topically exposed in industrial organic solvents. Toxicol Appl Pharmacol. 2010;242:29–37.
53. Dos Anjos JLV, Alonso A. Terpenes increase the partitioning and molecular dynamics of an amphipathic spin label in stratum corneum membranes. Int J Pharm. 2008;350(1–2):103–12.
54. Monteiro-Riviere NA, Inman AO, Erdmann D, Xia X, Riviere JE. Terpene effects on penetration of nanoparticles in human skin. Toxicologist. 2011;20 Suppl 2:462.
55. Tinkle SS, Antonini JM, Rich BA, Roberts JR, Salmen R, DePree K, et al. Skin as a route of exposure and sensitization in chronic beryllium disease. Environ Health Perspect. 2003;111:1202–8.
56. Rouse JG, Yang J, Ryman-Rasmussen JP, Barron AR, Monteiro-Riviere NA. Effects of mechanical flexion on the penetration of fullerene amino acid-derivatized peptide nanoparticles through skin. Nano Lett. 2007;7:155–60.
57. Gulson B, McCall M, Korsch M, Gomez L, Casey P, Oytam Y, et al. Small amounts of zinc from zinc oxide particles in sunscreens applied outdoors are absorbed through human skin. Toxicol Sci. 2010;118(1):140–9.
58. Monteiro-Riviere NA, Wiench K, Landsiedel R, Schulte S, Inman AO, Riviere JE. Safety evaluation of sunscreen formulations containing titanium dioxide and zinc oxide nanoparticles in UVB sunburned skin: an in vitro and in vivo study. Toxicol Sci. 2011;123(1):264–80.

7 Perspectives on Percutaneous Penetration of Nanomaterials

Mai A. Ngo, Michael O'Malley, and Howard I. Maibach

7.1 Introduction

Nanomaterials are defined as substances less than 100 nm in any one dimension [1]. Due to increased surface area, nanoparticles are generally more reactive and exhibition physiochemical properties which differ from that of the respective material at larger scales. Along with novel physiochemical properties, the toxicological profiles of nanomaterials may also be unanticipated and cannot be readily predicted from the known properties of the bulk material.

As nanomaterials have become increasingly prevalent in consumer products, their potential for human exposure has increased. While nanotechnology has important applications in many fields, the concern is that release of nanomaterials into the environment will increase more rapidly than scientific knowledge of possible exposure and toxicity. The extent of nanoparticle penetration into living tissue has not been determined conclusively and the latent risks and consequences of increased exposure to workers and consumers are largely unknown.

Commonly described nanoparticles include quantum dots (QDs), fullerenes, nanotubes, or nanowires. QDs are heterogenous nanoparticles that consist of a colloidal core surrounded by surface coatings. QD surface coatings or shells can prevent agglomeration or may be customized to specific applications. These surface coatings may affect absorption and toxicity. QDs have potential for use as diagnostic and imaging agents in biomedical applications and as semiconductors in the electronics industry. Fullerenes, used as a starting material in numerous industrial and pharmaceutical applications, are hydrophobic and have the tendency to form larger aggregates and could absorb small molecules. Carbon nanotubes (CNTs) can be found in several forms: single-wall (SWCNT), double-wall (DWCNT), and multi-wall carbon nanotubes (MWCNT), which can come in a variety of diameters and lengths.

The products into which nanomaterials are incorporated and the manner in which these products are intended for use, as well as the product life cycle will determine exposure to the environment and human health. Depending on the nanoproduct, people may be exposed to nanomaterials as single nanoparticles, agglomerates or aggregate particles, or embedded into other nanoscale matrix.

M.A. Ngo, Ph.D., (✉)
Department of Pesticide Regulation, California Environmental Protection Agency, 1001 "I" Street, P.O. Box 4015, Sacramento, CA 95812, USA
e-mail: mngo@cdpr.ca.gov

M. O'Malley
Center for Health and the Environment, University of California, Davis, CA, USA
e-mail: maomalley@ucdavis.edu

H.I. Maibach, M.D.
Department of Dermatology, University of California, 90 Medical Center Way, Surge 110, Box 0989, San Francisco, CA 94143-0989, USA
e-mail: maibachh@derm.ucsf.edu

A. Nasir et al. (eds.), *Nanotechnology in Dermatology*,
DOI 10.1007/978-1-4614-5034-4_7, © Springer Science+Business Media, LLC 2013

Possible exposure routes are thought to be respiratory an transdermal. Especially for epicutaneously applied nanoproducts, the concern is whether nanoparticles can penetrate skin. Most cosmetic applications of nanoparticles require the material to remain on the skin surface, the development and use of nanomaterials for biomedical applications may actually optimize for internalization by specific tissue or cell types and toxic outcome [2]. For example, there is active research in the field of topical vaccinations using nanoparticle-targeting of hair follicles [3, 4].

The skin may be an unintended route for localized and possibly systemic exposure to nanoparticles released during the manufacture, use, and disposal of nanomaterials. Cosmetics and personal care products are generally applied to human skin and are expected to result in only local exposure. Topical applications may also be a favorable route for pharmaceutical/therapeutic treatment of skin diseases because of the lower risk of systemic effects. In fact, the skin is a direct target for nanoparticle-based applications for transdermal drug delivery.

A dermally exposed or applied substance may penetrate the skin by three paths: transcellular, intercellular, and follicular. Percutaneous absorption is a dynamic process and there are many components with which a penetrant interacts before possibly gaining systemic access. Here, the classic ten steps of percutaneous penetration [5] are expanded to 15 perspectives: (1) physiochemical properties of the penetrant, (2) vehicle effects, (3) dose, duration, surface area, and frequency, (4) sub-anatomical pathways (skin appendages), (5) regional variability, (6) population variability, (7) surface conditions, (8) skin health and integrity, (9) substantivity and binding, (10) distribution, (11) loss from skin surface and exfoliation, (12) wash effect, (13) metabolic and photochemical transformation, (14) excretion kinetics, (15) method of determining absorption and toxicity. These are issues considered from the perspective of nanoparticles and the potential risks of dermal exposure. Understanding the potential for dermal penetration and possible toxicological outcomes is of great importance.

7.1.1 Physiochemical Properties of the Penetrant

Physiochemical properties of the penetrant may be the most pervasive factor influencing penetration, since these properties will influence interactions with the other skin components and pathways to be discussed. These properties of a nanoparticle will influence the manner in which it interacts with the skin and the extent to which it penetrates the skin layers or releases associated ingredients into the skin. Physiochemical properties influencing nanoparticle penetration resemble the major factors influencing the penetration of chemicals across the skin [6–11]. Factors influencing nanoparticle uptake and retention in living cells, including particle size, charge, surface characteristics, and mechanisms are reviewed in depth elsewhere [12].

For metal oxides nanoparticles such as titanium dioxide (TiO_2) and zinc oxide (ZnO), the greater specific surface area and chemical reactivity compared to larger particles potentially result in the generation of higher levels of reactive oxygen species (ROS) per unit mass. The ability of nanocrystalline titanium dioxide (nano-TiO_2) to efficiently generate ROS, particularly under ultraviolet (UV) illumination is utilized in applications such as self-cleaning glass and in water remediation and air purification [13]. Data suggests that nano-TiO_2 samples optimized for ROS production in photocatalysis are also more likely to generate damaging ROS species in cell culture [14]. Both nano-TiO_2 and nano-zinc oxide (nano-ZnO) were able to produce ROS in water suspension in addition to demonstrating antibacterial activity under light illumination [15].

It is generally viewed that nanoparticle toxicity and dermal penetration are a function of particle aggregate size. While Bos and Meinardi [7] argue that the molecular weight (MW) of a compound must be under 500 Da to be absorbed through skin, nanoparticles are estimated to have molar masses on the order of 10^6 Da and greater [16]. The penetration of 6 and 15 nm gold nanoparticles through human skin was studied and the results suggested that nanoparticles at this size range permeate the *SC* through intercellular pathways in a similar manner to drug molecules [17]. For larger nanoparticles, however, increasing evidence points to hair follicles as the main route by which they may cross the skin barrier (reviewed by Lademann et al. [18]). In another study, silica-coated iron oxide and dextran-coated nanoparticles were taken up by primary human macrophages and dendritic cells to a significantly higher degree, irrespective of size [19].

Although particle size may be an important factor in dermal penetration, particularly through hair follicles, the smallest particles do not have the greatest penetration. In an in vitro study of particles sized from 122 to 1,000 nm, Patzelt et al. suggested that particles in the size range of 400–700 nm had optimal penetration depth in porcine hair follicles compared to those that were smaller or larger [20]. Increasing penetration depth with increasing size was found for both poly-lactide-*co*-glycolide particles in the size range of 122–643 nm, and decreased significantly for 860 nm particles. For the silica oxide particles ranging in size from 300 to 1,000 nm, follicle penetration depth increased between 300 and 646 nm sized particles, and then decreased for particles larger than 646 nm. Since both particle types exhibited similar penetration depths with size, a mechanical, rather than chemical effect is presumed. The author suggests the optimal particle size for the deepest penetration corresponded approximately to the structure of the hair and follicles [21]. In a different study, microspheres in the size range of 750 and 6,000 nm were applied to freshly excised skin samples, 1,500 nm microspheres appeared to be the optimal size for hair follicles penetration [22].

Surface functionalization, such as for binding of specific ligands, allow nanoparticle targeting of specific cell populations and subcellular components. Because mammalian skin has a net negative charge under normal physiological conditions, it will attract positively charged molecules. Consistent with this concept, an in vitro study using anionic and cationic derivatives of nano-styrene found the cationic nanoparticles to have a clear affinity for the surface of excised porcine skin [23]. By contrast, in a comparison of in vitro penetration of spherical QDs (4.6 nm in diameter) and ellipsoid QDs (12×6 nm in dimensions) with neutral, anionic, and cationic different surface coatings, Ryman-Rasmussen et al. [24] reported the spherical QDs of all surface coatings were able to penetrate the SC and localize in the epidermis and dermis by 8 h. Similarly, the neutral and cationic-coated ellipsoid QDs localized within the epidermal layers by 8 h. However, ellipsoid QDs with anionic surface coatings penetrated only as far as the epidermis at 24 h.

Nevertheless, highly ionized nanoparticles may have decreased permeability through the lipid-rich *Stratum Corneum* (*SC*), as observed with some steroids of increased polarity [25]. The polarity of a steroid may increase its interaction with the relatively polar components of the *SC* and hinder its complete penetration through the skin. Ion pairing, where ions of opposite charge associate by electrovalent bonds, may facilitate dermal penetration of ionized molecules [26]. The pK_a of the nanoparticle or its chemical moieties will determine whether the ionized or non-ionized form of the nanoparticle will be favored under the moderately acidic pH of healthy skin. For a volatile chemical, the portion of the dose that does not volatilize or bind to the skin surface may be absorbed [27].

Relatively minor differences in physiochemical properties between two nanoparticles may have major toxicological consequences. For example, TiO_2 is a naturally occurring mineral found in two common forms: rutile and anatase. These forms differ in their crystal structures and optical and electrical properties [28]. In a study using nano-TiO_2 in cell culture, both smaller and larger nanoparticles were comparable in their effect on cellular behavior [14]. Although cytotoxicity was found only at relatively high concentrations of 100 μg/mL, anatase TiO_2 was found to be 100 times more toxic to an equivalent sample of rutile TiO_2. Cytotoxicity was attributed to efficiency of ROS generation and oxidative stress, which are thought to cause protein, DNA, membrane injury, mitochondrial damage, and inflammation. In another study investigating polyethylene glycol-functionalized (PEG) gold nanoparticles, Simpson et al. [29]

found short-chain PEG-gold nanoparticles were more water soluble and virtually nontoxic compared to long-chain PEG-nanoparticles. However, the long-chain PEG-gold nanoparticles had slower clearance rates.

7.1.2 Vehicle Effects

Vehicle-dependent effects on the penetration behavior of drugs following topical application have been widely examined [30]. Dermal absorption of a compound will be influenced by vehicle solubility and partitioning of the compound between the vehicle and the skin. Agents that are more soluble in an aqueous vehicle may tend towards limited absorption into the lipid-rich *SC*. Depending on the physiochemical properties of the penetrant, vehicle pH may also influence the ionization state of the compound and the rate of partitioning into the *SC*.

One general approach to overcoming the barrier properties of the skin is to use vehicles or other chemical compounds that help promote drug diffusion through the *SC* and epidermis. These substances, referred to as penetration enhancers, accelerants, adjuvants, or absorption promoters, act by reducing drug binding and interactions with skin components. Penetration enhancers, such as oil-based microemulsions, solvents or surfactants, are often added to topical drug formulations to increase dermal absorption. Overviews of particle-based formulations for transcutaneous applications in cosmetics and dermatology are provided elsewhere.

A frequent application of micro- and nanoemulsions and other colloidal drug carrier systems is towards enhancing penetration of hydrophilic as well as lipophilic substances (reviewed by Nohynek et al. [31]). Penetration of the lipophilic dye curcumin incorporated either in an oil-in-water microemulsion or in an amphiphilic cream, was investigated following application onto the skin of human volunteers [32]. Application of the microemulsion resulted in curcumin being found deeper in the *SC*, as well as in the follicular infundibula of hair follicles, compared to the cream formulation.

Nanoparticles can also be used to facilitate penetration of active drug ingredients. Since the epidermis is rich in lipids, lipid carriers for drugs may promote absorption into the outermost layer of the skin, the SC. Development of nanocarriers (NCs), colloidal systems such as microemulsions, vesicular (liposomes) and nanoparticular NCs, is discussed elsewhere [33, 34]. Similar to traditional emulsions, nanoemulsions are basically oil dispersed in water. Nanoemulsions are less visible than traditional emulsions, and their small size and hydrophilic exterior may facilitate transport of active ingredients across the horny layer of the epidermis. The lipophilic interior allows for transport of hydrophobic substances.

Solid lipid nanoparticles (SLN) and nanostructured lipid carriers (NLC) are alternative carrier systems of liposomes and emulsions. Both SLN and NLC are recognized for localization and targeting of active ingredients to the skin and controlled release of active ingredients. Due to the effects of occlusion and hydration, SLN and NLC may also enhance percutaneous penetration. SLN differ from NLC by composition of the solid particle matrix. Active ingredients may also be encapsulated in nanoparticles such as polymeric or lipid nanospheres and nanocapsules.

Sometimes, ingredients mean to be "inert," such as alcohols, ethers, glycols, or hydrocarbon oils, will act to augment transdermal penetration. Nanomaterials used in topical formulations, whether or not they are meant for promotion of drug uptake, may be absorbed and result in adverse reactions at the local as well as systemic level.

In some cases, the vehicle may change the integrity of the skin, such as by damage or altering skin hydration. Agents which cause skin irritation, inflammation, or are otherwise vasoactive, may also increase blood flow and systemic distribution of nanomaterials from the dermis. Organic solvents, such as acetone, may damage skin barrier properties and increase skin penetration [35]. One study reports oleic acid, ethanol, and oleic acid-ethanol were all capable of enhancing the transdermal delivery of ZnO nanoparticles by increasing the intercellular lipid fluidity or extracting lipids from the SC [36].

Another study reported chloroform, toluene, and cyclohexane to facilitate the skin penetration of pristine fullerenes [37]. A proposed mechanism for this facilitation of penetration is the super-saturation of fullerenes on the skin surface due to solvent evaporation. Consistently, when mineral oil was used as a vehicle, no penetration of fullerenes was detectable. However, in another study comparing toluene and water as vehicles found these vehicles to have minimal effect on the skin penetration of gold nanoparticles [17].

Vehicle ingredients may modulate the chemical and biological reactivity of nanomaterials. In one study investigating C60 fullerene energy transfer following UV irradiation, C60 embedded into micelles of surfactants generated ROS, but aggregated C60 did not [38]. This suggests that unlike C60 clusters, C60 in organic phase may contribute to redox reactions and oxidative damage in the natural environment and in biological systems. To counter such reactions, additives such as fillers, stabilizers, and antioxidants may be added to the C60 formulation. Yet again, these ingredients may have unanticipated influence on dermal absorption and toxicity. Degradation of the NCs and other additives such as in response to heat, hydrolysis, oxidation, or photo-irradiation, should also be considered.

In cases where skin penetration is not desired, penetration reducers can be used to prevent the compound from entering the systemic circulation [39]. Barrier-enhancing emulsions have reduced the penetration of exogenous proteins into the hair follicles to varying degrees [40].

7.1.3 Dose, Duration, Surface Area, and Frequency of Exposure

Following skin contact, the amount of chemical absorbed depends greatly on several conditions of exposure. Increasing the surface area over which a chemical is in contact will also increase the absorption. The tendency of a compound to spread laterally on the skin surface will influence measurements of dermal absorption, which is often expressed as dose per area (cm^2) [41]. The total mass absorbed into the body increases with dose. However, the efficiency of absorption or percent absorption may decrease.

Absorption may be viewed as flux (rate of dermal absorption measured in $mg/cm^2/h$), as total absorption, or as percent absorption of an applied dose. With flux, the absorption of a compound across a single barrier, the *SC*, is typically proportional to the dose under steady-state conditions [10]. Note that the percent absorption depends upon the concentration applied per unit area. For many compounds, the percent absorption at relatively high concentrations is inversely related to applied dose [42]. In other words, the percent absorbed increases as skin loading decreases, and the function is nonlinear. This is attributed to saturation of the absorptive capacity of the skin. Despite a decrease in efficiency of absorption, in terms of percent absorption, a continued increase in total penetration, in mass, can be observed with increased dose. Similarly, with multiple applications or exposures, skin may become saturated and resist penetration from subsequent doses. As shown for hydrocortisone and testosterone, absorption from a single application at a high concentration was greater than the absorption of the same concentration divided into equal doses and applied separately [42, 43].

Dermal penetration tends to increase with duration of contact or exposure, but chemical residues retained in the skin layers may continue to be released into the body even after the contact material had been removed from the skin surface. Experimental protocols often set periods of exposure and sample collection that may not be relevant. It is often necessary to extrapolate long-term exposure from short-term data, for example. In general, the total absorption of a compound appears to be linearly correlated with the amount of penetration into the *SC* observed shortly after application [44].

While many studies investigate the dermal absorption of a compound following a single dose, multiple or repeated exposures are more relevant to real-life. Although the experiment by Ryman-Rasmussen et al. [24] showed QDs of different sizes, shapes, and surface coatings reported penetration of intact skin at an occupationally relevant dose within the span of an average-length 8 h work day, it would also have been interesting to observe dermal penetration following repeated exposures.

7.1.4 Sub-anatomical Pathways (Skin Appendages)

Dermal penetration has been considered as the simple diffusion of chemicals through the outermost layer of the skin, the *SC*. Although the *SC* plays a critical role in the function of the permeability barrier, we now know that there are many components with which a penetrant interacts before entering the body.

Skin appendages include hair follicles, sebaceous glands, and sweat gland, which originate in the dermis. Diffusion of chemicals through skin appendages is also referred to as "shunt diffusion" from the perspective that they are "interruptions" in the skin barrier and act as portals through which topically applied chemicals or particles may be short-cut to the lower epidermis and the dermis.

In efforts to treat local reactions selectively or to reduce systemic side-effects, delivery of active compounds may be targeted to explicit components or structures within the skin, especially to distinct hair follicle compartments or cell populations. Molecular or particle size, chemical properties such as polarity, and the formulation will also influence delivery and retention of active ingredients from the infundibular reservoir.

Hair follicle parameters such as volume, density, and distribution on various body sites have been investigated [45]. Although the highest density of hair follicles was found on the forehead, and the lowest follicle density was found on the calf region, the follicular orifices of the calf region measured highest in average size. Overall, the highest infundibular volume and therefore the greatest potential follicular reservoir were calculated to be the forehead and the calf region. Various appendageal parameters have been correlated with anatomical variability in absorption of chemicals (reviewed by Ngo et al. [46]).

The hair follicle has a complex structure containing many cell types and specialized proteins. The follicular infundibulum increases the surface area through which chemicals may be absorbed, disrupts the epidermal barrier towards the lower parts of the follicle, and may serve as a reservoir (reviewed by Lademann et al. [18] and Vogt et al. [47]). Stem cells and a high density of antigen-presenting cells also surround hair follicles, making them an attractive target for research in regenerative medicine and immunomodulation (reviewed by Ohyama [48]). Additionally, the blood capillaries surrounding hair follicles may facilitate systemic uptake of drugs or chemicals brought to this area.

Transcutaneous penetration through the follicular pathway has been generally assumed to be minimal since hair follicles constitute only 0.1% of the total skin [49]. Nevertheless, numerous studies have ascribed significant contribution of the transfollicular route to the absorption of compounds through the skin [50–53]. Skin appendages, particularly hair follicles, are an important mechanism by which larger molecules, such as pollen proteins and TiO_2 particles, may be stored in the skin and a potential pathway by which they may gain access to the dermis [25, 54–56].

Studies have reported the localization of particles in the deeper layers of the SC, the viable epidermis, and deeper hair follicle structures (reviewed by Lademann et al. [18]). Larger particles will likely remain in the hair follicle until they are dislodged by hair growth or sebum flow. The ability of smaller particles to penetrate through the follicular epithelium and into the dermis and surrounding vasculature remains unclear. It may be that only a component of the nanoparticle will be absorbed systemically, such as zinc dissolved from nano-preparations of ZnO in sunscreen [57], or the metal core leached out of damaged QD, rather than the entire nanoparticle.

Particle size influences the depth of penetration, with different follicle structures interacting with particles of particular size [20, 22, 58]. However, only hair follicles exhibiting hair growth or sebum production may be "open" for particle penetration [59]. Without such activity, coverings consisting of desquamated corneocytes and dried sebum can block access to the follicle [60].

Consistent with this idea of "open" follicles, prior treatment of freshly excised skin samples with cyanoacrylate skin surface stripping-technique (CSSS) allowed for 1,500 nm microspheres to penetrate 55% of hair follicles, with a maximum penetration depth of >2,300 μm [22]. Without CSSS, microspheres were found to penetrate only 27% of hair follicles, and to a depth of approximately 1,000 μm. CSSS removes keratinized material, lipids, and other debris from follicular openings. In another study, 40 nm, but not 750 and 1,500 nm nanoparticles, were found to penetrate into vellus hair openings and through the follicular epithelium following CSSS and nanoparticle application to the skin [58]. Although nanoparticles of 750 and 1,500 nm size did not penetrate deeply into hair follicles, they aggregated superficially in the hair follicle openings. Depending on their size, microparticles may also be used to block hair follicles to prevent penetration of topically applied substances [61].

7.1.5 Regional Variability in Absorption

The extent of dermal absorption varies from one body region to another. With the exception of the genital area, body sites generally found to have the highest absorption are the forehead, scalp, and neck. This general pattern of regional variation in permeability appears to hold true for a wide variety of chemicals tested (reviewed by Ngo et al. [46]). Exceptions, such as the absorption of carbaryl and paraquat, have been noted. Regional variation has also been observed in animal models [42, 62].

Regional differences have traditionally been attributed to skin thickness and the number of cell layers in the *SC* [63]. For the scrotum, thinness of the *SC* as well as increased blood flow and subsequent clearance from this area may contribute to enhanced absorption at this site. Additional information on dermal permeability variations in the population and by anatomic site are provided elsewhere [64, 65].

Variability in absorption also appears to correlate with hair follicle volume and distribution on the different body sites (reviewed by Otberg et al. [45] and Ngo et al. [46]). Since hair follicles appear to facilitate particle penetration, such regional distribution will also be an important consideration when evaluating risk of nanoparticle exposure.

In undamaged skin, Tinkle et al. found that 50 and 100 nm particles penetrated the epidermis and dermis of skin that was flexed, but not skin that was flat [66]. Optical scanning laser confocal microscopy and size-selected fluorospheres were employed to demonstrate that motion, such as at the wrist, promotes particle penetration through the *SC*, into the epidermis and, occasionally, the dermis of human skin. In another study, skin flexing of dermatomed porcine samples for 60 and 90 min following application facilitated penetration of fullerene-peptide derivatives [67]. The nanoparticles were detected in the dermis 8 h after application to flexed skin, but were not found in the dermis of unflexed skin until 24 h after application. In another study, 320 nm particles were observed to penetrate deeper into the hair follicles when massage was applied to the skin samples [2]. It was assumed that massage caused movement of the hairs, a mechanism by which particles were pushed deep into the hair follicles. Mechanical forces and hair movements would be anticipated to occur in vivo, particularly for the limbs.

7.1.6 Population Variability in Absorption

Variability in skin properties reported for different groups in the human population is often attributed to age-related changes in the skin, general health, and genetic differences between individuals. Mangelsdorf et al. [68] demonstrated significant differences in follicular density and volume, depending on the body site examined, between volunteers of Asian, African, and Caucasian descent. Given the role hair follicles play in the penetration process in general, and for particles in particular, these results provide a potential basis for differences in the rate and extent of absorption between the various skin types. In addition to pigment quantity and distribution, racial differences have also been suggested for lipid content [69, 70] and the number of cell layers [71]. Studies addressing differences in skin properties based on race have been limited.

Changes in properties of human skin have been found with different developmental stages [64]. Although the structure of the epidermis and *SC* of newborn skin has been compared to that of adults, its function continues to develop for some time after birth [72, 73]. In the preterm neonate, thinness of the epidermis allows for enhanced percutaneous absorption [64]. A limited subcutaneous fat layer in the preterm infant presumably allows for direct systemic access for absorbed compounds. For 1- and 2-day-old infants, skin pH was found to be higher for all anatomic sites compared to adults [74].

It is generally proposed that diminishing barrier function with advanced age is attributable to various age-related changes in the skin [75]. Aging of the skin is associated with progressive atrophy of the dermis and thinning of epidermis, reduction in density of hair follicles and hair shaft diameter, slowed turnover rate of the epidermis, as well as other alterations in architectural organization and function [76]. Roskos and Maibach [77] reviewed the limited experimental data, suggesting that, in fact, penetration does not increase in the aged.

A study on ZnO particle absorption through human skin in vivo reported small, yet statistically significant differences in the absorption of zinc between male and female subjects [57]. Larger amounts of the tracer ^{68}Zn were detected in the blood of females than in males who received nanoparticle sunscreen. Interestingly, the subject with remarkably high levels of ^{68}Zn was also the subject with an unspecified "reaction" to the nanoparticle sunscreen, requiring discontinuation of use.

7.1.7 Skin Surface Conditions

Hydration appears to enhance the overall permeability of the skin. *SC* hydration can result in the swelling of corneocytes, and promotion of water uptake into intercellular regions. Skin moisture, influenced by relative humidity or perspiration, has been reported to increase the absorption of propoxur and chlorpyrifos [78, 79].

Occlusion can also increase *SC* hydration. An area of skin occluded for 24 h post-application of hydrocortisone resulted in a tenfold increase in absorption over non-occluded skin. Such enhancement of percutaneous absorption by occlusive conditions has been shown to enhance many, but not all topically applied compounds [80]. For example, occlusion increased the transdermal flux of the lipophilic citrophen, but did not change the delivery rate of caffeine, an amphiphilic molecule [81].

An area of skin may be occluded with a bandage, a layer of clothing, or with the application of a topical ointment with occlusive properties. By preventing the accidental removal of a chemical from the skin surface by wiping or evaporation of a volatile chemical, occlusion may serve to maintain a higher applied dose than might otherwise be experienced. In studying skin barrier enhancing formulations, Meinke and colleagues concluded that pretreatment of the skin with different emulsion formulations prevented pollen protein penetration into hair follicles due to the occlusive properties of the emulsions [3].

Increased temperature may also increase blood flow to an area, affecting clearance of the compound. In the absence of occlusion, increasing skin temperature may increase the evaporation of a volatile penetrant, altering the effective concentration of the compound on the skin surface. Moderate skin heating, to approximately 40 °C, has been shown to increase the transdermal delivery of both testosterone and fentanyl [82].

Additionally, occlusion may increase the pH and change the count and composition of microbial populations on the skin [83]. Occlusion may increase the local skin temperature. Depending on the penetrant, volatility and absorption may be affected by the moderate acidity of the skin surface. For example, D-methamphetamine hydrochloride has decreased volatility and consequently, greater dermal uptake potential under the moderately acidic conditions of the skin surface (pH 4–7.4). Using a simulated sweat solution as the donor fluid of an in vitro diffusion cell system,

Larese et al. [84] detected low-level penetration of silver nanoparticles through the *SC*. Although the role of the artificial sweat solution in facilitating penetration was not determined, it is conceivable that the ionization state of the nanoparticles may be altered by the saline properties of sweat.

7.1.8 Skin Health and Integrity

The structure and integrity of the skin will be compromised by damage or disease. Compounds that were most poorly absorbed through intact skin, generally, hydrophilic molecules [85], were found to have the greatest enhancement of penetration under conditions chemical or physical insult. Even relatively minor damage, such as by hypodermic needle abrasion and more substantial damage by tape-strip removal of *SC* layers, have been shown to increase permeability in various models [86]. Skin surface stripping with CSSS may also be considered a method by which the intact skin barrier is disrupted.

Diseases such as psoriasis, eczema, and dermatitis also compromise the skin's barrier function, resulting in the increased permeability of numerous compounds. Permeability and reservoir function of scar tissue was observed to differ from that of non-scarred skin [87, 88]. Diseased skin may also have differential drug-metabolizing potential which may alter the absorption of certain compounds.

At times, the penetrant itself may alter barrier function. For example, chronic application of hydrocortisone results in skin thinning and significantly enhances penetration [89]. Relatively low concentrations of gold nanoparticles were reported to enhance proliferation of keratinocytes [90], with subsequent modifications in barrier function suspected.

During the manufacture of nanomaterials, workers may have increased exposure to nanoparticles. Increased activity and mechanical stress may increase the risk of scratches or abrasion to the skin, particularly on their hands and face. Atopic dermatitis, eczema or other skin conditions, may also compromise the skin barrier to nanoparticle penetration.

In vitro studies using QDs suggested mechanical damage caused by abrasion allowed penetration of QDs into the viable dermal layers, whereas with flexing and tape stripping, QDs were only found as far as the uppermost *SC* layers [91]. In vitro experiments performed using human skin suggested polyvinylpyrrolidone-coated nano-silver penetration to be penetrate both intact and damaged skin [84], although this penetration was suggested to have been enhanced in damaged skin. Other research by this same group reported silver nanoparticles to inhibit cell growth and viability of the human keratinocyte HaCaT cell line, even at a relatively short exposure period of 24 h followed by 6 days of "recovery" in silver nanoparticle-free medium [92].

TiO_2 is used as a physical sunscreen because it reflects and scatters UV light. However, absorption of UV by TiO_2 in aqueous environments leads to the generation of hydroxyl radicals, which can initiate oxidative reactions and catalyze DNA damage with both an in vitro organic substrate and in human cells [93]. Oxidative stress responses and cytotoxicity as well as ultrastructural and morphological changes have been reported for human epidermal keratinocytes (HaCaT) exposed to single-walled CNTs in cell culture [94]. The oxidative damage and cytotoxicity was associated with the catalytic effects of iron within the SWCNT material.

7.1.9 Substantivity and Binding

The degree of interactions with the different skin components determines the time the chemical residues persist or accumulate in the skin as well as the time required to traverse the different layers and enter the systemic circulation. Chemicals may bind to the keratins within the corneocytes or to their envelope proteins, in the lipid milieu surrounding the corneocytes, or partition into the relatively aqueous dermis. As previously discussed, the structures of the various skin appendages may provide a compartment for accumulation and storage of particles. Chemical residues retained in the lower compartments of the skin may also act as long-term reservoirs [54, 55].

By contrast, substantivity is considered the non-penetrating absorption to the superficial layers of the skin. These residues have persistent therapeutic or biological activity, even after the vehicle removal. This persistence of effect of a topically applied drug or cosmetic is determined by the degree of physical and chemical bonding to the surface. Such resistance to removal or inactivation by sweating, swimming, bathing, and friction is important in the development of sunscreens or the dosing strategy for topical drugs such as antifungals or antibacterials.

Occlusion may inhibit shedding of dead cells and result in the longer-term retention of these residues. In risk assessment, there is also debate as to whether bound skin residues should be considered as part of the absorbed dose.

7.1.10 Distribution

Once the *SC* has been breached, interactions with the deeper components of the permeability barrier will again depend on the physiochemical properties of the penetrant in question. Proposed toxic effects of nanoparticles entering the systemic circulation include perturbation of the immune system, as well as other organ systems.

Once partitioned into the lower dermis or subcutaneous fat, affinity for components of blood, such as albumin, may help draw chemicals into the systemic circulation [95]. Thus, perfusion of the dermis promotes absorption and creates a sink for chemicals that have traversed the epidermis. Subsequent distribution of a chemical to target sites or organs within the body will also depend on blood flow. By reducing local blood flow, such as with use of a vasoconstrictor, systemic absorption can be inhibited. At the same time, drug penetration into deeper tissue layers adjacent to the application site may be enhanced by reduced blood flow [96]. Skin irritation, inflammation may also increase blood flow and systemic distribution of nanomaterials from the dermis.

Cases of local and systemic toxicity following dermal absorption of drugs or chemicals across the skin have been evaluated [97]. From research examining adverse health effects of inhaled or ingested nanoparticles, it has been shown that nanoparticles readily gain access to epithelial and interstitial sites in the lung and may be distributed to other organ systems via the circulatory system (reviewed by Borm et al. [98]). Wu et al. [99] reported TiO_2 nanoparticles to penetrate the skin and reach different tissues and organs after 60 days of application to hairless mice. In another study, the anti-inflammatory effects of nano-silver, but not the nanoparticles themselves, were suggested to be transferred from the section of treated skin to a distal site [100]. Nano-silver penetration was minimal and the authors suggested that systemic release of biological signals resulted from the skin surface interactions of the nano-silver. Immunohistochemical analysis supported the involvement of biological signaling molecules such as cytokines and growth factors.

Various transdermal drug delivery devices have been developed to bypass the permeation barrier of the *SC*. Transdermal delivery devices include non-needle syringes which inject a drug solution at high pressure directly into skin tissues or devices which introduce dissolved drugs through micro-scaled needles injections. Studies investigating the pharmacokinetics of such *SC*-bypassing devices have indicated blood flow at the injection site and diffusion through blood vessels were important factors in the local absorption and systemic migration of intracutaneously injected drugs [101, 102].

Following intracutaneous injection, such as those used to test for tuberculin antigen reactions, and subcutaneous injection, systemic absorption of salicylate was observed to be through the vasculature of the skin at the local site of injection [103]. Salicylate was also observed to migrate to local musculature from the injection site, and from there, enter the systemic circulation from capillary vessels. By contrast, there was little drug migration to muscle following topical application of salicylate. Additionally, intracutaneous injections of drugs in hypertonic solutions tended to have greater retention time in the skin and muscle [101]. Again, irritant, inflammatory, or otherwise vasoactive agents may increase blood flow and systemic distribution of nanomaterials from the dermis. Compounds with higher molecular weights, particularly above 10 kDa, were found to have limited systemic absorption following intracutaneous injection. Comparisons between in vitro and in vivo experiments also suggested vascular flow to be an important factor influencing systemic distribution following penetration of the *SC* [102].

Once nanomaterials penetrate into the dermis, they may enter the bloodstream and have the potential to accumulate or exert adverse effects on different organ systems. In a study evaluating the distribution of gold nanoparticles from 15 to 200 nm following intravenous injection in mice, Sonavane et al. [104] found that although the gold nanoparticles of all sizes were accumulated in organs like the liver, lung, and spleen, the smallest, 15 nm nanoparticles showed the most widespread organ distribution. The 15 and 50 nm nanoparticles appeared able to pass the blood–brain barrier while 200 nm nanoparticles showed only minute presence in organs. Translocation of nanoparticles along neuronal axons has also been reported elsewhere (reviewed by Oberdorster et al. [105]).

When small PEG-coated gold nanoparticles of 4 or 13 nm were compared to larger PEG-coated gold nanoparticles of 100 nm in mouse kinetic studies, the smaller nanoparticles showed high levels in blood for 24 h and were cleared by 7 days [106]. By contrast, the larger nanoparticles were completely cleared from the blood by 24 h. Interestingly, blood levels of the gold nanoparticles of all sizes increased again at 3 months, which correlated with organ levels. Liver and spleen levels of small gold nanoparticles peaked at 7 days and at 1 month in the mesenteric lymph node. By contrast, larger gold nanoparticles were taken up rapidly, approximately 30 min, into the liver, spleen, and mesenteric lymph nodes.

In other studies, the circulation half-life and tissue distribution of QD were determined by their specific surface coating [107, 108]. Evidence for oscillations in the QD blood flow patterns were observed during the arterial-venous extraction, a factor which could affect systemic disposition [108]. Direct migration of insulin into the lymphatic vessels has also been reported after application with a lipid-based transdermal delivery system [109].

Langerhans cells (LCs) are dendritic cells that reside in the epidermis and function as resident antigen-presenting cells in the skin (reviewed by Jakob et al. [110]). Their dendritic processes extend up into the *SC*. Because of their unique migratory ability, they can transport antigens from the epidermis to regional lymph nodes where they can initiate systemic immune responses. Many studies have shown that LCs are highly immunostimulatory and recent data suggests LCs have immunoregulatory properties [111].

Given the role of LC in adaptive immune response, uptake of nanoparticles by LC may theoretically prompt adverse immune responses, including autoimmunity. LCs have been targets of studies investigating transcutaneous vaccinations [112]. Other particle-based delivery systems targeting active compounds to the hair follicle for enhanced penetration efficiency and uptake by skin-resident cells have also been proposed [3]. The penetration of solid fluorescent 40 or 200 nm polystyrene nanoparticles as well as virus particles in murine skin were found to penetrate deeply into hair follicles and were internalized by perifollicular antigen-presenting cells. Particle transport to draining lymph nodes was also visualized.

7.1.11 Loss from Skin Surface and Exfoliation

The persistence of a chemical on the skin will also depend on its resistance to removal or inactivation. Mechanisms of removal from the skin surface include volatilization, sweating, washing, friction with or transfer to other surfaces, and exfoliation. The process of skin exfoliation constantly removes dead corneocytes from the surface of the epidermis, replacing them with newer cells emerging from the basal layer. Although often considered "absorbed" once bound to the *SC*, such chemical residues may eventually be lost by exfoliation. Deuterated water and urea have been used to measure skin cell turnover and loss by exfoliation [113, 114], which can be enhanced by washing/rubbing and contact with clothing. Theoretically, residues in the SC that are not lost by exfoliation may penetrate deeper into the skin or body with time. Ossadnik et al. [115] suggested nanoparticles which penetrate into hair follicles can only be removed by sebum flow and with hair growth. Because these are slow processes, in comparison to exfoliation, hair follicles are considered a long-term storage for mechanism for particles.

For volatile penetrants, evaporation will reduce the concentration of the chemical on the skin surface with time. Conditions such as occlusion or temperature may reduce the rate of volatilization. As basic protective barrier, clothing may reduce chemical exposures by preventing contact with the skin. Clothing may bind chemical residues and limit its bioavailability [116]. Clothing rubbing against the skin surface may rub away chemical residues or facilitate the exfoliation of skin cells and the subsequent loss of residues bound to them [114].

By contrast, studies have also described dermal absorption of chemicals from treated cloth fabric [117–119]. Although 3.2% of the available permethrin had leached from a fabric onto the skin after 7 days, most of that amount (2%) was absorbed and excreted, while the remaining amount (1.2%) was found on the skin surface. Permethrin residues remained in the cloth even after laundering with detergents [118]. The extent to which chemical residues are removed from clothing by laundering, and the retention of chemicals on different types of material, are detailed elsewhere [120, 121].

For the pesticides glyphosate and malathion, dermal absorption from solution placed directly on skin was found to be 1.42% and 8.77%, respectively [117], over a 24 h period. Absorption from freshly treated fabric approximated half the absorption from solution (0.74% for glyphosate and 3.92% for malathion). When the treated fabric was allowed to dry before being placed onto the skin, absorption decreased further to about 0.08% for glyphosate and 0.6% for malathion. When treated fabric was dried and then rehydrated, measured absorption for glyphosate increased again to almost half the absorption seen with freshly treated fabric (0.36%). However, absorption of malathion under conditions of rehydration approached the absorption measured when malathion solution was placed directly on the skin (7.34%). This rehydration simulates conditions of perspiration which are highly relevant to many of the physically demanding tasks in agricultural work. In another study, fabric discs were placed in sealed glass containers with ethylene oxide, a widely used fumigant [119]. Percutaneous absorption from treated fabric increased dramatically from 1.3 to 46% with occlusion with latex glove material.

Engineered nanomaterials that are incorporated into textile products, nanotextiles, include nano-silver, nano-TiO_2, nanosilica, nano-ZnO, nano-alumina (Al_2O_3), layered silica, carbon black, and CNTs (reviewed by Som et al. [122]). These nanomaterials are used to improve textile properties such as wrinkle resistance, water-repellency, antimicrobial and antistatic properties, and UV- and flammability resistance. Since there is continuous and intimate contact between clothing and skin, transfer and absorption of nanomaterials under these conditions should be considered.

The nanomaterial and the manner in which it is incorporated into the textile will have consequences on its bioavailability. The nanomaterial may be covalently or noncovalently bound directly to the fiber surface, incorporated in polymeric fiber coatings, or applied as single, agglomerated or aggregated nanoparticles. Apparently, it is possible to produce nanomaterial-free nanotextiles with nanoporous polymer or xero-gel coating. External factors such as abrasion, laundering, and heat exposure might result in the release of nanoparticles from nanotextiles.

7.1.12 Wash Effect

Washing and rubbing an area of urea-treated skin can remove upwards of 60% of the dose, even at 6 h after application [114]. Soap and water alone may not be effective in removing substances from the skin, particularly highly lipid-soluble chemicals [123]. Chemical residues, such as pesticides remaining in or on the skin after washing, may continue to be absorbed into the body [124]. Ironically, washing the application site between doses, a common step in experimental protocols of dermal absorption studies, has also been reported to enhance percutaneous absorption [125, 126].

In the occupational setting, washing of exposed skin for the purpose of decontamination can also enhance absorption [127]. This "wash-in" effect may be attributed to the influence of surfactants on the lipid properties of the *SC* and consequent alteration of barrier function [128, 129]. In fact, sodium lauryl sulfate, a common ionic surfactant and ingredient in liquid soaps and shampoos, is an irritant that can augment dermal penetration by producing skin irritation [130]. As noted previously, the generation of ROS and subsequent toxicity of fullerenes may be enhanced by surfactants [38]. These effects could facilitate the passage nanoparticles and of other co-exposed or subsequently exposed compounds.

Mechanical manipulations such as rigorous washing, has also suggested to be a mechanism by which hair follicle coverings, consisting of desquamated corneocytes and dried sebum, could be removed and particle penetration into follicles facilitated [60]. Another study proposed mineral oil as a decontamination solution after occupational exposure, based on their observation that mineral oil vehicle was able to prevent fullerene partitioning into the *SC* [37].

7.1.13 Metabolic and Photochemical Transformation

The skin is capable of a wide range of metabolic functions and is recognized as a significant site of biotransformation and photochemical transformation [131, 132]. The potential of a chemical compound to cause phototoxic or photoallergic reactions is of importance in both the occupational and the recreational setting where long hours are spent in the sun.

Depending on the properties of the chemical, photochemical degradation may also occur. In addition to cutaneous metabolism of endogenous substrates such as vitamin A and vitamin D, some xenobiotics must be metabolized into more water-soluble products before they can be excreted from the body [133]. Although some xenobiotic metabolism involves enzymes specific to the skin, the skin does not have the ability to metabolize all chemicals. For example, topically applied hexachlorophene does not appear to be readily metabolized. There are also population differences for enzymatic activity in the skin [134].

The observed attenuation in absorption of highly lipophilic compounds has been attributed to the relatively "aqueous" qualities of the lower layers of the epidermis and the dermis, compared to the lipid-rich *SC*. Cutaneous metabolism plays an essential role in the absorption of these highly lipophilic compounds by transforming them into more water soluble metabolites. The highly lipophilic butylparaben is metabolized to *p*-hydroxybenzoic acid which can more readily penetrate the skin. With the inhibition of cutaneous esterase activity with diisopropyl phosphorofluoridate, however, no *p*-hydroxybenzoic acid was detectable in the receptor fluid of the in vitro system [135]. A study using gold nanoparticles of 4–13 nm and 100 nm reported metabolic enzymes, namely CYP1A1 and 2B, to be transiently elevated in liver tissues from 24 h to 7 days [106]. This time frame mirrored the period of elevated gold levels in the liver. The larger gold nanoparticles did not appear to affect the metabolic enzymes.

Methodology for maintaining tissue viability and enzymatic activity for in vitro studies is discussed elsewhere [133, 136–139]. Nevertheless, some enzyme cofactors, such as glutathione, may be limited under in vitro conditions [140]. "Inert" ingredients or penetration enhancers in formulations may activate or inhibit xenobiotic metabolism, resulting in altered reactivity, retention, or subsequent toxicity of a dermally exposed compound. While in vivo metabolism often produces metabolites with physiochemical properties and penetration characteristics which differ from the parent compound, in vitro studies generally do not determine whether biotransformation and photochemical reactions occur during the absorption process.

The TiO_2 and ZnO have long been used in sunscreens as an alternative to existing chemical UV absorbers, such as *p*-aminobenzoic acid and benzophenones, which can cause allergic reactions in sensitive individuals. Nano-preparations of TiO_2 and ZnO are now incorporated into many sunscreens and cosmetics, yet studies have shown UV to induce free radical formation and subsequent phototoxicity from TiO_2 [141]. An investigation into carboxylated-QD penetration in mice in vivo suggested UV rays enhanced penetration into the skin [142]. Natural sunlight has been reported to break up aggregates of nanoparticles in aqueous medium, releasing primary particles of TiO_2 and MWCNT (Bennett, in press). The ability of ambient light to induce disaggregation of some nanoparticles may have significant consequences on toxicity and dermal penetration.

7.1.14 Excretion Kinetics

Depending on the compound, elimination may or may not be in proportion to the dose absorbed. For lipophilic compounds, elimination is dependent on metabolism. Lipophilic lindane, for example, is metabolized in the liver and its chlorophenol metabolites are readily excreted in urine. Hexachlorophene, on the other hand, is not readily metabolized and has limited excretion as a result of its lipophilicity [143].

Chemicals that are excreted largely unmetabolized have urinary concentrations that tend to be linear in relation to exposure. For chemicals that are extensively broken down into numerous metabolites, each metabolite may be difficult to quantify and to quantifiably relate to exposure or percutaneous penetration. There are still other chemicals for which metabolite excretion patterns are unknown. Similar to metabolism, excretion kinetics can differ greatly between even structurally related compounds. Xenobiotic metabolism and urine output are also known to vary greatly depending on age and activity level, and is subject to interindividual variability. Elimination routes between species may differ substantially.

Regarding in vivo studies, dermal absorption estimated from tissues, blood, or excreta sampling will be affected by the clearance rate of the compound and its metabolites from the biological media being tested (summarized by Nigg and Stamper [144]). A chemical may be slowly or rapidly eliminated into urine or feces, so the time of sampling is of great relevance in the design of such a study and sufficient collection period will differ depending on the chemical of interest. Additionally, when intravenous data is used for the correction of incomplete urinary recovery following dermal dosing, the underlying assumption is that the same metabolism and elimination profiles exist for both routes may be unsupported.

Yoshida et al. [102] found that following intracutaneous injection, the elimination from the skin was more rapid for smaller molecular weight compounds (0.160–3.8 kDa) in comparison to compounds of larger molecular weight (10–39.2 kDa). In a study evaluating the distribution of gold nanoparticles from 15 to 200 nm following intravenous injection in mice, Sonavane et al. [104] found that although the gold nanoparticles of all sizes were accumulated in organs like the liver, lung, and spleen, the smallest, 15 nm nanoparticles showed the most widespread organ distribution. The 15 and 50 nm nanoparticles were also appeared to be able to pass the blood–brain barrier while 200 nm nanoparticles showed only minute presence in organs.

While investigating in vivo disposition of the cadmium-based QD705 in mice, Yang et al. [145] found increasing tissue concentration and redistribution of QD705 for up to 28 days after a single intravenous dose. However, the plasma half-life of QD705 was relatively short (18.5 h). When QD705 was compared to cadmium ($CdCl_2$), they had very different kinetics in distribution and metabolism [146]. QD705 showed a longer plasma and body retention than $CdCl_2$ and could be detected in the brain during early exposure. The increased retention of QD705 in certain organs may result in higher toxicity compared to $CdCl_2$. In another study investigating the fate of 40 nm gold nanoparticles, Sadauskas et al. [147] were able to detect gold nanoparticles in liver tissue for greater than 6 months following intravenous injection.

7.1.15 Method of Determining Absorption and Toxicity

Dermal data are a central component in the assessment of risk from exposure to chemicals. In addition to dermal exposure and absorption, skin irritation and sensitization potential are also evaluated. When evaluating dermal data, it is important to keep in mind the controlled conditions under which these studies were done.

With the incorporation of nanomaterials into cosmetics and transdermal drug delivery devices, it is important to understand skin penetration and toxicity. A key advancement of methods for quantifying skin penetration came with the advent of radiolabeled compounds. Low doses of ^{14}C facilitated the tracking of chemical movement through the different layers of the skin. With this method, greater than 95% "mass balance" can typically be obtained, enabling excellent dose accountability and greatly augmenting the reliability of the data.

Tape stripping removes the *SC*, one layer at a time, allowing for the solubilization and analysis of residues in the *SC*. This can be done both in vivo and in vitro, however, the number of tape strips that are necessary to remove the *SC* will be influenced by such factors as species and vehicle used. Tape stripping has been combined with labeling of the *SC* with deuterated water or a minimally absorbed compound to measure skin cell turnover and loss by exfoliation [113, 114]. All skin residues that are not removed by tape stripping are assumed to have penetrated into the lower epidermis and dermis. Notably, the standard tape stripping method does not remove all residues, such as that from skin furrows or hair follicle openings [55, 148]. The cyanoacrylate skin surface biopsy method applies cyanoacrylate to the skin surface. Following polymerization, the layer of cyanoacrylate is removed from the skin along with corneocytes and follicular casts. With the combination method of differential stripping, tape stripping is used to remove the part of the *SC* that contains the topically applied substances, while subsequent cyanoacrylate skin surface biopsy removing and allowing for determination of follicular penetration [149]. These investigative methods also enlist spectrophotometry and confocal laser scanning microscopy. Specifics on the credibility of available methodologies and specialized chemical analyses, such as stable isotopes and accelerator mass spectroscopy (AMS), are provided elsewhere [150, 151].

Various factors will influence the final results and interpretations of studies designed to assess the ability of chemicals to penetrate the skin. Differences in experimental parameters such as animal species, anatomical site, vehicle, preparation of skin samples, and other exposure conditions affect the interpretation of and comparison between studies. Considerations for in vitro and in vivo methods used for studying hair follicle structure and for designing delivery systems using follicular penetration pathways are discussed elsewhere [152–154].

7.1.15.1 In Vivo

In vivo techniques usually involve the indirect measurements of compounds of interest in different body tissues and fluids following topical application [155]. The importance of "mass balance" and dose accountability is emphasized. Since complete recovery of the topically applied dose is not always achievable, adjustment for incomplete elimination may use recovery following parenteral dosing [156]. This type of adjustment assumes that the metabolism and elimination profile following topical administration is the same as that following parenteral administration. On the contrary, hepatic metabolism, which is experienced with oral or intravenous dosing, is known to be much more extensive than cutaneous metabolism [137]. Alternatively, dermal absorption may be considered as the sum of chemical residues from body fluids and tissues without the use of a correction factor to account for unknown excretion or storage pathways.

For ethical reasons, penetration potential of dermally exposed or applied chemicals is increasingly estimated from in vitro skin data. Interspecies differences in skin structure, such as epidermal or *SC* thickness, skin appendages, such as hair follicles and sweat glands, and metabolism have been documented [157–159]. However, evidence suggests that use of in vitro tests alone may not provide the best estimates for dermal penetration and systemic exposure.

Some dermatological effects, such as dermal sensitization and irritation, may be best verified with in vivo systems. Currently, data is limited for the ability of nanomaterials to elicit such effects. Efforts made to test toxicity of ultrafine TiO_2 by defined sets of tests have found limited toxicity [160]. These particles were not found to be skin irritants or dermal sensitizers in acute dermal irritation studies in rabbits and local lymph node assays in mice. However, short-term, reversible ocular conjunctival redness was demonstrated in rabbits. In another study, two products containing SWCNTs and two products containing MWCNTs were tested for acute dermal and acute eye irritation using rabbits, and skin sensitization using guinea pigs [161]. Only one MWCNT-containing product was found to cause very slight erythema at 24 h, but not at 72 h after patch removal in the dermal irritation and conjunctival redness tests. This product also caused blood vessel hyperemia at 1 h, but not at 24 h in the eye irritation experiments. Thus, only one of the four products was shown to be a very weak acute eye and skin irritant. None of the products tested exhibited skin-sensitization.

7.1.15.2 In Vitro

In vitro techniques typically involve placing a piece of excised skin into a diffusion chamber, applying a compound to one side of the skin, and then quantifying the amount of labeled material found in the collection fluid on the other side [162]. Fresh or frozen human skin, animal skin, or artificial membrane can be utilized in two basic types of diffusion chambers used: flow-through and static. In vitro techniques assume that the enzymatic activity present in viable epidermis, the vascularity and composition of the dermis, the surface conditions, and other processes of viable skin (such as a hydration and bacterial population) are little changed or have negligible influence on dermal absorption [5]. Although some of these conditions may be specifically controlled for, they are not commonly addressed in in vitro studies.

Advantages of in vitro methods include ease of use, with attainment of rapid results under controlled conditions. However, the very ability to control the numerous variables becomes a challenge when comparing results between studies. These variables include but are not limited to, receptor fluid composition, perfusion flow rate, air temperature and flow, and relative humidity [163–166].

Even with efforts made to standardize experimental protocol, standard deviations were high when comparing the mean absorption rates among different laboratories [167]. Interspecies differences in dermal absorption are well known, and no animal-to-human adjustment factors are available. Skin thickness, vascularization, and appendageal differences make extrapolation unreliable even between rodent species [168–170].

7.1.15.3 Mathematical Models

Quantitative structure activity relationships (QSARs) are mathematical models which use known molecular structures and their associated biological-activity to predict the behavior of other chemicals with similar structures. Analogous QSPRs have been derived to predict the capacity of chemical compounds to penetrate the *SC* [171]. These models also take into account the effect of vehicle and complex mixtures. Rather than using steady-state assumptions, current evidence supports the need to account for time-dependent diffusion with reversible binding in mathematical models of dermal absorption [172, 173]. Assessment of the correlations between a QSAR model based on in vitro flux data with one based on human in vivo data evaluates the strengths and weaknesses of predictive models [30].

The ability of any QSPR model to accurately make predictions is dependent on the quality of the data, specifically, dermal absorption values, from which the relationships are derived. Most of these QSPR models are based on data collected from various studies that use different concentrations, formulations, and vehicles, with differing experimental protocols, analysis, and interpretations. Dermal absorption values compiled in this manner may limit the ability to accurately make predictions about penetration properties.

When 12 QSPR models were evaluated for their ability to predict pharmacokinetic parameters for transdermal drug patches [6], most of these models underestimated the maximal plasma drug concentrations from observed in vivo data by as much as 10–100-fold. The greatest discrepancies between the predicted values from in vitro models and the in vivo data were for compounds with maximal or minimal values for molecular weight, hydrogen bonding, and $\log K_{o/w}$. The limited ability of these models to make accurate predictions for chemicals with peripheral values emphasizes the difficulty in developing generalized mathematical models to predict penetration of compounds across the spectrum of complex physiochemical characteristics. Although further improvement of predictive equations is necessary, QSPRs continue to be an extraordinary tool for estimating skin permeability.

Creating mathematical models of nanomaterials to predict their behavior and dermal penetration mandates basic knowledge of their properties and understanding the scientific principles behind their behavior. There is still much work needed for establishing models that will clarify the behavior of materials at this scale.

7.2 Summary

Transcutaneous vaccination, controlled drug release to skin and targeting of hair follicle-specific cell populations, and transdermal gene therapy are only a few of the new skin applications being investigated for nanomaterials. Novel carrier systems take advantage of improved skin penetration, specific targeting of tissue structures and cellular populations, deposition and sustained drug release can transform modern treatment for skin disorders. Nonetheless, there is concern over the safety of new cosmetics containing nanotechnology ingredients, although currently, there is little requirement for safety testing of cosmetic products prior to marketing.

For example, there has been a trend in the cosmetics industry to replace micronized TiO_2 and ZnO particles with nanoscale materials, owing to their ability to persist on the skin and adhere closely to the skin surface. Compared to traditional preparations, nano-TiO_2 and nano-ZnO are less chalky in appearance and accumulate more densely across the skin surface. Although the US Food and Drug Administration (FDA) has determined that nano-TiO_2 is a specific grade of the original product and not a new ingredient, concerns persist that the toxicological profiles for nanomaterials may not be equivalent to the respective bulk-form materials.

Optimally, potential toxicological and environmental outcomes should be evaluated prior to wide-spread implementation of nanomaterial use and incorporation into consumer products. There are limited data on nanoparticles that are increasingly being used in industry. Only a few specific nanoparticles have been investigated using a limited number of test systems [98]. Existing toxicoligcal data on nanoparticles suggest various adverse effects are possible.

Knowledge gaps include the behavior of nanoparticles once they enter the body. If nanoparticles enter the systemic circulation, the potential for bioaccumulation and the possible toxicological mechanisms outcomes are not known. Especially for nanoparticles designed to be cytotoxic, the same mechanisms with which these particles kill bacteria or cancer cells will also make them toxic to human cells and animals once they are released into the environment. Understanding mechanisms is needed in order to limit undesired effects and possibly correlate nanoparticle physiochemical properties to penetration potential. Correlating physiochemical properties of nanoparticles to penetration potential may be possible in the future.

A standardized test system for nanoparticles is necessary, including studies on the long-term effects of using nanomaterials in products. Further investigation is needed to clarify the potential for dermal penetration, irritation, and sensitization for nanomaterials, particularly for those incorporated into cosmetics and medical applications. Mechanistic studies on "representative" nanoparticles have been suggested [160]. Extrapolation of limited data to other newly developed nanomaterials does not seem advisable. However, these representative nanoparticle mechanisms may be combined with a standardized set of toxicity tests implemented by the companies that are developing nanomaterial-containing products.

Harmonization of safety requirements and testing for products containing nanomaterials is emphasized. The minimum criteria for such a set of testing could include substantial particle characterization, pulmonary toxicity studies, acute dermal toxicity and sensitization studies, acute oral and ocular toxicity studies, along with screening type genotoxicity, and aquatic toxicity studies [160]. With dermal studies, there is often limited data on the excretion and washout. The importance of mass balance must be stressed. Without this, it is difficult to interpret the data.

In the absence of experimental data for transdermal penetration in humans, in vitro, animal, or mathematical models are often used. Mathematical models would be particularly useful for predicting absorption of chemical mixtures since acquiring experimental data for all the possible combinations of chemicals would be impractical. Although valuable tools, the measurement of absorption with such methods may oversimplify transdermal penetration in humans, with a multitude of interactions at the molecular level being overlooked. Without extensive validation of highly standardized and well-controlled test systems, in vitro data alone cannot be used in place of in vivo data. In vivo verification of dermal penetration, preferably in humans, would add relevance to in vitro data.

Uncertainties regarding the impact of nanoscale materials would also be helped by greater consistency in methodologies used to assess health risks. As risk is a function of hazard and exposure, a basic set of hazard data are required for the development of risk management strategies for nanoscale products. Risk assessment under REACH guidance does not appear to be appropriate since this guidance was developed without considering nanomaterial mechanisms [174]. The main risks for human health may arise from chronic occupational exposure. At this time, available data for exposure and hazard is too limited for quantitative characterization of risk. Much remains to be done in the field of dermatotoxicology of nanomaterials. Utilizing the most sensitive analytical techniques and emphasizing in vivo human data should aid interpretative reliability and perspective.

References

1. ISO TS 27687:2008 – Nanotechnologies - Terminology and definitions for nano-objects—Nanoparticle, nanofibre and nanoplate.
2. Lademann J, Richter H, Teichmann A, Otberg N, Blume-Peytavi U, Luengo J, et al. Nanoparticles—an efficient carrier for drug delivery into the hair follicles. Eur J Pharm Biopharm. 2007;66(2):159–64.
3. Mahe B, Vogt A, Liard C, Duffy D, Abadie V, Bonduelle O, et al. Nanoparticle-based targeting of vaccine compounds to skin antigen-presenting cells by hair follicles and their transport in mice. J Invest Dermatol. 2009;129(5):1156–64.
4. Jung S, Patzelt A, Otberg N, Thiede G, Sterry W, Lademann J. Strategy of topical vaccination with nanoparticles. J Biomed Opt. 2009;14(2):021001.
5. Wester RC, Maibach HI. Cutaneous pharmacokinetics: 10 steps to percutaneous absorption. Drug Metab Rev. 1983;14(2):169–205.
6. Farahmand S, Maibach HI. Estimating skin permeability from physicochemical characteristics of drugs: a comparison between conventional models and an in vivo-based approach. Int J Pharm. 2009;375(1–2):41–7.
7. Bos JD, Meinardi MM. The 500 Dalton rule for the skin penetration of chemical compounds and drugs. Exp Dermatol. 2000;9(3):165–9.
8. Patel H, ten Berge W, Cronin MT. Quantitative structure-activity relationships (QSARs) for the prediction of skin permeation of exogenous chemicals. Chemosphere. 2002;48(6):603–13.
9. Potts RO, Guy RH. Predicting skin permeability. Pharm Res. 1992;9(5):663–9.
10. Scheuplein RJ, Blank IH. Mechanism of percutaneous absorption. IV. Penetration of nonelectrolytes (alcohols) from aqueous solutions and from pure liquids. J Invest Dermatol. 1973;60(5):286–96.
11. Scheuplein RJ. Mechanism of percutaneous absorption. II. Transient diffusion and the relative importance of various routes of skin penetration. J Invest Dermatol. 1967;48(1):79–88.
12. Thurn KT, Brown E, Wu A, Vogt S, Lai B, Maser J, et al. Nanoparticles for applications in cellular imaging. Nanoscale Res Lett. 2007;2(9):430–41.
13. Matilainen A, Sillanpaa M. Removal of natural organic matter from drinking water by advanced oxidation processes. Chemosphere. 2010;80(4):351–65.
14. Sayes CM, Wahi R, Kurian PA, Liu Y, West JL, Ausman KD, et al. Correlating nanoscale titania structure with toxicity: a cytotoxicity and inflammatory response study with human dermal fibroblasts and human lung epithelial cells. Toxicol Sci. 2006;92(1):174–85.
15. Lipovsky A, Gedanken A, Nitzan Y, Lubart R. Enhanced inactivation of bacteria by metal-oxide nanoparticles combined with visible light irradiation. Lasers Surg Med. 2011;43(3):236–40.

16. Schneider M, Stracke F, Hansen S, Schaefer UF. Nanoparticles and their interactions with the dermal barrier. Dermatoendocrinology. 2009;1(4):197–206.
17. Labouta HI, el-Khordagui LK, Kraus T, Schneider M. Mechanism and determinants of nanoparticle penetration through human skin. Nanoscale. 2011;3(12):4989–99.
18. Lademann J, Richter H, Schanzer S, Knorr F, Meinke M, Sterry W, et al. Penetration and storage of particles in human skin: perspectives and safety aspects. Eur J Pharm Biopharm. 2011;77(3):465–8.
19. Kunzmann A, Andersson B, Vogt C, Feliu N, Ye F, Gabrielsson S, et al. Efficient internalization of silica-coated iron oxide nanoparticles of different sizes by primary human macrophages and dendritic cells. Toxicol Appl Pharmacol. 2011;253(2):81–93.
20. Patzelt A, Richter H, Knorr F, Schafer U, Lehr CM, Dahne L, et al. Selective follicular targeting by modification of the particle sizes. J Control Release. 2011;150(1):45–8.
21. Lademann J, Patzelt A, Richter H, Antoniou C, Sterry W, Knorr F. Determination of the cuticula thickness of human and porcine hairs and their potential influence on the penetration of nanoparticles into the hair follicles. J Biomed Opt. 2009;14(2):021014.
22. Toll R, Jacobi U, Richter H, Lademann J, Schaefer H, Blume-Peytavi U. Penetration profile of microspheres in follicular targeting of terminal hair follicles. J Invest Dermatol. 2004;123(1):168–76.
23. Wu X, Landfester K, Musyanovych A, Guy RH. Disposition of charged nanoparticles after their topical application to the skin. Skin Pharmacol Physiol. 2010;23(3):117–23.
24. Ryman-Rasmussen JP, Riviere JE, Monteiro-Riviere NA. Penetration of intact skin by quantum dots with diverse physicochemical properties. Toxicol Sci. 2006;91(1):159–65.
25. Scheuplein RJ, Blank IH, Brauner GJ, MacFarlane DJ. Percutaneous absorption of steroids. J Invest Dermatol. 1969;52(1):63–70.
26. Hadgraft J, Valenta C. pH, pK(a) and dermal delivery. Int J Pharm. 2000;200(2):243–7.
27. Gilpin SJ, Hui X, Maibach HI. Volatility of fragrance chemicals: patch testing implications. Dermatitis. 2009;20(4):200–7.
28. Farrell KA. Synthesis effects of grain size and phase content in the anastase-rutile TiO2 system. Thesis, Worcester Polytechnic Institute. 2001. www.wpi.edu/Pubs/ETD/Available/etd-081699–142023.
29. Simpson CA, Agrawal AC, Balinski A, Harkness KM, Cliffel DE. Short-chain PEG mixed monolayer protected gold clusters increase clearance and red blood cell counts. ACS Nano. 2011;5(5):3577–84.
30. Smith EW, Maibach HI, editors. Percutaneous penetration enhancers. 2nd ed. Boca Raton: CRC Press; 2006.
31. Nohynek GJ, Lademann J, Ribaud C, Roberts MS. Grey goo on the skin? Nanotechnology, cosmetic and sunscreen safety. Crit Rev Toxicol. 2007;37(3):251–77.
32. Teichmann A, Heuschkel S, Jacobi U, Presse G, Neubert RH, Sterry W, et al. Comparison of stratum corneum penetration and localization of a lipophilic model drug applied in an o/w microemulsion and an amphiphilic cream. Eur J Pharm Biopharm. 2007;67(3):699–706.
33. Neubert RH. Potentials of new nanocarriers for dermal and transdermal drug delivery. Eur J Pharm Biopharm. 2011;77(1):1–2.
34. Schroeter A, Engelbrecht T, Neubert RH, Goebel AS. New nanosized technologies for dermal and transdermal drug delivery. A review. J Biomed Nanotechnol. 2010;6(5):511–28.
35. Moody RP, Wester RC, Melendres JL, Maibach HI. Dermal absorption of the phenoxy herbicide 2,4-D dimethylamine in humans: effect of DEET and anatomic site. J Toxicol Environ Health. 1992;36(3):241–50.
36. Kuo TR, Wu CL, Hsu CT, Lo W, Chiang SJ, Lin SJ, et al. Chemical enhancer induced changes in the mechanisms of transdermal delivery of zinc oxide nanoparticles. Biomaterials. 2009;30(16):3002–8.
37. Xia XR, Monteiro-Riviere NA, Riviere JE. Skin penetration and kinetics of pristine fullerenes (C60) topically exposed in industrial organic solvents. Toxicol Appl Pharmacol. 2010;242(1):29–37.
38. Lee J, Yamakoshi Y, Hughes JB, Kim JH. Mechanism of C60 photoreactivity in water: fate of triplet state and radical anion and production of reactive oxygen species. Environ Sci Technol. 2008;42(9):3459–64.
39. Trommer H, Neubert RH. Overcoming the stratum corneum: the modulation of skin penetration. A review. Skin Pharmacol Physiol. 2006;19(2):106–21.
40. Meinke MC, Patzelt A, Richter H, Schanzer S, Sterry W, Filbry A, et al. Prevention of follicular penetration: barrier-enhancing formulations against the penetration of pollen allergens into hair follicles. Skin Pharmacol Physiol. 2011;24(3):144–50.
41. Vieille-Petit A, Pirot F, Maibach HI. Lateral spreading from skin exposure: assessment with tape stripping. Submitted for publication.
42. Wester RC, Noonan PK, Maibach HI. Variations in percutaneous absorption of testosterone in the rhesus monkey due to anatomic site of application and frequency of application. Arch Dermatol Res. 1980;267(3):229–35.
43. Wester RC, Noonan PK, Maibach HI. Frequency of application on percutaneous absorption of hydrocortisone. Arch Dermatol. 1977;113(5):620–2.
44. Rougier A, Lotte C, Maibach HI. In vivo percutaneous penetration of some organic compounds related to anatomic site in humans: predictive assessment by the stripping method. J Pharm Sci. 1987;76(6):451–4.
45. Otberg N, Richter H, Schaefer H, Blume-Peytavi U, Sterry W, Lademann J. Variations of hair follicle size and distribution in different body sites. J Invest Dermatol. 2004;122(1):14–9.
46. Ngo MA, O'Malley M, Maibach HI. Percutaneous absorption and exposure assessment of pesticides. J Appl Toxicol. 2009;30(2):91–114.

47. Vogt A, Mandt N, Lademann J, Schaefer H, Blume-Peytavi U. Follicular targeting—a promising tool in selective dermatotherapy. J Investig Dermatol Symp Proc. 2005;10(3):252–5.
48. Ohyama M. Hair follicle bulge: a fascinating reservoir of epithelial stem cells. J Dermatol Sci. 2007;46(2):81–9.
49. Blank IH, Scheuplein RJ. Transport into and within the skin. Br J Dermatol. 1969;81 Suppl 4:4–10.
50. Feldmann RJ, Maibach HI. Regional variation in percutaneous penetration of 14C cortisol in man. J Invest Dermatol. 1967;48(2):181–3.
51. Maibach HI, Feldman RJ, Milby TH, Serat WF. Regional variation in percutaneous penetration in man. Arch Environ Health. 1971;23(3):208–11.
52. Kao J, Hall J, Helman G. In vitro percutaneous absorption in mouse skin: influence of skin appendages. Toxicol Appl Pharmacol. 1988;94(1):93–103.
53. Hueber F, Schaefer H, Wepierre J. Role of transepidermal and transfollicular routes in percutaneous absorption of steroids: in vitro studies on human skin. Skin Pharmacol. 1994;7(5):237–44.
54. Jacobi U, Engel K, Patzelt A, Worm M, Sterry W, Lademann J. Penetration of pollen proteins into the skin. Skin Pharmacol Physiol. 2007;20(6):297–304.
55. Lademann J, Richter H, Schaefer UF, Blume-Peytavi U, Teichmann A, Otberg N, et al. Hair follicles—a long-term reservoir for drug delivery. Skin Pharmacol Physiol. 2006;19(4):232–6.
56. Lademann J, Weigmann H, Rickmeyer C, Barthelmes H, Schaefer H, Mueller G, et al. Penetration of titanium dioxide microparticles in a sunscreen formulation into the horny layer and the follicular orifice. Skin Pharmacol Appl Skin Physiol. 1999;12(5):247–56.
57. Gulson B, McCall M, Korsch M, Gomez L, Casey P, Oytam Y, et al. Small amounts of zinc from zinc oxide particles in sunscreens applied outdoors are absorbed through human skin. Toxicol Sci. 2010;118(1):140–9.
58. Vogt A, Combadiere B, Hadam S, Stieler KM, Lademann J, Schaefer H, et al. 40 nm, but not 750 or 1,500 nm, nanoparticles enter epidermal CD1a+ cells after transcutaneous application on human skin. J Invest Dermatol. 2006;126(6):1316–22.
59. Lademann J, Otberg N, Richter H, Weigmann HJ, Lindemann U, Schaefer H, et al. Investigation of follicular penetration of topically applied substances. Skin Pharmacol Appl Skin Physiol. 2001;14 Suppl 1:17–22.
60. Otberg N, Richter H, Knuttel A, Schaefer H, Sterry W, Lademann J. Laser spectroscopic methods for the characterization of open and closed follicles. Laser Phys Lett. 2004;1(1):46–9.
61. Teichmann A, Ossadnik M, Richter H, Sterry W, Lademann J. Semiquantitative determination of the penetration of a fluorescent hydrogel formulation into the hair follicle with and without follicular closure by microparticles by means of differential stripping. Skin Pharmacol Physiol. 2006;19(2):101–5.
62. Mills PC, Magnusson BM, Cross SE. Investigation of in vitro transdermal absorption of fentanyl from patches placed on skin samples obtained from various anatomic regions of dogs. Am J Vet Res. 2004;65(12):1697–700.
63. Wester RC, Maibach HI. *In vivo* percutaneous absorption and decontamination of pesticides in humans. J Toxicol Environ Health. 1985;16(1):25–37.
64. Hoath SB, Maibach HI, editors. Neonatal skin: structure and function. 2nd ed. New York: Marcel Dekker; 2003.
65. Farage M, Maibach HI. The vulvar epithelium differs from the skin: implications for cutaneous testing to address topical vulvar exposures. Contact Dermatitis. 2004;51(4):201–9.
66. Tinkle SS, Antonini JM, Rich BA, Roberts JR, Salmen R, DePree K, et al. Skin as a route of exposure and sensitization in chronic beryllium disease. Environ Health Perspect. 2003;111(9):1202–8.
67. Rouse JG, Yang J, Ryman-Rasmussen JP, Barron AR, Monteiro-Riviere NA. Effects of mechanical flexion on the penetration of fullerene amino acid-derivatized peptide nanoparticles through skin. Nano Lett. 2007;7(1):155–60.
68. Mangelsdorf S, Otberg N, Maibach HI, Sinkgraven R, Sterry W, Lademann J. Ethnic variation in vellus hair follicle size and distribution. Skin Pharmacol Physiol. 2006;19(3):159–67.
69. Reinertson RP, Wheatley VR. Studies on the chemical composition of human epidermal lipids. J Invest Dermatol. 1959;32(1):49–59.
70. Wesley NO, Maibach HI. Racial (ethnic) differences in skin properties: the objective data. Am J Clin Dermatol. 2003;4(12):843–60.
71. Weigand DA, Haygood C, Gaylor JR. Cell layers and density of Negro and Caucasian stratum corneum. J Invest Dermatol. 1974;62(6):563–8.
72. Visscher MO, Chatterjee R, Munson KA, Pickens WL, Hoath SB. Changes in diapered and nondiapered infant skin over the first month of life. Pediatr Dermatol. 2000;17(1):45–51.
73. Nikolovski J, Stamatas GN, Kollias N, Wiegand BC. Barrier function and water-holding and transport properties of infant stratum corneum are different from adult and continue to develop through the first year of life. J Invest Dermatol. 2008;128(7):1728–36.
74. Yosipovitch G, Maayan-Metzger A, Merlob P, Sirota L. Skin barrier properties in different body areas in neonates. Pediatrics. 2000;106(1 Pt 1):105–8.
75. Farage MA, Miller KW, Elsner P, Maibach HI. Structural characteristics of the aging skin: a review. Cutan Ocul Toxicol. 2007;26(4):343–57.
76. Farage MA, Miller KW, Maibach HI, editors. Textbook of aging skin. 2nd ed. Heidelberg: Springer; 2010.
77. Roskos KV, Maibach HI. Percutaneous absorption and age. Implications for therapy. Drugs Aging. 1992;2(5):432–49.

78. Meuling WJ, Franssen AC, Brouwer DH, van Hemmen JJ. The influence of skin moisture on the dermal absorption of propoxur in human volunteers: a consideration for biological monitoring practices. Sci Total Environ. 1997;199(1–2):165–72.
79. Williams RL, Aston LS, Krieger RI. Perspiration increased human pesticide absorption following surface contact during an indoor scripted activity program. J Expo Anal Environ Epidemiol. 2004;14(2):129–36.
80. Zhai H, Maibach HI. Effects of skin occlusion on percutaneous absorption: an overview. Skin Pharmacol Appl Skin Physiol. 2001;14(1):1–10.
81. Treffel P, Muret P, Muret-D'Aniello P, Coumes-Marquet S, Agache P. Effect of occlusion on in vitro percutaneous absorption of two compounds with different physicochemical properties. Skin Pharmacol. 1992;5(2):108–13.
82. Shomaker TS, Zhang J, Ashburn MA. A pilot study assessing the impact of heat on the transdermal delivery of testosterone. J Clin Pharmacol. 2001;41(6):677–82.
83. Aly R, Shirley C, Cunico B, Maibach HI. Effect of prolonged occlusion on the microbial flora, pH, carbon dioxide and transepidermal water loss on human skin. J Invest Dermatol. 1978;71(6):378–81.
84. Larese FF, D'Agostin F, Crosera M, Adami G, Renzi N, Bovenzi M, et al. Human skin penetration of silver nanoparticles through intact and damaged skin. Toxicology. 2009;255(1–2):33–7.
85. Gattu S, Maibach HI. Modest but increased penetration through damaged skin: an overview of the in vivo human model. Skin Pharmacol Physiol. 2011;24(1):2–9.
86. Bronaugh RL, Stewart RF. Methods for *in vitro* percutaneous absorption studies V: permeation through damaged skin. J Pharm Sci. 1985;74(10):1062–6.
87. Hueber F, Besnard M, Schaefer H, Wepierre J. Percutaneous absorption of estradiol and progesterone in normal and appendage-free skin of the hairless rat: lack of importance of nutritional blood flow. Skin Pharmacol. 1994;7(5):245–56.
88. Hueber F, Wepierre J, Schaefer H. Role of transepidermal and transfollicular routes in percutaneous absorption of hydrocortisone and testosterone: in vivo study in the hairless rat. Skin Pharmacol. 1992;5(2):99–107.
89. Wester RC, Noonan PK, Maibach HI. Percutaneous absorption of hydrocortisone increases with long-term administration. In vivo studies in the rhesus monkey. Arch Dermatol. 1980;116(2):186–8.
90. Lu S, Xia D, Huang G, Jing H, Wang Y, Gu H. Concentration effect of gold nanoparticles on proliferation of keratinocytes. Colloids Surf B Biointerfaces. 2010;81(2):406–11.
91. Zhang LW, Monteiro-Riviere NA. Assessment of quantum dot penetration into intact, tape-stripped, abraded and flexed rat skin. Skin Pharmacol Physiol. 2008;21(3):166–80.
92. Zanette C, Pelin M, Crosera M, Adami G, Bovenzi M, Larese FF, et al. Silver nanoparticles exert a long-lasting antiproliferative effect on human keratinocyte HaCaT cell line. Toxicol In Vitro. 2011;25(5):1053–60.
93. Dunford R, Salinaro A, Cai L, Serpone N, Horikoshi S, Hidaka H, et al. Chemical oxidation and DNA damage catalysed by inorganic sunscreen ingredients. FEBS Lett. 1997;418(1–2):87–90.
94. Shvedova AA, Castranova V, Kisin ER, Schwegler-Berry D, Murray AR, Gandelsman VZ, et al. Exposure to carbon nanotube material: assessment of nanotube cytotoxicity using human keratinocyte cells. J Toxicol Environ Health. 2003;66(20):1909–26.
95. Menczel E, Maibach HI. In vitro human percutaneous penetration of benzyl alcohol and testosterone: epidermal-dermal retention. J Invest Dermatol. 1970;54(5):386–94.
96. Higaki K, Nakayama K, Suyama T, Amnuaikit C, Ogawara K, Kimura T. Enhancement of topical delivery of drugs via direct penetration by reducing blood flow rate in skin. Int J Pharm. 2005;288(2):227–33.
97. Alikhan FS, Maibach H. Topical absorption and systemic toxicity. Cutan Ocul Toxicol. 2011;30(3):175–86.
98. Borm PJ, Robbins D, Haubold S, Kuhlbusch T, Fissan H, Donaldson K, et al. The potential risks of nanomaterials: a review carried out for ECETOC. Part Fibre Toxicol. 2006;3:11.
99. Wu J, Liu W, Xue C, Zhou S, Lan F, Bi L, et al. Toxicity and penetration of TiO2 nanoparticles in hairless mice and porcine skin after subchronic dermal exposure. Toxicol Lett. 2009;191(1):1–8.
100. Nadworny PL, Landry BK, Wang J, Tredget EE, Burrell RE. Does nanocrystalline silver have a transferable effect? Wound Repair Regen. 2010;18(2):254–65.
101. Yoshida D, Todo H, Hasegawa T, Sugibayashi K. Dermatopharmacokinetics of salicylate following topical injection in rats: effect of osmotic pressure and injection volume on salicylate disposition. Int J Pharm. 2007;337(1–2):142–7.
102. Yoshida D, Todo H, Hasegawa T, Sugibayashi K. Effect of molecular weight on the dermatopharmacokinetics and systemic disposition of drugs after intracutaneous injection. Eur J Pharm Sci. 2008;35(1–2):5–11.
103. Yoshida D, Hasegawa T, Sugibayashi K. Targeting of salicylate to skin and muscle following topical injections in rats. Int J Pharm. 2002;231(2):177–84.
104. Sonavane G, Tomoda K, Makino K. Biodistribution of colloidal gold nanoparticles after intravenous administration: effect of particle size. Colloids Surf B Biointerfaces. 2008;66(2):274–80.
105. Oberdorster G, Elder A, Rinderknecht A. Nanoparticles and the brain: cause for concern? J Nanosci Nanotechnol. 2009;9(8):4996–5007.
106. Cho WS, Cho M, Jeong J, Choi M, Han BS, Shin HS, et al. Size-dependent tissue kinetics of PEG-coated gold nanoparticles. Toxicol Appl Pharmacol. 2010;245(1):116–23.

107. Ballou B, Lagerholm BC, Ernst LA, Bruchez MP, Waggoner AS. Noninvasive imaging of quantum dots in mice. Bioconjug Chem. 2004;15(1):79–86.
108. Lee HA, Imran M, Monteiro-Riviere NA, Colvin VL, Yu WW, Riviere JE. Biodistribution of quantum dot nanoparticles in perfused skin: evidence of coating dependency and periodicity in arterial extraction. Nano Lett. 2007;7(9):2865–70.
109. King MJ, Michel D, Foldvari M. Evidence for lymphatic transport of insulin by topically applied biphasic vesicles. J Pharm Pharmacol. 2003;55(10):1339–44.
110. Jakob T, Ring J, Udey MC. Multistep navigation of Langerhans/dendritic cells in and out of the skin. J Allergy Clin Immunol. 2001;108(5):688–96.
111. Igyarto BZ, Kaplan DH. The evolving function of Langerhans cells in adaptive skin immunity. Immunol Cell Biol. 2010;88(4):361–5.
112. Vogt A, Mahe B, Costagliola D, Bonduelle O, Hadam S, Schaefer G, et al. Transcutaneous anti-influenza vaccination promotes both CD4 and CD8 T cell immune responses in humans. J Immunol. 2008;180(3):1482–9.
113. Lindwall G, Hsieh EA, Misell LM, Chai CM, Turner SM, Hellerstein MK. Heavy water labeling of keratin as a non-invasive biomarker of skin turnover in vivo in rodents and humans. J Invest Dermatol. 2006;126(4):841–8.
114. Zheng Y, Vieille-Petit A, Chodoutaud S, Maibach HI. Dislodgeable stratum corneum exfoliation: role in percutaneous penetration? Cutan Ocul Toxicol. 2011;30(3):198–204.
115. Ossadnik M, Richter H, Teichmann A, Koch S, Schäfer U, Wepf R, et al. Investigation of differences in follicular penetration of particle-and nonparticle-containing emulsions by laser scanning microscopy. Laser Phys. 2006;16(5):747–50.
116. Hatch KL, Maibach HI. Textile chemical finish dermatitis. Contact Dermatitis. 1986;14(1):1–13.
117. Wester RC, Quan D, Maibach HI. In vitro percutaneous absorption of model compounds glyphosate and malathion from cotton fabric into and through human skin. Food Chem Toxicol. 1996;34(8):731–5.
118. Snodgrass HL. Permethrin transfer from treated cloth to the skin surface: potential for exposure in humans. J Toxicol Environ Health. 1992;35(2):91–105.
119. Wester RC, Hartway T, Serranzana S, Maibach HI. Human skin in vitro percutaneous absorption of gaseous ethylene oxide from fabric. Food Chem Toxicol. 1997;35(5):513–5.
120. Laughlin J. Decontaminating pesticide protective clothing. Rev Environ Contam Toxicol. 1993;130:79–94.
121. Fitzgerald RH, Manley-Harris M. Laundering protocols for chlorpyrifos residue removal from pest control operators' overalls. Bull Environ Contam Toxicol. 2005;75(1):94–101.
122. Som C, Wick P, Krug H, Nowack B. Environmental and health effects of nanomaterials in nanotextiles and facade coatings. Environ Int. 2011;37(6):1131–42.
123. Wester RC, Hui X, Landry T, Maibach HI. In vivo skin decontamination of methylene bisphenyl isocyanate (MDI): soap and water ineffective compared to polypropylene glycol, polyglycol-based cleanser, and corn oil. Toxicol Sci. 1999;48(1):1–4.
124. Zendzian RP. Pesticide residue on/in the washed skin and its potential contribution to dermal toxicity. J Appl Toxicol. 2003;23(2):121–36.
125. Bucks DA, Maibach HI, Guy RH. Percutaneous absorption of steroids: effect of repeated application. J Pharm Sci. 1985;74(12):1337–9.
126. Courtheoux S, Pechenot D, Bucks DA, Marty JP, Maibach HI, Wepierre J. Effect of repeated skin administration on in vivo percutaneous absorption of drugs. Br J Dermatol. 1986;115 Suppl 31:49–52.
127. Moody RP, Maibach HI. Skin decontamination: importance of the wash-in effect. Food Chem Toxicol. 2006;44(11):1783–8.
128. Ridout G, Hinz RS, Hostynek JJ, Reddy AK, Wiersema RJ, Hodson CD, et al. The effects of zwitterionic surfactants on skin barrier function. Fundam Appl Toxicol. 1991;16(1):41–50.
129. Fartasch M, Schnetz E, Diepgen TL. Characterization of detergent-induced barrier alterations—effect of barrier cream on irritation. J Investig Dermatol Symp Proc. 1998;3(2):121–7.
130. Marrakchi S, Maibach HI. Sodium lauryl sulfate-induced irritation in the human face: regional and age-related differences. Skin Pharmacol Physiol. 2006;19(3):177–80.
131. Kao J, Carver MP. Cutaneous metabolism of xenobiotics. Drug Metab Rev. 1990;22(4):363–410.
132. Oesch F, Fabian E, Oesch-Bartlomowicz B, Werner C, Landsiedel R. Drug-metabolizing enzymes in the skin of man, rat, and pig. Drug Metab Rev. 2007;39(4):659–98.
133. Kao J, Patterson FK, Hall J. Skin penetration and metabolism of topically applied chemicals in six mammalian species, including man: an in vitro study with benzo[a]pyrene and testosterone. Toxicol Appl Pharmacol. 1985;81(3 Pt 1):502–16.
134. Hikima T, Maibach HI. Gender differences of enzymatic activity and distribution of 17beta-hydroxysteroid dehydrogenase in human skin in vitro. Skin Pharmacol Physiol. 2007;20(4):168–74.
135. Bando H, Mohri S, Yamashita F, Takakura Y, Hashida M. Effects of skin metabolism on percutaneous penetration of lipophilic drugs. J Pharm Sci. 1997;86(6):759–61.
136. Kao J, Hall J, Shugart LR, Holland JM. An in vitro approach to studying cutaneous metabolism and disposition of topically applied xenobiotics. Toxicol Appl Pharmacol. 1984;75(2):289–98.
137. Bronaugh RL, Stewart RF, Storm JE. Extent of cutaneous metabolism during percutaneous absorption of xenobiotics. Toxicol Appl Pharmacol. 1989;99(3):534–43.

138. Kao J, Hall J. Skin absorption and cutaneous first pass metabolism of topical steroids: in vitro studies with mouse skin in organ culture. J Pharmacol Exp Ther. 1987;241(2):482–7.
139. Bronaugh RL. Methods for *in vitro* skin metabolism studies. In: Zhai H, Wilhelm KP, Maibach HI, editors. Marzulli and Maibach's dermatotoxicology. 7th ed. Boca Raton: CRC Press; 2008. p. 373–6.
140. Jewell C, Heylings J, Clowes HM, Williams FM. Percutaneous absorption and metabolism of dinitrochlorobenzene in vitro. Arch Toxicol. 2000;74(7): 356–65.
141. Popov AP, Zvyagin AV, Lademann J, Roberts MS, Sanchez W, Priezzhev AV, et al. Designing inorganic light-protective skin nanotechnology products. J Biomed Nanotechnol. 2010;6(5):432–51.
142. Mortensen LJ, Oberdorster G, Pentland AP, Delouise LA. In vivo skin penetration of quantum dot nanoparticles in the murine model: the effect of UVR. Nano Lett. 2008;8(9):2779–87.
143. Marzulli FN, Maibach HI. Relevance of animal models—the hexachlorophene story. In: Maibach HI, editor. Animal models in dermatology. Edinburgh: Churchill Living Stone; 1975. p. 156–67.
144. Nigg HN, Stamper JH. Biological monitoring for pesticide dose determination. In: Wang RGM, Franklin CA, Honeycutt RC, Reinert JC, editors. Biological monitoring for pesticide exposure: measurement, estimation, and risk reduction. Washington, DC: American Chemical Society; 1989. p. 6–27.
145. Yang RS, Chang LW, Wu JP, Tsai MH, Wang HJ, Kuo YC, et al. Persistent tissue kinetics and redistribution of nanoparticles, quantum dot 705, in mice: ICP-MS quantitative assessment. Environ Health Perspect. 2007;115(9):1339–43.
146. Yeh TK, Wu JP, Chang LW, Tsai MH, Chang WH, Tsai HT, et al. Comparative tissue distributions of cadmium chloride and cadmium-based quantum dot 705 in mice: safety implications and applications. Nanotoxicology. 2011;5(1):91–7.
147. Sadauskas E, Danscher G, Stoltenberg M, Vogel U, Larsen A, Wallin H. Protracted elimination of gold nanoparticles from mouse liver. Nanomedicine. 2009;5(2):162–9.
148. Mavon A, Miquel C, Lejeune O, Payre B, Moretto P. In vitro percutaneous absorption and in vivo stratum corneum distribution of an organic and a mineral sunscreen. Skin Pharmacol Physiol. 2007; 20(1):10–20.
149. Teichmann A, Jacobi U, Ossadnik M, Richter H, Koch S, Sterry W, et al. Differential stripping: determination of the amount of topically applied substances penetrated into the hair follicles. J Invest Dermatol. 2005;125(2):264–9.
150. Bronaugh RL, Maibach HI, editors. Percutaneous absorption: drugs, cosmetics, mechanisms, methodology. 4th ed. New York: Marcel Dekker; 2005.
151. Gilman SD, Gee SJ, Hammock BD, Vogel JS, Haack K, Buchholz BA, et al. Analytical performance of accelerator mass spectrometry and liquid scintillation counting for detection of 14C-labeled atrazine metabolites in human urine. Anal Chem. 1998;70(16): 3463–9.
152. Blume-Peytavi U, Vogt A. Human hair follicle: reservoir function and selective targeting. Br J Dermatol. 2011;165 Suppl 2:13–7.
153. Lademann J, Knorr F, Richter H, Blume-Peytavi U, Vogt A, Antoniou C, et al. Hair follicles—an efficient storage and penetration pathway for topically applied substances. Summary of recent results obtained at the Center of Experimental and Applied Cutaneous Physiology, Charite—Universitatsmedizin Berlin, Germany. Skin Pharmacol Physiol. 2008;21(3):150–5.
154. Knorr F, Lademann J, Patzelt A, Sterry W, Blume-Peytavi U, Vogt A. Follicular transport route—research progress and future perspectives. Eur J Pharm Biopharm. 2009;71(2):173–80.
155. OECD (2004) Guidelines for the testing of chemicals: test no. 427: skin absorption: in vivo method. Paris: Organisation for Economic Co-operation and Development.
156. Feldmann RJ, Maibach HI. Percutaneous penetration of steroids in man. J Invest Dermatol. 1969;52(1): 89–94.
157. Maibach HI, Wester RC. Percutaneous absorption: *in vivo* methods in humans and animals. J Am Coll Toxicol. 1989;8(5):803–13.
158. Bronaugh RL, Wester RC, Bucks D, Maibach HI, Sarason R. In vivo percutaneous absorption of fragrance ingredients in rhesus monkeys and humans. Food Chem Toxicol. 1990;28(5):369–73.
159. Wester RC, Maibach HI. *In vivo* animal models for percutaneous absorption. In: Bronaugh R, Maibach H, editors. Percutaneous absorption. New York: Marcel Dekker; 1985. p. 251–66.
160. Warheit DB, Hoke RA, Finlay C, Donner EM, Reed KL, Sayes CM. Development of a base set of toxicity tests using ultrafine TiO2 particles as a component of nanoparticle risk management. Toxicol Lett. 2007;171(3):99–110.
161. Ema M, Matsuda A, Kobayashi N, Naya M, Nakanishi J. Evaluation of dermal and eye irritation and skin sensitization due to carbon nanotubes. Regul Toxicol Pharmacol. 2011;61(3):276–81.
162. Tregear RT. Physical properties of the skin. New York: Academic; 1966.
163. Crutcher W, Maibach HI. The effect of perfusion rate on in vitro percutaneous penetration. J Invest Dermatol. 1969;53(4):264–9.
164. Chang SK, Brownie C, Riviere JE. Percutaneous absorption of topical parathion through porcine skin: in vitro studies on the effect of environmental perturbations. J Vet Pharmacol Ther. 1994;17(6):434–9.
165. Chang SK, Riviere JE. Percutaneous absorption of parathion in vitro in porcine skin: effects of dose, temperature, humidity, and perfusate composition on absorptive flux. Fundam Appl Toxicol. 1991;17(3): 494–504.

166. Reifenrath WG, Hawkins GS, Kurtz MS. Percutaneous penetration and skin retention of topically applied compounds: an in vitro-in vivo study. J Pharm Sci. 1991;80(6):526–32.
167. van de Sandt JJ, van Burgsteden JA, Cage S, Carmichael PL, Dick I, Kenyon S, et al. In vitro predictions of skin absorption of caffeine, testosterone, and benzoic acid: a multi-centre comparison study. Regul Toxicol Pharmacol. 2004;39(3):271–81.
168. Behl CR, Barrett M. Hydration and percutaneous absorption II: influence of hydration on water and alkanol permeation through Swiss mouse skin; comparison with hairless mouse. J Pharm Sci. 1981;70(11):1212–5.
169. Behl CR, El-Sayed AA, Flynn GL. Hydration and percutaneous absorption IV: influence of hydration on n-alkanol permeation through rat skin; comparison with hairless and Swiss mice. J Pharm Sci. 1983;72(1):79–82.
170. Behl CR, Flynn GL, Kurihara T, Harper N, Smith W, Higuchi WI, et al. Hydration and percutaneous absorption: I. Influence of hydration on alkanol permeation through hairless mouse skin. J Invest Dermatol. 1980;75(4):346–52.
171. Moss GP, Dearden JC, Patel H, Cronin MT. Quantitative structure-permeability relationships (QSPRs) for percutaneous absorption. Toxicol In Vitro. 2002;16(3):299–317.
172. Frasch HF, Barbero AM, Hettick JM, Nitsche JM. Tissue binding affects the kinetics of theophylline diffusion through the stratum corneum barrier layer of skin. J Pharm Sci. 2011;100(7):2989–95.
173. Anissimov YG, Roberts MS. Diffusion modelling of percutaneous absorption kinetics: 4. Effects of a slow equilibration process within stratum corneum on absorption and desorption kinetics. J Pharm Sci. 2009;98(2):772–81.
174. Aschberger K, Micheletti C, Sokull-Kluttgen B, Christensen FM. Analysis of currently available data for characterising the risk of engineered nanomaterials to the environment and human health—lessons learned from four case studies. Environ Int. 2011;37(6):1143–56.

8 Enhancement of Topical Delivery with Nanocarriers

Zoe Diana Draelos

8.1 Introduction

Dermatology is a unique area of medicine where topical delivery assumes primary importance in both the pharmaceutical and cosmeceutical realm. Delivery means putting an active ingredient with an intended purpose on the skin surface to produce some benefit. This benefit can occur on the skin surface, which is the case with sunscreen actives where the photoprotective material is designed to remain on the skin surface to either absorb or reflect UV radiation. More common in the pharmaceutical realm is the desire to place an active substance on the skin surface for penetration into the dermis for the modification of some key element of skin disease. This is the intended role for topical corticosteroids where the desire is to ameliorate dermal inflammation causing induration, scaling, and erythema in psoriatic plaques. Here the topical corticosteroid is carried into the dermis by stratum corneum disruption achieved through penetration enhancers, such as propylene glycol. Propylene glycol damages the stratum corneum lipid bilayer literally poking holes in this protective nonliving structure to allow free access of the corticosteroid to the viable epidermis and dermis. While this is an inexpensive well-studied method of drug delivery, over time propylene glycol-induced barrier damage may hamper barrier restoration, which is necessary for disease resolution. Further, the patient may experience the noxious sensory stimuli of stinging and burning from the propylene glycol-induced damage when applying the topical medication.

As well as allowing the active ingredient to enter the skin, delivery systems are also designed to maintain the active ingredient in a target location for a specified duration. Reaching the target location is necessary to achieve an optimal effect. For example, a drug designed to treat actinic keratoses should reside in the epidermis, while a drug designed to vasoconstrict blood flow should reside in the dermis. Drugs to modify hair growth require entry into the follicle while medications to decrease sebum production must enter the sebaceous gland. Targets within the skin are quite specific requiring drug to be present at the target long enough to achieve the desired effect. Sometimes delivery systems must provide slow sustained release to prolong efficacy.

Further delivery systems must allow the drug to reach the desired site for the required period of time in a biologically relevant form. The skin is an active metabolic structure able to esterify and enzymatically cleave drugs into alternate forms that may be more or less biologically active. This is the case with topically applied retinoids. Engineered cosmeceutical retinoid forms, such as biologically inactive retinyl esters, can be converted into the biologically active retinol through hydrolysis in the dermis. Topical acne medications, such as benzoyl

Z.D. Draelos, M.D. (✉)
Consulting Professor of Dermatology, Department of Dermatology, Duke University School of Medicine, 2444 North Main Street, High Point, NC, USA
e-mail: zdraelos@northstate.net

A. Nasir et al. (eds.), *Nanotechnology in Dermatology*,
DOI 10.1007/978-1-4614-5034-4_8, © Springer Science+Business Media, LLC 2013

peroxide, break down into benzoic acid and oxygen on the skin surface providing for the antibacterial, anti-inflammatory, and comedolytic effects valuable in acne treatment [1].

Unwanted side effects can also be minimized through careful delivery system selection. Retinoids are excellent agents for the treatment of acne, but exhibit unwanted dryness and irritation, especially when the concentration on the skin surface is high. A possible solution is the delivery of lower concentrations of retinoid for a prolonged time creating a time-released reservoir. This type of delivery increases efficacy by allowing skin retinoid exposure for more hours while minimizing side effects.

A variety of delivery systems have been developed to carry novel compounds to the skin. The simplest and oldest delivery systems are creams, lotions, and ointments. These represent delivery emulsions and have been the mainstay for the application of cosmetics, cosmeceuticals, and pharmaceuticals to the skin since antiquity, but new manufacturing techniques have created a nanosized world where particles smaller than 100 nm can be delivered to the skin surface, epidermis, and dermis. This chapter evaluates the present utility and future of nanocarriers in dermatology.

8.2 Anatomy of the Skin Relevant to Nanocarriers

The stratum corneum is uniquely designed to keep foreign materials from penetrating the skin. It does an amazing job of accomplishing this goal. The protein rich corneocytes surrounded by the covalently bound intercellular lipids form a waterproof barrier that is around 13 nm thick over most of the body. Any substance that damages protein or removes the intercellular lipids will increase penetration of substances into the skin. However, nanoparticles less than 13 nm can easily penetrate. 25 nm nanoparticles can penetrate into the upper 3–5 layers of the stratum corneum. Thus, the smaller the nanoparticle, the more the penetration and the larger the nanoparticle, the less penetration that can be expected. The access of nanocarriers into the skin can be modulated for ease of penetration and dose based on size. Overcoming the stratum corneum barrier is the biggest impediment to pharmaceutical and cosmeceutical delivery that might be overcome with nanocarrier technology.

8.3 Nanocarriers

Nanocarriers utilize nanoparticle technology to deliver active ingredients to areas inaccessible by standard delivery systems [1]. They are particularly apropos to dermatology where the stratum corneum, as discussed previously, is the major impediment to drug and cosmeceutical efficacy. The types of nanocarriers currently available are nanoemulsions, nanoliposomes, polymeric nanoparticle spheres, and quantum dots. Each of these nanocarriers is discussed in the context of current technology with attention to future possibilities.

8.3.1 Nanoemulsions

The most basic time-tested delivery system is the emulsion [2]. An emulsion is formed from oil and water, which are mixed and held in solution by an emulsifier. Most emulsifiers are surfactants, or soaps, which dissolve the two nonmiscible ingredients. The most common emulsions are oil-in-water, where the oil is dissolved in the water [3]. This emulsion is the most popular delivery system because the water evaporates leaving behind a thin film of oily ingredients. This is the basis for all moisturizers, the main method of transferring cosmeceuticals to the skin surface. If a large quantity of water is found in the emulsion, it is considered a lotion, while a thicker emulsion with less water is considered a cream. Creams typically contain a higher concentration of oily ingredients than lotions accounting for the thicker film produced.

Creams and lotions are the main cosmeceutical delivery systems because they are inexpensive to produce. They also impart moisturizing qualities to the skin surface, one of the main methods used by cosmeceuticals to improve skin texture and appearance. Thus, carefully constructed emulsions can accomplish moisturization and delivery of an active agent simultaneously. Creams and lotions are also the main delivery systems for topical medications, utilizing the same ingredients and technology as their OTC counterparts, to form a vehicle from which the drug is delivered to the skin.

Substances found in oil-in-water emulsions include oily substances such as petrolatum, mineral oil, vegetable oils, and dimethicone. The water evaporates leaving the oil behind to place a water impermeable barrier over the skin surface. Substances that attract water to the skin surface and function as humectants are found in the water portion of the emulsion. This is in contrast to water-in-oil emulsions, where the water-soluble substances are dissolved in the oil soluble substances. This emulsion produces an ointment, which may have a greasy feel leaving the skin warm and sticky.

A modern emulsion variant based on nanoparticles is known as a nanoemulsion [2]. Nanoemulsions, just like simple emulsions, are liquids with an oil-loving hydrophobic phase and a water-loving hydrophilic coexisting as a single phase with the aid of an emulsifier. The difference is that the droplets in these emulsions are on the nanoscale of 20–100 nm [3]. If the nano droplets are larger than 100 nm, the emulsion appears white, while nanoemulsions with droplets of 70 nm are transparent. Nanoemulsions offer the ability to deliver highly hydrophobic or lipophilic substances into the skin, which could not otherwise penetrate [4]. This offers unique nanocarrier opportunities because the stratum corneum is an excellent barrier to lipophilic pharmaceuticals and cosmeceuticals.

In the cosmeceutical realm, new nanoemulsions of ubiquinone have been developed. Ubiquinone, also known as coenzyme Q10, is an important antioxidant manufactured by the body and found in all skin cells. Topical penetration had been challenging; however, Hoppe was able to demonstrate coenzyme Q10 penetration into the viable epidermis and a reduction in oxidation with weak photon emission. They were also able to show a significant decrease in the expression of collagenase in human dermal fibroblasts following UVA radiation and improvement in orbital wrinkling with traditional emulsion formulations [5,6]. Because ubiquinone is found in both hydrophilic and hydrophobic cellular compartments, accounting for its name derivation from the world ubiquitous, enhanced topical delivery with nanoemulsions successfully produced higher concentrations of skin ubiquinone theoretically enhancing the skin's natural antioxidant capabilities.

Another current use for nanoemulsions is in the delivery of hyaluronic acid, a natural glycosaminoglycan found in the dermis. Hyaluronic acid has been popularized in dermatology as an injectable filler that can soften nasolabial and melolabial folds by attracting water in the dermis following injection. Topical hyaluronic acid similarly attracts water and can decrease skin surface fine lines and wrinkles through enhanced hydration. Nanoparticle hyaluronic acid can create an imperceptible hydration film over the skin surface creating the illusion of smoothness by filling in fine lines and increasing skin water content. This hyaluronic acid nanoemulsion is delivering a physical change to the skin surface while the ubiquinone nanoemulsion was delivering a chemical change. These examples highlight the utility of nanoemulsions in dermatology.

8.3.2 Nanoliposomes

Liposomes are a newer delivery system than the emulsions previously discussed; however, liposomes are usually suspended in emulsions for skin delivery. Liposomes are spherical vesicles with a diameter between 25 and 5,000 nm formed from membranes consisting of bilayer amphiphilic molecules, which possess both polar and nonpolar ends. The polar heads are directed toward the inside of the vesicle and toward its outer surface while the nonpolar, or lipophilic tails, are directed toward the middle of the bilayer. This delivery system was discovered by A.D. Bangham who reported his work in the 1965 Journal of Molecular Biology. His discovery was based on the observation that phospholipids could be dispersed in an aqueous solution to spontaneously form hollow vesicles, known as liposomes. Fig. 8.1.

Liposomes are based on the natural structure of the cell membrane, which has been highly

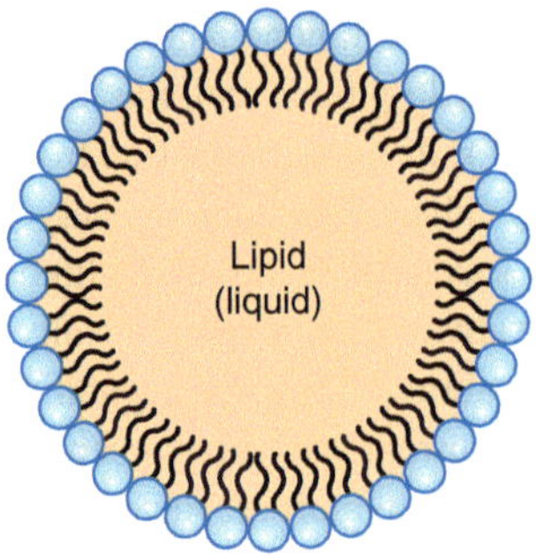

Fig. 8.1 Nanoliposome

conserved through evolutionary change. The name is derived from the Greek word "lipid" meaning fat and "soma" meaning body. Liposomes are primarily formed from phospholipids, such as phosphatidylcholine, but may also be composed of surfactants, such as dioleoylphosphatidylethanolamine. Their functionality may be influenced by chemical composition, vesicle size, shape, surface charge, lamellarity, and homogeneity.

The liposome is an extremely versatile structure. It can contain aqueous substances in its core, or nothing at all. Hydrophobic substances can dissolve in the phospholipid bilayer shell, which allows liposomes to deliver both oil-soluble and water-soluble substances. This characteristic is used in drug delivery where an oil-soluble drug can be dissolved in water if placed in the phospholipid shell. Similarly, water- and oil-soluble cosmeceutical ingredients can be dissolved in a stable solution without the use of an emulsifier with this technology.

It is unlikely that traditional liposomes diffuse across the stratum corneum barrier intact. The corneocytes are embedded in intercellular lipids, composed of ceramides, glycosylceramides, cholesterol, and fatty acids, which are structurally different from the phospholipids of the liposome. It is postulated that liposomes penetrate through the appendageal structures, but may also fuse with other bilayers, such as cell membranes, to release their ingredients. These penetration challenges are overcome with nanoliposomes that more readily traverse the stratum corneum due to their small size. Further, nanoliposomes can be specially designed to release their internal contents under certain conditions. For example, nanoliposomes can release when a desired pH or temperature is present in the stratum corneum. Nanoliposomes are also small enough to function as a natural target for macrophage phagocytosis, releasing their payload inside the macrophage phagosome and producing unique immune alterations. Nanoliposomes can also be coated with opsonins and ligands to encourage endocytosis in other cell types. Thus, future technology may utilize drug laden nanoliposomes to target cancer cells more specifically than the currently available chemotherapeutic agents.

One of the main challenges with nanoliposomes is their inherent instability. They are readily deformed and possibly lysed when compressed. They are subject to fusion, aggregation, and precipitation. This means that nanoliposomes can fuse together and combine their contents and bilayer shells in suspension, possibly rending the newly created liposome larger than 100 nm, technically outside the liposome range. More commonly nanoliposomes can physically stick together, again creating liposome clumps outside the nano range. Finally, the nanoliposomes can fall to the bottom of the emulsion under the effect of gravity and leave their evenly distributed state. In many situations, it is likely that nanoliposomes possess the properties of traditional liposomes due to their instability.

Nanoliposomes are further challenged by outside physical forces such as vigorous shaking or vibration. This type of movement is enough to physically lyse the nanoliposome destroying the discrete architecture and any benefits the liposomal structure might confer. The contents just become part of the traditional emulsion. A more stable type of nanoparticle is a polymeric nanoparticle sphere, discussed next.

8.3.3 Polymeric Nanoparticle Spheres

One of the major problems with nanoparticles is their tendency to stick together. Not only do nanoliposomes stick together, as previously mentioned, nanoparticles themselves tend to aggregate eliminating their unique size attributes. One method of preventing aggregration and imparting different properties to nanoparticles is to coat the nanoparticle with a polymer creating polymer nanoparticle spheres [7]. This is a valuable formulation technique because the nanoparticles can be charged, which may be of benefit in certain uses. Since charged particles do not readily penetrate the stratum corneum, anionically charged polymeric nanoparticle spheres are more likely to remain on the skin surface, a desirable quality for inorganic sunscreens. Additionally, polymeric nanoparticle spheres are smooth, creating a silky skin feel, which is not possible with the traditional rough-edged nanoparticles.

Nanoparticles also possess some unusual properties when struck by a photon of UV radiation [8]. Their small size allows the generation of secondary oxygen radicals, which is problematic in sunscreen formulations where the UV energy should be converted to something innocuous, such as heat, rather than oxygen radicals capable of causing DNA and collagen damage. While secondary oxygen radicals are less energetic than primary oxygen radicals, oxidative damage still occurs and should be minimized. The main use of polymeric nanoparticle spheres is in sunscreens where the zinc oxide nanoparticles can be coated to prevent secondary oxygen radicals.

8.3.4 Quantum Dots

The final presently available form of nanoparticles is known as quantum dots. Quantum dots are nanoparticles less than 10 nm, representing the smallest nanoparticles created. These extremely small nanoparticles possess interesting magnetic and electrical properties. For example, superparamagnetism is possible in nanoparticle magnetic materials. Ferromagnetic materials less than 10 nm can switch magnetization direction with room temperature thermal energy making them highly unstable, but opening up potential medical opportunities. These magnetic quantum dots could be applied to the skin surface and allowed to penetrate the skin, when the body is placed in a magnetic field the dots could change polarity creating the first real-time imaging technique for the skin. Something currently limited with computerized tomography (CT) or magnetic resonance imaging (MRI). Similarly, magnetic quantum dots could be injected into the body and a magnet placed over one area to deliver targeted drugs to cancerous tissue that preferentially absorbs and concentrates the quantum dots, such as cutaneous and metastatic melanoma. Quantum dots can even be created to carry two different drugs with the first drug released by a small remotely generated electromagnetic pulse and the second drug released by a stronger pulse. The unique dermatologic carrier possibilities for quantum dots are in their infancy.

8.4 Potential Applications and Challenges of Nanocarriers

Nanocarriers present numerous dermatologic application possibilities in both the cosmeceutical and pharmaceutical realms. For example, consider the ability of nanodelivered benzoyl peroxide to enter the pilosebaceous unit. This would allow very small concentrations of OTC acne actives to enter the follicular infindibulum to reach the location of P. acnes and more efficiently kill the organism. Nanoparticle topical antibiotics would create similar opportunities to decrease the incidence of skin infection with pathogens, such as staph aureus, and decrease skin colonization from organisms, such as Malassezia species, important in seborrheic dermatitis. Very small ingredient quantities with targeted delivery would increase efficacy while reducing toxicity and side effects in pharmaceutical dermatology.

In the realm of cosmeceuticals, nanocarriers have even more possibilities, since it would be easier to gain approval for the topical delivery of substances that are currently ingested. Topical nanocarriers could deliver antioxidants to the dermis where oxidative damage occurs in response to UV radiation. Presently, once endogenous Vitamin E and Vitamin C reserves are depleted in the dermis, UV damage progresses at a rapid rate. Nanosized antioxidants less than 13 nm could enter the dermis and augment endogenous antioxidants diminishing inflammatory cascade activations and supplementing the melanin cap over nuclear DNA. However, there are some important challenges that must be overcome in nanoparticle safety.

Nanocarriers theoretically could carry not only their desired payload, but also other unwanted substances into the skin. Imagine a situation where a nanoemulsion moisturizer is rubbed on the skin after walking in downtown New York City. Nanoparticles from internal combustion and pollution abound on the streets of New York City and land on the skin, perhaps even containing nanoparticle sized heavy metals. The subsequently applied nanoemulsion could carry the moisturizing ingredients, but also the heavy metals, past the stratum corneum and into the skin. It might also carry ingredients from a previously applied sunscreen into the skin that were not intended for penetration. In fact, the nanocarrier might draw everything on the surface of the skin into the body, including the insecticide droplets accidentally residing on the skin from a spray can be used to kill ants earlier in the day. Perhaps the biggest challenge for nanocarriers is the selectivity of the carrier system.

There are certain conditions under which nanocarriers might experience increased skin penetration [9]. For example, in areas where the skin is bent or flexed, such as in the antecubital and popliteal fossae, penetration is enhanced. Nanocarriers would also be expected to demonstrate altered absorption kinetics on wounded or abraded skin. Penetration enhanced products containing substances such as propylene glycol, a common ingredient in skin lotions, might also allow nanocarriers more rapid, enhanced access into the skin and ultimately the body through the dermal vasculature. These considerations are important as they govern the amount a nanocarrier can deliver over a given time period.

It is clear that a better understanding of nanocarriers is required, but there are other issues that need to be considered. Nanocarriers can only function as nanocarriers if they remain in the nanosize range. While a single nanoparticle less than 100 nm in one dimension can act as an effective nanocarrier, two or three or more nanoparticles stuck together might not demonstrate the same nanocarrier abilities. Indeed, aggregrated nanocarrier particles might be ineffective. It is challenging with present technology to maintain nanoparticles singly in solution and to prevent them from agglomerating, or accumulating in the bottom of the jar. Aggregrated or agglomerated nanoparticles are no longer nanoparticles. The biggest challenge in nanocarrier development may be the new emulsion technology required to keep the nanocarrier physically stable [10].

8.5 Conclusions

Nanotechnology as it applies to the topical delivery of active agents into the skin holds great promise as the next major advance in pharmaceutical and cosmeceutical dermatologics. In 2005, 54 topical nanoparticle products were in the worldwide marketplace compared with 1,015 in 2009. This number will increase as formulators recognize the benefits of nanocarriers and research elucidates some of the safety issues of present concern. For example, Sweden banned ten companies from marketing nanoparticle zinc oxide sunscreens and the European Union has passed legislation requiring the word "nano" placed in brackets on the label of products containing nanoparticles, which includes nanocarriers. Furthermore, the EU is requiring companies to notify the intent to release topical nanoparticle products as of November 2013. Once further research validates the safety and efficacy of nanocarrier formulations, the future of dermatology will be realized.

References

1. Tanghetti E. The evolution of benzoyl peroxide therapy. Cutis. 2008;82(5):5–11.
2. Solans C, Izquierdo P, Nolia J, Azemar N, Garcia-Celma MJ. Nanoemulsions. Curr Opin Colloid Interface Sci. 2005;10(3–4):103–10.
3. Tadros T, Izqulerdo P, Esquena J, Solans C. Formation and stability of nano-emulsions. Adv Colloid Interface Sci. 2004;108–109:303–18.
4. Sonneville-Aubrun O, Simonnet JT, L'Alloret F. Nanoemulsions: a new vehicle for skincare products. Adv Colloid Interface Sci. 2004;108–109:145–9.

5. Hoppe U, Bergemann J, Diembeck W, Ennen J, Gohla S, Harris I, et al. Coenzyme Q10, a cutaneous antioxidant and energizer. Biofactors. 1999;9(2–4):371–8.
6. Blatt T, Mundt C, Mummert C, Maksiuk T, Wolber R, Keyhani R, et al. Modulation of oxidative stresses in human aging skin. Z Gerontol Geriatr. 1999;32(2):83–8.
7. Schmidt G, Malwitz MM. Properties of polymer-nanoparticle composites. Curr Opin Colloid Interface Sci. 2003;8(1):103–8.
8. Rodriguez K, Fenollosa R, Meseguer F. Silicon microspheres as UV, visible and infrared filters for cosmetics. Cosmet Toiletries. 2010;125(9):42–50.
9. Harper A, Tonge S, Makein L, Kaszuba M, Connah M. Environmentally responsive nanoparticles for delivery as assessed via light scattering and near-infrared imaging. Cosmet Toiletries. 2010;125(9):52–7.
10. Staniland P, Hewitt J. Meeting the UVA challenge: inorganic-only sun protection. Cosmet Toiletries. 2010;125(9):32–40.

Hair Follicle Targeting with Nanoparticles

9

Fiorenza Rancan, Zahra Afraz, Behazine Combadiere, Ulrike Blume-Peytavi, and Annika Vogt

9.1 Introduction

Over the past years, hair follicles have gained a lot of interest in the field of skin penetration research as they represent an important penetration pathway for topically applied substances. They function as reservoirs [1, 2] and also as portals of entry to the viable skin layers [3]. Under physiological conditions the intact stratum corneum significantly impedes the skin penetration of large hydrophilic molecules and particulate structures. Hair follicles, in contrast, represent interruptions in this barrier and the importance of the transfollicular route of penetration has been demonstrated by several independent studies on various animal skin models, where hairy skin was compared with hair follicle-free skin areas [4, 5]. The role of the follicular penetration pathway compared to transepidermal route was also demonstrated by selective blockage of the follicular orifices in porcine ear skin [6] as well as human skin explants [7]. In vivo experiments with caffeine applied as shampoo formulations on skin with open or sealed hair follicles further underlined the importance of hair follicles for the transdermal permeation and systemic delivery of hydrophilic drugs [8]. The follicular route seems to be of special importance for particle penetration in skin. Particulate structures ranging from liposomes to solid inorganic particles and microspheres up to a diameter of 10 μm were shown to aggregate and remain in hair follicle openings. Such observations provided the basis for the idea of hair follicle targeting with particulate drug carriers [9]. Modifications of nanoparticle physicochemical properties, use of permeabilizing agents, as well as partial removal of the stratum corneum are some of the methods which were shown to increase the hair follicle penetration of drug-loaded nanoparticles as well as their targeting ability [10, 11]. In the following chapter, we outline how hair follicles act as entry pathway and reservoir structure for topically applied particles. We review the penetration properties of specific particle types and the influence of hair follicle parameters, e.g., hair follicle types, growth activity, and sebum production as modifying factors of particle–skin interactions. In fact, studies by our group and others strongly suggest that the hair follicle provides an important interface for such interactions. Nanoparticle-based hair follicle targeting aims at utilizing these interactions for therapeutic purposes, e.g., to deliver functionalized particles loaded with active compounds

F. Rancan (✉) • Z. Afraz • U. Blume-Peytavi • A. Vogt
Department of Dermatology and Allergy, Clinical Research Center for Hair and Skin Science, Charité-Universitätsmedizin Berlin,
Chariteplatz 1, 10117, Berlin, Germany
e-mail: fiorenza.rancan@charite.de

B. Combadiere
Institut National de la Santé et de la Recherche Médicale, INSERM UMR-S 945 and Université Pierre et Marie Curie (UPMC University Paris 06), Laboratory of Immunity and Infection,
83/91, Boulevard de l'Hôpital, 75013,
Paris Cedex 13, France

A. Nasir et al. (eds.), *Nanotechnology in Dermatology*,
DOI 10.1007/978-1-4614-5034-4_9, © Springer Science+Business Media, LLC 2013

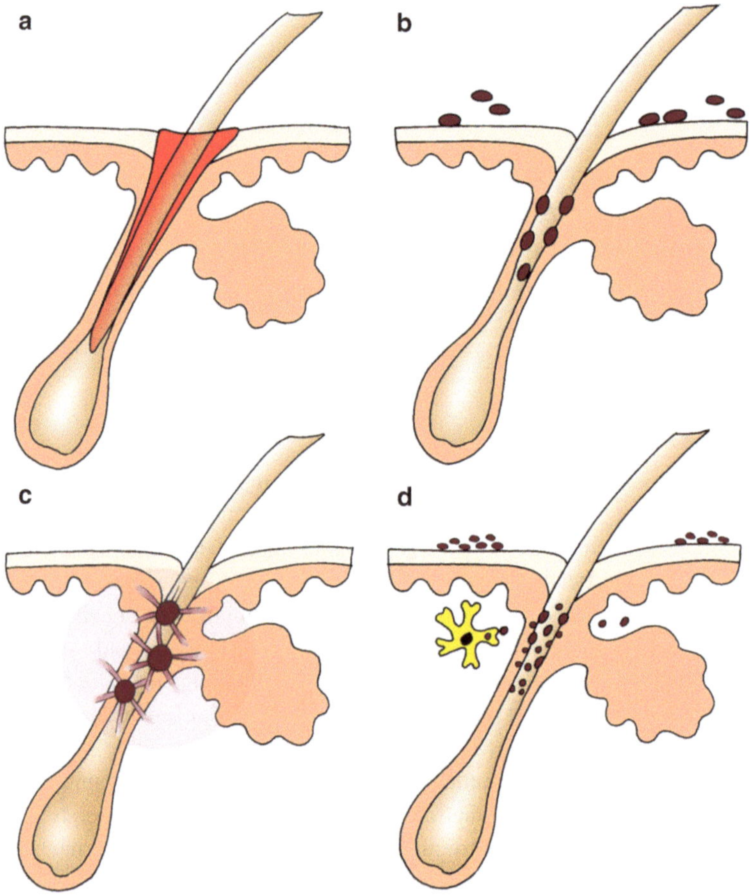

Fig. 9.1 Hair follicle orifices represent invaginations of the epidermis, which significantly increase the skin surface area available for penetration. As a consequence, the presence of large numbers of hair follicle significantly increases the penetration rates of topically applied compounds. With regard to topically applied particles and nanomaterials, there is strong experimental evidence that the hair follicle serves as reservoir (**a**), where the particles are partially protected from shedding and remain for prolonged time periods (**b**). During this time, prolonged release from such particle-based carrier systems could help create high perifollicular concentrations of active compounds which otherwise do not penetrate well, or which are being easily degraded (**c**) offering novel perspectives for the treatment of hair diseases. Similarly, prolonged release and protection from shedding could render dermatotherapeutics more effective. During the past years evidence emerged that translocation of nanomaterials across the skin barrier is possible, especially in barrier-disrupted skin and in hair follicles. Nano-sized carrier systems may hence be utilized to specifically target cell populations such as perifollicular dendritic cells (**d**), stem cells, or others

selectively to the hair follicles or even to specific skin compartments and cell populations. Particle-based targeting of hair follicles may include deposition of active compounds in the follicular reservoir, targeting of active compounds to specific compartments within the follicular duct, e.g., sebaceous gland, stem cell region, or even targeting of specific cell populations such as perifollicular antigen-presenting cells (Fig. 9.1).

9.2 Hair Follicle Physiology and Relevance for Particle-Based Drug Delivery

9.2.1 Anatomy

The pilosebaceous unit is a complex anatomical structure consisting of the hair shaft surrounded by epithelial and mesenchymal sheaths, the arrector pili muscle, and the sebaceous gland. Different regions can be identified within the hair follicle. The *infundibulum* is the region extending from the opening at the skin surface down to the sebaceous gland. In the intact pilosebaceous unit, the infundibulum is a reservoir where topically applied nanoparticles accumulate, mostly in the region between hair shaft and infundibulum walls but also on the hair shaft itself. In fact, there is evidence that some particle types get trapped underneath cuticle cells which may act as a pumping mechanism pushing the nanoparticles deep into the hair follicles [12]. In the lower part of the infundibulum the stratum corneum discontinues and provides a surface for nanoparticle–skin interactions [4, 7, 8]. Below the infundibulum the sebaceous duct opens into the hair follicle canal and connects the *sebaceous gland* with the hair follicle. The sebaceous gland releases sebum, a fungistatic and bacteriostatic mixture with protecting and hydrating function. In active hair follicles, the produced sebum flows through the duct and the hair follicle canal to the skin surface. The sebaceous gland is the main target for the treatment of acne-related disorders and was the main focus of early studies of particle-based drug delivery with microspheres [13]. The *bulge region* is located below the sebaceous gland. It is widely believed to be a major site of epithelial stem cells responsible for the regrowth of a new hair shaft during the hair cycle. Cells from this region also participate in wound healing [14]. Several other precursor cell populations including melanocyte precursors and Langerhans cell precursors reside in and around the bulge region as well [15]. Furthermore, the bulge region also marks the end of the permanent part of the hair follicle. The most prominent structure of the transient part, which extends and regresses according to the different hair cycle stages is the *hair bulb* with the *dermal papilla*, from where matrix cells form the hair shaft as a result of close epithelial–mesenchymal interactions [16]. Especially in the growth phase (anagen) where the follicle is fully developed, a rich network of microvessels surrounds the hair bulb. Targeting active compounds to the bulb and the suprabulbar region may open interesting new perspectives for the treatment of hair growth disorders. This region is not directly accessible to nanoparticles when the pilosebaceos unit is intact, because at the level of the sebaceous gland the hair shaft is tidily enveloped by the inner root sheath. However, drugs released from nanoparticles in the infundibulum can permeate the epithelial layers and reach the bulb.

9.2.2 Hair Follicle Types and Distribution

Different types of hair follicles can be differentiated in humans depending on their morphology and dimension. In general, it can be distinguished between terminal, vellus, and intermediate hair follicles. *Terminal hairs* are long, thick, pigmented and medullated, and can be found on the scalp and in hormone-dependent regions [17], such as beard, arm axillae, and pubic area. *Vellus hairs* are short, thin and unpigmented, and are found in all body regions, excluding palms, soles of the feet, lips, and portions of the external genitalia [17]. *Intermediate hairs* are produced by hair follicles with dimension and morphology in between vellus and terminal hairs. There are significant variations between different body regions with regard to the diameter of the follicular orifice, the thickness of the hair shaft, as well as the volume and surface area of the infundibula [18]. All these differences can affect the extent of nanoparticle penetration into the hair follicle canal, the surface area available for translocation and/or drug permeation, in case of nanoparticle-based drug delivery. For example, solid 750 nm microspheres penetrate deeply in human scalp terminal hair follicles [19], but do not enter into vellus hair follicles [20]. Besides those morphometric characteristics, the hair follicle density differs significantly among the various body regions and is a major determinant for different penetration rates. The lowest hair follicle density can be found in the calf (0.1 % of the total skin area), whereas scalp and the face are the regions with the highest hair follicle density (10 % of the total skin area) [21]. The finding that morphometric data are reproducible in specific body regions suggests that hair follicle targeting is a feasible approach for selective drug delivery [18]. However, it has to be taken into account that a large number of experimental studies on particle penetration and transcutaneous targeting is still being performed on animal skin, i.e., in mice, rats, hamsters, pig ears, and others. Careful interpretation of such results has to consider differences in hair follicle density and size among the different species.

9.2.3 Hair Cycling

Cyclic renewal is a characteristic unique feature of hair follicle biology. Each cycle consists of a growth phase (anagen) characterized by high proliferative and metabolic activity with subsequent hair fiber production followed by a regression phase (catagen), which is a short degenerative process associated with morphological changes. Until a new growing phase initiates, the hair follicle remains in a quiescent stage (telogen) [17]. Under physiological conditions, 85 % of scalp hair follicles are in the anagen phase and approximately 15 % is in the telogen phase, [17]. Changes in these percentages and alterations of the length of the hair cycle phases are characteristics of hair growth disorders [22, 23]. For example, in the case of androgenetic alopecia, shortening of the anagen phase is associated with continuous terminal hair follicle miniaturization to vellus hair follicles and hair loss. By contrast, lengthening of the anagen phase and transformation of vellus hair follicles into terminal hair has been reported for hypertrichosis and hirsutism [17, 22, 23]. Although most etiologies of hair growth disorders are still unknown improved knowledge on hair follicle physiology has resulted in a better understanding of hair growth disorders and identification of specific targets leading to new therapeutic approaches for hair growth regulation. For example, testosterone, estrogens, and their receptors are important hair cycle regulators and represent interesting candidates for nanoparticle-based therapeutic approaches [24, 25]. The complete hair follicle processes is only fully developed in anagen. During catagen and telogen, the cycling part undergoes regression, which also includes the microvasculature around the bulb. As part of the permanent portion, of the hair follicle, the hair follicle infundibulum is available during all hair follicle cycling phases. Although in vivo data as to whether during all hair follicle cycling does influence the accumulation of nanoparticles in the infundibulum itself is missing, the hair follicle infundibulum was shown to be more open for penetration either during active stages of hair growth and/or sebum production (see following section). Interestingly, the vascularization of the perifollicular epithelia is maximal during the anagen phase and the transdermal permeation of substances released within the hair follicle canals is probably higher during the growth phase.

9.2.4 Active and Inactive Hair Follicles

There is no available experimental data on nanoparticle penetration in different hair cycle phases. However, Lademann et al. [12] found that, under physiological conditions, not all hair follicles are open for penetration and defined "active" hair follicles according to hair growth ability and sebum production. In contrast, lack of hair growth and sebum production is found in inactive hair follicles and results in clogging of the hair openings which impedes the penetration of hydrophilic substances as well as nanoparticles [4, 7]. Opening of hair follicles by means of mechanical peeling effectively increases the hair follicle penetration rate [4]. In fact, our group used cyanoacrylate skin surface stripping as a technique to open hair follicles in a number of experimental studies [20, 26]. This method removes cellular debris, dried sebum, and a superficial layer of the stratum corneum from hair follicle openings making the infundibulum more accessible and facilitating follicular penetration. As active hair follicles are filled with sebum and because inactive hair follicles are closed by cell debris and hardened sebum, enhancement of the transfollicular penetration and drug delivery is feasible by using lipophilic vehicles compared to hydrophilic ones [27–29]. A study performed using ethanol-containing formulations has shown an increased follicular accumulation of minoxidil reflecting the partitioning of ethanol into the lipid-rich hair follicle canal [30]. Recently, 10 nm gold nanoparticles have been reported to function as an enhancer for the skin penetration of proteins [31]. This effect might be due to the surfactants used to stabilize the nanoparticles, which act as penetration enhancers.

9.3 Relevance of Physicochemical Properties of Particles for Hair Follicle Targeting

The physicochemical properties of the nanoparticles are major determinants in order to achieve a selective and efficient targeting of the different anatomical structures and cell populations in and around the hair follicle. The physicochemical characteristic of the nanoparticles can affect the penetration depth into the hair follicle canal, their release properties once penetrated in the hydrophobic environment of the infundibulum, their stability and biodegradability upon their deposition on skin, and their capacity to translocate to the skin viable layers.

9.3.1 Surface Properties

The *lipophilicity* of the carrier has a strong influence on the drug delivery properties of nanoparticles in the lipophilic environment of the hair follicle canals. Solid lipid nanoparticles, for example, were shown to improve the release of the loaded drug within the hair follicle canal [32, 33]. The disadvantage of these particle types is that they can be loaded with lipophilic substances but not with hydrophilic ones. The *surface charge* can influence the penetration within the hair follicle canal and even their uptake by cells. Amphoteric and cationic liposomes penetrated deeper compared to anionic ones in porcine hair follicle [34]. *Coating* of the nanoparticles with polymers, detergents, or proteins might also enhance the penetration of nanoparticles within the infundibulum. Moreover, the surface of nanoparticles can be functionalized with targeting units, like specific antibodies or substrates, in order to achieve an even more selective delivery of drugs to the desired target [35].

9.3.2 Size

Particle size is an important parameter for the penetration depth of nanoparticles in the hair follicle canal. Several studies performed on human skin models showed that particles with a diameter in the range of 10–500 nm accumulate in the hair follicle infundibulum, in contrast to microspheres (0.5–5 μm) which aggregate superficially in the hair follicle orifices. The smaller the nanoparticles are, the deeper they can penetrate into the hair follicle canal, so that the penetration depth into the infundibulum strictly depends on the particle size [19, 20, 36]. These results suggest that it is possible to target drugs to the different regions of the hair follicle by varying the size of nanoparticles [1, 9]. Size is also the major determinant for nanoparticle translocation across the skin barrier into the perifollicular epithelium. The translocation of nanoparticles in the skin is a key step for the targeting of specific cell populations, e.g., antigen-presenting cells for immunization purposes. Various investigations on the penetration of particles in human skin indicated that nanoparticles do not reach the epidermal and dermal layers when the skin barrier is intact [37, 38]. However, when the skin barrier is disrupted and skin immune system is activated, nanoparticles can have access to the living layers of the skin. In fact, nanoparticle translocation after the removal of superficial layers from the stratum corneum and of sebum from the hair follicle openings by means of cyanoacrylate biopsies was found to be size-dependent [20]. Translocated nanoparticles could also be detected in antigen-presenting cells, indicating that the stripping with cyanoacrylate enhanced the translocation of particles in the epidermis and the dermis, and allowed for targeting of the perifollicular antigen-presenting cells. Similarly, Mortensen et al. found that quantum dots accumulated in hair follicles and could translocate across the skin barrier after skin treatment with UV irradiation [39].

9.3.3 Form and Formability

Most of the hair follicle penetration data refer to particles as rigid spheres. Indeed, there is a bright spectrum of particle types with varying forms and different degrees of formability and flexibility. For example, the rigid inorganic gold nanoparticles can be manufactured as spheres, rods, ovals, and cubes. Also, carbon nanomaterials can form spheres as well as nanotubes. Further, inorganic metal and metal oxide particles are rigid, whereas particles made of organic polymers and lipids are soft and have different degrees of formability. At the moment, there are no studies, which compare particles with different forms or soft and rigid particles with regard to follicular penetration. However, from different distinct studies it is evident that particles with different plasticity can interact differently with skin and hair follicles. For example, whereas liposomes enhance skin penetration by releasing their load in the stratum corneum and in the hair follicle infundibulum [34], rigid nanoparticles of the same size remain on the skin surface and penetrate preferentially in the hair follicle openings [40].

9.3.4 Stability and Release Properties

Some particle types are not stable over time and disintegrate or loose stability and shape upon contact with the skin surface. Silver nanoparticles, for example, release silver ions over time and are therefore used as antibacterial additives in clothes. Liposomes are also not longer stable once they come in contact with skin lipophilic regions, e.g., the sebum-filled hair follicle canal. Therefore, environment-sensitive liposomes, which can disassemble and release their loads upon changes of lipophilicity, temperature or pH in the surroundings, are promising systems for drug delivery into the hair follicle infundibulum. Recently we identified a similar characteristics for polylactic acid (PLA) nanoparticles. Once accumulated in the hair follicle infundibulum, the polymer lost its supramolecular conformation and released over time the loaded fluorochrome, which could diffuse to the skin, especially to the sebaceous gland [26]. The stability of the PLA nanoparticles in the hair follicle canal depended on the hydrophobicity of the loaded dye [41].

9.3.5 Biodegradability

Among the various nanoparticle types which have been investigated in recent years, biodegradable nanoparticles are especially interesting for therapeutic use. This feature of nanoparticles is very important especially for those applications which require the translocation of nanoparticles to the perifollicular tissue or to other body compartments. Biodegradable nanoparticles are biologically compatible, undergo hydrolysis and/or can be metabolized to their constituting components which are not toxic and can easily be cleared from the body. PLA and polylactic-co-glycolic acid (PLGA) nanoparticles are the most investigated polymer-based biodegradable nanoparticles. They are degraded by hydrolysis within a few months. Biological nanomaterials like protein-based nanoparticles and virus-like particles (VLP) are also biodegradable and can be actively degraded after uptake into cells or processed by macrophages and antigen-presenting cells.

9.4 Particle-Based Hair Follicle Targeting for Dermatotherapy and Hair Diseases

As outlined above, the hair follicle does not only play a key role for particle penetration in human skin, it also contains several structures which could be very interesting targets for novel therapeutic approaches. Nanocarriers offer the unique opportunity to selectively target high concentrations of active compounds to the hair follicle openings and further on down to specific compartments along the follicular canal. In the following, we outline targets of interest and possible applications.

9.4.1 Sebaceous Gland

The sebaceous gland is one of the most attractive targets for particle-based drug delivery to hair follicles, especially for the therapy of acne vulgaris.

In fact, the delivery of acne-related therapeutics by means of liposomes or microspheres has been extensively studied and some formulations have already been translated into commercial products. Acne is a prevalent inflammatory disease of the pilosebaceous unit, which results from sebum production abnormalities of the sebaceous gland, increased proliferation and desquamation of the hair follicle epithelia, and bacterial proliferation [42]. Side effects and low efficacy are the main causes of patient incompliance in most of the available topical therapies [43, 44]. Improving hair follicle penetration of active substances and their targeting to the sebaceous gland by means of particle-based delivery systems results in higher local drug concentration at the site of action (enhanced efficacy), prolonged and sustained drug release and reduction of side effects [12, 13, 26, 45]. The ideal carrier for the treatment of acne should penetrate to the end of the infundibulum where the sebaceous gland duct opens to the hair follicle canal. However, acne is often characterized by clogged hair follicle orifices due to the thickening of the perifollicular epithelia and the excessive sebum production. Therefore, the choice of particle size, flexibility, and lipophilicity is crucial to reach an adequate penetration of the carriers into the infundibulum. In addition, mechanical peeling directed to open the hair follicle orifices could be helpful in order to maximize the targeting of nanoparticles to the hair follicles. Different types of lipid vesicles and liposomes have been investigated to deliver different types of anti-acne drugs. Liposomes and niosomes were found to enhance the drug accumulation at the administration site, as well as in the pilosebaceous unit, and were used to deliver antimicrobial agents, salicylic acid, retinoids as well as antiandrogens [46–50]. These carriers have sizes in the micrometer range, are formable and fuse with the lipid matrix and cell membranes once applied on skin. Among the particulate carriers in the nanometer range, solid lipid nanoparticles are the most investigated ones with regard to acne therapy [51–53]. They are biodegradable, do not fuse with skin components and once accumulated in the hair follicle infundibulum, allow for a sustained drug release from their solid matrix. Other valuable carriers for anti-acne drugs are polymeric nanoparticles. In fact, they are longer stable than liposomes, allowing for deposition in the hair follicle infundibulum and sustained drug release [26], and can be loaded with hydrophilic molecules, enabling the delivery of new developed drugs or even anti-acne vaccines [54].

9.4.2 Androgen Receptors

Androgens and androgen receptors in the pilosebaceous unit play an important role in acne pathophysiology. Androgens are regulating sebum production and sebocyte proliferation [55, 56] and influence keratinization in the upper part of the infundibulum, resulting in hair follicle obstruction. In this regard, acne therapy aims at regulating androgen synthesis and activity by using systemic antiandrogens or androgen blockers [57]. These drugs may have severe undesired effects, especially when administered systemically. In the past, topical formulations of antiandrogens failed to reach the target organs. Thus new approaches for topical applications minimizing systemic side effects are needed. Liposome-loaded antiandrogens resulted in lower serum drug level and better targeting to the site of action [57–59]. Therefore, the encapsulation of antiandrogens in particulate carriers which can be applied topically ensuring a constant release of drug to the site of action may improve the outcome of therapies based on antiandrogen approaches.

Antiandrogens are also used for the treatment of androgenetic alopecia, a hair growth disorder resulting from the shortening of the anagen phase, due to increased androgen receptor expression in the pilosebaceous unit [60, 61]. Systemic administration of finasteride, a 5a-reductase inhibitor, blocking the metabolization of testosterone to dihydrotestosterone, is one of the most common treatments available for androgenetic alopecia in men. However, no formulations have jet proven to be effective for topical applications. The results achieved with finasteride-loaded liposomes in acne treatments showed that particle-based formulations have a great potential to be effective also for the treatment of androgenetic alopecia.

In fact, it has been shown that solid lipid nanoparticles accumulate and remain in hair follicle canals for at least 24 h and that the encapsulation of an antiandrogen prodrug, RU 58841-myristate, in solid lipid nanoparticles decreased the transdermal permeation of the drug, which is responsible to side effects, and improved the cutaneous drug bioavailability [32].

9.4.3 Microorganisms

The hair follicle infundibulum represents an optimal niche for certain fungi or bacteria strains, not only because of the reservoir function, where they are less accessible for common antifungal and antibacterial formulation, but also due to the optimal sebum rich climate. Nano particles accumulate in the infundibula and act like a depot providing a constant and prolonged drug release influencing the local milieu and the colonization. Thus, particle-based delivery of compounds with antibacterial or antifungal properties to the hair follicle canals may help target colonization by bacteria or fungi [62]. Souto and Muller showed that the delivery of ketoconazol by solid lipid nanoparticles is feasible [63]. Solid lipid nanoparticles or nanostructured lipid carriers could increase the skin penetration of antifungal or antimicrobial agents [64, 65], and might therefore improve the topical therapy by reducing the application frequency and lowering the side effects. Sanna et al., demonstrated the follicular penetration and permeation of econazol to hair follicle epithelia delivered topically by solid lipid nanoparticles [66]. Polymer nanoparticles have also been investigated as carriers for azole antifungals. Also for this type of nanoparticles the follicular penetration pathway was predominant. Furthermore, their drug delivery properties were superior to those of a commercial liposomal preparation [67]. Targeting of active agents such as zinc pyrithione to the hair follicle infundibula by means of a particulate formulation of the active substance had positive impact to the treatment of dandruff and seborrhoeic dermatitis [68].

9.4.4 Systemic Delivery

The transdermal penetration data via the transfollicular pathway show the importance of hair follicles not only for the penetration of drugs to the skin but also as cutaneous shunts, which facilitate the translocation of drugs, and also of nanoparticles, to the blood stream. Although this possibility raises concerns about the body distribution of translocated nanoparticles and their chronic toxicity [69], it also opens new possibilities of alternative drug administration routes. The extensive capillary network surrounding the hair follicle infundibula makes hair follicles important entry points for the systemic delivery of topically applied drug formulations. The transdermal delivery can circumvent the first hepatic-pass and gastrointestinal digestion, resulting in lower but sustained doses of the actives and consequently in a lower toxicity risk. The use of biodegradable particles, which accumulate in the hair follicles canals and release their loads over time without crossing the skin barrier could make the transdermal delivery of high molecular weight hydrophilic substances like proteins (e.g., insulin) feasible.

9.4.5 Perifollicular Antigen-Presenting Cells

Studies by Lademann et al. [70] suggest that particles which have penetrated into the follicular duct are protected from shedding and remain in the hair follicle openings for several days, providing long time periods for intensive interactions with the hair follicle epithelium. Our own studies using in vivo confocal laser scanning microscopy in mice revealed that the hair follicle is the major site of nanoparticle translocation across the skin barrier. In fact, we were able to monitor particle diffusion into the perifollicular tissue in vivo and identified particles in Langerhans cells and perifollicular dendritic cells. Specific surface functionalization of such particles may allow to direct nanocarriers loaded with antigen and adjuvants directly to cell populations of interest, such as skin antigen-presenting cells. We recently explored possibilities to utilize this penetration pathway for transcutaneous vaccination approaches. In fact, the rationale for transcutaneous vaccination is strong because the skin is easily accessible, and there is evidence, that skin antigen-presenting cells can be targeted more effectively with less vaccine than conventional intramuscular injections. The use of nanocarriers for vaccine delivery would take advantage of penetration and storage in the hair follicle openings and could help increase vaccination efficacy via this route [61, 71, 72]. In previous studies we demonstrated that transcutaneous delivery of a licensed anti-influenza vaccine after cyanoacrylate hair follicle biopsies was safe and effective on skin surface areas as small as 16 cm^2 [63, 73]. Although nanoparticulate vaccine formulations are not available for applications in humans at this point, there is encouraging data available for nanocarrier-based model vaccines in animals. For example, transcutaneous application of biodegradable nanocarrier-based model vaccines was shown to be more effective than subcutaneous injections in triggering serum and mucosal antibodies as well as cellular immune responses [74].

9.4.6 Hair Bulb

Targeting of drugs to the follicular bulb region and the surrounding tissue is the aim of topical treatments of androgenetic alopecia and alopecia areata. Although not yet fully clarified, both hair diseases have different etiologies, thus requiring different therapeutic strategies. Androgenetic alopecia is a genetic predisposition with altered androgen receptor and enzymatic activity leading to alteration of the hair cycle. Current therapies aim at controlling hormone and prostaglandin metabolism. On the other hand, alopecia areata is an autoimmune disease characterized by CD4+ and CD8+ lymphocyte infiltration in the bulb region [75]. This autoimmune reaction directed against hair follicle-associated antigens breaks down the immune privilege of the hair bulb with invasion of inflammatory cells inside the hair follicle and interruption of hair shaft production. Interestingly, white hairs are spared from this process suggesting an involvement of melanocytic antigens in this process. Alopecia areata is regarded as an autoimmune disorder in genetically predisposed individuals. The most successful clinical therapeutic approaches in managing alopecia areata are based on immunosuppressant (cyclosporine A, prednisolone, photochemotherapy) and immunomodulation (diphencyprone, DPCP). However, a systemic long-term immunosuppression which may help to cope with an acute flare of the disease or maintenance of hair growth, cannot be the treatment of choice for an otherwise not life threatening disorder. The bulb region, located deep within the dermis, is not an easily accessible target for topically applied drugs. In addition, the morphology and function of the hair follicle in alopecia areata is altered. The targeting of the hair follicles by means of nanoparticles could therefore be of great advantage. In fact, the prolonged release of drugs in the hair follicles infundibulum could favor their permeation to the perifollicular tissue and to the dermal papilla. In order to target infiltrating lymphocytes it might be necessary to enhance hair follicle penetration or even skin permeability in order to target drugs or nanoparticle-loaded drugs to the bulb region. Recently it has been shown that the T-box21 small interfering RNA conjugated to cationized gelatine had mitigating effects on alopecia in C3H/HeJ mice [76]. Targeted, particle-mediated delivery of small interfering RNA could offer a more specific therapeutic approach. Such mode of delivery could improve the targeting of drugs to the hair follicle and may further increase the stability of such sensitive active molecules.

9.4.7 Stem Cells

The highly proliferative stem cells, located in the bulge region of the hair follicle, are responsible for hair growth, wound healing, and hair follicle reconstitution. They are capable of differentiating into other cell types like keratinocytes and melanocytes. Therefore, these cells are the attractive targets of gene delivery strategies for the permanent correction of genetic hair and skin disorders. Particles are valuable carriers for the delivery of DNA and si-RNA molecules in order to target them to the stem cells of the bulge area, whereby protecting them from degradation. By means of liposomes DNA and oligonucleotides have been targeted successfully to hair follicle progenitor cells [77], whereas Li and Hoffman [78] reported on the selective targeting of the lacZ reporter gene to hair follicles in mice. In another study, cationic liposomes were used to deliver gene encoding β-galactosidase to follicular stem cells at wound borders [79]. Nanoparticles have the potential to increase these results. In order to deliver genes, particulate carriers have not only to penetrate deep in the hair follicle infundibulum and reach the bulge region, but also to deliver their load into the cells and to the cell nucleus. For this purpose carriers equipped with penetrating peptides or viral vectors seems to be more adequate. For example, Hachiya et al. [80] evaluated the feasibility of gene transfer to follicular stem cells using HIV-based vectors and, in an experiment by Zhao et al. [81], the cutaneous delivery of the tyrosinase-gene pLme/SN by means of retrovirus resulted in melanin production in albino rats. Gene transfer to hair follicle cells has also been used to treat pigmentation and hair growth disorders [76].

9.5 Conclusions

Hair follicles can be considered as physiological breaks in the skin barrier and play a significant role as entry points for topically applied compounds and nanoparticles [9, 69]. The preferential accumulation of nanoparticles in the hair follicle infundibula and the fact that, depending on their physicochemical properties nanoparticles penetrate to different depths in the hair follicle canal and have different drug release properties, imply that particles could be important delivery systems to bring active compounds specifically to different hair follicle compartments [1, 2]. Depending on their application, particles could either accumulate in the infundibula and release their load over time, or cross the skin barrier and reach target cells in order to successfully fulfill their challenge. With regard to possible hazardous effects related to the translocation of nanomaterials or nano-sized degradation products from larger particles, the use of biodegradable particles currently represents one of the most interesting approaches in the field. The selective accumulation of nanoparticles in the hair follicle canals was confirmed for a wide range of particles as a result of their particulate nature and related physicochemical properties. Nevertheless, even more selective carrier systems can be manufactured, which might extremely improve nanoparticles selectivity for specific targets within the hair follicles. For example nanoparticles could be functionalized with antibodies or substrates specific for a certain surface cell marker or receptor [82]. Thus, it is to expect that multifunctional nanoparticles will be available, provided with targeting units, penetration enhancing units as well as environment-sensitive features designed for topical application and allowing a temporally and spatially controlled release of actives.

References

1. Blume-Peytavi U, Vogt A. Human hair follicle: reservoir function and selective targeting. Br J Dermatol. 2011;165 Suppl 2:13–7.
2. Papakostas D, Rancan F, Sterry W, Blume-Peytavi U, Vogt A. Nanoparticles in dermatology. Arch Dermatol Res. 2011;303(8):533–50.
3. Mitragotri S. Modeling skin permeability to hydrophilic and hydrophobic solutes based on four permeation pathways. J Control Release. 2003;86(1):69–92.
4. Lademann J, Knorr F, Richter H, et al. Hair follicles—an efficient storage and penetration pathway for topically applied substances. Summary of recent results obtained at the Center of Experimental and Applied Cutaneous Physiology, Charite-Universitatsmedizin Berlin, Germany. Skin Pharmacol Physiol. 2008;21(3):150–5.
5. Mahe B, Vogt A, Liard C, et al. Nanoparticle-based targeting of vaccine compounds to skin antigen-presenting cells by hair follicles and their transport in mice. J Invest Dermatol. 2009;129(5):1156–64.
6. Grams YY, Whitehead L, Cornwell P, Bouwstra JA. Time and depth resolved visualisation of the diffusion of a lipophilic dye into the hair follicle of fresh unfixed human scalp skin. J Control Release. 2004;98(3):367–78.
7. Lademann J, Otberg N, Richter H, et al. Investigation of follicular penetration of topically

applied substances. Skin Pharmacol Appl Skin Physiol. 2001;14 Suppl 1:17–22.
8. Otberg N, Teichmann A, Rasuljev U, Sinkgraven R, Sterry W, Lademann J. Follicular penetration of topically applied caffeine via a shampoo formulation. Skin Pharmacol Physiol. 2007;20(4):195–8.
9. Vogt A, Mandt N, Lademann J, Schaefer H, Blume-Peytavi U. Follicular targeting—a promising tool in selective dermatotherapy. J Investig Dermatol Symp Proc. 2005;10(3):252–5.
10. Prausnitz MR, Mitragotri S, Langer R. Current status and future potential of transdermal drug delivery. Nat Rev Drug Discov. 2004;3(2):115–24.
11. Morrow DI, Garland MJ, McCarron PA, Woolfson AD, Donnelly RF. Innovative drug delivery strategies for topical photodynamic therapy using porphyrin precursors. J Environ Pathol Toxicol Oncol. 2007;26(2):105–16.
12. Lademann J, Richter H, Teichmann A, et al. Nanoparticles—an efficient carrier for drug delivery into the hair follicles. Eur J Pharm Biopharm. 2007;66(2):159–64.
13. Rolland A, Wagner N, Chatelus A, Shroot B, Schaefer H. Site-specific drug delivery to pilosebaceous structures using polymeric microspheres. Pharm Res. 1993;10(12):1738–44.
14. Sellheyer K, Krahl D. Skin mesenchymal stem cells: prospects for clinical dermatology. J Am Acad Dermatol. 2010;63(5):859–65.
15. Nishikawa-Torikai S, Osawa M, Nishikawa S. Functional characterization of melanocyte stem cells in hair follicles. J Invest Dermatol. 2011;131(12):2358–67.
16. Wang X, Tredget EE, Wu Y. Dynamic signals for hair follicle development and regeneration. Stem Cells Dev. 2012;21(1):7–18.
17. Vogt A, McElwee K, Blume-Peytavi U. Biology of the hair follicle. In: Blume-Peytavi U, Tosti A, Whiting D, Trueb R, eds. *Textbook on Hair – From basic science to clinical application*. Germay: Springer; 2008:1–22.
18. Vogt A, Hadam S, Heiderhoff M, et al. Morphometry of human terminal and vellus hair follicles. Exp Dermatol. 2007;16(11):946–50.
19. Toll R, Jacobi U, Richter H, Lademann J, Schaefer H, Blume-Peytavi U. Penetration profile of microspheres in follicular targeting of terminal hair follicles. J Invest Dermatol. 2004;123(1):168–76.
20. Vogt A, Combadiere B, Hadam S, et al. 40 nm, but not 750 or 1,500 nm, nanoparticles enter epidermal CD1a+ cells after transcutaneous application on human skin. J Invest Dermatol. 2006;126(6):1316–22.
21. Knorr F, Lademann J, Patzelt A, Sterry W, Blume-Peytavi U, Vogt A. Follicular transport route—research progress and future perspectives. Eur J Pharm Biopharm. 2009;71(2):173–80.
22. Hordinsky M. Advances in hair diseases. Adv Dermatol. 2008;24:245–59.
23. Meidan VM, Touitou E. Treatments for androgenetic alopecia and alopecia areata: current options and future prospects. Drugs. 2001;61(1):53–69.
24. Ohnemus U, Uenalan M, Inzunza J, Gustafsson JA, Paus R. The hair follicle as an estrogen target and source. Endocr Rev. 2006;27(6):677–706.
25. Zouboulis CC, Bornstein SR. Endocrinology of the skin—a promising joint adventure. Horm Metab Res. 2007;39(2):69–70.
26. Rancan F, Papakostas D, Hadam S, et al. Investigation of polylactic acid (PLA) nanoparticles as drug delivery systems for local dermatotherapy. Pharm Res. 2009;26(8):2027–36.
27. Meidan VM, Bonner MC, Michniak BB. Transfollicular drug delivery—is it a reality? Int J Pharm. 2005;306(1–2):1–14.
28. Teichmann A, Heuschkel S, Jacobi U, et al. Comparison of stratum corneum penetration and localization of a lipophilic model drug applied in an o/w microemulsion and an amphiphilic cream. Eur J Pharm Biopharm. 2007;67(3):699–706.
29. Wu H, Ramachandran C, Weiner ND, Roessler BJ. Topical transport of hydrophilic compounds using water-in-oil nanoemulsions. Int J Pharm. 2001;220(1–2):63–75.
30. Grice JE, Ciotti S, Weiner N, Lockwood P, Cross SE, Roberts MS. Relative uptake of minoxidil into appendages and stratum corneum and permeation through human skin in vitro. J Pharm Sci. 2010;99(2):712–8.
31. Huang Y, Yu F, Park YS, et al. Co-administration of protein drugs with gold nanoparticles to enable percutaneous delivery. Biomaterials. 2010;31(34):9086–91.
32. Munster U, Nakamura C, Haberland A, et al. RU 58841-myristate—prodrug development for topical treatment of acne and androgenetic alopecia. Pharmazie. 2005;60(1):8–12.
33. Chen H, Chang X, Du D, et al. Podophyllotoxin-loaded solid lipid nanoparticles for epidermal targeting. J Control Release. 2006;110(2):296–306.
34. Jung S, Otberg N, Thiede G, et al. Innovative liposomes as a transfollicular drug delivery system: penetration into porcine hair follicles. J Invest Dermatol. 2006;126(8):1728–32.
35. Danhier F, Feron O, Preat V. To exploit the tumor microenvironment: passive and active tumor targeting of nanocarriers for anti-cancer drug delivery. J Control Release. 2010;148(2):135–46.
36. Alvarez-Roman R, Naik A, Kalia YN, Guy RH, Fessi H. Skin penetration and distribution of polymeric nanoparticles. J Control Release. 2004;99(1):53–62.
37. Zvyagin AV, Zhao X, Gierden A, Sanchez W, Ross JA, Roberts MS. Imaging of zinc oxide nanoparticle penetration in human skin in vitro and in vivo. J Biomed Opt. 2008;13(6):064031.
38. Baroli B, Ennas MG, Loffredo F, Isola M, Pinna R, Lopez-Quintela MA. Penetration of metallic nanoparticles in human full-thickness skin. J Invest Dermatol. 2007;127(7):1701–12.
39. Mortensen LJ, Oberdorster G, Pentland AP, Delouise LA. In vivo skin penetration of quantum dot nanoparticles in the murine model: the effect of UVR. Nano Lett. 2008;8(9):2779–87.
40. Graf C, Meinke M, Gao Q, et al. Qualitative detection of single submicron and nanoparticles in human skin

by scanning transmission x-ray microscopy. J Biomed Opt. 2009;14(2):021015.
41. Rancan F, Todorova A, Hadam S, et al. Stability of polylactic acid particles and release of fluorochromes upon topical application on human skin explants. Eur J Pharm Biopharm. 2012;80(1):76–84.
42. Zouboulis CC, Eady A, Philpott M, et al. What is the pathogenesis of acne? Exp Dermatol. 2005;14(2):143–52.
43. Gollnick H, Cunliffe W, Berson D, et al. Management of acne: a report from a Global Alliance to Improve Outcomes in Acne. J Am Acad Dermatol. 2003;49(1 Suppl):S1–37.
44. Leyden JJ. Topical treatment of acne vulgaris: retinoids and cutaneous irritation. J Am Acad Dermatol. 1998;38(4):S1–4.
45. Mordon S, Sumian C, Devoisselle JM. Site-specific methylene blue delivery to pilosebaceous structures using highly porous nylon microspheres: an experimental evaluation. Lasers Surg Med. 2003;33(2):119–25.
46. Honzak L, Sentjurc M. Development of liposome encapsulated clindamycin for treatment of acne vulgaris. Pflugers Arch. 2000;440(5 Suppl):R44–5.
47. Bhalerao SS, Raje Harshal A. Preparation, optimization, characterization, and stability studies of salicylic acid liposomes. Drug Dev Ind Pharm. 2003;29(4):451–67.
48. Manconi M, Sinico C, Valenti D, Lai F, Fadda AM. Niosomes as carriers for tretinoin: III. A study into the in vitro cutaneous delivery of vesicle-incorporated tretinoin. Int J Pharm. 2006;311(1–2):11–9.
49. Schafer-Korting M, Korting HC, Ponce-Poschl E. Liposomal tretinoin for uncomplicated acne vulgaris. Clin Investig. 1994;72(12):1086–91.
50. Bernard E, Dubois JL, Wepierre J. Importance of sebaceous glands in cutaneous penetration of an antiandrogen: target effect of liposomes. J Pharm Sci. 1997;86(5):573–8.
51. Mehnert W, Mader K. Solid lipid nanoparticles: production, characterization and applications. Adv Drug Deliv Rev. 2001;47(2–3):165–96.
52. Schafer-Korting M, Mehnert W, Korting HC. Lipid nanoparticles for improved topical application of drugs for skin diseases. Adv Drug Deliv Rev. 2007;59(6):427–43.
53. Date AA, Naik B, Nagarsenker MS. Novel drug delivery systems: potential in improving topical delivery of antiacne agents. Skin Pharmacol Physiol. 2006;19(1):2–16.
54. James KA, Burkhart CN, Morrell DS. Emerging drugs for acne. Expert Opin Emerg Drugs. 2009;14(4):649–59.
55. Stewart ME, Downing DT, Cook JS, Hansen JR, Strauss JS. Sebaceous gland activity and serum dehydroepiandrosterone sulfate levels in boys and girls. Arch Dermatol. 1992;128(10):1345–8.
56. Thiboutot D, Harris G, Iles V, Cimis G, Gilliland K, Hagari S. Activity of the type 1 5 alpha-reductase exhibits regional differences in isolated sebaceous glands and whole skin. J Invest Dermatol. 1995;105(2):209–14.
57. Stecova J, Mehnert W, Blaschke T, et al. Cyproterone acetate loading to lipid nanoparticles for topical acne treatment: particle characterisation and skin uptake. Pharm Res. 2007;24(5):991–1000.
58. Tabbakhian M, Tavakoli N, Jaafari MR, Daneshamouz S. Enhancement of follicular delivery of finasteride by liposomes and niosomes 1. In vitro permeation and in vivo deposition studies using hamster flank and ear models. Int J Pharm. 2006;323(1–2):1–10.
59. Kumar R, Singh B, Bakshi G, Katare OP. Development of liposomal systems of finasteride for topical applications: design, characterization, and in vitro evaluation. Pharm Dev Technol. 2007;12(6):591–601.
60. Hibberts NA, Howell AE, Randall VA. Balding hair follicle dermal papilla cells contain higher levels of androgen receptors than those from non-balding scalp. J Endocrinol. 1998;156(1):59–65.
61. Price VH. Testosterone metabolism in the skin. A review of its function in androgenetic alopecia, acne vulgaris, and idiopathic hirsutism including recent studies with antiandrogens. Arch Dermatol. 1975;111(11):1496–502.
62. Kaur IP, Kakkar S. Topical delivery of antifungal agents. Expert Opin Drug Deliv. 2010;7(11):1303–27.
63. Souto EB, Muller RH. SLN and NLC for topical delivery of ketoconazole. J Microencapsul. 2005;22(5):501–10.
64. Souto EB, Wissing SA, Barbosa CM, Muller RH. Development of a controlled release formulation based on SLN and NLC for topical clotrimazole delivery. Int J Pharm. 2004;278(1):71–7.
65. Souto EB, Muller RH. The use of SLN and NLC as topical particulate carriers for imidazole antifungal agents. Pharmazie. 2006;61(5):431–7.
66. Sanna V, Gavini E, Cossu M, Rassu G, Giunchedi P. Solid lipid nanoparticles (SLN) as carriers for the topical delivery of econazole nitrate: in-vitro characterization, ex-vivo and in-vivo studies. J Pharm Pharmacol. 2007;59(8):1057–64.
67. Bachhav YG, Mondon K, Kalia YN, Gurny R, Moller M. Novel micelle formulations to increase cutaneous bioavailability of azole antifungals. J Control Release. 2011;153(2):126–32.
68. Schwartz JR, Shah R, Krigbaum H, Sacha J, Vogt A, Blume-Peytavi U. New insights on dandruff/seborrhoeic dermatitis: the role of the scalp follicular infundibulum in effective treatment strategies. Br J Dermatol. 2011;165 Suppl 2:18–23.
69. Lademann J, Richter H, Schanzer S, et al. Penetration and storage of particles in human skin: perspectives and safety aspects. Eur J Pharm Biopharm. 2011;77(3):465–8.
70. Lademann J, Richter H, Schaefer UF, et al. Hair follicles—a long-term reservoir for drug delivery. Skin Pharmacol Physiol. 2006;19(4):232–6.
71. Combadiere B, Liard C. Transcutaneous and intradermal vaccination. Hum Vaccin. 2011;7(8):811–27.
72. Kanchan V, Panda AK. Interactions of antigen-loaded polylactide particles with macrophages and their correlation with the immune response. Biomaterials. 2007;28(35):5344–57.

73. Combadiere B, Vogt A, Mahe B, et al. Preferential amplification of CD8 effector-T cells after transcutaneous application of an inactivated influenza vaccine: a randomized phase I trial. PLoS One. 2010;5(5):e10818.
74. Liard C, Munier S, Arias M, et al. Targeting of HIV-p24 particle-based vaccine into differential skin layers induces distinct arms of the immune responses. Vaccine. 2011;29(37):6379–91.
75. Alkhalifah A. Topical and intralesional therapies for alopecia areata. Dermatol Ther. 2011;24(3):355–63.
76. Nakamura M, Jo J, Tabata Y, Ishikawa O. Controlled delivery of T-box21 small interfering RNA ameliorates autoimmune alopecia (Alopecia Areata) in a C3H/HeJ mouse model. Am J Pathol. 2008;172(3):650–8.
77. Domashenko A, Gupta S, Cotsarelis G. Efficient delivery of transgenes to human hair follicle progenitor cells using topical lipoplex. Nat Biotechnol. 2000;18(4):420–3.
78. Li L, Hoffman RM. The feasibility of targeted selective gene therapy of the hair follicle. Nat Med. 1995;1(7):705–6.
79. Branski LK, Masters OE, Herndon DN, et al. Preclinical evaluation of liposomal gene transfer to improve dermal and epidermal regeneration. Gene Ther. 2010;17(6):770–8.
80. Hachiya A, Sriwiriyanont P, Patel A, et al. Gene transfer in human skin with different pseudotyped HIV-based vectors. Gene Ther. 2007;14(8):648–56.
81. Zhao M, Saito N, Li L, et al. A novel approach to gene therapy of albino hair in histoculture with a retroviral streptomyces tyrosinase gene. Pigment Cell Res. 2000;13(5):345–51.
82. Franzen S. A comparison of peptide and folate receptor targeting of cancer cells: from single agent to nanoparticle. Expert Opin Drug Deliv. 2011;8(3):281–98.

Nano-Based Gene Therapy for Dermatologic Diseases

10

David L. Chen, Dan Zheng, and Amy S. Paller

10.1 Introduction

Nanotechnology examines and utilizes unique properties that arise in common materials when they are shrunk to sub-100 nm dimensions [1, 2]. At the nanoscale, the physical, chemical, and biological properties of materials differ from the properties of individual molecules or bulk matter [2]. Mixing and matching different properties at the nanoscales can lead to the creation of novel and useful materials in a wide range of fields. In fact, nanotechnology has already been incorporated into our daily life from consumer product goods, to medical device, to industrial products, and to scientific research tools. Engineered nanomaterials hold particularly great promise for studying biological processes and creating diagnostics and therapeutics, as their sizes are perfectly matched to the biological machines that orchestrate life. In the field of dermatology, nanoparticles have already been incorporated into sunscreens and cosmeceuticals for many years, for instance to control drug release to skin and skin appendages, and for targeting hair follicle-specific cell populations [3].

During the past two decades, our understanding of the underlying genetic basis of skin disease, whether the specific mutations in inherited skin disease or the ability for gene modulation of cutaneous neoplasia and inflammation, has dramatically increased. For example, the discovery of the alterations in gene expression and activation of signaling pathways in skin has revolutionized the treatment of psoriasis (e.g., small molecule and antibody inhibitors) and skin cancer (e.g., hedgehog pathway inhibitors for basal cell carcinoma and ^{V600E}BRAF inhibition for melanoma). In addition, the basis for more than 2,000 monogenic genodermatoses is now elucidated, leading to the capability of targeting through gene therapy and pathogenesis-based therapy [4, 5]. This chapter focuses on nanotechnology approaches that could be applied for topically applied gene therapy for skin disease, their potential pitfalls, and possibilities for the future.

10.2 Nanoparticles

Nanoparticles are currently the most commonly used vehicles for nanotechnology-based gene therapy for skin disorders. A broad spectrum of nanoparticles, ranging from soft nanostructures such as liposomes and polymer particles to inorganic solid metal and nanocrystalline particles, have already been tested as drug delivery systems

D.L. Chen • A.S. Paller (✉)
Department of Dermatology, Feinberg School of Medicine, Northwestern University,
676 N. St. Clair, Suite 1600, Chicago, IL 60611, USA
e-mail: dchen208@gmail.com; apaller@northwestern.edu

D. Zheng
Department of Chemistry, International Institute for Nanotechnology, Northwestern University,
2145 Sheridan Road, Evanston, IL 60208, USA
e-mail: danzheng2008@u.northwestern.edu

A. Nasir et al. (eds.), *Nanotechnology in Dermatology*,
DOI 10.1007/978-1-4614-5034-4_10, © Springer Science+Business Media, LLC 2013

in animal models with remarkable results. Nanoparticles can be categorized by their chemical composition, size and shape, and surface chemistry including charge and functionalization, all of which affect the interaction between nanotechnology and biological systems.

Nanoparticles can be distinguished by their chemical composition. In general, *soft particles* are made of organic materials (e.g., lipids, proteins, polymers) and can temporarily alter their shape by stress of contact with surfaces, while *rigid particles* are made of inorganic materials and are not deformable. *Liposomes* are among the most common soft nanomaterials and are composed of a phospholipid bilayer and a hydrophilic core. Hydrophilic substances can be integrated into the core of these nanoparticles, and lipophilic substances can be integrated into the shell. Drugs loaded on liposomes penetrate deeper into the hair follicle than drugs applied in other non-particulate formulations [6], leading to great potential utility for preferentially treated follicular disorders. *Solid lipid nanoparticles* (SLN) and *nanostructured lipid carriers* (NLC) have been engineered to create more stable nanosystems. They lack the micelle structure of the liposomes and are therefore more stable in hydrophilic as well as lipophilic environment, enabling the safe transportation and delivery of a variety of substances to skin [7–9]. *Virosomes* are hybrid nanostructures of liposomes and viral proteins. These virus-like particles mimic the nature of virus and have been used for vaccinations against HBV and HPV [10]. *Polymer-based nanoparticles*, such as those composed of polylactic acid (PLA) and polyglycolic-co-lactic acid (PLGA), have also been employed [11]. Their biodegradable properties have garnered much attention, since the accumulation and the potential cytotoxicity of nonbiodegradable materials are major health concerns. *Dendrimers* are highly branched structures with dimensions of a few nanometers. Their peripheral groups can be used to attach several drug molecules, as well as targeting and solubilizing moieties; thus, dendrimers are predominantly used as drug carrier systems [12]. The recently developed polyamidoamine (PAMAM) dendrimers have also been used as skin penetration enhancers [13]. Finally, *protein* and *peptide-based nanoparticles* such as gelatin nanoparticles are also biodegradable, nontoxic, and easy to cross-link and to modify chemically [14], leading to their potential to be used for drug delivery.

Typical rigid particles are colloidal structures of various shapes and made of different incompressible materials such as metals (e.g., gold or silver), metal oxides (e.g., iron oxide), or ceramics (e.g., silica). *Gold nanoparticles* exhibit unique properties, such as size- and shape-dependent optical and electronic features, have surfaces that can be readily modified with ligands, and have high surface area-to-volume ratio allowing for high drug payload [15, 16]. These properties enable a broad range of investigations and applications for detection and diagnostics [17–19], as well as for photothermal therapy [15]. Silver nanoparticles have been used widely for coating wounds and dressing burns because of their antibacterial effect [20]. Nanocrystalline materials, such as quantum dots (CdSe), are semiconductor materials with unique spectroscopic and optical properties. They can generate oxygen radicals after photosensitization and be used for the photodynamic therapy (PDT) [21]. The encapsulation of drugs in the inorganic nanoparticle core or their adsorption on the particle surface leads to longer circulation time by avoiding quick drug metabolism and degradation in tissue.

Size and shape of nanoparticles also influence their ability to interact with biological systems and their cellular internalization. Particles larger than 500 Da as single agents, for example, do not penetrate through intact epidermis well [22]. Functionalization of different moieties onto the particles can alter the particle surface charge, colloidal stability, bioavailability, as well as drug-carrying and targeting abilities, thus affecting the interaction of the particles with biological systems. As such, surfaces of the nanoparticles can act as a scaffold for the functionalization of different moieties in order to achieve optimal properties for interacting with the skin.

Particle surface charge can influence both particle transdermal penetration and the intracellular uptake. Jung et al. demonstrated that cationic liposomes could penetrate deeper into hair follicles than their anionic counterparts [6]. Positive charge influences cellular uptake since it allows interaction with the negatively charged cell membrane. On the other hand, a negative surface charge is important for nanocarriers of antigens and as adjuvants in vaccine formulations. For example, a negatively charged surface can adsorb HIV-1 p24 capsid peptide more favorably, thus increasing the stability against partial degradation [23].

Modification of the nanoparticle surface with covalent linkage to polyethylene glycol (PEGylation) or coating with polymers or silica can increase the particle stability against degradation and aggregation, which leads to a prolonged half-life time and a more controlled and constant drug release [24, 25]. Particle surface functionalization has also been shown to prevent recognition by the macrophages of the reticuloendothelial system, thus enhancing their bioavailability. Through functionalization of the nanoparticle surface with specific ligands such as monoclonal antibodies, peptides, or aptamers [16, 26, 27], an active targeting of a selected cell population or specific organ or tissues can be achieved. This is a major advantage in dermatology, given the many cell populations or structures within skin which may be specific targets for nanoparticle-based drug carriers. For example, functionalization may allow specific targeting of Langerhans cells or melanocytes in epidermis, stem cells in the follicular bulge region, or a specific population of infiltrating lymphocytes within the dermis [28].

10.3 Challenges and Limitation of Current Gene Therapy for Skin Diseases

The identification of pathogenic mutations that lead to inherited skin disease has revolutionized the ability to genotype and diagnose prenatally through DNA assessment. Many of the mutations that lead to skin disease are dominant negative mutations. In addition, many skin disorders have been shown to result from the inappropriate activation of signaling pathways or expression of gene, for example leading to inflammation or fibrosis. As a result, suppression of gene or protein expression through recent molecular strategies such as antisense DNA oligonucleotides and RNA interference with small interfering RNAs (siRNA) and microRNAs (miRNA) represents a targeted therapeutic approach that may be facilitated through nanotechnology. These gene-suppressing oligonucleotides are designed to be complementary to the genes of interest, and act as either physical impediments for protein translation or as triggers for RNA-induced silencing complexes (RISC) that directly cleave the genes of interest [29, 30]. These oligonucleotides are therapeutic options that potently suppress genetic targets at unprecedented specificity, including at the point mutation level, and are in ongoing trials as treatment for several noncutaneous disorders, among them diabetic macular edema, respiratory syncytial virus infection, hepatitis B, AIDS, and solid tumors [29, 31–33].

Recently, proof of concept that RNA interference can be used for suppressing genetic skin disease has been reported in a promising clinical trial for the treatment of pachyonychia congenita (PC). Individuals with PC have dominant negative mutations in one of four keratin genes (*KRT6a*, *KRT6b*, *KRT16*, or *KRT17*) that most commonly lead to characteristic nail dystrophy and, because of their plantar keratoderma, often tremendous pain resulting in incapacitation. siRNA targeting a specific *KRT6a* point mutation successfully cleared a zone of plantar keratoderma in a patient in a left-right double-blind controlled trial [5, 34, 35]. This preliminary study emphasizes the easy accessibility of skin and the potential to limit systemic side effects by localized delivery to skin. However, the siRNA in the PC trial was delivered by injection to bypass the epidermal barrier, limiting its applicability for drug development because of the associated pain of injection and the very limited treatment area.

Indeed, the primary technological challenge for delivery of genetic materials is the epidermal barrier. The stratum corneum and junctional regions between epidermal cells act as the major barrier to penetration of molecules across the skin. Large, negatively charged molecules such as antisense oligonucleotide, DNA plasmid, or siRNA do not easily penetrate the stratum corneum, as they have low lipophilicity [36]. Topical application of simple aqueous solutions of oligonucleotides results in little or no penetration into normal, viable human epidermis [37–39]. Physicochemical methods, such as mechanical approaches (e.g., local injection, microneedle arrays with and without dissolvable tips, iontophoresis, and electroporation), and chemical functionalization, such as addition of skin-penetrating peptides or emulsion with lipid carriers, have been used to facilitate epidermal penetration and cellular uptake [40–49]. However, these methods often require high concentrations of genetic material, and the delivery agents themselves may have unacceptable toxicity in in vivo models. For example, Takanashi et al. used a large amount of free siRNA (approximately 75 μM) in a lipid- and alcohol-containing "Gene Cream" to achieve gene knockdown in mouse skin [50]. None of these approaches employed nanotechnology.

Another major challenge of gene therapy in the skin is the instability of oligonucleotide in in vitro and in vivo biological systems. Nature has evolved an arsenal of enzymes, known as nucleases, to degrade foreign nucleic acids that enter cells, although nucleic acid stability is a key property of any system that will regulate genes therapeutically or be applicable for diagnostic testing [51]. Strategies to increase the stability of nucleic acids typically rely on chemical modification [52] or electrostatic interaction with a surface [53, 54]. However, such modification could be technically challenging and prohibitively expensive for commercial use.

Nanotechnology has helped in overcoming many of these barriers to delivering gene therapy for the skin. By adjusting the properties of nanoscale therapeutics, one can achieve a combination of successful gene transfer across the epidermis, while maintaining oligonucleotide stability for prolonged genetic knockdown. Although no FDA-approved gene delivery techniques using nanotechnology currently exist, tremendous progress has been made that promises to be translated in the near future to clinical applicability.

10.4 Nano-Based Genetic Therapy for Dermatologic Diseases: Current Status

Gene Knockdown: In their study to target the *KRT6a* mutation seen in PC, Jacobson et al. encapsulated the *KRT6a* site-specific siRNA in nanoparticles composed of a biodegradable copolymer of L-PLA-PEG [55]. When injected into the paws of mice, siRNAs encapsulated in nanoparticles retained biologic activity, and were released in a slower, sustained fashion. This finding is consistent with the ability of nanoparticle sequestration to protect siRNA from enzymatic degradation in vivo.

Nanoscale liposomes that encapsulate siRNA have been shown to improve the stability of siRNA, likely by preventing direct enzymatic degradation of siRNA [56, 57]. However, incomplete penetration occurs with topical application of nanoliposomes [58]. Tran et al. combined the protective properties of nanoliposome-encapsulation with the permeabilizing effects of low-frequency ultrasound in their mouse model to topically deliver siRNA targeting ^{T1799A}*BRAF*, the genetic basis for the ^{V600E}BRAF mutation found in a high percentage of melanomas, to the level of the epidermal–dermal junction, where melanocytes reside and melanoma arises [41, 59, 60]. By topically applying the nanoliposomes, and using low-frequency ultrasound at the application site, the group reduced the growth of ^{T1799A}*BRAF*+ melanomas in mice [41]. In the past few years, orally administered small molecule inhibitors that are more specific for the ^{V600E}BRAF mutation in melanoma have improved the survival of patients whose disease have this specific mutation [4, 61]. However, the impact of this new therapy is limited by the rapid development of resistance and the occurrence of often large numbers of skin tumors because of activation of alternative signaling pathways when BRAF is suppressed. The ability to target specifically to the point mutation, thus preventing suppression of the normal BRAF, and to concurrently target other signaling components through nano-delivery of more than one siRNA might further improve survival and decrease the risk of development of the cutaneous adverse effects. Systemic delivery of gene suppressing nanomaterials, ideally with functionalization that enables targeting to affected internal organs, would be optimal for metastatic disease, whereas topical delivery is advantageous for localized cutaneous involvement or a nonsurgical approach to the in-transit cutaneous metastases.

Nanoscale peptide-siRNA conjugates have also been used to deliver genetic therapy topically. Though peptides typically are unable to traverse the epidermis, recently engineered oligopeptides are able to interact and pass through tight junctions, or induce transfer across cellular membranes directly [62–64]. Uchida et al. conjugated RelA siRNA to AT1002, a synthetic oligomer capable of reversibly opening tight junctions, and Tat, a cell penetrating peptide [44]. Topical application of this siRNA-containing nanoconjugate to an atopic dermatitis mouse model reduced ear thickness, the severity of skin disease at the site of application, and cutaneous cytokine and serum IgE levels. Yi et al. applied a similar strategy in engineering a peptide-siRNA conjugate against microphthalmia-associated transcription factor (MITF) to reduce facial melasma in Asian patients [65].

Microneedle arrays, while not a nanotechnology approach, have been used to deliver siRNA to mouse skin, including the thick footpad region [49, 66]. With their micron-length projections, microneedles penetrate the epidermis and upper dermis painlessly, and can be coated with proteins, viruses, or oligonucleotides [46, 67]. Gonzalez-Gonzalez et al. have synthesized microneedles with a polyvinyl-alcohol copolymer ("protrusion array device") that dissolves with skin penetration, enabling both slow release of siRNA and clearance of the microneedles [49]. Such a system could theoretically be combined with nanotechnology approaches to improve cellular uptake and siRNA persistence after topical delivery to a precise skin depth.

Gene Correction: Despite considerable progress with gene knockdown nanotechnology approaches, introduction of larger amounts of DNA and gene splicing to replace mutant genetic material has primarily relied on transposons, lipid vectors, and viral vectors [68–71] that do not use nanotechnology. Given the potential to improve gene uptake at the cellular level with nanotechnology, early in vitro and in vivo investigations have used nanotechnology to facilitate the delivery of genes to noncutaneous sites [72–74]. For example, using a silica-based nanoparticle delivery system, Bharali et al. have achieved greater gene transfection in the central nervous system of mice compared to a viral transfection method [73]. Coated stainless steel microneedle arrays and protrusion array devices have allowed the delivery of plasmid GFP DNAs to mouse skin and human skin xenografts [49, 67], again raising the possibility of combining nanotechnological approaches with microneedles to improve gene transfection efficiency in the skin without the risks of viral/transposon based gene delivery systems [75–77].

Vaccination with Nano-Based Oligonucleotides: An interesting application of oligonucleotide delivery to skin through nanotechnology is for DNA vaccines. Oligonucleotide-based vaccines are theoretically safer than traditional alive or whole-killed viral vaccines, less expensive to manufacture, and able to elicit both humoral and cellular immune responses [78–81]. Intramuscular (IM) injections of DNA vaccines have only led to modest protective effects [80, 82, 83]. In comparison, delivery to skin can cause lasting immunity, because of the abundance of antigen-presenting dendritic cells in the epidermis and dermis that are able to take up genetic materials and trigger immune reactivity [82, 84, 85].

DNA vaccines delivered as particulates have been shown to improve efficiency, and nanoparticle and microspheres have thus been investigated as delivery systems [86]. Cui et al. delivered cationic nanoparticles engineered from oil-in-water microemulsions, coated with plasmid DNA (pDNA) containing the cytomegalovirus-beta-galactosidase reporter gene, intradermally with a needleless, CO_2 gas jet injection device. The result of this nanoparticle-pDNA delivery was a 200-fold greater immune response compared to naked pDNA alone, as measured by antigen-specific total IgG titer in the sera of the immunized mice [78, 87].

An alternative nanotechnology approach to deliver oligonucleotide vaccines is the use of nanoscale projections. Similar to the microneedle arrays, the Nanopatch™ system is a silicon-gold based patch with as many as 3,364 microprojections in a 4×4 mm area, but with nanoscale tips [88]. DNA vaccines are coated onto the projections, and by placing the coated Nanopatch™ onto intact skin, the vaccine can be simultaneously delivered to large numbers of epidermal APCs, the exact depth being controlled by the length of the projections. Chen et al. have shown that Nanopatch™ delivers DNA vaccine against HSV-2 painlessly and allows for higher titers of antigen-specific antibodies and increased vaccine stability compared to traditional IM injections of the same vaccine [88–90].

10.5 Concerns with Nanotechnology for the Delivery of Genetic Material

The toxicologic profiles and safety concerns of nanotechnology approaches must be addressed before successful human application can occur. These concerns can relate to either the nanomaterial or the genetic material, particularly when leading to gene interference. Due to their high surface area-to-volume ratios, nanomaterials may be chemically reactive [91]. Nanomaterials can cause inflammation in the pulmonary and cardiovascular system, can lead to increased blood clotting, deposit in the central nervous system, or even cause reactive oxidative stress regardless of location [3, 16, 92–96]. siRNA treatment in vitro and in vivo can cause off-target effects or activate components of the immune system, and this latter risk is theoretically greater when applied to the skin, given the many immune-mediating cells in human epidermis and dermis. Toxicity may result from the siRNA oligonucleotide used, the transfection agent, or a combination of both [97, 98]. Interestingly, the few nanotechnology-related studies with introduction of genetic material to the skin topically have shown little toxicity [16, 41, 99], although considerably more investigation is needed. The potential toxicity of nanomaterials will be covered elsewhere in the text, but it should be recognized that each nanomaterial with genetic conjugation has different properties, including toxicologic ones, that relate to the conjugate as whole.

10.6 Conclusions

Investigators studying gene therapy of dermatologic disorders have begun to take advantage of the unique properties of nanoscale therapeutics to overcome traditional barriers to the use of gene therapy, including the inefficient delivery of genetic material across the skin barrier and the instability of oligonucleotides in biological systems. Particularly exciting areas of application for topical therapy include vaccine delivery, inflammatory skin disease (e.g., suppressing cytokine function in psoriasis), cutaneous neoplasia (e.g., for more globally suppressing hedgehog signaling for basal cell carcinomas and Ras signaling for squamous cell carcinomas, or for point mutation suppression in melanomas with NRAS or BRAF mutations), or for treating the cutaneous manifestations of genetic skin disorders, either dominant negative (e.g., epidermolysis bullosa simplex or epidermolytic ichthyosis) or with activated signaling (e.g., suppression of mTOR in the angiofibromas or shagreen patches of tuberous sclerosis).

References

1. Drexler KE. Engines of creation. 1st ed. Garden City: Anchor Press/Doubleday; 1986.
2. Drexler KE. Nanosystems: molecular machinery, manufacturing, and computation. New York: Wiley; 1992.
3. Papakostas D, Rancan F, Sterry W, Blume-Peytavi U, Vogt A. Nanoparticles in dermatology. Arch Dermatol Res. 2011;303(8):533–50.
4. Flaherty KT, Puzanov I, Kim KB, et al. Inhibition of mutated, activated BRAF in metastatic melanoma. N Engl J Med. 2010;363(9):809–19.
5. Leachman SA, Hickerson RP, Hull PR, et al. Therapeutic siRNAs for dominant genetic skin disorders including pachyonychia congenita. J Dermatol Sci. 2008;51(3):151–7.
6. Jung S, Otberg N, Thiede G, et al. Innovative liposomes as a transfollicular drug delivery system: penetration into porcine hair follicles. J Invest Dermatol. 2006;126(8):1728–32.
7. Kristl J, Teskac K, Grabnar PA. Current view on nanosized solid lipid carriers for drug delivery to the skin. J Biomed Nanotechnol. 2010;6(5):529–42.
8. Saupe A, Wissing SA, Lenk A, Schmidt C, Muller RH. Solid lipid nanoparticles (SLN) and nanostruc-

tured lipid carriers (NLC)—structural investigations on two different carrier systems. Biomed Mater Eng. 2005;15(5):393–402.
9. Schafer-Korting M, Mehnert W, Korting HC. Lipid nanoparticles for improved topical application of drugs for skin diseases. Adv Drug Deliv Rev. 2007;59(6):427–43.
10. Ludwig C, Wagner R. Virus-like particles-universal molecular toolboxes. Curr Opin Biotechnol. 2007;18(6): 537–45.
11. Mahe B, Vogt A, Liard C, et al. Nanoparticle-based targeting of vaccine compounds to skin antigen-presenting cells by hair follicles and their transport in mice. J Invest Dermatol. 2009;129(5):1156–64.
12. Villalonga-Barber C, Micha-Screttas M, Steele BR, Georgopoulos A, Demetzos C. Dendrimers as biopharmaceuticals: synthesis and properties. Curr Top Med Chem. 2008;8(14):1294–309.
13. Venuganti VV, Perumal OP. Poly(amidoamine) dendrimers as skin penetration enhancers: influence of charge, generation, and concentration. J Pharm Sci. 2009;98(7):2345–56.
14. Zwiorek K, Kloeckner J, Wagner E, Coester C. Gelatin nanoparticles as a new and simple gene delivery system. J Pharm Pharm Sci. 2005;7(4):22–8.
15. Daniel MC, Astruc D. Gold nanoparticles: assembly, supramolecular chemistry, quantum-size-related properties, and applications toward biology, catalysis, and nanotechnology. Chem Rev. 2004;104(1):293–346.
16. Giljohann DA, Seferos DS, Daniel WL, Massich MD, Patel PC, Mirkin CA. Gold nanoparticles for biology and medicine. Angew Chem Int Ed Engl. 2010;49(19): 3280–94.
17. Rosi NL, Mirkin CA. Nanostructures in biodiagnostics. Chem Rev. 2005;105(4):1547–62.
18. Katz E, Willner I. Integrated nanoparticle-biomolecule hybrid systems: synthesis, properties, and applications. Angew Chem Int Ed Engl. 2004;43(45):6042–108.
19. Freeman RG, Raju PA, Norton SM, et al. Use of nanobarcodes particles in bioassays. Methods Mol Biol. 2005;303:73–83.
20. Chen X, Schluesener HJ. Nanosilver: a nanoproduct in medical application. Toxicol Lett. 2008;176(1):1–12.
21. Chatterjee DK, Fong LS, Zhang Y. Nanoparticles in photodynamic therapy: an emerging paradigm. Adv Drug Deliv Rev. 2008;60(15):1627–37.
22. Bos JD, Meinardi MM. The 500 Dalton rule for the skin penetration of chemical compounds and drugs. Exp Dermatol. 2000;9(3):165–9.
23. Ataman-Onal Y, Munier S, Ganee A, et al. Surfactant-free anionic PLA nanoparticles coated with HIV-1 p24 protein induced enhanced cellular and humoral immune responses in various animal models. J Control Release. 2006;112(2):175–85.
24. Mohamed F, van der Walle CF. PLGA microcapsules with novel dimpled surfaces for pulmonary delivery of DNA. Int J Pharm. 2006;311(1–2):97–107.
25. O'Reilly RK, Hawker CJ, Wooley KL. Cross-linked block copolymer micelles: functional nanostructures of great potential and versatility. Chem Soc Rev. 2006;35(11):1068–83.
26. Walther C, Meyer K, Rennert R, Neundorf I. Quantum dot-carrier peptide conjugates suitable for imaging and delivery applications. Bioconjug Chem. 2008;19(12):2346–56.
27. Cruz LJ, Tacken PJ, Fokkink R, et al. Targeted PLGA nano- but not microparticles specifically deliver antigen to human dendritic cells via DC-SIGN in vitro. J Control Release. 2010;144(2):118–26.
28. Vogt A, Mahe B, Costagliola D, et al. Transcutaneous anti-influenza vaccination promotes both CD4 and CD8 T cell immune responses in humans. J Immunol. 2008;180(3):1482–9.
29. Goodchild J. Therapeutic oligonucleotides. Methods Mol Biol. 2011;764:1–15.
30. Li L, Liu Y. Diverse small non-coding RNAs in RNA interference pathways. Methods Mol Biol. 2011;764:169–82.
31. Fire A, Xu S, Montgomery MK, Kostas SA, Driver SE, Mello CC. Potent and specific genetic interference by double-stranded RNA in Caenorhabditis elegans. Nature. 1998;391(6669):806–11.
32. Zamore PD. RNA interference: big applause for silencing in Stockholm. Cell. 2006;127(6):1083–6.
33. Stephenson ML, Zamecnik PC. Inhibition of Rous sarcoma viral RNA translation by a specific oligodeoxyribonucleotide. Proc Natl Acad Sci USA. 1978;75(1):285–8.
34. Leachman SA, Hickerson RP, Schwartz ME, et al. First-in-human mutation-targeted siRNA phase Ib trial of an inherited skin disorder. Mol Ther. 2010;18(2):442–6.
35. Lewin AS, Glazer PM, Milstone LM. Gene therapy for autosomal dominant disorders of keratin. J Investig Dermatol Symp Proc. 2005;10(1):47–61.
36. Kendall MA, Chong YF, Cock A. The mechanical properties of the skin epidermis in relation to targeted gene and drug delivery. Biomaterials. 2007;28(33):4968–77.
37. White PJ, Fogarty RD, Liepe IJ, Delaney PM, Werther GA, Wraight CJ. Live confocal microscopy of oligonucleotide uptake by keratinocytes in human skin grafts on nude mice. J Invest Dermatol. 1999;112(6):887–92.
38. White PJ, Fogarty RD, McKean SC, Venables DJ, Werther GA, Wraight CJ. Oligonucleotide uptake in cultured keratinocytes: influence of confluence, cationic liposomes, and keratinocyte cell type. J Invest Dermatol. 1999;112(5):699–705.
39. White PJ, Gray AC, Fogarty RD, et al. C-5 propyne-modified oligonucleotides penetrate the epidermis in psoriatic and not normal human skin after topical application. J Invest Dermatol. 2002;118(6):1003–7.
40. Lee WR, Shen SC, Zhuo RZ, Wang KC, Fang JY. Enhancement of topical small interfering RNA delivery and expression by low-fluence erbium:YAG laser pretreatment of skin. Hum Gene Ther. 2009;20(6):580–8.
41. Tran MA, Gowda R, Sharma A, et al. Targeting V600EB-Raf and Akt3 using nanoliposomal-small interfering RNA inhibits cutaneous melanocytic lesion development. Cancer Res. 2008;68(18):7638–49.

42. Garcia M, Larcher F, Hickerson RP, et al. Development of skin-humanized mouse models of pachyonychia congenita. J Invest Dermatol. 2011;131(5): 1053–60.
43. Ritprajak P, Hashiguchi M, Azuma M. Topical application of cream-emulsified CD86 siRNA ameliorates allergic skin disease by targeting cutaneous dendritic cells. Mol Ther. 2008;16(7):1323–30.
44. Uchida T, Kanazawa T, Kawai M, Takashima Y, Okada H. Therapeutic effects on atopic dermatitis by anti-RelA short interfering RNA combined with functional peptides Tat and AT1002. J Pharmacol Exp Ther. 2011;338(2):443–50.
45. Chabri F, Bouris K, Jones T, et al. Microfabricated silicon microneedles for nonviral cutaneous gene delivery. Br J Dermatol. 2004;150(5):869–77.
46. Gill HS, Prausnitz MR. Coated microneedles for transdermal delivery. J Control Release. 2007;117(2):227–37.
47. Prausnitz MR. Microneedles for transdermal drug delivery. Adv Drug Deliv Rev. 2004;56(5):581–7.
48. Gonzalez-Gonzalez E, Ra H, Hickerson RP, et al. siRNA silencing of keratinocyte-specific GFP expression in a transgenic mouse skin model. Gene Ther. 2009;16(8):963–72.
49. Gonzalez-Gonzalez E, Speaker TJ, Hickerson RP, et al. Silencing of reporter gene expression in skin using siRNAs and expression of plasmid DNA delivered by a soluble protrusion array device (PAD). Mol Ther. 2010;18(9):1667–74.
50. Takanashi M, Oikawa K, Sudo K, et al. Therapeutic silencing of an endogenous gene by siRNA cream in an arthritis model mouse. Gene Ther. 2009;16(8):982–9.
51. Gleave ME, Monia BP. Antisense therapy for cancer. Nat Rev Cancer. 2005;5(6):468–79.
52. Ng PS, Bergstrom DE. Alternative nucleic acid analogues for programmable assembly: hybridization of LNA to PNA. Nano Lett. 2005;5(1):107–11.
53. He XX, Wang K, Tan W, et al. Bioconjugated nanoparticles for DNA protection from cleavage. J Am Chem Soc. 2003;125(24):7168–9.
54. Han G, Martin CT, Rotello VM. Stability of gold nanoparticle-bound DNA toward biological, physical, and chemical agents. Chem Biol Drug Des. 2006;67(1):78–82.
55. Jacobson GB, Gonzalez-Gonzalez E, Spitler R, et al. Biodegradable nanoparticles with sustained release of functional siRNA in skin. J Pharm Sci. 2010;99(10): 4261–6.
56. Zhang C, Tang N, Liu X, Liang W, Xu W, Torchilin VP. siRNA-containing liposomes modified with polyarginine effectively silence the targeted gene. J Control Release. 2006;112(2):229–39.
57. Pal A, Ahmad A, Khan S, et al. Systemic delivery of RafsiRNA using cationic cardiolipin liposomes silences Raf-1 expression and inhibits tumor growth in xenograft model of human prostate cancer. Int J Oncol. 2005;26(4):1087–91.
58. El Maghraby GM, Williams AC, Barry BW. Can drug-bearing liposomes penetrate intact skin? J Pharm Pharmacol. 2006;58(4):415–29.
59. Yazdi AS, Palmedo G, Flaig MJ, et al. Mutations of the BRAF gene in benign and malignant melanocytic lesions. J Invest Dermatol. 2003;121(5):1160–2.
60. Davies H, Bignell GR, Cox C, et al. Mutations of the BRAF gene in human cancer. Nature. 2002;417(6892):949–54.
61. Chapman PB, Hauschild A, Robert C, et al. Improved survival with vemurafenib in melanoma with BRAF V600E mutation. N Engl J Med. 2011;364(26):2507–16.
62. Song KH, Fasano A, Eddington ND. Effect of the six-mer synthetic peptide (AT1002) fragment of zonula occludens toxin on the intestinal absorption of cyclosporin A. Int J Pharm. 2008;351(1–2):8–14.
63. Song KH, Fasano A, Eddington ND. Enhanced nasal absorption of hydrophilic markers after dosing with AT1002, a tight junction modulator. Eur J Pharm Biopharm. 2008;69(1):231–7.
64. Snyder EL, Dowdy SF. Recent advances in the use of protein transduction domains for the delivery of peptides, proteins and nucleic acids in vivo. Expert Opin Drug Deliv. 2005;2(1):43–51.
65. Yi X, Zhao G, Zhang H, et al. MITF-siRNA formulation is a safe and effective therapy for human melasma. Mol Ther. 2011;19(2):362–71.
66. Hickerson RP, Flores MA, Leake D, et al. Use of self-delivery siRNAs to inhibit gene expression in an organotypic pachyonychia congenita model. J Invest Dermatol. 2011;131(5):1037–44.
67. Gonzalez-Gonzalez E, Kim YC, Speaker TJ, et al. Visualization of plasmid delivery to keratincytes in mouse and human epidermis. Sci Rep. 2011;1(158):1–9.
68. Choate KA, Khavari PA. Direct cutaneous gene delivery in a human genetic skin disease. Hum Gene Ther. 1997;8(14):1659–65.
69. Siprashvili Z, Nguyen NT, Bezchinsky MY, Marinkovich MP, Lane AT, Khavari PA. Long-term type VII collagen restoration to human epidermolysis bullosa skin tissue. Hum Gene Ther. 2010;21(10):1299–310.
70. Ortiz-Urda S, Lin Q, Yant SR, Keene D, Kay MA, Khavari PA. Sustainable correction of junctional epidermolysis bullosa via transposon-mediated nonviral gene transfer. Gene Ther. 2003;10(13):1099–104.
71. Geusens B, Strobbe T, Bracke S, et al. Lipid-mediated gene delivery to the skin. Eur J Pharm Sci. 2011;43(4):199–211.
72. Suk JS, Suh J, Choy K, Lai SK, Fu J, Hanes J. Gene delivery to differentiated neurotypic cells with RGD and HIV Tat peptide functionalized polymeric nanoparticles. Biomaterials. 2006;27(29):5143–50.
73. Bharali DJ, Klejbor I, Stachowiak EK, et al. Organically modified silica nanoparticles: a nonviral vector for in vivo gene delivery and expression in the brain. Proc Natl Acad Sci USA. 2005;102(32):11539–44.
74. Roy I, Ohulchanskyy TY, Bharali DJ, et al. Optical tracking of organically modified silica nanoparticles as DNA carriers: a nonviral, nanomedicine approach for gene delivery. Proc Natl Acad Sci USA. 2005;102(2):279–84.
75. Yamamoto A, Kormann M, Rosenecker J, Rudolph C. Current prospects for mRNA gene delivery. Eur J Pharm Biopharm. 2009;71(3):484–9.

76. Baum C, Dullmann J, Li Z, et al. Side effects of retroviral gene transfer into hematopoietic stem cells. Blood. 2003;101(6):2099–114.
77. Belancio VP, Hedges DJ, Deininger P. Mammalian non-LTR retrotransposons: for better or worse, in sickness and in health. Genome Res. 2008;18(3):343–58.
78. Cui Z, Baizer L, Mumper R. Intradermal immunization with novel plasmid DNA-coated nanoparticles via a needle-free injection device. J Biotechnol. 2003;102:105–15.
79. Liu M, Hilleman M, Kurth R. DNA vaccines: a new era in vaccinology. Ann N Y Acad Sci. 1995;772:1–94.
80. Mumper R, Ledebur H. Dendritic cell delivery of plasmid DNA: application for controlled nucleic acid-based vaccines. Mol Biotechnol. 2001;19:79–95.
81. Gurunathan S, Klinman DM, Seder RA. DNA vaccines: immunology, application, and optimization*. Annu Rev Immunol. 2000;18:927–74.
82. Le TP, Coonan KM, Hedstrom RC, et al. Safety, tolerability and humoral immune responses after intramuscular administration of a malaria DNA vaccine to healthy adult volunteers. Vaccine. 2000;18(18):1893–901.
83. Haynes JR. Particle-mediated DNA vaccine delivery to the skin. Expert Opin Biol Ther. 2004;4(6):889–900.
84. Hunsaker BD, Perino LJ. Efficacy of intradermal vaccination. Vet Immunol Immunopathol. 2001;79(1–2):1–13.
85. Raz E, Carson DA, Parker SE, et al. Intradermal gene immunization: the possible role of DNA uptake in the induction of cellular immunity to viruses. Proc Natl Acad Sci USA. 1994;91(20):9519–23.
86. O'Hagan DT. Recent advances in vaccine adjuvants for systemic and mucosal administration. J Pharm Pharmacol. 1998;50(1):1–10.
87. Banchereau J, Steinman RM. Dendritic cells and the control of immunity. Nature. 1998;392(6673):245–52.
88. Chen X, Kask AS, Crichton ML, et al. Improved DNA vaccination by skin-targeted delivery using dry-coated densely-packed microprojection arrays. J Control Release. 2010;148(3):327–33.
89. Wermeling DP, Banks SL, Hudson DA, et al. Microneedles permit transdermal delivery of a skin-impermeant medication to humans. Proc Natl Acad Sci USA. 2008;105(6):2058–63.
90. Chen X, Prow TW, Crichton ML, et al. Dry-coated microprojection array patches for targeted delivery of immunotherapeutics to the skin. J Control Release. 2009;139(3):212–20.
91. Nasir A. The future of nanotechnology in dermatology. US Dermatol. 2009;3(1):11–3.
92. Oberdorster G, Finkelstein JN, Johnston C, et al. Acute pulmonary effects of ultrafine particles in rats and mice. Res Rep Health Eff Inst. 2000;(96):5–74; discussion 75–86.
93. Apopa PL, Qian Y, Shao R, et al. Iron oxide nanoparticles induce human microvascular endothelial cell permeability through reactive oxygen species production and microtubule remodeling. Part Fibre Toxicol. 2009;6:1.
94. Li N, Xia T, Nel AE. The role of oxidative stress in ambient particulate matter-induced lung diseases and its implications in the toxicity of engineered nanoparticles. Free Radic Biol Med. 2008;44(9):1689–99.
95. Klaper R, Crago J, Barr J, Arndt D, Setyowati K, Chen J. Toxicity biomarker expression in daphnids exposed to manufactured nanoparticles: changes in toxicity with functionalization. Environ Pollut. 2009;157(4):1152–6.
96. Donaldson K, Stone V, Clouter A, Renwick L, MacNee W. Ultrafine particles. Occup Environ Med. 2001;58(3):211–6.
97. Fedorov Y, Anderson EM, Birmingham A, et al. Off-target effects by siRNA can induce toxic phenotype. RNA. 2006;12(7):1188–96.
98. Lin X, Ruan X, Anderson MG, et al. siRNA-mediated off-target gene silencing triggered by a 7 nt complementation. Nucleic Acids Res. 2005;33(14):4527–35.
99. Chen D, Zheng D, Giljohann D, et al. Topically-delivered polyvalent siRNA-gold nanoparticles traverse epidermis to cause gene suppression without toxicity. J Invest Dermatol. 2011;131 Suppl 1:S64.

Emerging Nanomedicine for Skin Cancer

11

Puiyan Lee, Adnan Nasir, and Kenneth K.Y. Wong

11.1 Basic Facts on Skin Cancer

Skin cancer is one of the most common cancers seen worldwide. Around 20 % of the population gets some form of skin cancer in the world. In the United States, the incidence rate of skin cancer is even higher, representing the top five cancers reported by CDC (Center for Disease Control and prevention) [1]. High mortality and morbidity have been associated with melanoma and it is indeed the second leading cause of death in women aged between 19 and 25 in the United States according to world skin cancer foundation [2]. Since skin is the largest organ in the body, it is sometimes associated with the metastasis from other cancers. In fact, the cutaneous metastasis is a common manifestation from patients with breast cancer [3].

P. Lee
Department of Surgery, The University of Hong Kong, Laboratory Block,
Sassoon Road, Hong Kong SAR, China
e-mail: puiyanrlee@gmail.com

A. Nasir
Department of Dermatology, UNC School of Medicine, 3100 Thurston Bowles Building, Chapel Hill, NC 27544, USA
e-mail: anasir@live.com

K.K.Y. Wong(✉)
Department of Surgery, The University of Hong Kong, Queen Mary Hospital,
Pokfulam Road, Hong Kong SAR, China
e-mail: kkywong@hku.hk

11.2 Topical Delivery

Topical application of chemotherapy is an attractive option for skin cancer since the skin is easily accessible. This modality also allows a higher concentration of drugs at the local site and also a reduction of systemic drug load required for sufficient therapeutic effect, when compared to oral or parenteral drug administration. The topical route also reduces the need for frequent, repeated administration for the maintenance of therapeutic drug level. Thus, tremendous research has been focused on the design of topical treatment, especially in the field of skin cancer.

11.3 Nanotechnology and Topical Treatments

It has long been known that particles less than 600 Da (~0.5 nm^3) can be well absorbed into skin [4]. Thus, formulation of drugs into nanoparticle size as compared to microparticle size seems to be a promising direction in topical delivery. 5-Flurouracil (so called Efudex®, Carac®, Fluoroplex®) is an FDA approved drug for the treatment of suprabasal carcinoma and has been commercially available as a topical cream [5]. Numerous novel strategies to try to improve the anticancer action of 5-fluorouracil have been reported, one example of which is the assembly of drugs into a nanoparticle drug delivery system such as chitosan [6].

A. Nasir et al. (eds.), *Nanotechnology in Dermatology*,
DOI 10.1007/978-1-4614-5034-4_11, © Springer Science+Business Media, LLC 2013

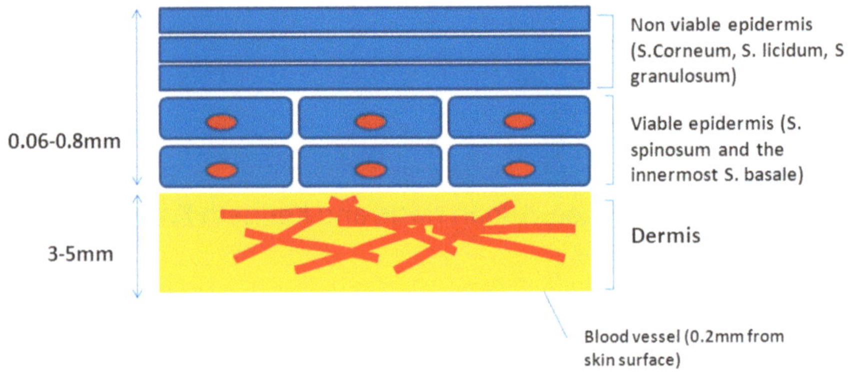

Fig. 11.1 Schematic diagram showing basic anatomy of the epidermal and dermal layers of the skin

In order to be well absorbed via the topical route, adequate skin penetration and permeation is required and is still a challenge. Nanotechnology enables the investigation of particle distribution in skin. The likelihood of skin penetration and permeation by a nanoparticle was studied originally to address the safety concern. It was shown that nanoparticles <10 nm could not penetrate through viable epidermis by a very commonly used nanoparticle available in commercial market—zinc oxide (ZnO) in sunblock. ZnO with the size of 10 nm was shown to have penetration across the intact skin [7] while an imaging study using SEM combined with multiphoton microscopy showed that topically applied ZnO measuring 20–30 nm lacked penetration and remained in the outermost layer of skin, stratum corneum (SC) [8].

Further studies confirmed that nanoparticles with diameter ranged from 6 to 15 nm could indeed penetrate through intact skin but it was highly variable according to surface polarity and the nature of nanoparticles [9]. This was corroborated by another study which demonstrated that the penetration and permeation were highly dependent on the hydrophilicity, shape, and morphology of nanoparticles.

The passage of larger nanoparticles (>20 nm) however, was controversial, due to differences in the design and methodology of the models in various studies. Some studies did report that >20 nm penetrating into the SC, but many others showed that >20 nm nanoparticles was not able to penetrate into the viable epidermis in human full-thickness skin [10, 11]. For transdermal drug to be well-distributed and well-absorbed in deeper skin such as the hypodermis, it is important to reach the dermis rich in blood vessel for better circulation, as seen in Fig. 11.1. The depth of penetration from SC into viable epidermis is highly dependent on the hydrophobicity. One study showed that removal of SC facilitates the uptake of hydrophilic drugs at higher level than lipophillic drugs [12]. This suggests that the SC prevents the penetration of hydrophillic drugs at a very efficient level.

All these studies on size, polarity, and charge selectivity, suggest the presence of barrier the passage of particles across the skin. Understanding the nature and physiology of skin barrier will aid the design and development of better strategy for skin drug delivery.

11.4 Skin as Site for Topical Delivery

Skin is the largest natural barrier that covers the entire body. It is the first line of defense to maintain immunological, physical, and UV sunlight barrier based on its integrity from structure and function, thus to prevent the entry of microbes and bacteria into the body [13]. Despite these problems, the skin remains the most attractive site for drug delivery due to the large area available.

11.4.1 Skin Barrier

Skin is composed of epidermis, dermis, and hypodermis. The epidermis consists of five layers of cells

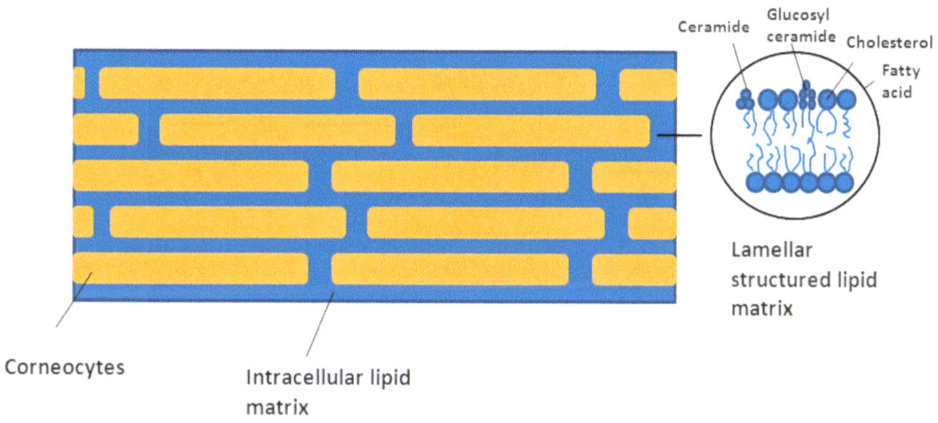

Fig. 11.2 A simplified diagram showing lipophillic nature of the stratum corneum in the skin

packed as the outermost stratum corneum (SC), stratum licidum, stratum granulosum, stratum spinosum, and the innermost stratum basale. The outermost stratum corneum acts like a natural barrier for blocking the entry of most molecules. The impermeability of the skin to a wide range of drugs is primarily associated with lipophillic nature of the stratum corneum (Fig. 11.2). Stratum corneum is composed of a multilayered "brick and mortar" structure where the brick is the corneocytes (80 % of the mass) and the mortar is the intracellular lipid matrix (15 % of mass) mainly of ceramides, fatty acids, triglycerides, cholesterol, and wax [14]. This would explain the impermeability to hydrophilic molecules since the intracellular lipid matrix limits the entry of hydrophilic molecules while the lipophillic molecules can travel freely within the matrix. The situation is more complicated than the description since the molecules travel transversely across the heterogenous lipidic complex within SC. The particles have to partition through hydrophobic lipid matrix and relative hydrophilic corneocytes. The partition and diffusion between hydrophilic and hydrophobic environment are very unfavorable to most drugs [15].

11.4.2 Tight Junction

It is located at stratum granulosum and some found in follicles and epithelial layers. Tight junction is responsible for the passage of nutrients. It forms a barrier for the movement of molecules as small as ions and restricted the passage between cell sheets. It is densely packed with affiliated proteins with space between 10 and 30–50 Å. That can explain why the space can allow the 15 Å (~3.5 kDa) but not the larger ions to get through.

11.4.3 Skin Barrier in Abnormal Condition

The question was originated by the study showing that the zinc nanoparticles could penetrate the skin more effectively when the skin was physically or UV damaged [16]. Apart from physical damage, disruption of the stratum corneum caused by diseases or genetic mutation can lead to the damage in structure and organization. Several genes and promoters regulate the organization of stratum corneum. TALE is a commonly found homeodomain protein in the epidermal barrier. The modification in expression is associated with skin disease and characterized by the thick and non-compact morphology of the stratum corneum [17]. Furthermore, transglutaminase is a calcium-dependent enzyme that is closely related to the stratum corneum. Any defect of this enzyme will result in early murine neonatal death, suggesting that the SC is a crucial protective layer for the body [18]. It is this reason that makes it challenging for any uptake of drug through the SC.

The question of whether cancer will affect the nature of SC has been investigated in one study where the structure integrity of the skin barrier was associated with a cancer gene. In this study, tumor *MT-sp-1* gene knockout caused the malformation of SC [19]. However, further studies are necessary to examine the unknown morphology, permeability, and the penetration depth of drug uptake through the skin. By far, whether cancer will cause the SC disruption is questionable. That can explain why there are extensive research into different strategies to improve the efficiency of topical delivery for skin cancer therapy.

11.5 Strategies for Enhancing the Permeation

Many strategies have been developed and explored to facilitate permeation and access of drugs across the skin for decades. Some strategies using physical or chemical methodology directly enhance the transport into SC. It has long been advocated on theoretical grounds that water hydration can open up the structure of SC, resulting in an increase in the permeability of SC since additional permeation content (normally 15–20 %) cause the modification of partitioning and permeation solubility [20, 21]. But the feasibility is questionable. Other strategy will be discussed as follows.

11.5.1 Permeation Enhancer by Organic Solvent

This is the simplest strategy. The solvent provides one or several functions as follows: (1) it can increase the solubility; (2) the partition coefficient; and (3) the diffusibility of the carrier. The chemical enhancer is mostly an organic solvent such as DMSO, propylene glycol or ethanol. The disadvantage of this method is that the solvent can interact with proteins in SC and also the lipids, resulting in a disruption of the organized structure of SC. This can be associated with irritation or irreversible damage of skin. Furthermore, as the drug then travels through mainly by diffusion, the uptake dose cannot be well controlled.

11.5.2 Physical Methods

For several decades, transdermal delivery has been a very common treatment administration option when there is necessity of long-term drug delivery since it is less invasive. One example is the transdermal patch for analgesic treatment. However, delivery efficiency across the intact skin is not high. For the delivery of insulin for diabetes, even a small protein like insulin (51 kDa), cannot be well absorbed via the skin due to inability to penetrate into skin. With improvement in recent technology, only around 10 % permeation efficiency can be achieved by the transdermal patch per day [22]. Many researchers have attempted to use alternative energy to facilitate the permeation across the skin and allow the passage for a wider range of molecules.

Transdermal ionophoresis has been reported to be useful in aiding the penetration of hydrophillic peptides into the skin using electric current. However, it is highly dependent on the conductivity of the particles [23]. This method has been shown to be particularly useful to deliver the relative small, highly mobile, and mostly charged or some neutral particles [23, 24]. In contrast, delivery of larger particles (>7,000 Da, ~5 nm) are not effective [23].

A newer technique is sonophoresis, utilizing low-frequency ultrasound to aid the transdermal passage of particles across and into the skin. It involves the disruption of lipid bilayer on the skin surface, thus enabling reduced size selectivity of the skin barrier [25]. Furthermore, the use of ultrasound also increases the skin permeability of skin barrier to the passage of hydrophilic molecules. As a result, it alleviates the limitation to the delivery of more hydrophilic natural drugs such as proteins, peptides, and hormones. Indeed, this method has also been tested on commercial synthetic drugs such as ketoprofen [26]. Nonetheless, one drawback of this is the generation of heat, which can cause a rise of skin temperature by 7–10 °C [26]. Thus, sonophoresis is associated with tissue damage during the disruption of barrier properties [26].

Tape stripping comparatively is a less invasive method to expose the dermal tissue by peeling off the SC (horny layer). However, removal of epithelium is associated with inflammation [27]. A study has shown that drug molecules of ~250 nm incorporated into nanoparticle carrier can penetrate the SC. Depending on the hydrophobicity of the drug, removal of the honry layer by tape stripping facilitates the uptake. While uptake of hydrophobic methoxsalen was increased threefold, the more hydrophillic penciclovir increased 1,300-fold [12, 28]. The permeation study with commonly used technique using confocal microscopy performed in porcine skin showed that the permeation across the skin is at greater depth up to 60 μm at 2 h post-application. Despite this, no evidence has reported that nanoparticle formulation showed an enhancement in permeation compared with control [29].

11.5.3 Mechanical Massage

These strategies utilize simple physical technique to enhance topical delivery, aiming to create less tissue damage with minimal toxicity concerns. Mechanical massage has long been known to have the capability of facilitating effective molecular transfer into tissue. Liu and Huang has shown a highly reproducible enhancement of HGF plasmid DNA transfer by tenfold into the liver when receiving intravenous injection of HGF plasmid DNA followed by four repeated manually mechanical massage at 1 s apart [30].

Stimulating mechanical stress has also been shown to increase penetration of the particles into skin [31, 32]. Tinkle et al. investigated the potential for mechanical stress to enhance the dermal penetration. Results showed that 1 μm of the beads could penetrate SC and reach the dermal layer after the skin had been treated with topical application of beads followed by 60 min flex. This indicated that the duration and number of repetition was crucial in affecting the depth of skin penetration. Besides, frequency could be a factor. Machida's group showed that low-frequency massage pulse enhanced both skin permeation and transdermal absorption of drugs [33].

11.6 Solid Lipid Nanoparticles

Encapsulation of drugs into nanoparticle formulation has been demonstrated to have the potential to aid skin permeation, due to the flexibility of the carrier material that would be compatible with the drug payload, as well as the environment [34]. Many of the current formulation have been tailored to reach more depth, so the carrier can release the drug and diffuse into the target.

On the basis of the lipophillic property of barrier in the skin, solid lipid nanoparticles (SLN) favors skin penetration and permeation. Furthermore, SLN has been reported with targeting potential leading to high accumulation in skin [35–39]. Recent technique has been used to incorporate or encapsulate drugs within the lipophillic carrier for enhanced topical delivery. Visualization of the nanoparticle carrier distribution using confocal microscopy showed that the penetration reached as deep as 60 μm.With the combination of the tape-stripping technique for removal of SC [29], it is possible to allow the penetration of large size of the nanoparticle carrier (>200 nm) while normally SC can allow the penetration of 600 Da size. The increase in penetration indicates the increased availability of drug into the skin tissue.

11.6.1 Formulations

The composition of drug with optimal lipophilicity is also crucial for facilitating skin penetration of drug. Since melanoma and basal cell carcinoma are two common skin carcinoma, deeper penetration indicates the possibility of delivery effectively into the tumor site. In permeation study, solid lipidic nanoparticle consisting of either tristearin glyceride or steric acid with size dimension ranging from 123 to 173 nm showed that nanoparticle at 123 nm was the most optimum and there was the highest concentration reaching the dermis. Although the size dimension of SLN is dominantly critical in determining skin permeation, as compared to the lipid composition, the skin retention of SLN seems to depend on the composition. Nanoparticles with smallest composition of long hydrocarbon chain lecithin (4 %) but moderate size showed significantly higher skin accumulation compared with the smallest size [40].

11.6.2 Toxicity

To reduce the concern of toxicity, the material for making solid lipidic nanoparticles should be biocompatible and biodegradable. We have synthesized a lipidic nanoparticle formulation consisting of Cetyl alcohol and Brij 78. The material has been reported to have excellent biocompatibility. We have demonstrated that encapsulation of toxic cancer therapeutics within our lipophilic nanoparticles could reduce the systemic toxicity. Thus it would be a great potential in developing our lipophilic nanoparticle carrier for topical use [41].

References

1. Incidence and mortality web-based report. Human Services, Centers for Disease Control and Prevention, and National Cancer Institute. Atlanta, GA: Department of Health and United States Cancer Statistics: 1999–2007; 2010. http://www.cdc.gov/uscs.
2. Krathen RA, Orengo IF, Rosen T. Cutaneous metastasis: a meta-analysis of data. South Med J. 2003;96(2):164–7.
3. Spencer PS, Helm TN. Skin metastases in cancer patients. Cutis. 1987;39(2):119–21.
4. Barry BW. Breaching the skin's barrier to drugs. Nat Biotechnol. 2004;22:165–7.
5. Micali G, Lacarrubba F, Dinotta F, Massimino D, Nasca MR. Topical cream for treating cancer. Expert Opin Pharmacother. 2010;11(9):1515–27.
6. Arias JL. Novel strategies to improve the anticancer action of 5-fluorouracil by using drug delivery systems. Molecules. 2008;13(10):2340–69.
7. On regulatory aspects of nanomaterials. In: Scientific Committee on Consumer Products, European Commission; 2008. http://www.europarl.europa.eu/sides/getDoc.do?pubRef=-//EP//NONSGML+REPORT+A6-2009- 0255+0+DOC+PDF+V0//EN.
8. Zvyagin AV, Zhao X, Gierden A, Sanchez W, Ross JA, Roberts MS. Imaging of zinc oxide nanoparticle penetration in human skin in vitro and in vivo. J Biomed Opt. 2008;13(6):064031.
9. Labouta HI, El-Khordagui LK, Kraus T, Schneider M. Mechanism and determinants of nanoparticle penetration through human skin. Nanoscale. 2011;3:4989–99.
10. Ryman-Rasmussen JP, Riviere JE, Monteiro-Riviere NA. Variables influencing interactions of untargeted quantum dot nanoparticles with skin cells and identification of biochemical modulators. Nano Lett. 2007;7:1344–8.
11. Baroli B, Ennas MG, Loffredo F, Isola M, Pinna R, Lopez-Quintela MA. Penetration of metallic nanoparticles in human full-thickness skin. J Invest Dermatol. 2007;127:1701–12.
12. Anigbogu AN, Williams AC, Barry BW. Permeation characteristics of 8-methoxypsoralen through human skin; relevance to clinical treatment. J Pharm Pharmacol. 1996;48:357–66.
13. Moens E, Veldhoen M. Epithelial barrier biology: good fences make good neighbours. Immunology. 2012;135:1–8.
14. Nino M, Calabrò G, Santoianni P. Topical delivery of active principles: the field of dermatological research. Dermatol Online J. 2010;16(1):4.
15. Dhiman S, Singh TG, Rehni AK. Transdermal patches: a recent approach to new drug delivery system. Int J Pharm Pharm Sci. 2011;3(5):26–34.
16. Gamer AO, Leibold E, Ravenzwaay VB. The in vitro absorption of microfine zinc oxide and titanium dioxide through porcine skin. Toxicol In Vitro. 2006;20(3):301–7.
17. Jackson B, Brown SJ, Avilion AA, O'Shaughnessy RF, Sully K, Akinduro O, Murphy M, Cleary ML, Byrne C. TALE homeodomain proteins regulate site-specific terminal differentiation, LCE genes and epidermal barrier. J Cell Sci. 2011;124(Pt 10):1681–90.
18. Matsuki M, Yamashita F, Ishida-Yamamoto A, Yamada K, Kinoshita C, Fushiki S, et al. Defective stratum corneum and early neonatal death in mice lacking the gene for transglutaminase 1 (keratinocyte transglutaminase). Proc Natl Acad Sci U S A. 1998;95(3):1044–9.
19. Scheuplein RJ, Blank IH. Permeability of the skin. Physiol Rev. 1971;51(4):702–47.
20. Scheuplein RJ, Blank IH. Mechanism of percutaneous absorption: IV. Penetration of nonelectrolytes (alcohols) from aqueous solutions and from pure liquids. J Invest Dermatol. 1973;60(5):286–96.
21. List K, Haudenschild CC, Szabo R, Chen W, Wahl SM, Swaim W, Engelholm LH, Behrendt N, Bugge TH. Matriptase/MT-SP1 is required for postnatal survival, epidermal barrier function, hair follicle development, and thymic homeostasis. Oncogene. 2002; 21(23):3765–79.
22. Tiwary AK, Sapra B, Jain S. Innovations in transdermal drug delivery: formulations and techniques. Recent Pat Drug Deliv Formul. 2007;1:23–36.
23. Kanikkannan N. Iontophoresis-based transdermal delivery systems. BioDrugs. 2002;16(5):339–47.
24. Batheja P, Thakur R, Michniak B. Transdermal iontophoresis. Expert Opin Drug Deliv. 2006;3(1):127–38.

25. Pahade A, Jadhav VM, Kadam VJ. Sonophoresis: an overview. Int J Pharm Sci Rev Res. 2010;3(2):1–32.
26. Herwadkar A, Sachdeva V, Taylor LF, Silver H, Banga AK. Low frequency sonophoresis mediated transdermal and intradermal delivery of ketoprofen. Int J Pharm. 2012;423(2):289–96.
27. Holzmann S, Tripp CH, Schmuth M, Janke K, Koch F, Saeland S, Stoitzner P, Romani N. A model system using tape stripping for characterization of Langerhans cell-precursors in vivo. J Invest Dermatol. 2004;122(5):1165–74.
28. Morgan CJ, Renwick AG, Friedmann PS. The role of stratum corneum and dermal microvascular perfusion in penetration and tissue levels of water-soluble drugs investigated by microdialysis. Br J Dermatol. 2003;148:434–43.
29. Alvarez-Román R, Naik A, Kalia YN, Guy RH, Fessi H. Enhancement of topical delivery from biodegradable nanoparticles. Pharm Res. 2004;21(10):1818–25.
30. Liu F, Huang L. Noninvasive gene delivery to the liver by mechanical massage. Hepatology. 2002;35(6):1314–9.
31. Tinkle SS, Antonini BA, Rich JR, et al. Skin as a route of exposure and sensitization in chronic beryllium disease. Environ Health Perspect. 2003;111:1201–8.
32. Schneider M, Stracke F, Hansen S, Schaefer UF. Nanoparticles and their interactions with the dermal barrier. Dermatoendocrinology. 2009;1(4):197–206.
33. Sakurai H, Takahashi Y, Machida Y. Influence of low-frequency massage device on transdermal absorption of ionic materials. Int J Pharm. 2005;305(1–2):112–21.
34. Baroli B, Ennas MG, Loffredo F, Isola M, Pinna R, López-Quintela MA. Penetration of metallic nanoparticles in human full-thickness skin. J Invest Dermatol. 2007;127:1701–12.
35. Maia CS, Mehnert W, Sch¨afer-Korting M. Solid lipid nanoparticles as drug carriers for topical glucocorticoids. Int J Pharm. 2000;196:165–7.
36. Sivaramakrishnan R, Nakamura C, Mehnert W, Korting HC, Kramer KD, Sch¨afer-Korting M. Glucocorticoid entrapment into lipid carriers—characterization by parelectic spectroscopy and influence on dermal uptake. J Control Release. 2004;97:493–502.
37. Souto EB, Wissing SA, Barbosa CM, M¨uller RH. Development of a controlled release formulation based on SLN and NLC for topical clotrimazole delivery. Int J Pharm. 2004;278:71–7.
38. Song C, Liu S. A new healthy sunscreen system for human: solid lipid nanoparticles as carrier for 3,4,5-trimethoxybenzoylchitin and the improvement by adding vitamin E. Int J Biol Macromol. 2005;36:116–9.
39. Chen HB, Chang XL, Yang XL, Du DR, Liu W, Liu J, Weng T, Yang YJ, Xu HB. Podophyllotoxin-loaded solid lipid nanoparticles for epidermal targeting. J Control Release. 2006;110:296–306.
40. Liu J, Hu W, Chen H, Ni Q, Xu H, Yang X. Isotretinoin-loaded solid lipid nanoparticles with skin targeting for topical delivery. Int J Pharm. 2007;328(2):191–5.
41. Lee PY, Zhu Y, Tan J, Sun RWY, Hao W, Liu X, Che CM, Wong KK. The cytotoxic effects of lipidic formulated gold-porphyrin nanoparticles for the treatment of neuroblastoma. Nanotechnol Sci Appl. 2010;3:23–8.

Nanotechnology and the Diagnosis of Cutaneous Malignancies

12

Julie K. Kim, Adnan Nasir, and Kelly C. Nelson

12.1 Introduction

The application of nanotechnology to dermatology has opened new doors in the diagnosis and treatment of cutaneous malignancies in the twenty-first century. Noninvasive diagnosis of cutaneous malignancies has traditionally been limited by the confines of what can be seen with visual inspection. Despite the availability of adjunctive tools, such as the dermatoscope, to increase accuracy in correct identification of lesions, skin biopsy with histopathologic confirmation is frequently necessary to ascertain a diagnosis. Thus, the possibility of imaging potentially malignant lesions noninvasively by coupling nanotechnology with various imaging techniques is very exciting. Much of this new technology still remains under investigation. This chapter will provide an overview of cutaneous malignancies, as well as an overview of the traditional methods and current developments in cutaneous imaging on the nanotechnology front.

12.2 Overview of Cutaneous Malignancies

Nonmelanoma skin cancers (NMSCs) represent the most commonly diagnosed and curable of all cancers in the United States with more than two million cases documented each year [1]. The two major types of NMSCs are basal cell carcinomas (BCCs) and squamous cell carcinomas (SCCs) which arise from dysplasia of the epidermal keratinocytes. UV radiation has been established as the greatest risk factor for the development of these lesions. Approximately 80 % of NMSCs occur on the head and neck region, followed by the trunk and upper extremities, indicating the predilection of these cutaneous malignancies for sun-exposed skin. BCCs comprise approximately 80 % of all NMSCs and maintain cure rates of 96–99 % with early detection. However, they are associated with a high rate of recurrence (approximately 20 %) within 1 year of therapy if not treated adequately [2, 3]. SCCs, on the other hand, may not encompass the majority of NMSCs but carry a greater potential for invasion and metastases that accounts for most of NMSC-related mortality [4]. There appears to be a rising trend in NMSC incidence which may be accredited to greater UV exposure from ozone depletion and sun-seeking behavior, longer life spans, and increased detection of skin cancer stemming from improved public awareness [5].

J.K. Kim, MD (✉)
Department of Dermatology, Duke University Medical Center, DUMC Box 3135, 200 Trent Drive, Room 4033, Purple Zone, Duke South, Durham, NC, 27710, USA
e-mail: Julie.k.kim@duke.edu

A. Nasir, MD
Department of Dermatology, UNC Chapel Hill, Chapel Hill, NC, USA
e-mail: anasir@live.com

K.C. Nelson, MD
Department of Dermatology, Duke University Medical Center,Durham, NC, USA
e-mail: Kelly.nelson@duke.edu

A. Nasir et al. (eds.), *Nanotechnology in Dermatology*,
DOI 10.1007/978-1-4614-5034-4_12, © Springer Science+Business Media, LLC 2013

Melanoma is one of the most aggressive skin cancers arising from the malignant transformation of melanocytes at the dermal–epidermal junction. UV radiation (both natural, from sun exposure, and artificial, from tanning beds) is a well-documented risk factor for melanoma. Other risk factors include being of a fair skin phototype, having a personal or family history of melanoma, the presence of atypical nevi, and the presence of certain genetic disorders (such as xeroderma pigmentosum) [6]. The overall incidence of melanoma has increased rapidly in the past few decades [7, 8], attributable to both increased surveillance and a rise in high-risk behaviors such as tanning bed use [9]. Approximately, 70 % of melanomas arise de novo, whereas the remaining 30% are thought to originate from preexisting nevi [10]. Potential precursors of melanoma include atypical melanocytic nevi which are pigmented lesions that represent disordered proliferations of variably atypical melanocytes. It is difficult to predict whether a given atypical nevus will evolve into a melanoma, as it may very well stay quiescent and remain benign during the individual's lifetime. Adding to the challenge of melanoma diagnosis are amelanotic melanomas which present as pink lesions that can often be confused with other cutaneous neoplasms, such as BCCs, or may even be overlooked.

12.3 Traditional Methods for the Diagnosis of Skin Malignancies

Traditional methods for the diagnosis of cutaneous malignancies include visual inspection with total body skin examination. For patients who have a large number of atypical nevi or considerable anxiety about their lesions, total body photography may be used as an adjunct tool to capture pigmented lesions at their baseline and to track potentially malignant changes [11]. This can ultimately minimize the number of unnecessary biopsies. Dermoscopy, the examination of skin lesions using a handheld magnifier called a dermatoscope, has also become increasingly popular and has improved the diagnostic accuracy of visual inspection. This has given dermatologists well trained in this technique a powerful and reliable tool to further aid in the diagnosis of melanocytic lesions [12].

Visual inspection certainly has its drawbacks as the total body skin examination is very subjective, and whether a given lesion appears concerning for a malignancy may be open to interpretation. As a result, skin biopsy with histopathologic confirmation still remains the gold standard of diagnosis. Histopathologic analysis provides information on specific high-risk skin cancer features such as the depth of invasion and perineural involvement to aid in the stratification of patients into the appropriate treatment trajectory.

The disadvantages of biopsy with histopathologic confirmation include time, cost, discomfort, and inconvenience to the patient. Another nuance to consider is that despite these shortcomings, a biopsy (even as the current gold standard) still does not absolutely clinch a diagnosis, as histopathology may be indeterminate for complex melanocytic lesions. In situations where the diagnosis is equivocal, various immunohistochemical tumor markers may be helpful in differentiating benign melanocytic lesions from those that are malignant (See Table 12.1) [13, 14]. However, even these stains are subject to interpretation and do not necessarily provide a definitive answer.

In vivo imaging technologies offer noninvasive alternatives to further visualize lesions of concern. Reflectance confocal microscopy (RCM) and MelaFind are currently the only two FDA-approved imaging devices designed with the goal of increasing the diagnostic accuracy of melanocytic lesions while reducing the number of unnecessary biopsies. One of the greatest advantages of RCM is its ability to image with cellular resolution, allowing for visualization of vascular and tumor nests that are otherwise not noticeable to the naked eye. However, RCM's cellular resolution diminishes at the dermal layer, which is one limitation of this technology [13]. MelaFind is a computerized system coupled with a handheld scanner that uses diagnostic imaging algorithms to predict the likelihood of a pigmented lesion to be a melanoma. When considering borderline lesions (such as severely atypical nevi), MelaFind has demonstrated an

Table 12.1 Immunohistochemical tumor markers of melanoma [13]

S-100	• 21-Kd protein that functions with intracellular calcium trafficking and microtubular assay • Sensitivity: 97–100 % • Low specificity (stains wide variety of lesions including almost all benign and malignant melanocytic lesions and their metastases) • Staining is nuclear and cytoplasmic • Most sensitive marker for desmoplastic melanomas
HMB-45 (human melanoma, black-45)	• Mouse monoclonal antibody • Sensitivity: 69–93 % ; decreased sensitivity in metastatic lesions • Staining is cytoplasmic and granular • Helpful in distinguishing nevi from melanoma
Melan-A/MART-1 (melanoma antigen recognized by T cells)	• Melanocyte differentiation antigen • Sensitivity: 75–92 %, specificity: 95–100 % • More intense and diffuse staining than HMB-45 • May be easier to interpret, especially in metastatic melanomas
MEL-5 (aka TA-99)	• Mouse monoclonal antibody • Sensitivity: 95 % but also stains normal epidermal melanocytes • Stains melanosome-containing cells • May be negative in amelanotic, desmoplastic, or dermal melanomas
Ki67	• Proliferation marker • Positive in less than 5 % of tumor cells in most benign nevi • Positive in 13–30 % of cases of cells in malignant melanomas • Increase in recurrence rate and mortality with increasing positivity found to be correlated

impressive sensitivity of greater than 95%; however, this is countered by a relatively lower specificity of close to 10 %. Another limitation of Melafind may be its efficacy in detecting less common subtypes of melanomas such as amelanotic, acral lentiginous, and spitzoid melanomas [15]. Other imaging techniques currently under investigation include Raman spectroscopy, high-frequency ultrasound, and optical computed tomography [13].

12.4 Nanotechnology for the Diagnosis of Skin Malignancies

The application of nanotechnology to medicine (or nanomedicine) has brought a number of innovative projects dealing with diagnosis and therapy. In dermatology, nanomedicine has particularly intriguing applications in the diagnosis of cutaneous malignancies. Several different types of nanoparticles are currently under investigation. Quantum dots, gold nanoparticles, magnetic nanoparticles, and hydrogel core shell nanoparticles are specifically being studied for their use in molecular in vivo skin imaging [16].

12.4.1 Quantum Dots

Quantum dots are solid semiconductor particles that have been widely studied due to their unique optical properties. These particles fluoresce when excited by a light source and are particularly useful in molecular imaging given their ability to absorb light over a wide range of wavelengths but reemit in a narrow spectra dependent on particle size. Quantum dots emit bright fluorescence signals that remain stable for hours, which allows for continuous imaging of biological samples over extended periods of time. This stability may also lead to a resistance to degradation, leading to cellular toxicity due to exposure of the core metals (quantum dots are composed of a metallic core shielded by a shell) to physiological fluids, particularly when the core/shell/coating complex is unstable. Other limiting factors to the use of quantum dots include their unknown long-term toxicity and unknown optimal levels of cellular uptake necessary for quality imaging.

The imaging potential of quantum dots has been demonstrated, particularly in conjunction with fluorescence microscopy. Quantum dot nanoprobes have been used to tag biomarkers that are overexpressed in tumor vasculature and are considered key receptors for tumor targeting. The result is an enhanced fluorescence response compared to untagged peptides, enabling in vivo tumor imaging of melanoma cells and vasculature [17]. To further enhance their uptake, quantum dots have also been encapsulated with hydrogel nanoparticles which allow enhanced accumulation within tumor tissues rather than in the periphery. However, this additional enhancement may later impede excretion from the body, which could be potentially harmful to the renal system [18].

12.4.2 Gold Nanoparticles

Gold nanoparticles also emit stable and intense optical signals, functioning as good labels for sensors in various analytic techniques used for detection. Their size and shape can be further fine tuned to be efficiently integrated into cells. One advantage of gold nanoparticles is their relative nontoxicity to human cells compared to quantum dots because they do not utilize heavy metals and are considered to be bioinert [19]. They have also been used for systemic therapeutics in other applications without significant toxicity [20]. However, their use may be limited by the inherent difficulties in noninvasive monitoring of these particles [19].

The tumor-imaging potential of gold nanoparticles has been explored in several studies investigating its application in photoacoustic tomography (PAT). In one study, injection of gold nanoparticles into mice with breast cancer tumors demonstrated an accumulation of these particles within the tumor; in comparison to mice who received injections of saline, the gold nanoparticle-injected mice demonstrated a substantial increase in image clarity on PAT [19]. This idea has been further extrapolated to the arena of melanoma diagnosis through the use of bioconjugated gold nanocages which, when coupled to melanocyte stimulating hormone (MSH), have been shown to provide enhanced signaling compared to nanocages coupled to polyethylene glycol when imaged with PAT. The theory behind this application is that melanomas tend to overexpress alpha-MSH receptors, and the use of bioconjugated gold nanocages coupled with PAT may allow for the detection of early-stage melanomas and metastatic lymph nodes [21].

12.4.3 Magnetic Nanoparticles

Magnetic nanoparticles may be coupled with magnetic resonance imaging (MRI) technology to further enhance the detection of biomarkers, especially in smaller or deeper tumors which are not often picked up by standard imaging techniques. This was found to be especially useful in one study investigating the detection of angiogenesis as a way to determine the presence of small tumors in the body by targeting melanoma biomarkers such as $\alpha v\beta 3$ (alpha nu beta 3) integrin which are not normally expressed on quiescent endothelial cells. Such nanoparticles facilitate localization of angiogenic vasculature when used in conjunction with MRI technology and improve tumor recognition by differentiating them from surrounding tissues. This technology may be utilized to help risk-stratify patients in terms of surgical management to minimize locoregional recurrences and to aid in delineating metastases. A disadvantage of this particular application is that it still requires injection of contrast for optimal localization and thus may not be a completely noninvasive procedure. However, compared to the gold standard technique of biopsy and histopathologic confirmation, it is still far less invasive and allows visualization of tumor bulk before any surgical measures are taken [22].

12.4.4 Hydrogel Core Shell Nanoparticles

Nanoparticles can also be used in the detection of serum biomarkers that aid in the detection of cutaneous malignancies. The detection of serum biomarkers can be particularly challenging as there may be a low abundance of these particles expressed in smaller sized lesions. Proteins more commonly found in the serum, such as albumin, may also interfere with detection, and whatever is successfully detected may be degraded quickly after collection. This is where hydrogel core shell nanoparticles may prove useful. By entrapping smaller and less abundant biomarkers, these nanoparticles have the ability to offer protection from enzymatic degradation, thereby improving detection rates of smaller cutaneous malignancies. This novel application has been studied in the pro-apoptotic protein Bak which may be a candidate melanoma biomarker and negative prognostic factor for melanoma. In one study, serum Bak levels were found to be more elevated in thin melanomas compared to thick melanomas. This may imply that lower levels of Bak correlate with a poorer prognosis due to downregulation of pro-apoptotic activity and increased survival of melanoma cells. Utilizing this technology may ultimately result in improved risk stratification of patients with melanoma into more appropriate treatment trajectories [23].

12.5 Advantages of Nanotechnology

There are many benefits of nanotechnology as applied to medicine. Diagnostic devices are postulated to have more rapid processing speeds with increased sensitivity and specificity compared to conventional, gold standard devices. Treatments utilizing nanotechnology for drug delivery are anticipated to be more efficacious, customized, and target specific given the small size of the particles. Nanoparticles may also be engineered for combined therapeutic and diagnostic applications (theranostic uses). Such a combined approach may allow for simultaneous monitoring of therapeutic efficacy and toxicity of the treatment drug through imaging [24, 25].

12.6 Future Directions

The ability to image and treat diseases simultaneously is a unique feature of nanomedicine. Theranostic formulations may allow for therapeutic agents to become more personalized and individualized for the patient. The potential to monitor drug response and efficacy through these applications will help to ensure that the appropriate treatments are being provided to patients and provide an indication of when alternative therapeutic options should be considered. However, further investigation into the toxicities associated with nanoparticles is warranted to ensure that these materials can be incorporated into medical practice without endangering patient safety. As a discipline that relies heavily upon visual inspection for diagnosis, the field of dermatology can benefit greatly from these advances in noninvasive cutaneous imaging and therapeutics.

References

1. Jemal A, Siegel R, Xu J, Ward E. Cancer statistics, 2010. CA Cancer J Clin. 2010;60:277–300.
2. Kim RH, Armstrong AW. Nonmelanoma skin cancer. Dermatol Clin. 2012;30:125–39, ix
3. Diepgen TL, Mahler V. The epidemiology of skin cancer. Br J Dermatol. 2002;146 Suppl 61:1–6.
4. Weinstock MA, Bogaars HA, Ashley M, Litle V, Bilodeau E, Kimmel S. Nonmelanoma skin cancer mortality. A population-based study. Arch Dermatol. 1991;127:1194–7.
5. Rogers HW, Weinstock MA, Harris AR, et al. Incidence estimate of nonmelanoma skin cancer in the United States, 2006. Arch Dermatol. 2010;146:283–7.
6. Tuong W, Cheng LS, Armstrong AW. Melanoma: epidemiology, diagnosis, treatment, and outcomes. Dermatol Clin. 2012;30:113–24, ix.
7. Erickson C, Driscoll MS. Melanoma epidemic: facts and controversies. Clin Dermatol. 2010;28:281–6.
8. Linos E, Swetter SM, Cockburn MG, Colditz GA, Clarke CA. Increasing burden of melanoma in the United States. J Invest Dermatol. 2009;129:1666–74.
9. Robinson JK, Rigel DS, Amonette RA. Trends in sun exposure knowledge, attitudes, and behaviors: 1986 to 1996. J Am Acad Dermatol. 1997;37:179–86.
10. Wolff K, Johnson RA. Cutaneous melanoma. In: Sydor AM, Englis MR, editors. Fitzpatrick's color atlas & synopsis of clinical dermatology. 6th ed. New York, NY: McGraw Hill Medical; 2009. p. 308–27.
11. Malvehy J, Puig S. Follow-up of melanocytic skin lesions with digital total-body photography and digital dermoscopy: a two-step method. Clin Dermatol. 2002;20:297–304.
12. Argenziano G, Ferrara G, Francione S, Di Nola K, Martino A, Zalaudek I. Dermoscopy–the ultimate tool for melanoma diagnosis. Semin Cutan Med Surg. 2009;28:142–8.
13. Ohsie SJ, Sarantopoulos GP, Cochran AJ, Binder SW. Immunohistochemical characteristics of melanoma. J Cutan Pathol. 2008;35:433–44.

14. El Tal AK, Abrou AE, Stiff MA, Mehregan DA. Immunostaining in Mohs micrographic surgery: a review. Dermatol Surg. 2010;36:275–90.
15. Smith L, Macneil S. State of the art in non-invasive imaging of cutaneous melanoma. Skin Res Technol. 2011. doi:doi: 10.1111/j.1600-0846.2011.00503.x.
16. Monheit G, Cognetta AB, Ferris L, et al. The performance of MelaFind: a prospective multicenter study. Arch Dermatol. 2011;147:188–94.
17. Zuo L, Wei W, Morris M, Wei J, Gorbounov M, Wei C. New technology and clinical applications of nanomedicine. Med Clin North Am. 2007;91:845–62.
18. Li Z, Huang P, Lin J, et al. Arginine-glycine-aspartic acid-conjugated dendrimer-modified quantum dots for targeting and imaging melanoma. J Nanosci Nanotechnol. 2010;10:4859–67.
19. Nair A, Shen J, Thevenot P, et al. Enhanced intratumoral uptake of quantum dots concealed within hydrogel nanoparticles. Nanotechnology. 2008;19: 485102.
20. Zhang Q, Iwakuma N, Sharma P, et al. Gold nanoparticles as a contrast agent for in vivo tumor imaging with photoacoustic tomography. Nanotechnology. 2009;20:395102.
21. Aillon KL, Xie Y, El-Gendy N, et al. Effects of nanomaterial physicochemical properties on in vivo toxicity. Adv Drug Deliv Rev. 2009;61(6):457–66.
22. Kim C, Cho EC, Chen J, et al. In vivo molecular photoacoustic tomography of melanomas targeted by bio conjugated gold nanocages. ACS Nano. 2010;4(8):4559–64.
23. Schmieder AH, Winter PM, Caruthers SD, et al. Molecular MR imaging of melanoma angiogenesis with alphanubeta3-targeted paramagnetic nanoparticles. Magn Reson Med. 2005;53:621–7.
24. Longo C, Gambara G, Espina V, et al. A novel biomarker harvesting nanotechnology identifies Bak as a candidate melanoma biomarker in serum. Exp Dermatol. 2011;20:29–34.
25. Nasir A. Nanotechnology and dermatology: part I–potential of nanotechnology. Clin Dermatol. 2010;28:458–66.

13 The Skin Immune System

Adnan Nasir and Anthony Gaspari

13.1 Introduction

The skin is the largest organ in the body. It serves many functions, including thermoregulation, endocrine homeostasis, and transduction of environmental stimuli. The latter can be from simple registering of heat and cold to photoreception [1] or to the complex haptic processing required to read Braille. The skin actively and passively defends against chemical, thermal, electrical, radioactive, physical, and other environmental and microbial insults. The latter defenses fall under the broad purview of the skin immune system. Nanotechnology exploits the unique properties of matter on the nanoscale to selectively target the skin immune system, either for the purposes of augmenting immunity, in the case of immunodeficiency or in generating an immune response against a tumor or pathogen, or for the purposes of selectively inhibiting the immune system, for example, to treat autoimmune disease or prevent the rejection of a grafted organ.

A. Nasir, M.D., Ph.D. (✉)
Department of Dermatology, UNC Chapel Hill, 3100 Thurston Bowles Building, Chapel Hill, NC 27544, USA
e-mail: anasir@live.com

A. Gaspari, M.D.
Department of Dermatology, University of Maryland Medical Center,
Baltimore, MD 21201, USA

The components of the skin immune system are multifaceted, broad, and involve interactions of a variety of cell types and systems (Fig. 13.1). The skin immune system is finely tuned. Normal skin lacks any inflammation yet is teeming with cells involved in the immune response, suggesting tight regulatory control.

The skin immune system is broadly divided into innate and adaptive immunity. The former is considered evolutionarily ancient, rapid, and relatively nonspecific. The latter is viewed as a more recent evolutionary development, and is characterized by its power, specificity, adaptive nature, and memory. The innate immune system senses the presence of pathogen-specific molecules using pattern recognition receptors. When bound to their ligands, these receptors trigger responses leading to pathogen neutralization, for example, phagocytosis and remodeling by tissue-resident macrophages. They also trigger the adaptive immune system. The adaptive immune system relies on the presentation of antigen to T cells, activation of effector T cells, and on the stimulation of antibody-producing B cells. Feedback loops prevent responses from exceeding desired limits. These divisions, however, are arbitrary, and there is considerable overlap and interplay between the innate and adaptive components of the immune system. The influence of epigenetic changes on both innate and adaptive immunity is considerable. These include gene rearrangement of immunoglobulin genes and T cell receptor genes as well as regulatory RNAs such as miRNA.

A. Nasir et al. (eds.), *Nanotechnology in Dermatology*,
DOI 10.1007/978-1-4614-5034-4_13, © Springer Science+Business Media, LLC 2013

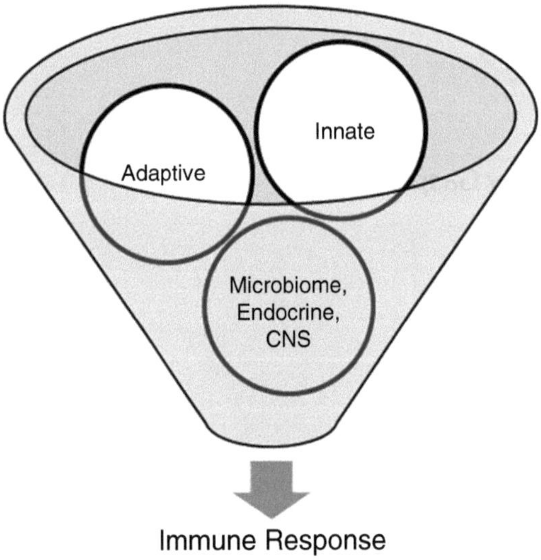

Fig. 13.1 The totality of the immune response is a summation of the activity of the innate immune system, the adaptive immune system, and related tissues and organ systems (adipose tissue, the central nervous system, the peripheral nervous system), and the microbial community of the host (microbiome).

The barrier function of the skin is an important aspect of the innate immune system. The majority of the stratum corneum on non-intertriginous and non-mucosal skin is dry. Cells of the stratum corneum are shed daily at a significant rate. Vital minerals such as iron are trapped in inaccessible forms on the skin [2]. Desiccation, shedding, and mineral sequestration deny invaders and irritants purchase. Nanoparticles which sequester allergens may, in the same fashion, reduce irritant and allergic potential. For example, topically applied calcium phosphate nanoparticles, which bind nickel, have been shown to reduce cutaneous persistence and penetration of nickel, and to reduce allergic contact dermatitis in animal models [3, 4]. Natural moisturizing factor and filaggrin, among other compounds, prevent transepidermal water loss and have antimicrobial properties [5]. Filaggrin is a highly soluble, histidine-rich 37 kD protein which is cleaved from a 400 kD precursor. Its gene is located in the Epidermal Differentiation Complex (EDC), a cluster of 60 genes on 1q21. Histidine serves as a humectant, pH regulator, and is antimicrobial. The altered pH from lack of filaggrin activates serine proteases like kallikrein 5, which favors Th2 cytokines. Nearly 10 % of Europeans harbor a loss-of-function mutation in filaggrin. Mutations are associated with an increased risk of atopic dermatitis, asthma, hay fever, peanut allergy, nickel contact dermatitis, and influence the severity of asthma, alopecia areata, and susceptibility to herpetic infection. Filaggrin deficiency permits passive protein allergen transfer across the epidermis (e.g., squinlan). Disruption of the barrier by physical or mechanical damage of keratinocytes, whether endogenous (from scratching) or exogenous (from trauma), elicits the release of defensive inflammatory cytokines [6].

Antimicrobial defenses include defense proteins called antimicrobial peptides. Examples include cathelicidin and human beta defensin [7]. The expression of these proteins is augmented in inflammatory skin diseases such as acne rosacea and psoriasis. Lack of antimicrobial peptides may lead to increased risk of infection or colonization with pathogens such as *Staphylococcus aureus*. Diminished levels of antimicrobial peptides are seen in diseases such as atopic dermatitis.

In the epidermis, Langerhans cells interdigitate with one another and with keratinocytes in a dense arbor of dendrites. En face sections of epidermis staining for Langerhans cells reveal a rich network of dendrites with an extensive reach over multiple planes of the epidermis, virtually assuring contact with any potential invaders in transit. Antigens coming into contact with Langerhans cells are endocytosed and presented on the cell surface in the context of major histocompatibility complex (MHC). These Langerhans cells exit the epidermis and migrate to draining lymph nodes to present the antigens to the T cells.

Components of the immune system in the dermis include recently described resident dermal dendritic cells, as well as any circulating T cells and B cells. Dermal dendritic cells (BDCA-1+), only recently described, are present in at least three subtypes, are potent antigen-presenting cells, and can become Langerhans cells (CD1a+). In fact, mononuclear cells of the skin may form a continuum between macrophages (CD163+) and dendritic cells depending on the microenvironment and may function in one of several modes: antigen-presenting mode, migratory mode, and phagocytic mode. One subset of dermal dendritic cells has been known to accumulate in psoriasis. Targeting different modes or subtypes may be a useful target for the treatment of skin diseases. Recent estimates suggest that nearly two-thirds of the circulating T cells in the body reside in the skin. Under the influence of cytokines and homing receptors, T cells can be stimulated to exit the circulation and accumulate in the dermis or the epidermis.

Subcutaneous fat is no longer regarded as a passive insulator or energy storage component of the skin [8, 9]. Adipose tissue is rich in T cells, B cells, and macrophages, and adipose tissue is one of the largest components of the skin endocrine immune system. Fat cells secrete adipokines such as leptin and adiponectin, and pro-inflammatory cytokines such as IL-6 and TNF-α which have a variety of effects on inflammation. Leptin is a 16-kD protein hormone, a product of the OB gene, and binds its receptor, a member of the type I cytokine superfamily. Receptor binding activates the Jak/AP-3 pathway. The receptor is expressed in injured keratinocytes. Leptin on T cells prevents apoptosis. And it lines endothelial vessels, where it upregulates the expression of adhesion molecules such as ICAM-1 and cadherin. Adiponectin is a secreted protein which can trimerize and oligomerize. It inhibits TNF-α and downregulates ICAM-1 in endothelial vessels. In psoriasis, the adipose tissue compartment of the hypodermis is often expanded, the levels of adiponectin are decreased, and the levels of leptin are often increased. The levels of IL-23 and IL-17 are increased in obese individuals. Angiotensinogen is made in the adipose tissue and is converted to angiotensin I and II. Angiotensin II has been shown to stimulate T cell proliferation, homing receptor expression, and the production of IFN-γ and TNF-α. Agents which target adipose tissue may be useful in the treatment of some inflammatory skin diseases as well as their systemic complications and comorbidities (e.g., cardiovascular disease, diabetes).

The role of diet in influencing the immune system has also been suggested. Studies of mice have shown that augmenting the diet with eicosapentaenoic acid (EPA) can increase adiponectin levels, decrease insulin resistance, and reduce TLR-2 expression and NFκB activation and translocation. The active bioflavinoid in green tea, epigallocatechin-3-gallate (EGCG), has been shown to reduce the signs of inflammation in experimental autoimmune encephalomyelitis in mice by decreasing Th1 and Th17 cells and increasing the so-called Tregs. TLR activated by pathogens has been shown to lead to the hydroxylation of 25(OH) Vitamin D, leading to an increase in cathelicidin expression. In this instance, nutrition and innate immunity converge to neutralize a pathogen in a positive feedback loop. Curcumin has been used to treat a variety of skin diseases including psoriasis and scleroderma. Variable results may have to do with bioavailability. Nanoemulsions are being developed which enhance topical delivery of curcumin. In animal studies, topical curcumin nanoemulsions have been shown to reduce ear swelling in a TPA assay and to inhibit cyclin D1 expression. Intravenous ω-3 fatty acids and oral inositol have shown some benefit in psoriasis. Alcohol may have a negative effect on psoriasis disease severity and expression. Patients with psoriasis can have elevated anti-gliadin IgG and IgA antibodies. In at least one study, patients with psoriasis and anti-gliadin antibodies placed on a gluten-free diet for 3 months noted symptomatic improvement in their disease. A variety of chemicals can exacerbate pemphigus including thiols, thiocyanates, phenols, and tannins. The high tannin content of Brasilian river water, which is neutralized by chlorination, may account for the discrepancy in pemphigus prevalence between rural and urban areas. Other environmental toxins affecting the skin immune system include tobacco. Smoking may be associated with rheumatoid arthritis, with the relative risk increasing nearly twelvefold with certain subtypes of HLA and PTPN22, possibly through the induction of antibodies to citrullinated peptides. Tobacco may also affect the efficacy of medications such as antimalarials.

The nervous system interacts with the skin immune system in a variety of ways. Noxious stimuli can be avoided, swatted, or scratched away. Sensations such as heat, pain, or itch can trigger healing behaviors such as rest and wound care. Pathogenic scratching or self-injurious behavior can trigger inflammation. Stress has been shown to alter the immune response to a variety of pathogens, including herpes simplex virus, as well as lower the threshold for itch. A number of receptors for neurotransmitters have been identified on the cells of the immune system and neuronal dendrites have been observed in the synaptic contact with lymphocytes [10].

The microbiome is the sum total of microorganisms residing in an individual. With the advent of genomics, it is now possible to sequence the entire genome of a bacterium (4 MBp) within 1 day. Partial sequences have revealed that we are hosts to many thousands of microbial species, including bacteria, viruses, yeast, and fungi [11, 12]. Some skin viruses have been associated with disease, such as the recently described association with Polyomavirus and Merkel cell carcinoma; a number of viruses have been discovered in skin from the polyoma family and other families which are not known to be associated with any disease, such as HPyV6, HPyV7, and Siphoviruses. Even within a single species, there is considerable genetic variation. For example, for E. coli, the variation in genetic material between pathogenic and nonpathogenic strains is up to 36 %. While microbial genomes are small, a substantial proportion of genetic material is functionally expressed. For example, for E. coli, about 90 % of the genome is coding; for *Mycobacterium leprae*, the number is closer to 50 %; and for the human genome, it is just under 1 %. Current estimates are that many thousands of species of microbes reside in the human body, with a significant number on the skin, including bacteria, fungi, arthropods, yeast, and viruses. While many of these organisms may be commensals, some, as symbionts, may actively alter host defense, host nutrient uptake, or host disease activity [13]. For example, the role of *Malassezia* species in the activity of psoriasis and atopic dermatitis and the role of *S. aureus* in the activity of atopic dermatitis are well known. The utility of probiotics in the prevention and management of atopic dermatitis remains unclear. Some species have been shown to contain genomic islands of 100–200 kb. Known as CRISPRs (contiguous regularly spaced interspersed short palindromic sequences), these may protect microbes from phage or virus attack or from predation by nematodes and protozoa.

Some may produce antimicrobial compounds, which may be useful for research, or may protect the host from pathogens, for example, nanoparticulate Siphoviruses which are bacteriophage and the potential pathogen of Propionibacterium acnes [14]. Another organism, *Staphylococcus epidermis*, generates phenol-soluble molecules such as PSM and γδ [15]. These form alpha helical domains which penetrate lipid membranes causing cell death via a mechanism analogous to the innate immune system's antimicrobial peptide LL-37. The discovery of new organisms in the epidermal microbiome may be helpful in future therapy and in the study of host defense.

Metagenomics is the analysis of the genome of the microbiome, and has led to important disease associations, for example, with Crohn's disease and with multiple sclerosis. The association of certain strains of pathogens and psoriasis has recently been established. The role of parasitic infection and immune regulation has also recently been described. In at least one study, administering eggs of *Trichinosis suis* has alleviated the symptoms of inflammatory bowel disease [16, 17].

Regulatory RNAs such as miRNAs have been found to be involved in feedback loops for the epigenetic regulation of gene activity. Specific miRNAs have been found to be important in skin diseases such as psoriasis and atopic dermatitis [18]. Small RNAs may be excellent targets for specific nanoparticle-based therapies such as siRNAs and aptamers [19].

As already mentioned, despite the abundance of lymphocytes, macrophages, dendritic cells, and monocytes in at least one layer of the skin, normal skin is not inflamed. Furthermore, despite the presence of a number of microorganisms on the stratum corneum, the skin is again not inflamed. Finally, if the innate or adaptive immune system reacts to an environmental assault, the response is tailored and self-limited. Thus, the skin immune system possesses a number of active and passive properties which maintain homeostasis.

Promotion of defense against pathogens. The innate immune system, lacking "danger" signals, remains ready but quiescent. If pathogens enter a defect in the skin barrier, they will activate pathogen-associated molecule receptors, such as TLR, and trigger an innate response, which may include antimicrobial peptides or phagocytosis. Furthermore, the innate immune system may activate antigen-presenting cells to begin the process of triggering adaptive immunity. Feedback loops eliminate further reaction once the threat has been neutralized.

The principles of nanotherapy typically involve targeted delivery using nanocarriers [20, 21]. The carriers themselves may have no intrinsic activity, their dimensions are in the nanoscopic size range, and they carry active ingredients in their interior or surface. They may be capable of releasing their contents in response to a variety of physiologic, pharmacologic, or physical stimuli (Fig. 13.2a–c). Examples of naturally occurring nanocarriers include transport proteins such as lipoproteins; carriers of cellular information such as exosomes; and viruses, which are nucleic acid carriers. Typical carriers circulate in the tissue by means of diffusion, and this is size dependent. For example, 50 nm particles diffuse at the rate of 3,000 nm/s. Diffusion can also be enhanced by temperature, charge, electricity, and magnetism. Carriers must be small enough to not be obstructed by the extracellular matrix. For example, albumin is around 10 nm, and easily moves in lymph. Nanocarriers in circulation should not inadvertently adhere to vessel walls, trigger clotting cascades, or embolize. Carriers should avoid undesirable consequences, such as active ingredient denaturation and inadvertent triggering of an immune response. Carriers should not accidentally release their contents in nontarget cells/tissues.

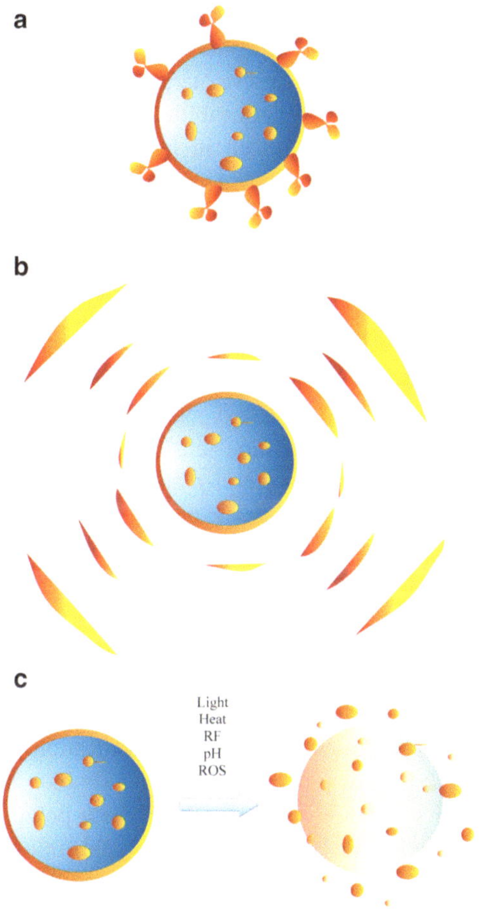

Fig. 13.2 (**a**) Nanoparticle delivery can be targeted using cell-specific high-affinity molecules such as ligands, aptamers, receptors, and monoclonal antibodies. (**b**) Nanoparticles can emit radiation in the form of light, which can be detected for imaging purposes, or heat, which can be used for therapeutic ablation. (**c**) Nanopolymers can be sensitive to triggers such as heat, magnetism, radiofrequency, laser light, or enzymatic activity to target release.

Nanocarriers should be designed to precisely regulate the uptake and release of their contents under reproducible conditions. Carriers can transport drugs which are poorly soluble, or protect drugs from a harsh extracellular or intracellular environment. They can be modified to remain in the circulation longer, or catabolized and excreted rapidly. Nanocarriers come in four varieties: hydrophilic short-lived payload, hydrophobic payload, soluble carriers, and field-responsive carriers. Examples of carriers with fragile hydrophilic payloads include liposomes, polymer vesicles, niosomes, and cubosomes. Carriers with hydrophobic cargo include microemulsions, polymer micelles, solid lipid nanoparticles, and polymer nanoparticles. Hydrophilic carriers include dendrimers, PEGylated proteins, and polymer conjugates. And field-responsive carriers include metal nanoparticles, magnetic nanoparticles, and quantum dots.

To direct specific localization of carriers, they can be coated in homing molecules. Examples include antibodies, receptor proteins, aptamers, ligands, and specific peptides. This is a type of active targeting. Passive targeting relies on enhanced plasma permeability and retention (EPPR), commonly seen in leaky vasculature, such as that encountered in areas of inflammation or neoangiogenesis.

Once the nanocarrier arrives at its destination, its payload needs to be released or activated. Sometimes this release can be automatic and responsive to particular microenvironmental conditions. For example, inflammation may entail the generation of reactive oxygen species. Carriers made with an oxidation-sensitive matrix may degrade in the presence of reactive oxygen species, releasing embedded active ingredients only at the sites of inflammation and not around healthy tissue. Oxidation-responsive polymers may, for example, have hydrophobic disulfide bonds, which degrade and disperse into smaller soluble chains containing hydrophilic sulfhydryl groups in the presence of ROS. These types of polymers are easily made through the emulsification and photooxidation of block copolymers which are either symmetric or asymmetric. These polymer carriers are stealthy as well, and are able to repel surface protein adhesion. If polymers are coupled to an enzyme, such as glucose oxidase, their oxidation-dependent release requires glucose. This type of enzymatically coupled release offers still more flexibility in carrier design.

Carriers can enter cells via passive permeation, caveolin-dependent endocytosis, clathrin-dependent endocytosis, macropinocytosis, and clathrin/caveolin-independent endocytosis. Carriers with a polycationic core can transport nucleic acids such as RNA. When these polymers are taken up into cells, the endosomal pH drops, the nucleic acid becomes protonated, and is released from the particles. Furthermore, polymers increase osmotic swelling of the endosome and become released into the cytoplasm.

Carriers can be thermosensitive. Certain polymers can be intact below a critical threshold temperature and dissolve into monomers above a critical threshold temperature. This type of temperature-dependent release of active ingredients can be controlled externally or physiologically (through heat generated via inflammation, or infection).

Some carriers can be made to respond to electromagnetic fields using a property of metallic nanoparticles known as surface plasmon resonance. Surface plasmon resonance can enormously and constructively amplify entrained wavelengths of light on a metallic nanoparticle whose size matches a resonant frequency of incident monochromatic light. Metallic particles exhibiting SPR can be coupled to targeting molecules such as ligands or antibodies. In the presence of their resonant wavelength of light, they emit heat or light of significant amplitude to initiate selective photothermolysis or bright beacons of light.

Nanomaterials can be attached to sensors and arrays for in vitro diagnosis, therapy, or simultaneous diagnosis and therapy (theranostics) [22–26].

While a comprehensive discussion of diseases involving the immune system is beyond the scope of this review, a few notable ones are worth mentioning.

Alopecia Areata is an autoimmune disorder of the hair follicles. Severe early onset is associated with HLA (DRB*0401; DQB1*0301). The immune etiology has been further established by grafting experiments of affected scalp onto nude mice, which show hair regrowth. Classically, hair may be a site of immune privilege. It does not express MHC-I or MHC-II molecules, and has a low LC, DC density. It is abundant in inhibitory cytokines such as MSH-α, TFG-β, and IGF-1. The target of the immune response in alopecia areata is unknown, but it may be pigment as the disease process tends to spare white hairs. Treatments vary and include topical steroids, systemic steroids, cyclosporine, minoxidil, biologic agents, anthralin, and the induction of allergic contact dermatitis using DNCB, DPCP, and SADBE. Nanoformulations of minoxidil, cyclosporine, and steroids may enhance topical delivery and minimize systemic side effects [27–30].

13.2 Contact Dermatitis

Contact dermatitis is associated with significant morbidity and mortality. Nickel is the most common cause of allergic contact dermatitis in the United States, accounting for nearly 10 % of the cases. Common sources of nickel include jewelry, metal snaps in clothing, coins, and even faceplates of cell phones. Systemic sources of nickel include foods such as chocolate. The effects of contact dermatitis due to nickel range from localized dermatitis to widespread disease requiring hospitalization. Nickel can be persistent on the skin, and the effects of dermatitis can persist for days to weeks beyond avoidance of the allergen. Strategies for the management of nickel contact dermatitis include avoidance, coating of metals containing nickel, and reducing or eliminating nickel in the diet. Nanoparticle barrier creams have shown significant affinity for nickel and diminished responses to topical nickel in a porcine model [4].

13.3 Atopic Dermatitis

Atopic dermatitis is extremely common with studies showing nearly 15–30 % of children and 2–10 % of adults in developed countries are affected. Nearly half of cases are diagnosed before the age of 6 months, and up to 85 % are diagnosed by the age of 5 years. Nearly 70 % of the cases undergo remission before adolescence.

The distribution of lesions may change over the course of a lifetime, with the cheeks and scalp dominating in infancy, and the flexures and dorsal limbs in childhood and adulthood, with lichenification more common in adults. Itching can be severe, chronic, and debilitating, often being worse at night and disrupting sleep.

Atopic dermatitis may be a predisposing factor to other diseases including asthma, allergic rhinitis, as well as contact dermatitis, such as nickel dermatitis. Patients with atopic dermatitis may be at greater risk for viral skin infections including molluscum contagiosum, verruca vulgaris, and disseminated herpes simplex. Patients with atopic dermatitis may lack natural defenses against bacterial pathogens such as *S. aureus* and may be predisposed to *S. aureus* colonization.

There is a strong concordance among monozygotic twins, and a number of genes have been identified which are associated with atopic dermatitis. These include the EDC cluster on 1q21, as well as genes which favor IgE production (e.g., cytokines IL-4,5,12,13; and GM-CSF). Genes encoding barrier function proteins have also been implicated in atopic dermatitis. These include filaggrin (FLG), which is associated with early onset atopic dermatitis associated with asthma. Filaggrin variants which resist proteolytic cleavage tend to be associated with atopic dermatitis, as the products of proteolysis contribute to the stratum corneum water-binding capacity and reduce transepidermal water loss. The expression of FLG is regulated and can be altered or reduced by inflammation. Other components of the barrier implicated in atopic dermatitis include involucrin, loricrin, and skin lipids. Disruptions in the skin's barrier may permit allergens and irritants to penetrate the skin, promoting an inflammatory or allergic response.

The skin contains roughly one million T cells per square centimeter of BSA. Keratinocytes in atopic dermatitis patients produce high levels of IL-7 like TSLP, which drives Th2 differentiation. This Th2 polarization yields high levels of GM-CSF and PGE2, which lead to skin sensitization, as well as sensitization of the lungs and the GI tract. This type of sensitization can be adoptively transferred.

T regs are abundant in the circulation of patients with atopic dermatitis, but their numbers are reduced or absent in the skin.

Dendritic cells contain IgE receptors and promote Th2 polarization after antigen capture.

Defects in innate immunity may contribute to the pathogenesis of atopic dermatitis. Cells in the epidermis, including keratinocytes, contain receptors for pathogens. These include TLR's, C-type lectins, NOD receptors, and PR receptors. At least ten different TLR's have been identified, capable of recognizing bacterial, viral, and fungal structures. Upon stimulation, TLRs can activate antimicrobial peptides including cathelicidins and defensins, such as cathelicidin LL-37 (which has potent antiviral activity); human beta defensins, HBD-1,2,3; and dermicidin. Antimicrobial peptide levels are downregulated by cytokines such as IL-4, IL-10, and IL-13, which may account for the colonization by *S. aureus*, *Malassezia*, and increased susceptibility to viral infection (from reduced cathelicidin LL-37).

Pathogens can exacerbate atopic dermatitis in several ways. For example, *S. aureus* ceramidase disrupts the skin's barrier. *S. aureus* enterotoxin promotes the formation of enterotoxin-specific

IgE. Enterotoxins act as superantigens, directly binding MHC and TCR. They induce T cell homing receptors such as CLAA; induce beta isoforms of glucocorticoid receptors (contributing to resistance to corticosteroid therapy); and induce TNF receptors on APC, preventing suppression from T regs.

Pruritus is a significant cause of morbidity in atopic dermatitis. Pruritus is mediated by inflammation, which may itself be triggered by scratching, which releases pro-inflammatory cytokines, and may also be mediated by neuropeptides, proteases, and kinins. IL-31 is highly pruritogenic and is upregulated by *S. aureus* exotoxins. Scratching also exposes keratinocytes and facilitates adhesion of *S. aureus*.

Patients with atopic dermatitis may express antibodies to foods and aeroallergens. About 25 % of patients express autoantibodies to manganese SOD and calcium-binding proteins. These intracellular KC proteins are released upon scratching. They may have antigenic characteristics which mimic microbial structures.

Strategies for the care of atopic dermatitis include prevention, genetic therapy, reducing colonization, reducing allergic contact dermatitis, reducing itch, and reducing inflammation. Personalized medicine may help tailor treatment to the disease state of the individual.

13.4 Autoimmune Diseases

A number of skin diseases have an autoimmune basis. These include lupus and psoriasis. Some of these diseases overlap with other disorders of autoimmunity, including type I diabetes mellitus, inflammatory bowel disease, and multiple sclerosis. These diseases share several features including chronic and intermittent bouts of inflammation, destruction of target organs, and immune complex deposition. Genetic and environmental factors may contribute to these diseases. Some observers believe that immune cell signaling is a balance of positive and negative regulation, and that in autoimmune diseases a quantitative threshold is required to tip the balance in favor of autoimmunity. Thus, a combination of genetic predisposition and epigenetic and environmental triggers precipitate autoimmunity. Genome-wide surveys have revealed more than 200 loci common to autoimmune disorders. These involve the MHC, the innate, and adaptive immune systems.

The immediate response of the immune system occurs via innate immunity. It is the phylogenetically older ancestral arm of the immune system and relies on pattern-associated receptors. The prevalence of autoimmune diseases may be increasing because of increased hygiene and lack of external pathogens to trigger the innate pathway. Intracellular sensors for bacterial peptidoglycan (such as NOD2) and genes which target microbes to lysosomes (such as ATG16L1) are associated with Crohn's. Eradication of intestinal parasites has been shown to increase the risk of Crohn's. Similar findings have been described for mouse models of autoimmune diabetes. Viral pathogens have been implicated in diabetes (picornavirus, which induces interferon-induced helicase C domain containing protein 1). Missense mutations in interferon-induced helicase C domain containing protein 1 protect against autoimmune diabetes. Furthermore, interferon-induced helicase C domain containing protein 1 is associated with psoriasis and IgA deficiency. Increased IRF5 (interferon regulatory factor) is associated with lupus erythematosus, rheumatoid arthritis, and primary biliary cirrhosis.

The adaptive immune system consists of many cells including T cells and B cells which contain highly diverse, specific, and powerful receptors. These are somatically generated, selected for affinity and avidity, and display memory. Problems in the adaptive immune system arise from faulty control of self-reactivity. Self-reactive T and B cells are kept in check by a variety of mechanisms. These include the so-called central mechanisms (elimination of autoreactive clones in the bone marrow and the thymus) and peripheral mechanisms (suppression of autoreactive clones by T regs). Genes involved in central control include TRAR for T cells and PTPN22 for pre-B cells. Genes involved in peripheral mechanisms include FOXP3. Ubiquitin is involved in immune signaling pathways and may quantitatively influence the threshold for autoimmunity.

Genes involved in this pathway include TNFAIP3, TNIP1, UBASH3A, and UBEL3. IL-2, IL-23, and IL-23R have been implicated in celiac disease, ulcerative colitis, rheumatoid arthritis, ankylosing spondylitis, and psoriasis.

While genetics and the environment both contribute to the development of autoimmune diseases, genetics are the basis of therapeutic avenues including: TNF inhibitors, B cell inhibitors such as rituximab, B7-CTLA4 inhibitors such as abatacept involved in T cell activation, anti-p40 antibodies in patients with psoriasis (with a strong IL-23 genetic signature), and agents which block lymphocyte trafficking, or kinases. Combination therapies may be required given the complexity of autoimmune diseases.

13.4.1 Bullous Dermatoses

Pemphigus vulgaris is caused by antibodies to desmoglein. The exact molecular basis of disruption of the epidermis is not exactly known. Therapies include immunosuppressive medication and plasmapheresis, which attempt to reduce the levels of circulating autoantibodies. Systemic therapies include corticosteroids; antimetabolites such as azathioprine, methotrexate, mycophenolate mofetil; alkylating agents such as cyclophosphamide; and calcineurin inhibitors such as cyclosporine. IVIg has shown benefit in the management of pemphibus vulgaris. Antibodies to the B cell marker CD 20 (rituximab) have also proven beneficial [31]. This is a chimeric monoclonal antibody which has been approved for the treatment of rheumatoid arthritis, and for B cell lymphoma. In a small study of patients with pemphigus vulgaris, 18 of the 21 recipients had remission, with nearly half, that is, eight of 18, experiencing disease-free intervals of up to 3 years. Treatment with rituximab resulted in a substantial drop in the circulating B cells, by up to 99 % for up to 6 months. These coincided with marked decreases in autoantibody titers of anti-desmoglein 1 and anti-desmoglein 3. No changes in the circulating T cells were noted, nor were there any changes in the IgG levels, in antibodies to pneumococcus, or tetanus toxoid. Side effects included pyelonephritis, sepsis, and death.

13.4.2 Psoriasis

Psoriasis has been reclassified as an immune-mediated inflammatory disease (IMID) [32]. Psoriasis is an autoimmune disorder. The initial trigger of immune activation in psoriasis is not clear, but the antigen is believed to be an epidermal antigen. Evidence includes adoptive transfer with BMT and adoptive cure with BMT. The best candidate antigen is β-hemolytic Streptococcus; oligoclonal T cells in patients with guttate psoriasis reacting to BHS cross react with keratins. There is recognition that psoriasis is associated with comorbidities in other organ systems, including cardiovascular disease, metabolic syndrome, and diabetes mellitus. Psoriasis is associated with Th1, Th17, and Th22 expansion and activation, leading to the production of cytokines such as IFN-γ, TNF-α, IL-17, and IL-22. Innate immunity in psoriasis is also in overdrive, classically through TNF-α/iNOS producing dendritic cells. By producing TNF, and IL-20, these dendritic cells drive keratinocytes and other cells to produce inflammatory cytokines. These cytokines can affect the liver (via CRP), the joints (via MMP's), adipose tissue, and blood vessels. The microbiome of psoriatic skin differs from normal skin, and whether this difference contributes to the disease state is not yet clear [12].

13.4.3 Urticaria

Chronic urticaria can be associated with a variety of causes (physical factors, drugs, foods, food additives, infections, systemic disease, and stress) in 20–30 % of patients, but for the vast majority it is idiopathic. Nanofiltered C1 has been used to treat hereditary angioedema [33]. Cyclosporine has been used in the treatment of urticaria, and nanoliposomal cyclosporine has shown greater efficacy and reduced toxicity [34, 35]. A significant proportion of patients with chronic idiopathic urticaria have antibodies to IgE or the high-affinity receptor to IgE. Patients with autoimmune urticaria tend not to respond to antihistamines. Nanoparticulate biologic antibodies such as omalizumab, rituximab, ofatumumab, TNF-α inhibitors, and immunosuppressive drugs such as methotrexate may be useful in the management of refractory

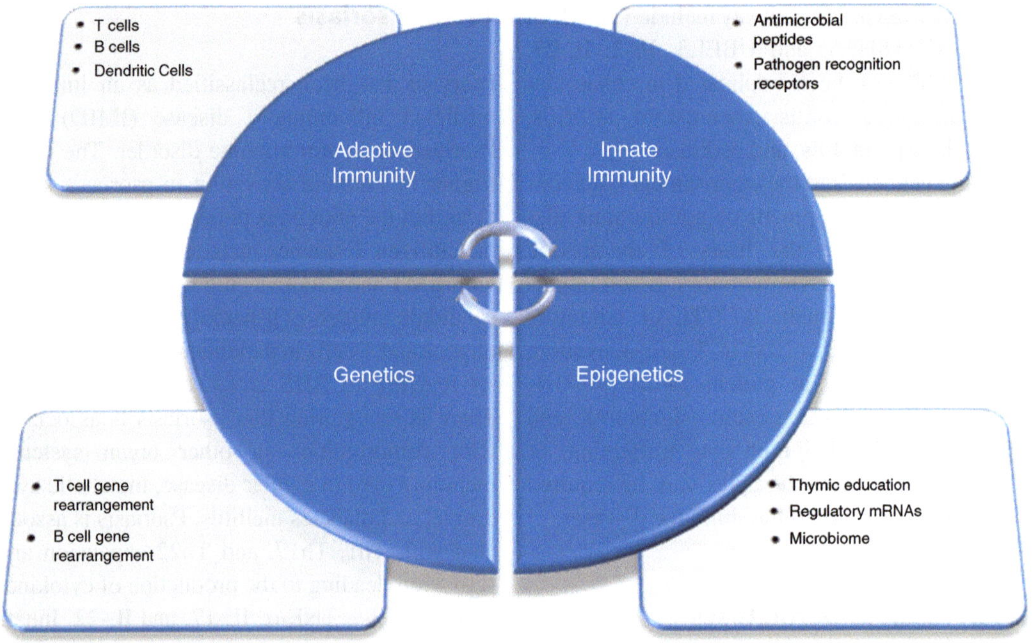

Fig. 13.3 The skin immune system is the result of close coordination and integration of multiple components. Each of these can serve as a target for the diagnosis of skin disease and for the selective enhancement or suppression of immunity.

urticaria [36–40]. Nanoformulations of methotrexate have been developed, including those for topical delivery [41–43]. Leukotriene inhibitors, such as 5-lipoxygenase inhibitors, have also shown some benefit in the management of urticaria [44]. Dapsone has been used in the treatment of urticaria, and nanoliposomal carriers of dapsone have been developed [45, 46]. Plasmapheresis has been used in the management of refractory urticaria [47, 48], and the use of magnetic nanofiltration may make this process more efficient with fewer side effects [49]. The symptoms of urticaria may be ameliorated by antihistamines, and controlled-release antihistamine nanodelivery systems have been developed with improved taste and desirable pharmacokinetics [50].

13.4.4 Vitiligo

Vitiligo vulgaris is caused by the autoimmune loss of melanocytes. Patients with vitiligo vulgaris are at slightly greater risk for other autoimmune disorders including: autoimmune thyroid disease, rheumatoid arthritis, psoriasis, adult-onset type 1 diabetes, pernicious anemia, systemic lupus erythematosus, and Addison's disease [51]. Genome-wide analysis shows links to HLA, especially HLA-A O2, -DR B1 04, and -DQ A1. HLA-A2 01 is able to present tyrosinase to autologous cytotoxic T lymphocytes, and is a good prognostic marker for melanoma. IL2-R A variants have been associated with vitiligo and other autoimmune diseases including type I diabetes, rheumatoid arthritis, multiple sclerosis, Grave's disease, and lupus. One variant of the tyrosinase protein associated with vitiligo is thermosensitive. The 402Q substitution is misfolded and easily degraded compared to the wild-type 402R enzyme [52].

In summary, the skin immune system comprises the innate immune system, the adaptive immune system, and the peripheral components including the endocrine system and the nervous system (Fig. 13.3). Genetic and epigenetic mechanisms dictate control of the skin immune system, and alterations can lead to disease. A variety of strategies to restore the skin immune system to health using nanotechnology are being explored.

References

1. Xiang Y, et al. Light-avoidance-mediating photoreceptors tile the Drosophila larval body wall. Nature. 2010;468(7326):921–6.
2. Dahl MV. Dermatophytosis and the immune response. J Am Acad Dermatol. 1994;31(3 Pt 2):S34–41.
3. Vemula PK, Anderson RR, Karp JM. Animal models for nickel allergy. Nat Nanotechnol. 2011;6(9):533.
4. Vemula PK, Anderson RR, Karp JM. Nanoparticles reduce nickel allergy by capturing metal ions. Nat Nanotechnol. 2011;6(5):291–5.
5. Garcia Bartels N, et al. Effect of standardized skin care regimens on neonatal skin barrier function in different body areas. Pediatr Dermatol. 2010;27(1):1–8.
6. Oyoshi MK, et al. Cellular and molecular mechanisms in atopic dermatitis. Adv Immunol. 2009;102:135–226.
7. Nakatsuji T, Gallo RL. Antimicrobial peptides: old molecules with new ideas. J Invest Dermatol. 2012;132(3 Pt 2):887–95.
8. Ozdemir M, et al. Serum leptin, adiponectin, resistin and ghrelin levels in psoriatic patients treated with cyclosporin. J Dermatol. 2012;39(5):443–8.
9. Gerdes S, et al. Leptin, adiponectin, visfatin and retinol-binding protein-4—mediators of comorbidities in patients with psoriasis? Exp Dermatol. 2012;21(1):43–7.
10. Arck PC, et al. Neuroimmunology of stress: skin takes center stage. J Invest Dermatol. 2006;126(8):1697–704.
11. Cho I, Blaser MJ. The human microbiome: at the interface of health and disease. Nat Rev Genet. 2012;13(4):260–70.
12. Kong HH, Segre JA. Skin microbiome: looking back to move forward. J Invest Dermatol. 2012;132(3 Pt 2):933–9.
13. Bilbo SD, et al. Reconstitution of the human biome as the most reasonable solution for epidemics of allergic and autoimmune diseases. Med Hypotheses. 2011; 77(4):494–504.
14. Lood R, Collin M. Characterization and genome sequencing of two Propionibacterium acnes phages displaying pseudolysogeny. BMC Genomics. 2011;12:198.
15. Otto M. Molecular basis of Staphylococcus epidermidis infections. Semin Immunopathol. 2012;34(2):201–14.
16. Motomura Y, et al. Helminth antigen-based strategy to ameliorate inflammation in an experimental model of colitis. Clin Exp Immunol. 2009;155(1):88–95.
17. Wolff MJ, Broadhurst MJ, Loke P. Helminthic therapy: improving mucosal barrier function. Trends Parasitol. 2012;28(5):187–94.
18. Bak RO, Mikkelsen JG. Regulation of cytokines by small RNAs during skin inflammation. J Biomed Sci. 2010;17:53.
19. Ng EW, et al. Pegaptanib, a targeted anti-VEGF aptamer for ocular vascular disease. Nat Rev Drug Discov. 2006;5(2):123–32.
20. Nasir A. Nanodermatology: a bright glimpse just beyond the horizon—part I. Skin Therapy Lett. 2010;15(8):1–4.
21. Ulbrich W, Lamprecht A. Targeted drug-delivery approaches by nanoparticulate carriers in the therapy of inflammatory diseases. J R Soc Interface. 2010;7 Suppl 1:S55–66.
22. Kunjachan S, et al. Theranostic systems and strategies for monitoring nanomedicine-mediated drug targeting. Curr Pharm Biotechnol. 2012;13(4):609–22.
23. Amiri H, Mahmoudi M, Lascialfari A. Superparamagnetic colloidal nanocrystal clusters coated with polyethylene glycol fumarate: a possible novel theranostic agent. Nanoscale. 2011;3(3):1022–30.
24. Caldorera-Moore ME, Liechty WB, Peppas NA. Responsive theranostic systems: integration of diagnostic imaging agents and responsive controlled release drug delivery carriers. Acc Chem Res. 2011;44(10):1061–70.
25. Puri A, Blumenthal R. Polymeric lipid assemblies as novel theranostic tools. Acc Chem Res. 2011;44(10):1071–9.
26. Yoo D, et al. Theranostic magnetic nanoparticles. Acc Chem Res. 2011;44(10):863–74.
27. Kim ST, et al. Topical administration of cyclosporin A in a solid lipid nanoparticle formulation. Pharmazie. 2009;64(8):510–4.
28. Baspinar Y, Keck CM, Borchert HH. Development of a positively charged prednicarbate nanoemulsion. Int J Pharm. 2010;383(1–2):201–8.
29. Padois K, et al. Solid lipid nanoparticles suspension versus commercial solutions for dermal delivery of minoxidil. Int J Pharm. 2011;416(1):300–4.
30. Gelfuso GM, et al. Chitosan microparticles for sustaining the topical delivery of minoxidil sulphate. J Microencapsul. 2011;28(7):650–8.
31. Ahmed AR, et al. Treatment of pemphigus vulgaris with rituximab and intravenous immune globulin. N Engl J Med. 2006;355(17):1772–9.
32. Zippin JH. The genetics of psoriasis. J Drugs Dermatol. 2009;8(4):414–7.
33. Hofstra JJ, et al. Treatment of hereditary angioedema with nanofiltered C1-esterase inhibitor concentrate (Cetor(R)): multi-center phase II and III studies to assess pharmacokinetics, clinical efficacy and safety. Clin Immunol. 2012;142(3):280–90.
34. Hollander SM, Joo SS, Wedner HJ. Factors that predict the success of cyclosporine treatment for chronic urticaria. Ann Allergy Asthma Immunol. 2011;107(6):523–8.
35. Czogalla A. Oral cyclosporine A—the current picture of its liposomal and other delivery systems. Cell Mol Biol Lett. 2009;14(1):139–52.
36. Sanchez-Machin I, et al. T cell activity in successful treatment of chronic urticaria with omalizumab. Clin Mol Allergy. 2011;9:11.
37. Sagi L, et al. Evidence for methotrexate as a useful treatment for steroid-dependent chronic urticaria. Acta Derm Venereol. 2011;91(3):303–6.
38. Czuczman MS, et al. Ofatumumab monotherapy in rituximab-refractory follicular lymphoma: results from a multicenter study. Blood. 2012;119(16):3698–704.
39. Taylor PC, et al. Ofatumumab, a fully human anti-CD20 monoclonal antibody, in biological-naive,

rheumatoid arthritis patients with an inadequate response to methotrexate: a randomised, double-blind, placebo-controlled clinical trial. Ann Rheum Dis. 2011;70(12):2119–25.
40. Ivyanskiy I, Sand C, Francis ST. Omalizumab for chronic urticaria: a case series and overview of the literature. Case Rep Dermatol. 2012;4(1):19–26.
41. Concannon C, et al. Nanoemulsion encapsulation and in vitro SLN models of delivery for cytotoxic methotrexate. Curr Drug Discov Technol. 2010;7(2):123–36.
42. Corem-Salkmon E, et al. Convection-enhanced delivery of methotrexate-loaded maghemite nanoparticles. Int J Nanomedicine. 2011;6:1595–602.
43. Lin YK, et al. Combination of calcipotriol and methotrexate in nanostructured lipid carriers for topical delivery. Int J Nanomedicine. 2010;5:117–28.
44. Di Lorenzo G, et al. Leukotriene receptor antagonists in monotherapy or in combination with antihistamines in the treatment of chronic urticaria: a systematic review. J Asthma Allergy. 2008;2:9–16.
45. Chougule M, Padhi B, Misra A. Development of spray dried liposomal dry powder inhaler of Dapsone. AAPS PharmSciTech. 2008;9(1):47–53.
46. Noda S, Asano Y, Sato S. Long-term complete resolution of severe chronic idiopathic urticaria after dapsone treatment. J Dermatol. 2012;39(5):496–7.
47. Grattan CE, et al. Plasmapheresis for severe, unremitting, chronic urticaria. Lancet. 1992;339(8801):1078–80.
48. Jiang X, et al. A case report of double-filtration plasmapheresis for the resolution of refractory chronic urticaria. Ther Apher Dial. 2008;12(6):505–8.
49. Herrmann IK, et al. Device for continuous extracorporeal blood purification using target-specific metal nanomagnets. Nephrol Dial Transplant. 2011;26(9):2948–54.
50. Li FQ, et al. Cetirizine dihydrochloride loaded microparticles design using ionotropic cross-linked chitosan nanoparticles by spray-drying method. Arch Pharm Res. 2010;33(12):1967–73.
51. Spritz RA. Six decades of vitiligo genetics: genome-wide studies provide insights into autoimmune pathogenesis. J Invest Dermatol. 2012;132(2):268–73.
52. Berson JF, et al. A common temperature-sensitive allelic form of human tyrosinase is retained in the endoplasmic reticulum at the nonpermissive temperature. J Biol Chem. 2000;275(16):12281–9.

14 Diagnosis and the Skin Immune System

Adnan Nasir and Anthony Gaspari

14.1 Introduction

The traditional means of diagnosis of inflammatory skin diseases include physical examination, diagnostic tests such as skin biopsy for histology and immunofluorescence, and a variety of special serologic assays. Tools and techniques utilizing nanotechnology are designed to utilize less material; be minimally invasive; and offer rapid, sensitive, and specific results.

Skin biopsies are painful, requiring the injection of local anesthetic; are invasive, requiring a scalpel or punch tool; and often leave scars. Some diseases may require multiple biopsies for proper diagnosis. For example, autoimmune bullous diseases often require lesional tissue for routine histology, and perilesional tissue for immunofluorescence.

Using nanomethodologies, tissue assessment can be noninvasive or minimally invasive, and has the potential to offer rapid and reliable results. Visualization of the skin can be enhanced with optical methods that rely on the metabolic activity of the skin or the fine structure of the skin. For example, near-infrared imaging techniques can enhance visualization of carbon nanotubes, quantum dots or gold particles with near-infrared surface plasmon resonance [1, 2]. Quantum dots coupled to cell specific ligands can allow for the detection of cell types in vivo [3–5]. Optical methods can be used to diagnose inflammatory diseases as well as assess response to therapy [6–9]. In the case of gold nanoparticle ablation using surface plasmon resonance, the technique is theranostic, leading to identification and simultaneous selective photothermolysis of ligand-bearing cells and tissues. The latter has been demonstrated successfully for melanoma, but the principal can be applied to augment or suppress immunity [10, 11]. For example, rituximab binds CD20 leading to B cell depletion and has shown benefit in some cases of lupus. Gold nanoparticles coupled to rituximab have been used for surface-enhanced Raman scattering for the visualization of the CD20-bearing chronic lymphocytic leukemia cells and may be useful for immunophenotyping [12]. Microneedles and carbon nanotubes can be used for transdermal delivery and sampling [13–16]. Microneedles have been used in a variety of other topical delivery and sensing studies [13, 17–27].

Serologic assays vary in their sensitivity and processing time. Serologic assays can require significant quantities of sample, especially if an extensive panel is ordered. Optical analysis using gold nanorods has single-molecule detection capability, and modified assays using this technique would require extremely small amounts of

A. Nasir, MD, PhD (✉)
Department of Dermatology, UNC Chapel Hill,
3100 Thurston Bowles Bldg, Chapel Hill, NC, 27544, USA
e-mail: anasir@live.com

A. Gaspari, MD
Department of Dermatology, University of Maryland Medical Center,
Baltimore, MD, 21201, USA

A. Nasir et al. (eds.), *Nanotechnology in Dermatology*,
DOI 10.1007/978-1-4614-5034-4_14, © Springer Science+Business Media, LLC 2013

material [28]. Binding molecules can include biotin, antibodies, receptors, ligands, and aptamers.

Aptamers can be made of single-stranded DNA or RNA. Aptamers fold into unique three-dimensional conformations based on their sequence and length. The resulting macromolecule can be selected for a strong dissociation constant for a desired compound (**Fig.** 14.1).

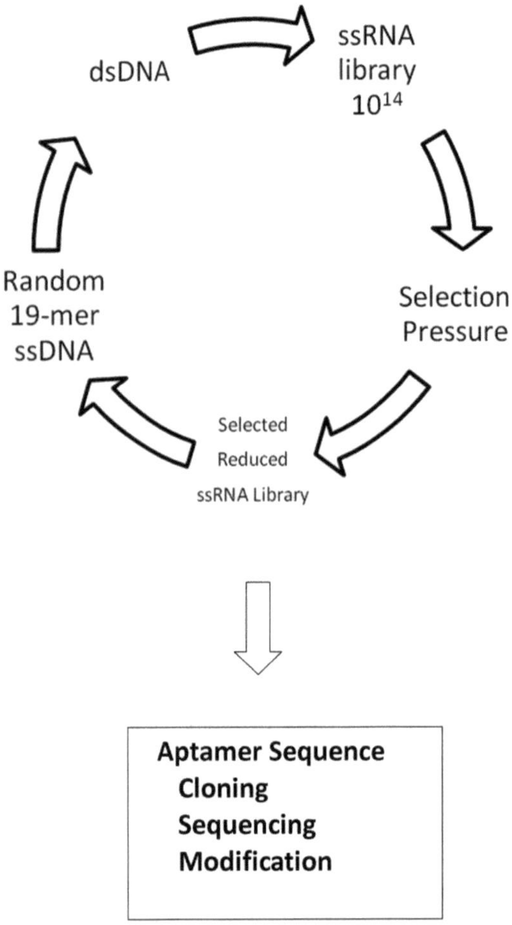

Fig. 14.1 Aptamers are single-stranded nucleic acids (DNA or RNA) that form unique secondary and tertiary structures. These structures can be iteratively selected from a large library of random oligomers for binding affinity to a desired target (metal ion, drug, carbohydrate moiety, lipid, protein, nucleic acid). The aptamer with the greatest binding affinity can then be cloned and synthesized in large quantities. Aptamers can be functionalized with organic molecules (antibodies, ligands, carbon nanotubes, toxins) and inorganic molecules (gold nanoparticles) to add to their capabilities as sensors or bioactive compounds.

Aptamers are generated from an oligonucleotide library comprising random 25-nucleotide-long oligomers of random composition. This allows for the generation of 10^{15} different and unique oligonucleotides. A 40-nucleotide aptamer library could generate a repertoire of 10^{24} unique sequences. Large families of aptamers can be generated rapidly and be selected for binding to a variety of molecules including organic compounds, inorganic compounds, ions, proteins, and drugs. Iterative rounds of selection can yield aptamers with high selectivity and affinity for their target [29]. The utility of aptamers has been demonstrated for plasma protein detection in the picomolar range [30]. Using T4 ligase, aptamers can be bound to chips to make arrays and biosensors [29–32]. Examples of molecules recognized by customers include proteins such as the HIV tat and rev proteins, vEGF, basic FGF, and IgE. Peptides, drugs such as neomycin, vitamins, and inorganic compounds such as metal ions have also been bound with high specificity, selectivity, and affinity to aptamers. Unlike monoclonal antibodies, aptamers do not require animals, can be generated with high purity, have little or no batch-to-batch variability, are highly stable, and are very small in size. Arrays can be used for multiple diagnoses, for example, in assessing cytokine profiles [31, 33–35], larger proteins and macromolecules [36], or in distinguishing subsets of autoimmune diseases [37] as well as infections [38].

Arrays have also been used for the identification of genes associated with psoriasis [39]. Chromatin immunoprecipitation techniques have explored networks of transcription pathways associated with squamous cell carcinoma [40]. Newer methods of analysis of expression profiles (RNAseq) and rapid next-generation, automated DNA-sequencing technology can allow for high output analysis of mutations and may have implications for prognosis, identifying subsets of disease, and for pharmacogenomics [41–43]. Microchip arrays using nanoparticles have also demonstrated utility in the diagnosis of allergic disease and for the detection and diagnosis of HPV [36, 44]. Theranostic aptamers for simultaneous imaging and targeting of tumors have been developed [45]. Rapid high-throughput screening methods can facilitate the rapid identification and development of aptamers [46]. The hedgehog pathway is regulated by miRNAs, and peptides or aptamers could be used for therapy [47].

Drug eruptions are typically caused by immune-mediated hypersensitivity reactions. In many instances, specific drug-reactive T cells are associated with them. Certain HLA types may be associated with susceptibility to reactions by some drugs, and HLA screening may be cost-effective in pretreatment pharmacogenomics [48, 49]. For example, HLA-B*5701 is associated with hypersensitivity reactions against abacavir in HIV patients, and HLA-B*1502 is associated with carbamazepine hypersensitivity in Han Chinese [50, 51].

14.2 Theranostics

The simultaneous diagnosis and therapy of disease using nanoparticles is termed theranostics (Fig. 14.2). For example, quantum dot-based multiplexed array technology has been used for the ex vivo detection of low-affinity epitope autoreactive T cells in diabetes [52]. This type of assay could be adapted for use in the detection of drug-reactive T cells. Nanomolecular mimics of autoreactive epitopes have been developed, and these can be targeted to autoreactive lymphocytes for simultaneous diagnosis and therapy of vulnerability to drug hypersensitivity [53]. Other examples can be found in imaging studies as well as in transdermal drug delivery devices [54–58]. In the case of malignant tumors, metabolic utilization of folate differs from their normal counterparts. A malignant epidermal cell line (KB) was targeted in this fashion using a dual-ligand theranostic gold nanoparticle [59]. Melanocyte-stimulating hormone coupled to gold nanoshells has been used for the ablation of melanoma with sparing of surrounding tissue in murine models [10].

In summary, the diagnosis of the skin immune system using nanotechnology involves specific markers, which deliver a high signal-to-noise ratio, and can be assessed visually, using special optics, or through the use of microarrays. The hallmarks of these technologies are compactness, sensitivity, speed, minimal specimen quantity, high throughput, and multiplexing. In many instances, therapy is directly coupled to diagnosis (theranostics).

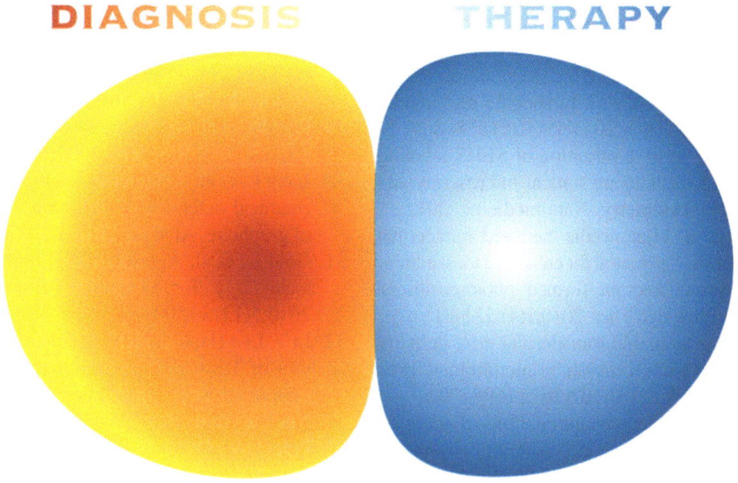

Fig. 14.2 Theranostic nanoparticles typically have dual functionality. One component (orange on the left) is used to target a specific cell or tissue type for diagnosis (e.g., MSH for melanocytes), and another (blue on the right) is used to treat the disease (e.g., a gold nanoshell heated by surface plasmon resonance using near infrared light).

References

1. Wang J, et al. Fluorophore-gold nanoparticle complex for sensitive optical biosensing and imaging. Nanotechnology. 2012;23(9):095501.
2. Welsher K, et al. A route to brightly fluorescent carbon nanotubes for near-infrared imaging in mice. Nat Nanotechnol. 2009;4(11):773–80.
3. Erogbogbo F, et al. Bioconjugation of luminescent silicon quantum dots for selective uptake by cancer cells. Bioconjug Chem. 2011;22(6):1081–8.
4. Haas P, et al. Fast quantitative single-molecule detection at ultralow concentrations. Anal Chem. 2010;82(14):6299–302.
5. Mogensen M, et al. Optical coherence tomography for imaging of skin and skin diseases. Semin Cutan Med Surg. 2009;28(3):196–202.
6. Zulfakar MH, et al. In vivo response of GsdmA3Dfl/+ mice to topically applied anti-psoriatic agents: effects on epidermal thickness, as determined by optical coherence tomography and H&E staining. Exp Dermatol. 2011;20(3):269–72.
7. Aydin SZ, et al. Optical coherence tomography: a new tool to assess nail disease in psoriasis? Dermatology. 2011;222(4):311–3.
8. Buder K, Knuschke P, Wozel G. Evaluation of methylprednisolone aceponate, tacrolimus and combination thereof in the psoriasis plaque test using sum score, 20-MHz-ultrasonography and optical coherence tomography. Int J Clin Pharmacol Ther. 2010;48(12):814–20.
9. Egawa M, et al. In vivo characterization of the structure and components of lesional psoriatic skin from the observation with Raman spectroscopy and optical coherence tomography: a pilot study. J Dermatol Sci. 2010;57(1):66–9.
10. Lu W, et al. Targeted photothermal ablation of murine melanomas with melanocyte-stimulating hormone analog-conjugated hollow gold nanospheres. Clin Cancer Res. 2009;15(3):876–86.
11. Pastor F, et al. Targeting 4-1BB costimulation to disseminated tumor lesions with bi-specific oligonucleotide aptamers. Mol Ther. 2011;19(10):1878–86.
12. Maclaughlin CM. et al. Evaluation of SERS labeling of CD20 on CLL cells using optical microscopy and fluorescence flow cytometry. Nanomedicine. 2012
13. Donnelly RF, et al. Microneedle-mediated intradermal nanoparticle delivery: potential for enhanced local administration of hydrophobic pre-formed photosensitisers. Photodiagnosis Photodyn Ther. 2010;7(4):222–31.
14. Wu J, et al. Programmable transdermal drug delivery of nicotine using carbon nanotube membranes. Proc Natl Acad Sci U S A. 2010;107(26):11698–702.
15. Singh R, Nalwa HS. Medical applications of nanoparticles in biological imaging, cell labeling, antimicrobial agents, and anticancer nanodrugs. J Biomed Nanotechnol. 2011;7(4):489–503.
16. Degim IT, Burgess DJ, Papadimitrakopoulos F. Carbon nanotubes for transdermal drug delivery. J Microencapsul. 2010;27(8):669–81.
17. Al-Qallaf B, Das DB. Optimizing microneedle arrays to increase skin permeability for transdermal drug delivery. Ann N Y Acad Sci. 2009;1161:83–94.
18. Badran MM, Kuntsche J, Fahr A. Skin penetration enhancement by a microneedle device (Dermaroller) in vitro: dependency on needle size and applied formulation. Eur J Pharm Sci. 2009;36(4–5):511–23.
19. Bal SM, et al. Influence of microneedle shape on the transport of a fluorescent dye into human skin in vivo. J Control Release. 2010;147(2):218–24.
20. Chen H, et al. Iontophoresis-driven penetration of nanovesicles through microneedle-induced skin microchannels for enhancing transdermal delivery of insulin. J Control Release. 2009;139(1):63–72.
21. Donnelly RF. Re: microneedle-mediated intradermal delivery of 5-aminolevulinic acid. J Control Release. 2008;129(3):153.
22. Donnelly RF, et al. Microneedle arrays permit enhanced intradermal delivery of a preformed photosensitizer. Photochem Photobiol. 2009;85(1):195–204.
23. Donnelly RF, et al. Microneedle-mediated intradermal delivery of 5-aminolevulinic acid: potential for enhanced topical photodynamic therapy. J Control Release. 2008;129(3):154–62.
24. Donnelly RF, et al. Microneedle arrays allow lower microbial penetration than hypodermic needles in vitro. Pharm Res. 2009;26(11):2513–22.
25. Farnworth TK, et al. Comparison of skin necrosis in rats by using a new microneedle electrocautery, standard-size needle electrocautery, and the Shaw hemostatic scalpel. Ann Plast Surg. 1993; 31(2):164–7.
26. Miller PR, et al. Multiplexed microneedle-based biosensor array for characterization of metabolic acidosis. Talanta. 2012;88:739–42.
27. Windmiller JR, et al. Microneedle array-based carbon paste amperometric sensors and biosensors. Analyst. 2011;136(9):1846–51.
28. Zijlstra P, Paulo PM, Orrit M. Optical detection of single non-absorbing molecules using the surface plasmon resonance of a gold nanorod. Nat Nanotechnol. 2012;7(6):379–82.
29. Platt M, et al. Aptamer evolution for array-based diagnostics. Anal Biochem. 2009;390(2):203–5.
30. Jiang L, et al. Aptamer-based highly sensitive electrochemical detection of thrombin via the amplification of graphene. Analyst. 2012;137(10):2415–20.
31. Gold L, et al. Aptamer-based multiplexed proteomic technology for biomarker discovery. PLoS One. 2010;5(12):e15004.
32. Li Y, Lee HJ, Corn RM. Fabrication and characterization of RNA aptamer microarrays for the study of protein-aptamer interactions with SPR imaging. Nucleic Acids Res. 2006;34(22):6416–24.
33. Potuckova L, et al. Rapid and sensitive detection of cytokines using functionalized gold nanoparticle-based immuno-PCR, comparison with immuno-PCR and ELISA. J Immunol Methods. 2011;371(1–2):38–47.
34. Jensen GC, et al. Inkjet-printed gold nanoparticle electrochemical arrays on plastic. Application to

immunodetection of a cancer biomarker protein. Phys Chem Chem Phys. 2011;13(11):4888–94.
35. Liu Y, et al. Micropatterned aptasensors for continuous monitoring of cytokine release from human leukocytes. Anal Chem. 2011;83(21):8286–92.
36. Simion M, et al. Detection of human papilloma viruses using nanostructurated silicon support in microarray technology. J Nanosci Nanotechnol. 2011;11(10):9102–9.
37. Kreuter A, et al. Expression of antimicrobial peptides in different subtypes of cutaneous lupus erythematosus. J Am Acad Dermatol. 2011;65(1):125–33.
38. Zelada-Guillen GA, et al. Label-free detection of Staphylococcus aureus in skin using real-time potentiometric biosensors based on carbon nanotubes and aptamers. Biosens Bioelectron. 2012;31(1):226–32.
39. Ellinghaus D, et al. Combined analysis of genome-wide association studies for Crohn disease and psoriasis identifies seven shared susceptibility loci. Am J Hum Genet. 2012;90(4):636–47.
40. Botti E, et al. Developmental factor IRF6 exhibits tumor suppressor activity in squamous cell carcinomas. Proc Natl Acad Sci U S A. 2011;108(33):13710–5.
41. Mellmann A, et al. Prospective genomic characterization of the German enterohemorrhagic Escherichia coli O104:H4 outbreak by rapid next generation sequencing technology. PLoS One. 2011;6(7):e22751.
42. Rothberg JM, et al. An integrated semiconductor device enabling non-optical genome sequencing. Nature. 2011;475(7356):348–52.
43. Elliott AM, et al. Rapid detection of the ACMG/ACOG-recommended 23 CFTR disease-causing mutations using ion torrent semiconductor sequencing. J Biomol Tech. 2012;23(1):24–30.
44. Teste B, et al. Microchip integrating magnetic nanoparticles for allergy diagnosis. Lab Chip. 2011;11(24):4207–13.
45. Kim JK, et al. Molecular imaging of a cancer-targeting theragnostics probe using a nucleolin aptamer- and microRNA-221 molecular beacon-conjugated nanoparticle. Biomaterials. 2012;33(1):207–17.
46. Dausse E, et al. HAPIscreen, a method for high-throughput aptamer identification. J Nanobiotechnology. 2011;9:25.
47. Katoh Y, Katoh M. Hedgehog signaling, epithelial-to-mesenchymal transition and miRNA (review). Int J Mol Med. 2008;22(3):271–5.
48. Bharadwaj M, et al. Drug hypersensitivity and human leukocyte antigens of the major histocompatibility complex. Annu Rev Pharmacol Toxicol. 2012;52:401–31.
49. Wei CY, et al. A recent update of pharmacogenomics in drug-induced severe skin reactions. Drug Metab Pharmacokinet. 2012;27(1):132–41.
50. Norcross MA, et al. Abacavir induces loading of novel self-peptides into HLA-B*57: 01: an autoimmune model for HLA-associated drug hypersensitivity. AIDS. 2012;26(11):F21–9.
51. Chung WH, Hung SI. Recent advances in the genetics and immunology of Stevens-Johnson syndrome and toxic epidermal necrosis. J Dermatol Sci. 2012;66(3):190–6.
52. Unger WW, et al. Discovery of low-affinity preproinsulin epitopes and detection of autoreactive CD8 T-cells using combinatorial MHC multimers. J Autoimmun. 2011;37(3):151–9.
53. DeNardo GL, et al. Nanomolecular HLA-DR10 antibody mimics: a potent system for molecular targeted therapy and imaging. Cancer Biother Radiopharm. 2008;23(6):783–96.
54. Amiri H, Mahmoudi M, Lascialfari A. Superparamagnetic colloidal nanocrystal clusters coated with polyethylene glycol fumarate: a possible novel theranostic agent. Nanoscale. 2011;3(3):1022–30.
55. Caldorera-Moore ME, Liechty WB, Peppas NA. Responsive theranostic systems: integration of diagnostic imaging agents and responsive controlled release drug delivery carriers. Acc Chem Res. 2011;44(10):1061–70.
56. Gittard SD. et al. Multiphoton microscopy of transdermal quantum dot delivery using two photon polymerization-fabricated polymer microneedles. Faraday Discuss. 2011;**149**:171–85; discussion 227–45
57. Puri A, Blumenthal R. Polymeric lipid assemblies as novel theranostic tools. Acc Chem Res. 2011;44(10):1071–9.
58. Yoo D. et al. Theranostic magnetic nanoparticles. Acc Chem Res. 2011
59. Li X, et al. Enhancement of cell recognition in vitro by dual-ligand cancer targeting gold nanoparticles. Biomaterials. 2011;32(10):2540–5.

15 Augmenting the Skin Immune System

Adnan Nasir and Anthony Gaspari

15.1 Introduction

Manipulation of the immune system for clinical dermatology can be divided into three categories: augmentation of immunity, suppression of immunity, or alteration of immunity such as induction of tolerance. Current methods for augmenting immunity include vaccination, medications, or interventions such as stem cell transplantation. Traditional vaccination relies upon administration of either antigens, inactivated or attenuated, or live organisms, with or without adjuvants. In the case of inherited immune deficiencies, gene therapy or bone marrow transplants have been performed. In the case of acquired immune deficiencies due to infection (such as human immunodeficiency virus), drugs directed at the causative organism have been used. In the case of acquired immune deficiency due to malignancy, antitumor medications have been indicated. While these therapies tend to be effective for their respective indications, they may lack efficiency, may lack a robust and sustained change in immune status, or may lack specificity and entail significant side effects or unintended effects.

There are many classifications of the immune system, but the broadest one divides it into two arms: innate and adaptive immunity. The innate immune system is triggered by molecular signals of pathogens binding pathogen recognition receptors located in all compartments of a cell from the cell membrane to the cytosol. Members of the receptor family include TLR, CLR, NLR, and DAI. Innate responses tend to be rapid and less specific, but recruit components of the adaptive immune system for a delayed, more specific, and more sustained response with memory.

Activation of the immune system requires three signals: the antigen, a co-stimulatory signal, or a third or danger signal (Fig. 15.1). Examples of danger signals include CpG, LPS, or I:C, which activate the pathogen recognition receptors TLR9, TLR4, andTLR3, respectively. Pathogen recognition receptors are molecular sensors that detect a variety of pathogens including bacteria, viruses, fungi, and parasites. They are located at diverse sites outside the cell, on the cell surface, and intracellularly. These receptors mediate innate immune activation. TLR families include the toll-like receptors, C-type lectin receptors, nucleotide oligomerization domain-like receptors, and retinoic acid-inducible gene-based-like receptors. Their interactions can be complex and either synergistic or antagonistic.

Examples of TLR ligands include bacterial cell wall components such as peptidoglycan, lipoprotein,

A. Nasir, M.D., Ph.D. (✉)
Department of Dermatology, UNC Chapel Hill, 3100 Thurston Bowles Building, Chapel Hill, NC 27544, USA
e-mail: anasir@live.com

A. Gaspari, M.D.
Department of Dermatology, University of Maryland Medical Center,Baltimore, MD 21201, USA

A. Nasir et al. (eds.), *Nanotechnology in Dermatology*,
DOI 10.1007/978-1-4614-5034-4_15, © Springer Science+Business Media, LLC 2013

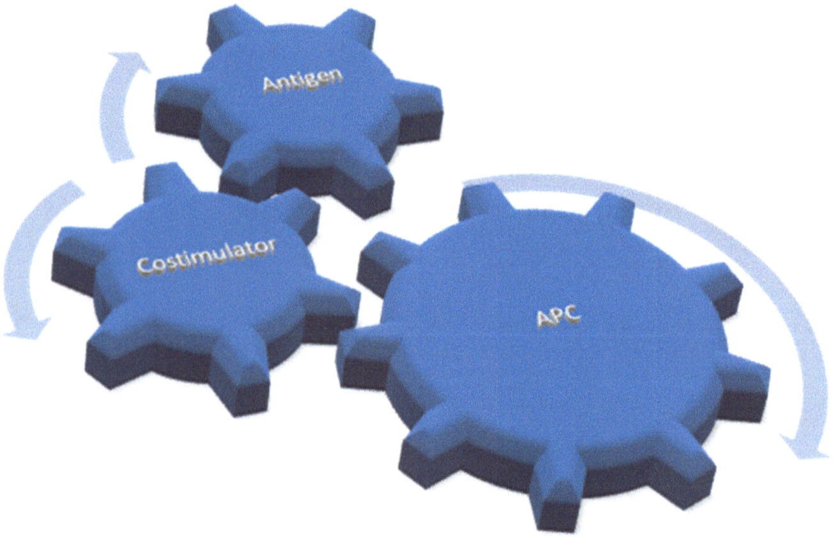

Fig. 15.1 The immune system requires three signals presented in the context of MHC in order to elicit an immune response. Absence of a co-stimulatory signal or a third signal can lead to anergy or tolerance induction

lipoteichoic acid, lipopolysaccharide, MPLA (mono phosphoryl lipid A), and flagellin. Fungal wall components such as zymosan, parasitic proteins such as *Toxoplasma gondii* Profilin, carbohydrates, and bacterial nucleic acids such as CpG, as well as viral nucleic acids incorporating DNA and RNA are other targets for TLRs. TLRs can be found ubiquitously on the cell surface, in the cell interior, in endosomes, and in the endoplasmic reticulum. And they can be found in many cell types including DCs, T cells, B cells, natural killer cells, fibroblasts, and epithelial cells.

PRRs create an elaborate surveillance network which surveys distinct cellular components on the cell surface in endosomes and in the cytosol and are capable of recognizing a diverse array of pathogens in every conceivable cellular compartment. These pathogens have the potential for triggering one or more than one TRR.

For nearly two centuries, vaccines have been one of the most successful and economical methods of preventing infectious disease, superseded only by sanitation methods. The successful elimination of polio has been nearly achieved, and strides are being made to develop vaccines for malaria. Classical vaccination involves exposing a small sample of a disease-causing agent to patients to elicit a response which has specificity and memory. Classical vaccination is not effective in all individuals or in all diseases; this is especially true in the elderly and in organisms which have developed unique methods of evading the immune system such as trypanosomiasis and HIV. The adjuvant MF59 consists of squalene and surfactants and improves influenza responsiveness in the elderly, possibly via MCP1, MIP1-α,β. Future vaccines may be for cancer, allergy, or diseases such as Alzheimer's, which the immune system would normally not target [1, 2].

Vaccines can be live attenuated vaccines, inactivated vaccines, and subunit vaccines (Fig. 15.2). The latter tend to be less effective because they do not always trigger danger signals. Live attenuated vaccines have limited utility in patients with weakened immune systems or when deadly pathogens are used. Inactivated vaccines may require frequent booster shots to reinforce the immune response. The best vaccines are rationally designed and ideally targeted to antigen-presenting cells such as Langerhans cells and dendritic cells. The ideal vaccine should be easy to manufacture, inexpensive, portable, stable, effective, safe, and easy to administer.

The purpose of nanovaccine delivery is to stabilize antigens, target them appropriately, and control immunologic functionality to either

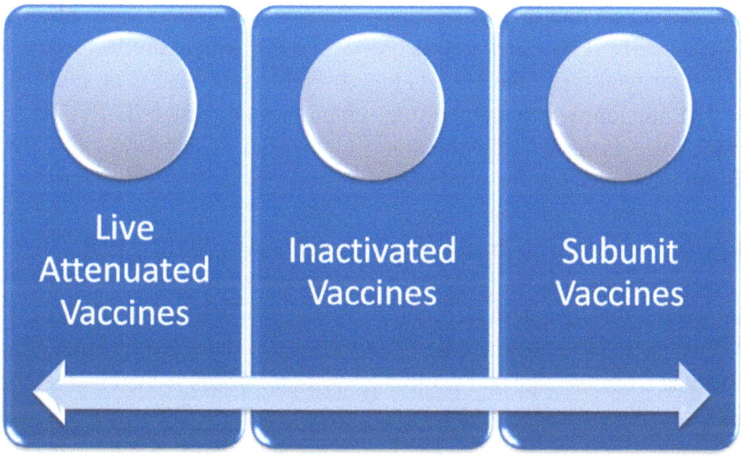

Fig. 15.2 Potential vaccine substrates can include live attenuated vaccine, killed or inactivated vaccine, and subunit vaccine. Subunit vaccines tend to be less immunogenic unless administered with an adjuvant

enhance immune activation or specifically inhibit immunity. Nanoparticle vaccines target antigens to DCs. They also target TLR ligand to DCs in order to induce DC maturation and activation, induce NK activation, break self-tolerance, and reverse Treg suppression.

Using nanotechnology, a number of agents are employed to modify the immune system. These include fullerenes, polymers, dendrimers, and nanoshells. Products incorporating nanotechnology have to be tested for biocompatibility. They must not interact with red blood cells, must not promote coagulation, and must have no protein-binding properties which may interfere with serum enzymes. Products must be tested for bio-accumulation, such as in the reticuloendothelial system. Products should be readily metabolized and easily cleared either by the kidneys, liver, or the gastrointestinal tract.

Vaccine efficiency has been shown to be enhanced with adjuvant use. Some synthetic adjuvants for tumor antigens, such as α-Galactosylceramide, induce NKT anergy. When conjugated with polylactic acid-based nanoparticles, these adjuvants stimulate NKT and do not induce anergy [3].

Nanoparticulate biologic agents such as ipilimumab and tremelimumab activate CTLA-4, resulting in T cell activation. Ipilimumab has recently shown benefit in the treatment of metastatic melanoma [4, 5].

TLRs are activated in pathogen defense. Dermatophyte exposure leads to co-stimulation via the myd88 pathway. Leprosy activates TLR1 and 2 and promotes Th1 immunity, and inhibits Th2 activity. The lack of innate immunity may contribute to vulnerability to colonization or infection. In atopic dermatitis, antimicrobial peptides LL-37 and HBD-2 are reduced, possibly accounting for increased susceptibility to staphylococcal colonization.

CpG induces TLR9 and has been shown to serve as a useful adjuvant. The binding of CpG to TLR9 is inhibited by antimalarials such as hydroxychloroquine, and may be responsible for their effectiveness in SLE. Lysosomotropic nanoparticles of antimalarials have shown benefit in tumor models and may have potential in the therapy of autoimmune diseases [6].

Acne severity may be related to the host inflammatory response. This is clearly seen in the use of topical and intralesional steroids for the treatment of inflammatory and cystic lesions. *Propionibacterium acnes* expresses TLR2 ligand, activates TLR2, and host reactivity is associated with antibodies against carbohydrate moieties in its cell wall [7, 8].

There is good evidence that the immune system is involved in skin cancer surveillance and elimination. Longitudinal observations demonstrate spontaneous involution of actinic keratoses and nonmelanoma skin cancer [9, 10]. Histology and cytokine profiles suggest that involuting cancers contain Th1 CD4 cells; IFN-γ; and are likely killed by perforins, granzymes, and Fas/FasL [11, 12]. Enhancing Th1 immunity can involve the induction of contact hypersensitivity, inflammation caused by excision, administration of interferon regulators such as imiquimod, and photodynamic therapy [13, 14]. In patients with malignancy from faulty DNA repair, topical delivery of a nanoencapsulated liposomal *Escherichia coli*-derived repair enzyme, T4 endonuclease V, reverses UVB-induced DNA damage and reduces the incidence of basal cell carcinomas and actinic keratoses [15].

Melanoma evades host immune response by inhibiting TLR4. TLR agonists combined with nanoparticulate CD40L agonists are being tested for the augmentation of anti-melanoma immunity [16]. Vaccines containing CpG (a TLR9 agonist), poly I:C (a TLR3 agonist), or nanoparticulate-modified adenovirus overcome this inhibition [17–21].

Malignancies such as mycosis fungoides may respond to Toll receptor agonists. Imiquimod and adjuvants such as CpG ODN increase IFN-γ and cell-mediated immunity and can be used as single-agent subcutaneous therapy. Nanoparticulate imiquimod has been used with antigen-loaded flexible liposomes. These may be adapted for the therapy of mycosis fungoides [22]. DC vaccines using adjuvants such as IL-12, -15, -18, -21; GCSF (increasing function and number of DC); imidazoquinoline; anti-CD40; and CTLA4-Ig have been tried [23, 24]. Monoclonal antibodies against CCR4 lead to T reg death; and, Bexarotene reduces CCR4, which may explain its mechanism of action [25]. Histone deacetylase inhibitors (HDACi) such as vorinostat and romidepsin increase IL-2R and enhance susceptibility to denileukin diftitox [26].

T cells can be found in melanoma lesions. Tumors evade immunity by downregulating MHC [27]. Melanomas, particularly advanced melanomas, release soluble factors which diminish levels of co-stimulatory molecules CD80/86 on the surfaces of dendritic cells causing them to induce tolerance [28, 29]. Adhesion molecules allow metastasis, and MUC18 may be a target for therapy (ABX-MA1) [30, 31]. They may secrete cytokines which are autocrine growth factors (b-FGF, IL-6, IL-8), or resist TFGβ, or make immunosuppressive IL-10 [12].

Vaccine candidates for melanoma have included a variety of antigens including peptides (MAA-CTL), proteins (tyrosinase), and surface gangliosides (GM2, GM3, GD3). The latter has included an engineered anti-idiotypic antibody BEC-2 which mimics GD2 [32, 33]. Results with these candidates have been equivocal. Adjuvants such as QS21 have been shown to augment responsiveness [34]. Adoptive transfer studies of tumor-infiltrating lymphocytes suggest a role for CD4 and CD8 T cells [11]. The use of phage-expressing HSP for melanoma vaccination has resulted in mean remission time increases of 0.6–1.5 years in patients with melanoma [35]. Viral oncolysates using vaccinia virus and Newcastle disease virus have shown benefit in the vaccination of melanoma [36, 37]. Peptide vaccines alone do not work as well because of tumor escape, neutralizing anti-peptide antibodies, and selection of tumor variants with loss of target peptide. Heteroclytic peptide vaccines, which bind HLA with higher affinity, are more effective. Dendritic cell vaccines obtained by either pulsing dendritic cells with antigens or cell lysates or transfecting cells with antigens have had modest success in the treatment of melanoma [38]. There is some evidence to suggest that nanoliposomally encapsulated antigens yield a more effective immune response [39]. Direct targeting of angiogenesis with immunoliposomes is another strategy being used for tumor control [40].

Special vaccines include glycoconjugates. These have been used for tetanus and diphtheria and are showing promise for the prophylaxis of viral and parasitic infections [41] as well as fungal infections. More recently, glycoconjugate vaccines have been developed for skin cancer. Tumor-associated carbohydrate antigens are overexpressed on some cancers and their expression may represent multiple cancer-associated genetic abnormalities [42]. Target moieties include MUC1, GM2, GD2, GD3, Lewis-Y, STn, TF, and Tn. These have been coupled to carriers such as KLH and used in clinical trials [43].

Nanoliposomal conjugates have been used for the elicitation of antitumor immunity against a malignant adenocarcinoma cell line MCF7 by combining tumor-associated glycopeptides MUC1, a T- and B cell peptide epitope, and a TLR2 ligand with resultant high antibody titer responses [44–46].

15.2 Adjuvants

Adjuvants were initially described in the last century [47] and were used to describe a vaccine additive which resulted in a stronger immune response. Adjuvants have the potential to create vaccines from hitherto impossible compounds. They can also maximize the mean response to the existing vaccines. In 1881, WB Coley found that by injecting live bacteria into cancer patients, he was able to increase their ability to fight off tumor. Presumably, the bacteria were able to augment host immunity by generating a danger signal. Danger signals such as CpG have been used to augment the effectiveness of antitumor vaccines.

Adjuvants enhance antibody titers, reduce antigen doses, reduce the number of vaccinations needed for a successful immunization, stabilize antigens, and can overcome weak immune responses in the elderly and in infants. Examples of traditional adjuvants include insoluble aluminum salts (referred to as alum). Adjuvants in nanoparticles capitalize on size and design to enhance immune responses [48]. The exact mechanism of action of Alum is unknown, but it may work by retaining antigens in depot form at the injection site, increasing uptake by APC, recruiting and activating APC, and by directly activating the inflammasome complex (NLRP3) [49, 50]. Alum forms loose aggregates 1–20 μm in diameter and may make antigens less thermally stable, thus unmasking epitopes. Alum adjuvants are not effective elicitors of cell-mediated immunity. The incorporation of TLR agonists has been shown to partially overcome this limitation. For example, nanoemulsion adjuvants containing MF59 are being tested for vaccines against HSV and have been successful in vaccines against HPV [51–53].

Adjuvants may work by activating the innate immune system. The use of adjuvants requires a better understanding of their immune mechanisms, a better ability to activate immunity, and a better ability to deactivate tolerogenicity and to reduce toxicity. Adjuvants may work by slowing antigen release (depot effect), targeting delivery to APC, and directly activating the immune system. Small differences in nanoparticle antigen properties such as their size, structure, and manufacturing impurities can result in significant differences in the immune response.

Reverse vaccinology is a high-throughput method of generating effective, targeted vaccines [54]. It is useful for the identification of vaccines for which traditional methods have failed, either because of antigen variability or because of immune evasion. In the first stage, bioinformatic tools are used to identify potential antigens (Fig. 15.3). The bioinformatic revolution has brought together a number of tools for a better understanding of protein and antigen conformation. These include databases of antigen–antibody complexes, typically involving 20 aa residues occupying 60–100 nm [55]. The surfaceome, or surfome, of the target is identified with proteomic techniques, including 2D gel protein electrophoresis, elution, and analysis of proteins by MALDI-TOF MS. In cases of hydrophobic membrane proteins, which are harder to study, gel-free methods such as w-D liquid chromatography are used. Antigens can also be expressed on gene chip microarrays. Once candidate antigens are identified, they are mass produced by PCR, PIPE cloning, gene synthesis, culture, or cell-free synthesis. Candidate proteins tagged with hexahistidine are robotically purified. Vaccine antigenicity is assessed by molecular epidemiology. Immunoreactivity against human sera is tested by ELISA, immuno dot blot, and protein microarray chips. Convalescent human sera provide the best reagent for these tests. Immunogenicity is tested in vitro and in vivo. For example, there is currently no vaccine for *Staphylococcus aureus*, which elicits high antibody titers, because antibodies are not protective against disease. In the final step, antigen toxicity is reduced by site-directed mutagenesis. Reverse

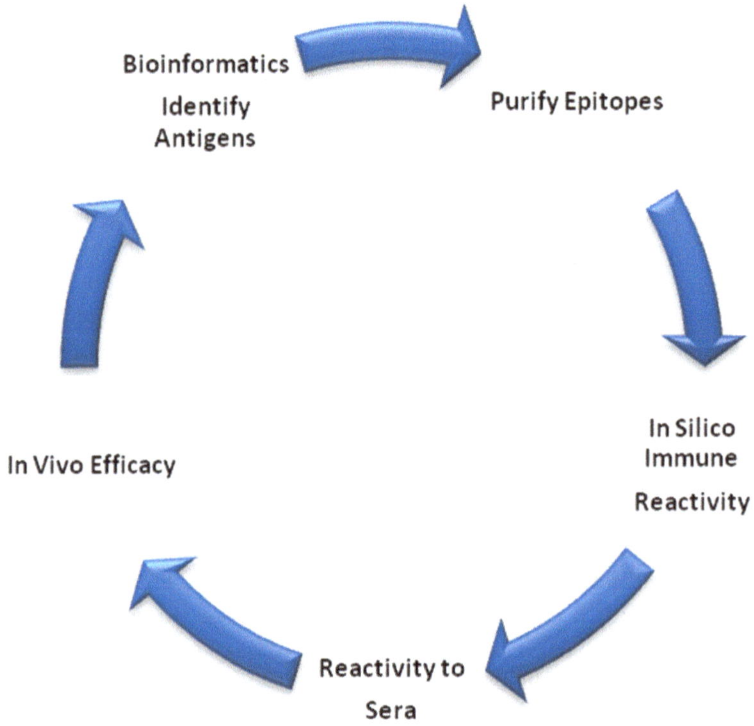

Fig. 15.3 In reverse vaccinology, bioinformatics is used to identify potential epitopes or vaccine candidates, which are then isolated and mapped. Candidate antigens are purified and selected by ELISA and tested for immune reactivity in vitro against convalescent sera and in vivo. Toxicity is reduced by rational site-directed mutagenesis

vaccinology has proven successful in identifying candidate vaccines for bacterial, viral, and fungal pathogens [56, 57].

Special routes of vaccination include cardiac, mucosal, oral, and intralymphatic (Fig. 15.4). Exosomes are essentially extracellular phospholipid carriers of cellular contents including proteins and nucleic acids. They range in size from 30 to 100 nm and are secreted by cells into the surrounding environment. Dendritic cells and other cells of the myeloid lineage have a high capacity for taking up exosomes by phagocytosis. This uptake has been exploited in the management of melanoma in clinical trials [53, 58]. Nanomimetics of exosomes are being developed in the laboratory [58, 59].

The route of vaccination may determine the immunogenicity of various vaccines. Mucosal administration is needle-free, does not require skilled personnel, and can result in mucosal and systemic immunity. Mucosal vaccines must be stable to gastric pH, intestinal enzymes, and bile salts. Polymeric nanodelivery systems can shield antigens and overcome these obstacles [60]. Antigen-presenting cell in the gut may be programmed for the induction of tolerance [61]. Mucosal vaccines have been used to reduce DTH and result in oral tolerance [62]. Carriers for the induction of an immune response have also included transgenic plants, for example, for protection against poxviruses [63, 64]. Intranasal viral-like nanoparticles have been used for the elicitation of neutralizing antibodies against HPV [65]. The TLR ligand CpG is stable in the gut and has been used for the enhancement of Th1 and Th2 responses via oral, intranasal, intravaginal, and intrarectal routes, and has elicited responses to Chlamydia and HSV [66, 67]. Topical vaccination using microneedles and nanovehicles can lead to a significant immune response [68–72].

The site of immunization on the skin has been shown to correlate with immune response. Superficial scarification is the most effective method [73]. Subcutaneous, intradermal, and intra-

Fig. 15.4 There are a number of routes to immunization. Topical delivery utilizes nanoparticles, hydrogel patches, or microneedles. Intralymphatic vaccination is very efficient and requires very little antigen. Mucosal delivery can be used both to augment immunity and to induce tolerance

muscular methods are less effective. Microneedles and carbon nanotube-based needles may be able to deliver antigen superficially to the epidermis, or to the dermal–epidermal junction, and may be able to maximize immune responsiveness while minimizing antigen dose [74–76]. Arrays of microneedles or carbon nanotubes may be able to deliver multiple antigens simultaneously.

Topical vaccines have been developed. These have been used for investigation of influenza vaccines, HIV vaccines, and VEGFR2 [77–80]. The latter is a DNA vaccine containing murine beta defensin 2, murine VEGFR2 (mFlk-1), and a linkage peptide [(G4S)3] in a cationic liposome. This vaccine was shown to reduce LL tumor burden and increase survival in treated vs. control (VEGFR2 alone or MBD2 alone) animals. Patches have been developed for the cutaneous delivery of antigens in vaccine studies. One example is a dry transdermal patch that becomes hydrated by transepidermal water loss. The low-level of hydration results in a highly concentrated antigen mixture which is able to penetrate the epidermis. This patch has been used to successfully immunize recipients against *E. coli*, tetanus, influenza, and anthrax [81]. Heterologous vaccination relies on using two different modalities of immunization to elicit immunity. An influenza patch has been shown to heterologously augment responses in elderly patients treated by intramuscular, subcutaneous, or intradermal flu vaccination [82, 83]. Ceramic needles have been used for the delivery of topical vaccines [84]. Needles can be manufactured to deliver a single antigen or an array of antigens. As opposed to traditional intramuscular vaccination, which yield a CD4 response, topical cyanoacrylate stripping or ceramic needle vaccination elicits a CD4 CD8 response. Intraepidermal vaccination has been shown to reduce the dosage vaccine by four-fifths. It may also directly target epidermal LC and DCs [73].

Vaccination via the intralymphatic route has certain advantages over traditional methods. It has been estimated that antigens need to be presented to at least 10^7 T and B cells to elicit an immune response. In typical vaccination procedures, 0.0001–0.1 % of the antigen reaches lymph nodes. Direct intralymphatic vaccination leads to enhanced immunity against a variety of antigens, including tumor antigens, and requires nanogram quantities of antigen [85]. Particles in the 20–200 nm range enter directly into lymphatics, while larger particles are transported by APC [86, 87]. Intralymphatic vaccination has also been used for the induction of tolerance [88]. Lymph nodes are readily visualized by ultrasound. The procedure of injection is rapid and simple. Because of the poor innervation of lymph nodes, the procedure causes minimal discomfort.

15.3 Dendritic Cell Delivery

For dendritic cells, antigen uptake occurs via several mechanisms (Fig. 15.5). In vivo, cells take up materials either by endocytosis, phagocytosis, or pinocytosis [86, 89–92]. Pinocytosis can be mediated either by clathrin, caveolin, or neither. Clathrin-mediated endocytosis works for particles 120 nm and smaller and does not require actin. Caveolin-mediated endocytosis captures particles 60 nm and smaller and does not require actin. Cell uptake which bypasses both mechanisms is useful for particles 90 nm and smaller.

Each lymph node has approximately 5,000 antigen-bearing DCs which persist for 2 weeks. In the absence of antigen, each DC contacts about 500 unique T cells per hour. In the presence of antigen, DCs are highly effective at recruiting

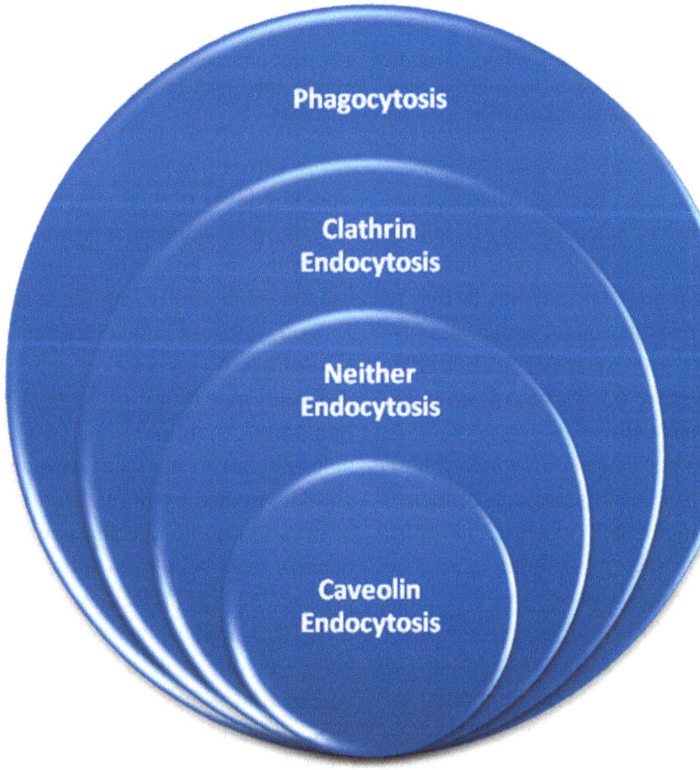

Fig. 15.5 Dendritic cell uptake can be either by phagocytosis, endocytosis, or pinocytosis. The latter can be mediated by clathrin, caveolin, or neither. The clathrin-mediated mechanism is suitable for particles ≤120 nm, the caveolin mediated mechanism is for particles ≤60 nm, and the mechanism which utilizes neither is effective for particles ≤90 nm

antigen-specific T cells. T cells circulate to a high degree, and in adoptive transfer studies, 44 % of the circulating T cells were found to interact with DCs in vivo. Strategies to deliver antigen to DCs include ex vivo and in vivo strategies. Ex vivo strategies involve peptide pulsing, protein pulsing, and genetic modification. In vivo strategies targeting antigens to DCs include the use of proteins, peptides, DNA, and nanoparticulate delivery (NPD). NPD delivery includes polymeric systems and lipid systems. Polymeric systems include PLGA, starch, and other polymers. Lipid systems include exosomes, liposomes, and nanoemulsions. The benefits of NPD include enhanced DC internalization. This can be manipulated through control of the biomaterial vehicle, the size of particles, their surface characteristics, and their shape. The dimensions of NPD systems are comparable to microorganisms.

When delivering particles to DCs, the use of particles is more efficient than solute [48, 86, 93]. Enhanced phagocytosis of NPD is 1,000–10,000 times more efficient than the uptake of soluble antigen. The use of receptors on nanoparticles also enhances DC delivery. Particles use receptors such as the mannose receptor (DEC-205), DC-SIGN, and nanoparticles can be functionalized with ligands such as mannose, fucose, and N-acetylglucosamine. Escape of antigens from the endosome to the cytosol promotes the MHC-I pathway. Intracellular reservoirs of antigens can be sustained for many days and allow for efficient immune responses with minute levels of antigens.

TLR4 ligand MPLA incorporated into nanoparticles enhanced DC response by three orders of magnitude when compared to soluble Ag [94]. TLR9 responsiveness is maximized in endosomes, where optimal signaling requires low pH. PLGA nanoparticles contain lactic acid and glycolic acid, thus acidifying endosomes. CpG, taken up in endosomes, is also released in an acidic environment.

Poly I:C is a synthetic dsRNA. It induces stable maturation of DC, lasts up to 48 h, and induces p70 of IL-12 which has been associated with shock, renal failure, coagulopathy, and hypersensitivity reactions [18]. Poly I:C12U results mismatch dsRNA, accelerated hydrolysis, and is less toxic. No dose-limiting toxicity has been seen in HIV. In one study, immature DCs were isolated from patients with advanced ovarian cancer ascites fluid. The cells were harvested and treated with Poly IC 12 U, LPS, TNF, and PGE2. This led to an activation of the immature DCs into mature DCs as measured by morphology and activation markers such as MHC II, CD 40, and CD 86. Cytokines such as IL-12 and IL-10 decreased and IL-6 increased. Supernatants from the mature DCs generated high levels of interferon gamma in purified CD4 T cells. CD8 cells primed with mature DCs induced efficient tumor lysis of autologous ovarian tumor cells.

Sustained release of antigens may also augment immunity. Cubosomes, liposomes, and hydrogels, by delaying release and maximizing retention of antigens, may increase vaccine efficacy and decrease the need for boosting. This approach is being used for KLK peptide CpG ODN – TB HBV [95, 96]. Poly (I:C) 31 may have a depot effect on antigens. A polymer gel is being used for HPV viral-like particles in nanoliposomes in a nasal powder form called Gelsite, which results in a good response and which is antigen sparing. The GSK vaccine Cervarix consists of a virus like-particle made of the L1 proteins do HPV 16, 18, and adjuvants of alum, QS21, and MPL packaged in a liposome [52]. Their malaria vaccine contains MPL, QS21, and antigen RTS-S. In small clinical trials in Africa, it has shown protective benefit [97].

Gadolinium in contrast dye is associated with nephrogenic fibrosis. Gadolinium in fullerene nanoparticles has been shown to induce dendritic cell maturation in murine models. DCs exposed to the particles express IL-12p70, co-stimulatory molecules, MHC molecules, and change from a CCL5 to a CCL19-responsive phenotype. The mature DCs promote a Th1 phenotype in naïve T cells [98].

Virosomes comprising empty influenza envelopes have been shown to generate antibodies, delay parasitemia, and decrease hepatic parasites in phase I studies of malaria [99]. Studies have shown that the shape of the antigen can determine the immune response. In general, antigens which are cubical or short squat cylinders 'generate negative immune responses, and antigens which are hexagonal or long and cylindrical generate positive immune responses [100].

Currently, a vaccine for malaria containing a partial parasite surface protein, hepatitis B surface antigen, a protein link, and RTS-S in conjunction with two adjuvants is in phase 3 clinical trials in infants [101]. A vaccine for influenza using AS03 oil in water emulsion yields protective antibodies in the majority of elderly subjects, immunocompromised subjects, and subjects with egg allergy [102–104]. AS03 avian H5N1 flu vaccine yields protective antibodies using one-third antigen amounts

Polymers containing ISS (immunostimulatory oligonucleotide) have been used to augment the immune response and to generate the same antibody titer from antigen concentrations reduced 100-fold. Linkage to ISS increases antigen uptake by antigen-presenting cells, increases the Th1 response, or blocks IgE epitopes, and has been successful in treating ragweed allergy [105].

An adjuvant oligonucleotide PF-3512676 stimulates TRL9 and increases Th1 immunity. This has been used to reduce antigen concentrations ×10 to ×1,000. It may enhance the maturation of plasmacytoid and myeloid dendritic cells in lymph nodes and has been used in the treatment of melanoma and basal cell carcinoma in small clinical trials [106, 107].

Oxidation-responsive nanoparticles made of polymers such as PPS (polypropylene sulfide) dissociate upon oxidation, releasing contents in their capsules [108]. Such responsive release may target therapy to sites of inflammation, or may target intracellular organisms.

Antimicrobial peptides come in three varieties: alpha, beta, and theta (not found in humans) and are widely distributed. They are abundant in granules and act to permeabilize membranes. Cathelicidin is found in the nail bed and eccrine glands and has activity against a variety of bacteria, *Candida*, and HSV. The expression of some antimicrobial peptides can be altered by vitamin D [109, 110]. The proteolytic subunits including RK-31 and KS-30 have more potent activity. Granulysin and RNAase-7 are active against *P. acnes* [111, 112]. The latter is also active against *S. aureus* and vancomycin-resistant *Enterococcus*. There is some evidence that antimicrobial peptide levels are reduced in diseases such as hidradenitis suppurativa, atopic dermatitis, and psoriasis [113–115]. The use of nanoparticles for the delivery of antimicrobial peptides is being investigated [116, 117]. A synthetic antimicrobial peptide, KSLW, coupled to a pluronic nanocarrier has shown benefits in animal models of wound healing [118].

In summary, the skin immune system has a number of components that are involved in the upregulation of immunity. Activation of the immune system using nanotechnology involves exploitation of innate immunity, encapsulation of antigens, epicutaneous delivery, dendritic cell targeting, and the use of high-throughput techniques for the generation of vaccines.

References

1. Imbimbo BP, et al. Solanezumab for the treatment of mild-to-moderate Alzheimer's disease. Expert Rev Clin Immunol. 2012;8(2):135–49.
2. Arigoni M, et al. A vaccine targeting angiomotin induces an antibody response which alters tumor vessel permeability and hampers the growth of established tumors. Angiogenesis. 2012;15(2):305–16.
3. Thapa P, et al. Nanoparticle formulated alpha-galactosylceramide activates NKT cells without inducing anergy. Vaccine. 2009;27(25–26):3484–8.
4. Hodi FS, et al. Improved survival with ipilimumab in patients with metastatic melanoma. N Engl J Med. 2010;363(8):711–23.
5. Ascierto PA, Marincola FM, Ribas A. Anti-CTLA4 monoclonal antibodies: the past and the future in clinical application. J Transl Med. 2011;9:196.
6. Marin GH, et al. Exploratory study on the effects of biodegradable nanoparticles with drugs on malignant B cells and on a human/mouse model of Burkitt lymphoma. Curr Clin Pharmacol. 2010;5(4):246–50.
7. Nakatsuji T, Gallo RL. Antimicrobial peptides: old molecules with new ideas. J Invest Dermatol. 2012;132(3 Pt 2):887–95.
8. Nakatsuji T, et al. Vaccination targeting a surface sialidase of P. acnes: implication for new treatment of acne vulgaris. PLoS One. 2008;3(2):e1551.
9. Wong DA, et al. Cytokine profiles in spontaneously regressing basal cell carcinomas. Br J Dermatol. 2000;143(1):91–8.

10. Criscione VD, et al. Actinic keratoses: natural history and risk of malignant transformation in the Veterans Affairs Topical Tretinoin Chemoprevention Trial. Cancer. 2009;115(11):2523–30.
11. Friedman KM, et al. Tumor-specific CD4+ melanoma tumor-infiltrating lymphocytes. J Immunother. 2012;35(5):400–8.
12. Lazar-Molnar E, et al. Autocrine and paracrine regulation by cytokines and growth factors in melanoma. Cytokine. 2000;12(6):547–54.
13. Macpherson N, Lamrock E, Watt G. Effect of inflammation on positive margins of basal cell carcinomas. Australas J Dermatol. 2010;51(2):95–8.
14. Maguire Jr HC, et al. Phase I study of R24 in patients with metastatic melanoma including evaluation of immunologic parameters. Cancer Biother Radiopharm. 1998;13(1):13–23.
15. Cafardi JA, Elmets CA. T4 endonuclease V: review and application to dermatology. Expert Opin Biol Ther. 2008;8(6):829–38.
16. Stone GW, et al. Nanoparticle-delivered multimeric soluble CD40L DNA combined with toll-Like Receptor agonists as a treatment for melanoma. PLoS One. 2009;4(10):e7334.
17. Tittarelli A, et al. Toll-like receptor 4 gene polymorphism influences dendritic cell in vitro function and clinical outcomes in vaccinated melanoma patients. Cancer Immunol Immunother. 2012 May 3. [Epub ahead of print] PubMed PMID: 22552381.
18. Cheng YS, Xu F. Anticancer function of polyinosinic-polycytidylic acid. Cancer Biol Ther. 2011;10(12): 1219–23.
19. Tormo D, et al. Therapeutic efficacy of antigen-specific vaccination and toll-like receptor stimulation against established transplanted and autochthonous melanoma in mice. Cancer Res. 2006;66(10):5427–35.
20. Tarhini AA, et al. Safety and immunogenicity of vaccination With MART-1 (26–35, 27L), gp100 (209–217, 210M), and tyrosinase (368–376, 370D) in adjuvant with PF-3512676 and GM-CSF in metastatic melanoma. J Immunother. 2012;35(4):359–66.
21. Speiser DE, et al. Memory and effector CD8 T-cell responses after nanoparticle vaccination of melanoma patients. J Immunother. 2010;33(8):848–58.
22. Li N, et al. Effective transcutaneous immunization by antigen-loaded flexible liposome in vivo. Int J Nanomedicine. 2011;6:3241–50.
23. Ni X, Duvic M. Dendritic cells and cutaneous T-cell lymphomas. G Ital Dermatol Venereol. 2011;146(2):103–13.
24. Kim YH, et al. In situ vaccination against mycosis fungoides by intratumoral injection of a TLR9 agonist combined with radiation: a phase 1/2 study. Blood. 2012;119(2):355–63.
25. Richardson SK, et al. Bexarotene blunts malignant T-cell chemotaxis in Sezary syndrome: reduction of chemokine receptor 4-positive lymphocytes and decreased chemotaxis to thymus and activation-regulated chemokine. Am J Hematol. 2007;82(9):792–7.
26. Prince HM, Dickinson M. Romidepsin for cutaneous T-cell lymphoma. Clin Cancer Res. 2012;18(13):3509–15.
27. Degenhardt Y, et al. Distinct MHC gene expression patterns during progression of melanoma. Genes Chromosomes Cancer. 2010;49(2):144–54.
28. Baumgartner JM, et al. DC maturation and function are not altered by melanoma-derived immunosuppressive soluble factors. J Surg Res. 2012;176(1):301–8.
29. McCarter M, et al. Melanoma skews dendritic cells to facilitate a T helper 2 profile. Surgery. 2005;138(2):321–8.
30. McGary EC, Lev DC, Bar-Eli M. Cellular adhesion pathways and metastatic potential of human melanoma. Cancer Biol Ther. 2002;1(5):459–65.
31. Melnikova VO, Bar-Eli M. Bioimmunotherapy for melanoma using fully human antibodies targeting MCAM/MUC18 and IL-8. Pigment Cell Res. 2006;19(5):395–405.
32. Bradbury PA, Shepherd FA. Immunotherapy for lung cancer. J Thorac Oncol. 2008;3(6 Suppl 2):S164–70.
33. Wondimu A, et al. Peptides mimicking GD2 ganglioside elicit cellular, humoral and tumor-protective immune responses in mice. Cancer Immunol Immunother. 2008;57(7):1079–89.
34. Deng K, et al. Synthesis of QS-21-xylose: establishment of the immunopotentiating activity of synthetic QS-21 adjuvant with a melanoma vaccine. Angew Chem Int Ed Engl. 2008;47(34):6395–8.
35. Randazzo M, et al. Active-specific immunotherapy of human cancers with the heat shock protein Gp96-revisited. Int J Cancer. 2012;130(10):2219–31.
36. Hersey P. Active immunotherapy with viral lysates of micrometastases following surgical removal of high risk melanoma. World J Surg. 1992;16(2):251–60.
37. Hersey P. Evaluation of vaccinia viral lysates as therapeutic vaccines in the treatment of melanoma. Ann N Y Acad Sci. 1993;690:167–77.
38. Van Nuffel AM, et al. Epitope and HLA-type independent monitoring of antigen-specific T-cells after treatment with dendritic cells presenting full-length tumor antigens. J Immunol Methods. 2012;377(1–2):23–36.
39. van Broekhoven CL, et al. Targeting dendritic cells with antigen-containing liposomes: a highly effective procedure for induction of antitumor immunity and for tumor immunotherapy. Cancer Res. 2004;64(12):4357–65.
40. Wicki A, et al. Targeting tumor-associated endothelial cells: anti-VEGFR2 immunoliposomes mediate tumor vessel disruption and inhibit tumor growth. Clin Cancer Res. 2012;18(2):454–64.
41. Debierre-Grockiego F. Glycolipids are potential targets for protozoan parasite diseases. Trends Parasitol. 2010;26(8):404–11.
42. Kannagi R, et al. Current relevance of incomplete synthesis and neo-synthesis for cancer-associated alteration of carbohydrate determinants—Hakomori's concepts revisited. Biochim Biophys Acta. 2008;1780(3):525–31.
43. Livingston PO. Approaches to augmenting the immunogenicity of melanoma gangliosides: from whole

melanoma cells to ganglioside-KLH conjugate vaccines. Immunol Rev. 1995;145:147–66.
44. Ingale S, Buskas T, Boons GJ. Synthesis of glyco(lipo) peptides by liposome-mediated native chemical ligation. Org Lett. 2006;8(25):5785–8.
45. Ingale S, et al. Robust immune responses elicited by a fully synthetic three-component vaccine. Nat Chem Biol. 2007;3(10):663–7.
46. Ingale S, et al. Increasing the antigenicity of synthetic tumor-associated carbohydrate antigens by targeting Toll-like receptors. Chembiochem. 2009;10(3):455–63.
47. Ebisawa I. The encounter of Gaston Ramon (1886–1963) with formalin: a biographical study of a great scientist. Kitasato Arch Exp Med. 1987;60(3):55–70.
48. Oyewumi MO, Kumar A, Cui Z. Nano-microparticles as immune adjuvants: correlating particle sizes and the resultant immune responses. Expert Rev Vaccines. 2010;9(9):1095–107.
49. Schijns VE, Lavelle EC. Trends in vaccine adjuvants. Expert Rev Vaccines. 2011;10(4):539–50.
50. Sharp FA, et al. Uptake of particulate vaccine adjuvants by dendritic cells activates the NALP3 inflammasome. Proc Natl Acad Sci U S A. 2009;106(3):870–5.
51. Stanberry LR. Clinical trials of prophylactic and therapeutic herpes simplex virus vaccines. Herpes. 2004;11 Suppl 3:161A–9.
52. Schauner S, Lyon C. Bivalent HPV recombinant vaccine (Cervarix) for the prevention of cervical cancer. Am Fam Physician. 2010;82(12):1541–2.
53. Tan A, De La Pena H, Seifalian AM. The application of exosomes as a nanoscale cancer vaccine. Int J Nanomedicine. 2010;5:889–900.
54. Bowman BN, et al. Improving reverse vaccinology with a machine learning approach. Vaccine. 2011;29(45): 8156–64.
55. Dormitzer PR, Ulmer JB, Rappuoli R. Structure-based antigen design: a strategy for next generation vaccines. Trends Biotechnol. 2008;26(12):659–67.
56. Palumbo A, et al. A chemically modified antibody mediates complete eradication of tumours by selective disruption of tumour blood vessels. Br J Cancer. 2011;104(7):1106–15.
57. Palumbo E, et al. Antigen identification starting from the genome: a "Reverse Vaccinology" approach applied to MenB. Methods Mol Biol. 2012;799:361–403.
58. Andre F, et al. Exosomes for cancer immunotherapy. Ann Oncol. 2004;15 Suppl 4:iv141–4.
59. Kooijmans SA, et al. Exosome mimetics: a novel class of drug delivery systems. Int J Nanomedicine. 2012;7:1525–41.
60. Mahapatro A, Singh DK. Biodegradable nanoparticles are excellent vehicle for site directed in-vivo delivery of drugs and vaccines. J Nanobiotechnology. 2011;9:55.
61. Kagnoff MF. Oral tolerance: mechanisms and possible role in inflammatory joint diseases. Baillieres Clin Rheumatol. 1996;10(1):41–54.
62. Pabst O, Mowat AM. Oral tolerance to food protein. Mucosal Immunol. 2012;5(3):232–9.
63. Salyaev RK, Rigano MM, Rekoslavskaya NI. Development of plant-based mucosal vaccines against widespread infectious diseases. Expert Rev Vaccines. 2010;9(8):937–46.
64. Rigano MM, et al. Plants as biofactories for the production of subunit vaccines against bio-security-related bacteria and viruses. Vaccine. 2009;27(25–26):3463–6.
65. Balmelli C, et al. Nasal immunization of mice with human papillomavirus type 16 virus-like particles elicits neutralizing antibodies in mucosal secretions. J Virol. 1998;72(10):8220–9.
66. Cheng C, et al. Induction of protective immunity by vaccination against Chlamydia trachomatis using the major outer membrane protein adjuvanted with CpG oligodeoxynucleotide coupled to the nontoxic B subunit of cholera toxin. Vaccine. 2009;27(44):6239–46.
67. Tengvall S, et al. CpG oligodeoxynucleotide augments HSV-2 glycoprotein D DNA vaccine efficacy to generate T helper 1 response and subsequent protection against primary genital herpes infection in mice. J Reprod Immunol. 2005;68(1–2):53–69.
68. Gittard SD, et al. Multiphoton microscopy of transdermal quantum dot delivery using two photon polymerization-fabricated polymer microneedles. Faraday Discuss 2011;149:171–85; discussion 227–45.
69. Nasir A. Nanotechnology in vaccine development: a step forward. J Invest Dermatol. 2009;129(5):1055–9.
70. Liard C, et al. Targeting of HIV-p24 particle-based vaccine into differential skin layers induces distinct arms of the immune responses. Vaccine. 2011;29(37):6379–91.
71. Mahe B, et al. Nanoparticle-based targeting of vaccine compounds to skin antigen-presenting cells by hair follicles and their transport in mice. J Invest Dermatol. 2009;129(5):1156–64.
72. Badran MM, Kuntsche J, Fahr A. Skin penetration enhancement by a microneedle device (Dermaroller) in vitro: dependency on needle size and applied formulation. Eur J Pharm Sci. 2009;36(4–5):511–23.
73. Kupper TS. Old and new: recent innovations in vaccine biology and skin T cells. J Invest Dermatol. 2012;132(3 Pt 2):829–34.
74. Degim IT, Burgess DJ, Papadimitrakopoulos F. Carbon nanotubes for transdermal drug delivery. J Microencapsul. 2010;27(8):669–81.
75. Wu J, et al. Programmable transdermal drug delivery of nicotine using carbon nanotube membranes. Proc Natl Acad Sci U S A. 2010;107(26):11698–702.
76. Im JS, Bai B, Lee YS. The effect of carbon nanotubes on drug delivery in an electro-sensitive transdermal drug delivery system. Biomaterials. 2010;31(6): 1414–9.
77. Miyazawa M, et al. Phase I clinical trial using peptide vaccine for human vascular endothelial growth factor receptor 2 in combination with gemcitabine for patients with advanced pancreatic cancer. Cancer Sci. 2010;101(2):433–9.
78. Mochimaru H, et al. Suppression of choroidal neovascularization by dendritic cell vaccination targeting VEGFR2. Invest Ophthalmol Vis Sci. 2007;48(10):4795–801.
79. Lori F. DermaVir: a plasmid DNA-based nanomedicine therapeutic vaccine for the treatment of HIV/AIDS. Expert Rev Vaccines. 2011;10(10):1371–84.

80. Wang YS, et al. Immunity against tumor angiogenesis induced by a fusion vaccine with murine beta-defensin 2 and mFlk-1. Clin Cancer Res. 2007;13(22 Pt 1):6779–87.
81. Frech SA, et al. Use of a patch containing heat-labile toxin from Escherichia coli against travellers' diarrhoea: a phase II, randomised, double-blind, placebo-controlled field trial. Lancet. 2008;371(9629):2019–25.
82. Frolov VG, et al. Transcutaneous delivery and thermostability of a dry trivalent inactivated influenza vaccine patch. Influenza Other Respi Viruses. 2008;2(2):53–60.
83. Frech SA, et al. Improved immune responses to influenza vaccination in the elderly using an immunostimulant patch. Vaccine. 2005;23(7):946–50.
84. Quan FS, et al. Dose sparing enabled by skin immunization with influenza virus-like particle vaccine using microneedles. J Control Release. 2010;147(3):326–32.
85. Kreiter S, et al. Intranodal vaccination with naked antigen-encoding RNA elicits potent prophylactic and therapeutic antitumoral immunity. Cancer Res. 2010;70(22):9031–40.
86. Manolova V, et al. Nanoparticles target distinct dendritic cell populations according to their size. Eur J Immunol. 2008;38(5):1404–13.
87. Kobiasi MA, et al. Control of size dispersity of chitosan biopolymer microparticles and nanoparticles to influence vaccine trafficking and cell uptake. J Biomed Mater Res A. 2012;100(7):1859–67.
88. Senti G, et al. Intralymphatic immunotherapy for cat allergy induces tolerance after only 3 injections. J Allergy Clin Immunol. 2012;129(5):1290–6.
89. Besterman JM, Low RB. Endocytosis: a review of mechanisms and plasma membrane dynamics. Biochem J. 1983;210(1):1–13.
90. Lim JP, Gleeson PA. Macropinocytosis: an endocytic pathway for internalising large gulps. Immunol Cell Biol. 2011;89(8):836–43.
91. Meyer C, et al. Interleukin-6 receptor specific RNA aptamers for cargo delivery into target cells. RNA Biol. 2012;9(1):67–80.
92. Zaki NM, Tirelli N. Gateways for the intracellular access of nanocarriers: a review of receptor-mediated endocytosis mechanisms and of strategies in receptor targeting. Expert Opin Drug Deliv. 2010;7(8): 895–913.
93. De Temmerman ML, et al. Particulate vaccines: on the quest for optimal delivery and immune response. Drug Discov Today. 2011;16(13–14):569–82.
94. Cluff CW. Monophosphoryl lipid A (MPL) as an adjuvant for anti-cancer vaccines: clinical results. Adv Exp Med Biol. 2009;667:111–23.
95. Barry M, Cooper C. Review of hepatitis B surface antigen-1018 ISS adjuvant-containing vaccine safety and efficacy. Expert Opin Biol Ther. 2007;7(11):1731–7.
96. Vandepapeliere P, et al. Vaccine adjuvant systems containing monophosphoryl lipid A and QS21 induce strong and persistent humoral and T cell responses against hepatitis B surface antigen in healthy adult volunteers. Vaccine. 2008;26(10):1375–86.
97. Lell B, et al. A randomized trial assessing the safety and immunogenicity of AS01 and AS02 adjuvanted RTS, S malaria vaccine candidates in children in Gabon. PLoS One. 2009;4(10):e7611.
98. Yang D, et al. [Gd@C(82)(OH)(22)](n) nanoparticles induce dendritic cell maturation and activate Th1 immune responses. ACS Nano. 2010;4(2):1178–86.
99. Cech PG, et al. Virosome-formulated Plasmodium falciparum AMA-1 & CSP derived peptides as malaria vaccine: randomized phase 1b trial in semi-immune adults & children. PLoS One. 2011;6(7):e22273.
100. Garcia A, et al. Microfabricated engineered particle systems for respiratory drug delivery and other pharmaceutical applications. J Drug Deliv. 2012;2012: 941243.
101. Agnandji ST, et al. First results of phase 3 trial of RTS, S/AS01 malaria vaccine in African children. N Engl J Med. 2011;365(20):1863–75.
102. Gagnon R, et al. Safe vaccination of patients with egg allergy with an adjuvanted pandemic H1N1 vaccine. J Allergy Clin Immunol. 2010;126(2): 317–23.
103. Rubinstein E, et al. The responses of Aboriginal Canadians to adjuvanted pandemic (H1N1) 2009 influenza vaccine. CMAJ. 2011;183(13):E1033–7.
104. Cooper C, et al. High-level immunogenicity is achieved vaccine with adjuvanted pandemic H1N1(2009) and improved with booster dosing in a randomized trial of HIV-infected adults. HIV Clin Trials. 2012;13(1):23–32.
105. Tulic MK, et al. Local induction of a specific Th1 immune response by allergen linked immunostimulatory DNA in the nasal explants of ragweed-allergic subjects. Allergol Int. 2009;58(4):565–72.
106. Hofmann MA, et al. Phase 1 evaluation of intralesionally injected TLR9-agonist PF-3512676 in patients with basal cell carcinoma or metastatic melanoma. J Immunother. 2008;31(5):520–7.
107. Molenkamp BG, et al. Intradermal CpG-B activates both plasmacytoid and myeloid dendritic cells in the sentinel lymph node of melanoma patients. Clin Cancer Res. 2007;13(10):2961–9.
108. Cerritelli S, Velluto D, Hubbell JA. PEG-SS-PPS: reduction-sensitive disulfide block copolymer vesicles for intracellular drug delivery. Biomacromolecules. 2007;8(6):1966–72.
109. Hegyi Z, et al. Vitamin D analog calcipotriol suppresses the Th17 cytokine-induced proinflammatory S100 "alarmins" psoriasin (S100A7) and koebnerisin (S100A15) in psoriasis. J Invest Dermatol. 2012;132(5):1416–24.
110. Deguchi E, et al. Topical vitamin D3 analogues induce thymic stromal lymphopoietin and cathelicidin in psoriatic skin lesions. Br J Dermatol. 2012;167(1):77–84.
111. McInturff JE, et al. Granulysin-derived peptides demonstrate antimicrobial and anti-inflammatory effects against Propionibacterium acnes. J Invest Dermatol. 2005;125(2):256–63.
112. Simanski M, et al. Antimicrobial RNases in cutaneous defense. J Innate Immun. 2012;4(3):241–7.
113. Schittek B. The multiple facets of dermcidin in cell survival and host defense. J Innate Immun. 2012;4(4): 349–60.

114. Hofmann SC, et al. Expression of innate defense antimicrobial peptides in hidradenitis suppurativa. J Am Acad Dermatol. 2012;66(6):966–74.
115. Dombrowski Y, Schauber J. Cathelicidin LL-37: a defense molecule with a potential role in psoriasis pathogenesis. Exp Dermatol. 2012;21(5):327–30.
116. Brandelli A. Nanostructures as promising tools for delivery of antimicrobial peptides. Mini Rev Med Chem. 2012;12(8):731–41.
117. Nguyen DN, et al. Lipid-derived nanoparticles for immunostimulatory RNA adjuvant delivery. Proc Natl Acad Sci USA. 2012;109(14): E797–803.
118. Williams RL, et al. Synthetic decapeptide reduces bacterial load and accelerates healing in the wounds of restraint-stressed mice. Brain Behav Immun. 2012;26(4):588–96.

16 Nanoparticle-Based Epidermal and Dermal Vaccination

Angèle Soria, Annika Vogt, Adnan Nasir, and Béhazine Combadière

Abbreviations

APC	Antigen-presenting cell
CSSS	Cyanoacrylate tape-surface stripping
DMF	Dermal macrophages
(D) (p) DC	(Dermal) (plasmacytoid) dendritic cell
HFT	Hair follicular targeting
ID	Intradermal
IM	Intramuscular
LC	Langerhans cell
LN	Lymph node
MVA	Modified Vaccinia Ankara virus
NP	Nanoparticle
PLA	Poly (lactic) acid nanoparticles
SC	Subcutaneous
TC	Transcutaneous
TCM	Central memory T cells
TEM	Effector memory T cells
TLR	Toll-like-receptor
VLP	Virus-like-particle

A. Soria
Institut National de Santé et de Recherche Médicale, INSERM UMR-S 945 and Université Pierre et Marie Curie (UPMC—Paris 6), Laboratory of Immunity and Infection, 91, boulevard de l'Hôpital, 75013, Paris, Cedex 13, France and Department of Dermatology and Allergy, Hôpital Tenon—Assistance Publique Hôpitaux de Paris (AP-HP), 4 rue de la Chine, 75020, Paris, France
e-mail: angele.soria@tnn.aphp.fr

A. Vogt
Clinical Research Center for Hair and Skin Science, Department of Dermatology and Allergy, Charité-Universitätsmedizin Berlin, Chariteplatz 1, 10117, Berlin, Germany
e-mail: annika.vogt@charité.de

A. Nasir
Department of Dermatology, UNC Chapel Hill, Chapel Hill, NC, 67550, USA
e-mail: anasir@live.com

B. Combadière (✉)
Institut National de Santé et de Recherche Médicale, INSERM UMR-S 945 and Université Pierre et Marie Curie (UPMC—Paris 6), Laboratory of Immunity and Infection, 91, boulevard de l'Hôpital, 75013, Paris, Cedex 13, France
e-mail: behazine.combadiere@upmc.fr

16.1 Introduction

Over the past two decades, interest in the topical delivery of vaccinations has gained extraordinary interest. This surge is a direct result of the evolving understanding of the high immune potential of skin-resident and inflammatory cells, and importantly, the urgent need to improve the immunogenicity of innovative vaccines against infectious diseases. To address this approach, a variety of needle-free methods of vaccination based on nanotechnology have emerged with the potential to improve immunity to the delivered vaccines. Nanotechnology has allowed the development of vaccines based on solid or biodegradable nanoparticles (NPs), virus-like particle, virosomes, and liposomes. Translation to the bedisde has to date been limited however, due to

A. Nasir et al. (eds.), *Nanotechnology in Dermatology*,
DOI 10.1007/978-1-4614-5034-4_16, © Springer Science+Business Media, LLC 2013

the rigorous criteria of GMP production, and therefore safety, immunogenicity, and efficacy for future strategies in humans needs to be demonstrated. Here, we review our current knowledge on two directions in topical delivery of vaccinations that are progressing simultaneously; one, the skin immunization method; and two, NP-based vaccines. When combined, these two approaches could bring a new insight into vaccination.

16.2 Nanoparticle-Based Vaccines

The rationale for the use of particle-based vaccines is to optimize vaccine uptake by antigen-presenting cells (APCs) [1–3]. Thus, their delivery into the skin where a high density of APC can be found might offer a means to instigate immunity against a broad range of infectious diseases [4, 5]. This is of particular importance when considering infectious agents for which current vaccines are rather limited or vaccines are poorly immunogenic. Variables, such as the material composition, charge, size, cell-attaching surface antigens, and routes of administration are likely to be important determinants of the subseuqent immunological outcome, and therefore different approaches are currently under investigation to optimize this result.

Among such novel delivery strategies are polymer microparticles with encapsulated antigen that slow and limit degradation of antigenic proteins, peptides, or DNA, with the ability to prolong their persistence over a period of time in vivo [5, 6]. However, several classes of vaccines based on smaller particles in the nanometer range could be used to target vaccine compounds specifically to skin immune cells. Interesting candidates include virus-like particles (VLP), DNA and attenuated small viral vectors, and a family of polymeric NPs, such as poly-D, L-lactide-*co*-glycolide and poly-D, L-*co*-glycololic acid (PLGA), poly-D, L-lactic acid (PLA) and others, which can carry proteins, peptides, or DNA. Particles ranging from 20 to 200 nm in diameter can be internalized by professional APCs such as dendritic cells (DC) in clathrin-dependent internalization manner, whereas bigger particles bewteen 0.5 and 5 μm are taken up by phagocytosis or macropinocytosis by macrophages. There is evidence that NPs show increased uptake by APCs [7–10], which suggests that they may be better inducers of immune responses than microparticles. In addition, localization of DC and macrophages in lymphoid tissues where the immune response is initiated, is also distinct and might then differentially influence the polarity of the immune responses [11, 12]. It has been recently shown that submicrospheres can be modeled after mast cell granules. A major role of this type of particles is that they can incorporate antigen and cytokines. Furthermore, their localization in the lymph nodes (LNs) promotes and polarizes antigen-specific immune responses [13].

Formulations of NP-based vaccines can be modified to increase APC interaction. For instance, because of high mannose receptor expression on DC, formulations based on mannolysation of antigen-delivery system have been proposed for more efficient targeting and activating of APC such as DC and macrophages [14]. These authors showed that ovalbumine/mannan–PLGA–NPs induced stronger in vitro antigen-specific CD4+ and CD8+ T-cell responses than ovalbumin/PLGA–NPs without mannolysation [15, 16]. Another strategy to improve the efficiency of NP-based vaccines is endocytic pathway targeting such as association with C-type lectin receptors (CLRs). Cruz et al. demonstrated that targeting the DC-SIGN improved antigen processing by human DC, which resulted in increased antigen-specific T-cell activation with 10- to100-fold lower concentrations of antigen [17, 18]. Similarly, several toll-like receptors (TLRs) ligands have been incorporated into PLGA–NPs, such as TLR-4 ligand (7-acyl lipid A) [19] or TLR-9 ligand (CpG), both of which enhanced the immunogenicity of particle-based vaccines by increasing APC activation [20].

VLPs, liposomes, and virosomes are also considered as particle-based vaccine, but are non-infective viral or biological products consisting of assembled viral-envelope proteins into lipid layers without genetic material [21]. They have also been shown to stimulate both cellular and humoral immunity (Grgacic Methods 2006, de Bruijn

Vaccine 2005). Recent successes of VLP vaccine development include the Human Papilloma Virus vaccine (Gardasil®, Cervarix®) and HBV (Engerix-B®, Recombivax-HB®) administrated by intramuscular (IM) route [22, 23]. VLP vaccines containing influenza hemagglutinin (HA) and neuraminidase (NA) antigens are also under development for future influenza vaccination strategies [24] as well as in other diseases [25, 26].

In addition, DNA-encoding immunogens can be coated on NPs or be delivered using polylysine formulation. Singh et al. have shown, using pDNA-coated poly (D,L-lactic-coglycolic acid) (PLGA) microspheres, that with smaller microspheres (300 nm > 1 μm > 30 μm), there is a stronger-induced immune response [27], suggesting that smaller and more readily engineered NPs would preferentially target APCs. In order to increase the efficacy of in vivo transfection of skin cells and antigen presentation, delivery methods (bio-injectors, gene gun, or electroporation) have been developed [28]. DNA injection followed by electroporation dramatically increases the amount of antigen at the site of immunization and resulted in both increased humoral and cellular immune responses in human [29–31] and mice models [32]. Although many electroporation methods have been used for muscle tissues in the past, both electroporation and bio-injector methods have recently been applied to the skin tissue and shown to induce direct-presentation of the antigen by skin APCs [33] or cross-presentation after uptake by keratinocytes [34–38].

Other non-degradable particles of variable or undefined size have been used to pre-activate DCs after being mixed with or conjugated to antigens. However, it was shown that nondegradable particles were much less effective at cross-presenting antigens than degradable particles [39–41].

16.3 Skin Immunology

16.3.1 Innate Immunity in the Skin

The skin is a major immunological organ with a dense network of professional APCs consisting of mainly Langerhans cells (LCs) [42, 43] in the epidermis and various subpopulation of dermal dendritic cells (DDC) and dermal macrophages (DMF) in the dermis [44, 45] (see also Fig. 16.1 for additional references).

The epidermis is composed of stratified layers, including squamous and keratinized covering, contains the stratum basal, the stratum spinosum, the stratum granulosum, and the stratum corneum. The stratum corneum lipids are composed primarily of ceramides, cholesterol, and fatty acids that are assembled into multilamellar bilayers. The stratum corneum is pierced by several appendages, i.e., sweat glands and hair follicles (HFs). Other specialized cells in the epidermis include melanocytes, which produce pigment (melanin), and adaptive immune memory cells such as antigen-spepcific CD8+ cytotoxic T-cells that can be found in the stratum basal and spinosum [46, 47]. Overall, 2–3 % of LCs circulate naturally from the skin to the LNs even in the absence of inflammatory signals. The homeostatic migration of LCs from the epidermis to the lymphoid tissues is due to CCR7 expression on LCs and the chemoattractive chemokine CCL20 chemokine that is constitutively expressed by lymphatic vessel and by endothelial cells.

The dermis is composed of collagen and elastic tissue and reticula fibers. It contains many specialized cells, such as dendritic cells (DC) including dermal DCs (DDCs) [48], dermal MF, and plasmacytoid DCs (pDCs), memory T-cells (including CD4 and CD8), γδ T-cells, natural killer (NK), and NKT cells. In healthy human normal skin, γδ CD3 T cells, NK, NKT cells, and pDCs constitute the innate immune compartments. Their role in vaccination using particle-based vaccine has not yet been elucidated. Mast cells and fibroblasts are also present. Fibroblasts are mesenchymal cells that synthesize the extra cellular matrix components. They also secrete cytokines that play an essential role in the mechanisms of cutaneous inflammation such as IL-6 (involved in DC activation) and TGF-β (role in the maturation of LC precursors). Blood and lymphatic vessels and nerves are present throughout the dermis, facilitating recruitment of immune cells (neutrophils, monocytes, monocytes derived DCs, memory T cells, and pDC) [49].

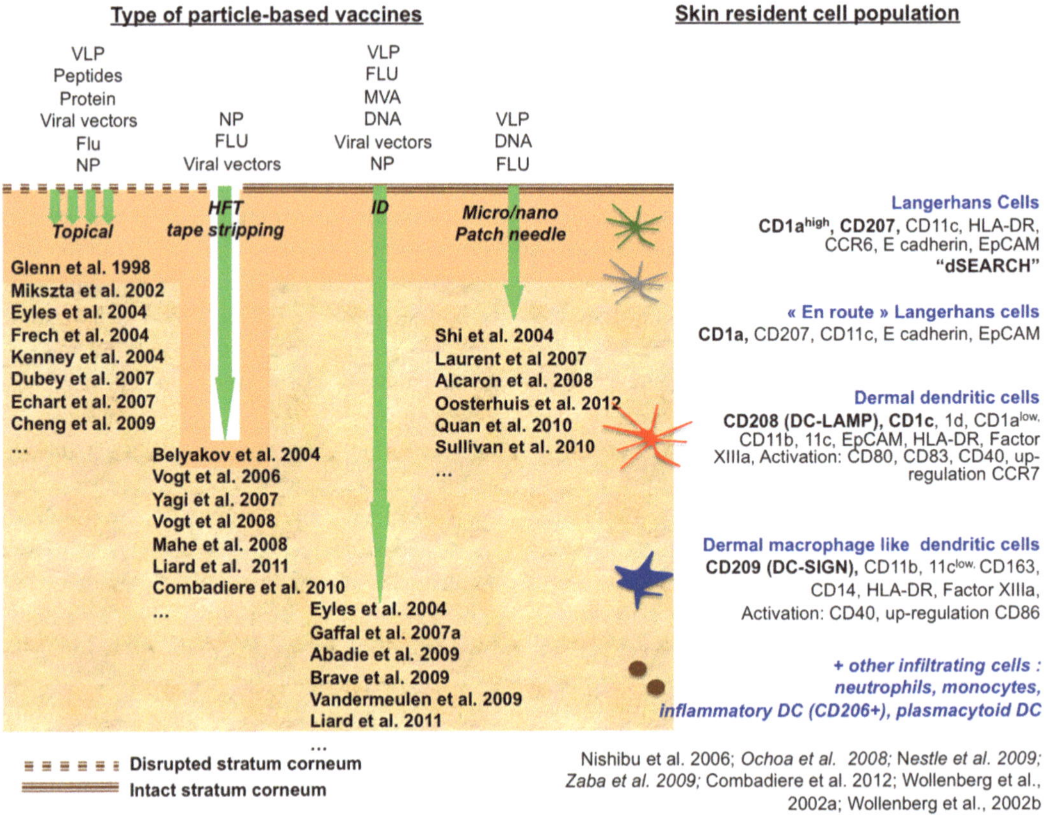

Fig. 16.1 Skin modes of particle-based vaccine penetration. Four distinct penetration pathways of vaccine targeting various skin layers. Each penetration route preferentially allows the uptake by antigens by distinct subpopulations of skin antigen-presenting cells (APC). *NP* nanoparticle; *VLP* virus-like particles; *MVA* modified vaccinia Ankara; *Flu* influenza; *LC* langerhans cells; *DDC* dermal dendritic cells; *DMF* dermal macrophages; *HFT* hair follicular targeting; *ID* intra-demal

The hypodermis is a layer of white fat also called as adipose tissue, connected to the lower part of the dermis by extensive collagen fibers. It contains only fibroblasts and adipocytes. Vaccine injected subcutaneously is more prompted to remain in the fatty layers and are uptaken by monocytes and transported at low rate to peripheral lymphoid tissues.

16.3.2 Adaptive Immunity in the Skin

The concept of the skin-associated lymphoid tissue (SALT) was first described in 1983 by Streilein [50] and later Bos and col. introduced the skin immune system (SIS) [51–53]. These concepts introduced the idea of circulating immune cells that continuously traffic between the skin and the draining LNs, providing optimal immunosurveillance. Moreover, the distal part of the follicular immune system appears to represent a specialized compartment of the SIS. There is an interaction between intra-epithelial T cells and LCs, hich generates a particularly effective anti-infectious defense system [54–56].

Indeed, the skin tissue contains nearly 2×10^{10} resident T cells (1×10^{6} cells per cm^2)—twice as many as the total number of circulating T cells in the blood [46, 57, 58]. When confronted with an infectious pathogen, the immune system responds (amplification phase), controls (effector and contraction phases), and remembers (memory phase). In parallel to B cells that produce antibodies (Abs), two different types of CD4 and

CD8 T cells are generated [59]: effector memory T cells (TEM), circulating and homing to peripheral tissues and central memory T cells (TCM) mostly in lymphoid tissues [60]. Puissant-Lubrano et al. have shown that the residual IFNγ+ TNFα+ and IFNγ+IL-2+ long-term effector/memory CD4 lymphocytes generated in response to smallpox vaccination (dermal injection) expressed skin-homing marker (CLA+ for Cutaneous-Leukocyte Antigen). The presistence of vaccinia-virus memory cells correlates with a reduction in clinical disease after viral challenge in humans. These results suggest a curical role for circulating TEM expressing a skin-homing marker on the control of virus infection in the skin [61].

Histologic studies of the skin have shown that more than 95 % of the T cells are CD45RO memory T cells and can respond rapidly to antigen restimulation [62]. Less than 5 % of T cells are CD45ROneg expressing the CLA along with P- and E-selectin ligands that allow them to bind to vascular endothelial cells [63]. Skin T cells also co-express CCR4 and CCR10 receptors, which interact with the CCL17 and CCL27 chemokines expressed in the dermis, and for 50 % of them CCR8 [64–67]. In normal skin, in the absence of inflammatory stimuli, the majority (>90 %) of CLA+ skin-homing TEM are resident in skin. As they do not express receptors for CCR7, these TEM cells cannot enter into the LNs and reside in skin [57, 58]. The skin TEM CD4+ and CD8+ T lymphocytes are grouped in clusters, on the post-capillary venules under the dermal–epidermal junction. Adaptive T cells in the skin are also of importance because they constitute the major first line of defense for "recall" antigen that might be re-introduced into the skin during re-infection. So far, the interaction of particle-based vaccine containing a "recall" antigen during re-vaccination with antigen-specific T cells has not been studied.

The follicular canal is a major port of entry for microbial invasion; therefore, the distal outer root sheat is specialized in microbial antigen recognition [54, 68, 69]. This area of the human hair follicle contains many more CD4+ and CD8+ T cells than the interfollicular epidermis [70]. During the growth phase (anagen), the most prominent immune cells located in or around the hair follicle are LC, CD4+, CD8+ T cells, macrophages, and mast cells whereas B cells, natural killer cells, and γδ T cells are sparse [71]. During anagen, they can mostly be found in the distal, i.e., superficial compartments of the hair follicle, while the proximal bulb and the peribulbar region are characterized by a paucity of immune cells ("immune privilege of the hair follicle"). In the SIS, LC and DC are found in distal hair follicle epithelium whereas macrophages and mast cells are distributed in the perifollicular connective tissue sheath [70].

The skin has essential immunological functions, both during tissue homeostasis and in various pathological conditions due to continous trafficking of circulating immune cells between the skin, the draining LNs but also mucosal sites. Further studies are needed to explore possible strategies to utilize those interactions for particle-based vaccination strategies [12].

16.4 Skin Immunization: Can We Reach an Optimal Delivery of Nanoparticle-Based Vaccines?

16.4.1 Passive Transcutaneous Vaccination

Passive transcutaneous (TC) vaccination consists of a topical application of vaccine compounds. The major limitation unfortunately is the relative impermeability of the horny layer for all particle-based vaccines. The skin barrier, with its highly lipophilic stratum corneum, limits the skin penetration of exogenous substances. The stratum corneum lipids are composed primarily of ceramides, cholesterol, and fatty acids that are assembled into multilamellar bilayers. Generally, a molecular mass less than 500 Da and high lipophilicity favor passive penetration across the horny layer. Passive penetration of hydrophilic molecules such as peptides, proteins, DNA, or particle-based vaccine is however difficult through intact stratum corneum [72].

16.4.2 Epidermal Penetration of Particle-Based Vaccines

In the literature, the terms epidermal, TC, or transdermal vaccination are all being used for different approaches, which aim at targeting vaccine across the skin barrier to epidermal or dermal DC. Obviously, vaccination technologies which include topical application of vaccines on the skin surface have the potential of being non-invasive, needle-free, and painless [4, 12]. Because intact stratum corneum significanlty impairs the penetration of larger compounds, numerous technologies have been proposed to improve penetration and overcome the skin barrier, in most cases by transient disruption of its structure [73, 74]. Many of these techniques, albeit based on topical vaccine applications, include forced penetration, significant skin surface abrasion by chemical or physical means, or barrier disruption by microneedle systems, which frequently target considerble amounts of the vaccine to the dermis rather than the epidermis.

One widely used technique to effectively deliver larger compounds into the skin is gene-gun technology, which permits targeting of both epidermis and dermis and uses compressed helium to propel micrometer-sized particles. This technique successfully induced humoral and cellular immune responses [75–77] and has been investigated in various different animal models against several diseases, including influenza, Hepatitis B virus, Marburg fever, Malaria, Ebola, and Hanta viruses [78]. Penetration enhancement can also be achieved by application of electric fileds to move both charged and uncharged species across the skin, or by acoustical methods with ultrasound or ultrasonic waves at various frequencies [79]. High-voltage pulses transiently increase skin permeability by a mechanism called electroporation mostly used for gene expression using plamid DNA vectors but also peptides and proteins [80, 81]. It has been reported that electroporation technique produces a 10- to 1,000-fold increase in the expression of transgene after direct tissue injection of naked plasmid DNA [82]. The electroporation for DNA vaccine is a highly efficient method of DNA delivery that induces both humoral and cellular immune responses [83–85]. Thus, human clinical trials using DNA vaccines with electroporation are widely developed today against various infectious diseases (HIV, HCV, influenza, and malaria) [86] and cancers [38]. Finally, the use of chemical compounds causes transient disruption of the horny layer structure and helps augment NP penetration [4, 72].

The presented technologies aim at delivering the complete vaccine contstructs, e.g., particle-bound vaccine or vaccine compounds with a particulate structure into the skin. Even though these approaches are mostly considered non-invasive and are usually well tolerated, it has to be taken into account that many of these techniques cause inflammation or cell death in the epidermis, which may influence the immunological outcomes. In contrast, vesicular particles of the liposome-type, including niosomes, ultradeformable transferosomes, and others, are promising tools, which act as penetration enhancers. Such carrier systems shuttle vaccine through/across the skin barrier and release their incorporated compounds into the viable tissue [87–89].

16.4.3 Immunization Through Hair Follicles

Unlike vesicular particles, e.g., lipsosomes, solid particles do not fuse with the horny layers and usually remain on the surface of intact stratum corneum. Based on multiple experimental studies using different model systems and optical methods ranging from conventional microscopy techniques to X-ray microscopy, there is a consensus in the field of nanosciences that most types of NPs do not penetrate the skin barrier as long as it is intact. Even under physiological conditions, however, natural interruptions occur in the horny layer mainly due to sweat gland and hair follicle openings. In particular, the invaginations caused by hair follicles result in a significant increase in the skin surface area available for penetration. Moreover, the penetration pathway into and along hair follicles seems to be especially relevant for large molecules and particles, and there is evidence that such particles remain in the hair follicle openings much longer, i.e., for up to 10 days rather than on the epidermis where they are being shed with the corneocytes (for more details, refer to Chapter 9, "Hair Follicle Targeting with Nanoparticles"). Such extended time periods offer the opportunity for both resident and infiltrated APCs to take up the vaccine compounds and process the antigens [90].

During the past years, multiple experimental studies in mice and in human skin confirm that the hair follicle plays a major role for the penetration of topically applied NPs. In fact, it seems to be a major site of translocation for particles and particulate vaccines up to a diameter of approximately 300 nm. For example, topical application of several sizes of nanoparticles on the skin showed that 40 nm polystyrene particles penetrated into hair follicles openings and were uptaken by LCs [91, 92]. It has to be noted that those studies were conducted using a tape-stripping technique, which results in the opening of hair follicles by removing cellular debris and dried sebum and thereby increases the hair follicles available for penetration. This technique was brought forward by our groups as a novel approach for TC vaccination also in clinical trials on humans (see later). However, because hair follicle penetration in skin explants was also shown to be approximately on the order of ten times less than under in vivo conditions, it lent itself as an improtant experimental step to study hair follicle penetration ex vivo. In mice we monitored particle penetration in hair follicles and translocation into the perifollicular tissue in vivo using 200 nm NPs or modified vaccinia Ankara (MVA) virus (approximatively 290 nm) [93, 94]. Hair follicle penetration of either 40–50 nm (latex beads) or 200 nm (PLA fluorescent or protein-coated NP) or MVA (290 nm) showed preferential uptake by LCs and transport to lymphoid tissues, where both humoral and cellular responses were initiated [93, 95, 96]. It was also demonstrated that 40–49 nm NPs can stimulate CD8+ T-producing IFNg effector cells (Th1 response) whereas bigger particles (93–123 nm) will direct the immune response toward the production of IL-4 secreting CD4+ T cells (Th2 response) [97].

The follicular duct epidermal sheath is also associated with an environment with high immune potential. It is surrounded by LCs and closely related to DDCs. LCs are the main DC subset in the epidermis and constitutively reside in the suprabasal layers and are regularly spaced among keratinocytes. The main surface markers are CD1a, CD207 (or Langerin) in humans and in mice, a C-lectin type II receptor internalized in Birbeck granules. When LCs mature, they express E-cadherin, an adhesion marker in epithelial cells (Fig. 16.1), and are among the first DCs to come in contact with microbial antigens [98–101]. Several functions were shown in vitro; LCs takes up and process antigen and microbial fragment for presentation to effectors T cells [102]. Importantly, there is a regular extension of LC dendrites between keratinocytes: the "dSEARCH" (dendrite surveillance extension and retraction cycling habitude) that might facilitate particle uptake [103]. LCs do indeed function as epidermal sentinels surveying their environment and possibly communicating with each other by "touching" after activation [104]. Another subtype of epidermal DCs, known as inflammatory dendritic epidermal cells (IDECs), are distinguished from LC by the expression of the macrophage mannose receptor CD206. This receptor was found in the inflamed epidermis of patients with atopic dermatitis [105, 106]. Their role and origin remain to be studied.

16.4.4 Intradermal Delivery

Antigen can be directly delivered into the dermis by conventional intradermal (ID) needles, microneedles, or pressure-injector. Charles Mantoux invented the standard ID immunization technique in 1910, which consists in inserting a 26–27 Gauge needle in the skin with the needle at a 45° angle. An adverse event is that ID vaccination is more painful than conventional IM or subcutaneous (SC) injections and can cause local inflammatory reactions [107]. The benefit here, however, is that this technique allows for an injection with less vaccine solution (about 100 μL, i.e.,

1/5th of the classical volume injected with the IM or SC route) [12, 107]. Because of the difficulty to succeed in ID delivery by mantoux injection method, standardized micron-scale needles have also been used to pierce the stratum corneum and create holes large enough for molecules to enter and small enough to avoid pain [108, 109]. This method can also be used to deliver proteins [110, 111] and DNA [112], as well as miniaturized and applied in drug-coated patch. In a rat model, influenza vaccine administered with microneedles induced similar antibody responses compared to intra-muscular (IM) delivery with fivefold less vaccine dose [113]. In addition, a novel dissolving polymer-microneedle patch for influenza VLP vaccination in mice has been developed. This technique showed similar Abs titers (hemagglutinin inhibition assays) in the microneedle group compared to IM-vaccinated group [114].

ID immunization with PLA or MVA in human skin explants induced respectively after 2 and 4 h morphological modifications of epidermal LCs; decrease dendrite length, dendrite number, and LC area. Moreover, the density of LCs was decreased in epidermis between 2 and 4 h after PLA or MVA ID immunization [96]. In a Langerin-diphteria toxin receptor (DTR) transgenic mice conditional depletion of LCs in epidermis, we showed that LCs are necessary to induced a MVA-specific IFNg-secreting CD8+ T cells, but not of CD4 T cells or humoral response [96]. ID injection of recombinant HCV-VLPs induced strong and broad humoral and cellular immune responses in mice [115, 116]. Similar approaches induced a protective immunity in chimpanzees [117].

In conclusion, ID delivery induced superior or comparable (for lower doses by ID route) of both humoral and cellular protection compared to IM vaccination [12, 118, 119].

16.4.5 Subcutaneous Delivery

The hypodermis is a layer of adipose tissue, connected to the lower part of the dermis by extensive collagen fibers. It contains only fibroblasts and adipocytes. But in contrast to the epidermis or the dermis, it is naturally devoid of resident immune cells. SC administration of smaller NP (20–200 nm) and virus-like-particles (VLPs) about 30 nm showed that particles are in resident skin DCs and under capsulate macrophages in draining LNs. Transcutaneous (TC), intradermal (ID), and subcutaneous (SC) routes of immunization using PLA-NP coated with HIV p24 antigen were compared for the quality of immune responses. We demonstrated that SC immunization induced HIV-1 p24-specific IgG in the absence of antigen-specific CD8 T cells, whereas the ID route induced both cellular and humoral responses [95]. Interestingly, TC application through hair follicular ducts, targeting epidermal LCs, induced major antigen-specific CD8 effector cells in the absence of IgG in the serum. In addition, ID and TC routes induced IgA in vaginal lavage when SC did not [95]. These results showed the major difference in the quality of immune responses induced using TC, ID, and SC immunization.

In conclusion, passive penetration, topical application, hair follicle penetration, forced injection (macro, micro, nanoneedle, or with pressure) are different modes of vaccination that when combined with different types of particle-based compounds induce changes to the micro-environement of the skin that are not well characterized or compared in the literature. Most authors aim to validate their mode of administration or vaccine compounds by studying the final outcome of the immune responses and protection in murin models. However, these first steps in vaccination including antigen-uptake, transport to the lymphoid organs, localization of antigen and the amount presented in the LNs, and proper processing and presentation by APCs to the T and B cell that are in specific zones of lymphoid tissues are essential to the quality and intensity of immune responses [11, 12]. Furthermore, recent advances in systems biology studies have shown the crucial importance of innate immunity that dictates adaptive immunity to viral infection or after vaccination [120–122] and need be investigated in the future for nantechnology-based vaccines.

16.5 Human Clinical Trials: How Far Are We?

16.5.1 Epidermis as Target Skin Layer for Vaccination in Human

The potential for TC vaccination with patch or tape-stripping has been validated with several vaccine models in animals studies and human clinical trials. Transcutanoeus routes have been studied in various diseases among which diphtheria [123], cholera toxine [124–127], tetanus [128], yersiniosis [129], herpes virus simplex 1 [130], HIV [131], influenza [92, 132], vaccinia virus [93], anthrax [133], and tumoral peptides [134, 135].

Various vaccine models are being studied in phase-I and -II human clinical trials. For immunization against *Escherichia coli* (*E. coli*), three clinical trials with *E. coli* lymphotoxin applied in patch form [127] demonstrated that patch application promotes a high rate of antibodies anti-lymphotoxin. Other studies with seasonal influenza vaccine in the elderly resulted in better seroconversion with patch than with IM injection [136, 137]. TC vaccination using cyanoacrylate tape-surface stripping (CSSS) and cutaneous application of trivalent inactivated influenza vaccine induced both CD4+ and CD8+ T cell response whereas IM injection of vaccine induced only strong effectors CD4+ T cell response but not influenza-specific CD8 cells [92, 132]. A clinical trial phase I/II with live attenuated measles vaccine (Rouvax®) patch had showed the production of IgA in saliva and measles-specific CD8+ T cell immune response [138]. Lastly, in a clinical trial with seven patients with advanced melanoma, following five CSSS applications of a mixture of melanoma and HIV peptides, melanoma- and HIV-specific cytotoxic T-cell were noted in vivo [135]. There are clearly multiple studies which suggest that targeting of mostly epidermal DCs, e.g., by vaccine delivery via hair follicles, induces CD8+ T cell immune response whereas IM route did not [139–151].

16.6 Conclusion

Targeting skin APCs for vaccination and using nano-based vaccines for optimal delivery will likely have significant benefits in enhancing the immune responses, polarizing T cell immunity, reducing vaccine doses as well as the number of injection if the delivery method is reproducible and acceptable. Human clinical trials have shown that skin-immunization with conventional vaccine is comparable and sometimes even superior when compared to conventional vaccination. More human clinical trials are necessary to evalaute the efficacy of particle-based vaccines (NPs, VLP, virosomes) in the future and should also be evaluated in immunocompromised populations (elderly, HIV infected, immunosuppressed patients). Further investigations are needed, especially with respect to their tolerance, safety, feasibility, and immunogenicity in human.

Acknowledgments This work has been funded by the EU-FP7 health program CUT'HIVAC "Cutaneous and Mucosal HIV vaccination" (No. 241904), Fondation pour la Recherche Medicale (FRM France), and Agence National de Recherche (ANR-bio-mergence). B. Combadière is an awardee of the INSERM-Interface AP/HP program.

References

1. Jewell CM, Lopez SC, Irvine DJ. In situ engineering of the lymph node microenvironment via intranodal injection of adjuvant-releasing polymer particles. Proc Natl Acad Sci U S A. 2011;108:15745–50.
2. Klippstein R, Pozo D. Nanotechnology-based manipulation of dendritic cells for enhanced immunotherapy strategies. Nanomedicine. 2010;6:523–9.
3. Rice-Ficht AC, Arenas-Gamboa AM, Kahl-McDonagh MM, Ficht TA. Polymeric particles in vaccine delivery. Curr Opin Microbiol. 2010;13:106–12.
4. Combadiere B, Mahe B. Particle-based vaccines for transcutaneous vaccination. Comp Immunol Microbiol Infect Dis. 2008;31:293–315.
5. Gupta RK, Chang AC, Griffin P, Rivera R, Guo YY, Siber GR. Determination of protein loading in biodegradable polymer microspheres containing tetanus toxoid. Vaccine. 1997;15:672–8.
6. Langer R, Cleland JL, Hanes J. New advances in microsphere-based single-dose vaccines. Adv Drug Deliv Rev. 1997;28:97–119.

7. Foged C, Brodin B, Frokjaer S, Sundblad A. Particle size and surface charge affect particle uptake by human dendritic cells in an in vitro model. Int J Pharm. 2005;298:315–22.
8. O'Hagan DT, Rappuoli R. Novel approaches to vaccine delivery. Pharm Res. 2004;21:1519–30.
9. O'Hagan DT, Singh M, Dong C, Ugozzoli M, Berger K, Glazer E, et al. Cationic microparticles are a potent delivery system for a HCV DNA vaccine. Vaccine. 2004;23:672–80.
10. O'Hagan DT, Singh M, Ulmer JB. Microparticles for the delivery of DNA vaccines. Immunol Rev. 2004;199: 191–200.
11. Abadie V, Bonduelle O, Duffy D, Parizot C, Verrier B, Combadiere B. Original encounter with antigen determines antigen-presenting cell imprinting of the quality of the immune response in mice. PLoS One. 2009;4:e8159.
12. Combadiere B, Liard C. Transcutaneous and intradermal vaccination. Hum Vaccin. 2011;7:811–27.
13. St John AL, Chan CY, Staats HF, Leong KW, Abraham SN. Synthetic mast-cell granules as adjuvants to promote and polarize immunity in lymph nodes. Nat Mater. 2012;11(3):250–7.
14. Ghotbi Z, Haddadi A, Hamdy S, Hung RW, Samuel J, Lavasanifar A. Active targeting of dendritic cells with mannan-decorated PLGA nanoparticles. J Drug Target. 2011;19:281–92.
15. Hamdy S, Haddadi A, Hung RW, Lavasanifar A. Targeting dendritic cells with nano-particulate PLGA cancer vaccine formulations. Adv Drug Deliv Rev. 2011;63:943–55.
16. Hamdy S, Haddadi A, Shayeganpour A, Samuel J, Lavasanifar A. Activation of antigen-specific T cell-responses by mannan-decorated PLGA nanoparticles. Pharm Res. 2011;28:2288–301.
17. Cruz LJ, Rueda F, Cordobilla B, Simon L, Hosta L, Albericio F, et al. Targeting nanosystems to human DCs via Fc receptor as an effective strategy to deliver antigen for immunotherapy. Mol Pharm. 2011;8:104–16.
18. Cruz LJ, Tacken PJ, Fokkink R, Figdor CG. The influence of PEG chain length and targeting moiety on antibody-mediated delivery of nanoparticle vaccines to human dendritic cells. Biomaterials. 2011;32:6791–803.
19. Hamdy S, Molavi O, Ma Z, Haddadi A, Alshamsan A, Gobti Z, et al. Co-delivery of cancer-associated antigen and Toll-like receptor 4 ligand in PLGA nanoparticles induces potent CD8+ T cell-mediated anti-tumor immunity. Vaccine. 2008;26:5046–57.
20. Fischer S, Schlosser E, Mueller M, Csaba N, Merkle HP, Groettrup M, et al. Concomitant delivery of a CTL-restricted peptide antigen and CpG ODN by PLGA microparticles induces cellular immune response. J Drug Target. 2009;17:652–61.
21. Jennings GT, Bachmann MF. The coming of age of virus-like particle vaccines. Biol Chem. 2008;389:521–36.
22. Block SL, Brown DR, Chatterjee A, Gold MA, Sings HL, Meibohm A, et al. Clinical trial and post-licensure safety profile of a prophylactic human papillomavirus (types 6, 11, 16, and 18) l1 virus-like particle vaccine. Pediatr Infect Dis J. 2010;29:95–101.
23. Buonaguro FM, Tornesello ML, Buonaguro L. Virus-like particle vaccines and adjuvants: the HPV paradigm. Expert Rev Vaccines. 2009;8:1379–98.
24. Haynes JR. Influenza virus-like particle vaccines. Expert Rev Vaccines. 2009;8:435–45.
25. Wagner R, Deml L, Teeuwsen V, Heeney J, Yiming S, Wolf H. A recombinant HIV-1 virus-like particle vaccine: from concepts to a field study. Antibiot Chemother. 1996;48:68–83.
26. Yang C, Ye L, Compans RW. Protection against filovirus infection: virus-like particle vaccines. Expert Rev Vaccines. 2008;7:333–44.
27. Singh M, Kazzaz J, Ugozzoli M, Chesko J, O'Hagan DT. Charged polylactide co-glycolide microparticles as antigen delivery systems. Expert Opin Biol Ther. 2004;4:483–91.
28. Oosterhuis K, van den Berg JH, Schumacher TN, Haanen JB. DNA vaccines and intradermal vaccination by DNA tattooing. Curr Top Microbiol Immunol. 2012;351:221–50.
29. Murakami T, Sunada Y. Plasmid DNA gene therapy by electroporation: principles and recent advances. Curr Gene Ther. 2011;11:447–56.
30. Sardesai NY, Weiner DB. Electroporation delivery of DNA vaccines: prospects for success. Curr Opin Immunol. 2011;23:421–9.
31. van Drunen Littel-van den Hurk S, Hannaman D. Electroporation for DNA immunization: clinical application. Expert Rev Vaccines. 2010;9:503–17.
32. Abdulhaqq SA, Weiner DB. DNA vaccines: developing new strategies to enhance immune responses. Immunol Res. 2008;42:219–32.
33. Lee PW, Hsu SH, Tsai JS, Chen FR, Huang PJ, Ke CJ, et al. Multifunctional core-shell polymeric nanoparticles for transdermal DNA delivery and epidermal Langerhans cells tracking. Biomaterials. 2010;31: 2425–34.
34. Cheng JY, Huang HN, Tseng WC, Li TL, Chan YL, Cheng KC, et al. Transcutaneous immunization by lipoplex-patch based DNA vaccines is effective vaccination against Japanese encephalitis virus infection. J Control Release. 2009;135:242–9.
35. Kim D, Hoory T, Wu TC, Hung CF. Enhancing DNA vaccine potency by combining a strategy to prolong dendritic cell life and intracellular targeting strategies with a strategy to boost CD4+ T cell. Hum Gene Ther. 2007;18:1129–39.
36. Mkrtichyan M, Ghochikyan A, Movsesyan N, Karapetyan A, Begoyan G, Yu J, et al. Immunostimulant adjuvant patch enhances humoral and cellular immune responses to DNA immunization. DNA Cell Biol. 2008;27:19–24.
37. Rice J, Dossett ML, Ohlen C, Buchan SL, Kendall TJ, Dunn SN, et al. DNA fusion gene vaccination mobilizes effective anti-leukemic cytotoxic T lymphocytes from a tolerized repertoire. Eur J Immunol. 2008;38:2118–30.

38. Rice J, Ottensmeier CH, Stevenson FK. DNA vaccines: precision tools for activating effective immunity against cancer. Nat Rev Cancer. 2008;8:108–20.
39. Kwon YJ, James E, Shastri N, Frechet JM. In vivo targeting of dendritic cells for activation of cellular immunity using vaccine carriers based on pH-responsive microparticles. Proc Natl Acad Sci U S A. 2005;102:18264–8.
40. Kwon YJ, Standley SM, Goh SL, Frechet JM. Enhanced antigen presentation and immunostimulation of dendritic cells using acid-degradable cationic nanoparticles. J Control Release. 2005;105:199–212.
41. Kwon YJ, Standley SM, Goodwin AP, Gillies ER, Frechet JM. Directed antigen presentation using polymeric microparticulate carriers degradable at lysosomal pH for controlled immune responses. Mol Pharm. 2005;2:83–91.
42. Igyarto BZ, Kaplan DH. The evolving function of Langerhans cells in adaptive skin immunity. Immunol Cell Biol. 2010;88:361–5.
43. Streilein JW, Lonsberry LW, Bergstresser PR. Depletion of epidermal langerhans cells and Ia immunogenicity from tape-stripped mouse skin. J Exp Med. 1982;155:863–71.
44. Romani N, Flacher V, Tripp CH, Sparber F, Ebner S, Stoitzner P. Targeting skin dendritic cells to improve intradermal vaccination. Curr Top Microbiol Immunol. 2012;351:113–38.
45. Teunissen MB, Haniffa M, Collin MP. Insight into the immunobiology of human skin and functional specialization of skin dendritic cell subsets to innovate intradermal vaccination design. Curr Top Microbiol Immunol. 2012;351:25–76.
46. Clark RA. Skin-resident T cells: the ups and downs of on site immunity. J Invest Dermatol. 2010;130:362–70.
47. Rupec RA, Boneberger S, Ruzicka T. What is really in control of skin immunity: lymphocytes, dendritic cells, or keratinocytes? facts and controversies. Clin Dermatol. 2010;28:62–6.
48. Nestle FO, Nickoloff BJ. Dermal dendritic cells are important members of the skin immune system. Adv Exp Med Biol. 1995;378:111–6.
49. Nestle FO, Di Meglio P, Qin JZ, Nickoloff BJ. Skin immune sentinels in health and disease. Nat Rev. 2009;9:679–91.
50. Streilein JW. Skin-associated lymphoid tissues (SALT): origins and functions. J Invest Dermatol. 1983;80(Suppl):12s–6.
51. Bos JD, Kapsenberg ML. The skin immune system: progress in cutaneous biology. Immunol Today. 1993;14:75–8.
52. Bos JD, Luiten RM. Skin immune system. Cancer Treat Res. 2009;146:45–62.
53. Bos JD, Zonneveld I, Das PK, Krieg SR, van der Loos CM, Kapsenberg ML. The skin immune system (SIS): distribution and immunophenotype of lymphocyte subpopulations in normal human skin. J Invest Dermatol. 1987;88:569–73.
54. Paus R, Christoph T, Muller-Rover S. Immunology of the hair follicle: a short journey into terra incognita. J Investig Dermatol Symp Proc. 1999;4:226–34.
55. Paus R, Ito N, Takigawa M, Ito T. The hair follicle and immune privilege. J Investig Dermatol Symp Proc. 2003;8:188–94.
56. Ruckert R, Hofmann U, van der Veen C, Bulfone-Paus S, Paus R. MHC class I expression in murine skin: developmentally controlled and strikingly restricted intraepithelial expression during hair follicle morphogenesis and cycling, and response to cytokine treatment in vivo. J Invest Dermatol. 1998;111:25–30.
57. Clark RA, Chong B, Mirchandani N, Brinster NK, Yamanaka K, Dowgiert RK, et al. The vast majority of CLA+ T cells are resident in normal skin. J Immunol. 2006;176:4431–9.
58. Clark RA, Chong BF, Mirchandani N, Yamanaka K, Murphy GF, Dowgiert RK, et al. A novel method for the isolation of skin resident T cells from normal and diseased human skin. J Invest Dermatol. 2006;126:1059–70.
59. Sallusto F, Lenig D, Forster R, Lipp M, Lanzavecchia A. Two subsets of memory T lymphocytes with distinct homing potentials and effector functions. Nature. 1999;401:708–12.
60. Mackay CR, Marston WL, Dudler L. Naive and memory T cells show distinct pathways of lymphocyte recirculation. J Exp Med. 1990;171:801–17.
61. Puissant-Lubrano B, Bossi P, Gay F, Crance JM, Bonduelle O, Garin D, et al. Control of vaccinia virus skin lesions by long-term-maintained IFN-gamma+TNF-alpha+effector/memory CD4+ lymphocytes in humans. J Clin Invest. 2010;120:1636–44.
62. Wherry EJ, Teichgraber V, Becker TC, Masopust D, Kaech SM, Antia R, et al. Lineage relationship and protective immunity of memory CD8 T cell subsets. Nat Immunol. 2003;4(3):225–34.
63. Picker LJ, Terstappen LW, Rott LS, Streeter PR, Stein H, Butcher EC. Differential expression of homing-associated adhesion molecules by T cell subsets in man. J Immunol. 1990;145:3247–55.
64. Koelle DM, Huang J, Hensel MT, McClurkan CL. Innate immune responses to herpes simplex virus type 2 influence skin homing molecule expression by memory CD4+ lymphocytes. J Virol. 2006;80:2863–72.
65. Scala E, Guerra EC, Giani M, Pirrotta L, Locanto M, Mondino C, et al. Delayed allergic reaction to suxamethonium driven by oligoclonal Th1-skewed CD4+CCR4+IFN-gamma+memory T cells. Int Arch Allergy Immunol. 2006;141:24–30.
66. Schaerli P, Ebert LM, Moser B. Comment on "The vast majority of CLA+ T cells are resident in normal skin". J Immunol 2006;177:1375–6; author reply 1376–7.
67. Seneviratne SL, Black AP, Jones L, di Gleria K, Bailey AS, Ogg GS. Interleukin-4 promotes human CD8 T cell expression of CCR7. Immunology. 2007;120:66–72.

68. Paus R, Eichmuller S, Hofmann U, Czarnetzki BM, Robinson P. Expression of classical and non-classical MHC class I antigens in murine hair follicles. Br J Dermatol. 1994;131:177–83.
69. Paus R, van der Veen C, Eichmuller S, Kopp T, Hagen E, Muller-Rover S, et al. Generation and cyclic remodeling of the hair follicle immune system in mice. J Invest Dermatol. 1998;111:7–18.
70. Christoph T, Muller-Rover S, Audring H, Tobin DJ, Hermes B, Cotsarelis G, et al. The human hair follicle immune system: cellular composition and immune privilege. Br J Dermatol. 2000;142:862–73.
71. Paus R, Hofmann U, Eichmuller S, Czarnetzki BM. Distribution and changing density of gamma-delta T cells in murine skin during the induced hair cycle. Br J Dermatol. 1994;130:281–9.
72. Bos JD, Meinardi MM. The 500 Dalton rule for the skin penetration of chemical compounds and drugs. Exp Dermatol. 2000;9:165–9.
73. Prausnitz MR. Microneedles for transdermal drug delivery. Adv Drug Deliv Rev. 2004;56:581–7.
74. Prausnitz MR, Mitragotri S, Langer R. Current status and future potential of transdermal drug delivery. Nat Rev Drug Discov. 2004;3:115–24.
75. Gaffal E, Schweichel D, Tormo D, Steitz J, Lenz J, Basner-Tschakarjan E, et al. Comparative evaluation of CD8+CTL responses following gene gun immunization targeting the skin with intracutaneous injection of antigen-transduced dendritic cells. Eur J Cell Biol. 2007;86:817–26.
76. Porgador A, Irvine KR, Iwasaki A, Barber BH, Restifo NP, Germain RN. Predominant role for directly transfected dendritic cells in antigen presentation to CD8+ T cells after gene gun immunization. J Exp Med. 1998;188:1075–82.
77. Torres CA, Iwasaki A, Barber BH, Robinson HL. Differential dependence on target site tissue for gene gun and intramuscular DNA immunizations. J Immunol. 1997;158:4529–32.
78. Li N, Peng LH, Chen X, Nakagawa S, Gao JQ. Transcutaneous vaccines: novel advances in technology and delivery for overcoming the barriers. Vaccine. 2011;29:6179–90.
79. Mitragotri S, Blankschtein D, Langer R. Ultrasound-mediated transdermal protein delivery. Science (New York, NY). 1995;269:850–3.
80. Prausnitz MR, Bose VG, Langer R, Weaver JC. Electroporation of mammalian skin: a mechanism to enhance transdermal drug delivery. Proc Natl Acad Sci U S A. 1993;90:10504–8.
81. Prausnitz MR, Edelman ER, Gimm JA, Langer R, Weaver JC. Transdermal delivery of heparin by skin electroporation. Biotechnology (NY). 1995;13:1205–9.
82. Somiari S, Glasspool-Malone J, Drabick JJ, Gilbert RA, Heller R, Jaroszeski MJ, et al. Theory and in vivo application of electroporative gene delivery. Mol Ther. 2000;2:178–87.
83. Brave A, Hallengard D, Gudmundsdotter L, Stout R, Walters R, Wahren B, et al. Late administration of plasmid DNA by intradermal electroporation efficiently boosts DNA-primed T and B cell responses to carcinoembryonic antigen. Vaccine. 2009;27:3692–6.
84. Brave A, Hallengard D, Malm M, Blazevic V, Rollman E, Stanescu I, et al. Combining DNA technologies and different modes of immunization for induction of humoral and cellular anti-HIV-1 immune responses. Vaccine. 2009;27:184–6.
85. Vandermeulen G, Richiardi H, Escriou V, Ni J, Fournier P, Schirrmacher V, et al. Skin-specific promoters for genetic immunisation by DNA electroporation. Vaccine. 2009;27:4272–7.
86. Liu MA. DNA vaccines: an historical perspective and view to the future. Immunol Rev. 2011;239:62–84.
87. Abraham W, Downing DT. Preparation of model membranes for skin permeability studies using stratum corneum lipids. J Invest Dermatol. 1989;93:809–13.
88. Dubey V, Mishra D, Nahar M, Jain NK. Vesicles as tools for the modulation of skin permeability. Expert Opin Drug Deliv. 2007;4:579–93.
89. Matsuzaki K, Imaoka T, Asano M, Miyajima K. Development of a model membrane system using stratum corneum lipids for estimation of drug skin permeability. Chem Pharm Bull (Tokyo). 1993;41:575–9.
90. Knorr F, Lademann J, Patzelt A, Sterry W, Blume-Peytavi U, Vogt A. Follicular transport route—research progress and future perspectives. Eur J Pharm Biopharm. 2009;71:173–80.
91. Vogt A, Combadiere B, Hadam S, Stieler KM, Lademann J, Schaefer H, et al. 40 nm, but not 750 or 1,500 nm, nanoparticles enter epidermal CD1a+ cells after transcutaneous application on human skin. J Invest Dermatol. 2006;126:1316–22.
92. Vogt A, Mahe B, Costagliola D, Bonduelle O, Hadam S, Schaefer G, et al. Transcutaneous anti-influenza vaccination promotes both CD4 and CD8 T cell immune responses in humans. J Immunol. 2008;180:1482–9.
93. Mahe B, Vogt A, Liard C, Duffy D, Abadie V, Bonduelle O, et al. Nanoparticle-based targeting of vaccine compounds to skin antigen-presenting cells by hair follicles and their transport in mice. J Invest Dermatol. 2009;129:1156–64.
94. Nasir A. Nanotechnology in vaccine development: a step forward. J Invest Dermatol. 2009;129:1055–9.
95. Liard C, Munier S, Arias M, Joulin-Giet A, Bonduelle O, Duffy D, et al. Targeting of HIV-p24 particle-based vaccine into differential skin layers induces distinct arms of the immune responses. Vaccine. 2011;29:6379–91.
96. Liard C, Munier S, Joulin-Giet A, Bonduelle O, Hadam S, Duffy D, et al. Intradermal immunization triggers epidermal langerhans cell mobilization required for CD8 T-cell immune responses. J Invest Dermatol. 2012;132(3 Pt 1):615–25.
97. Mottram PL, Leong D, Crimeen-Irwin B, Gloster S, Xiang SD, Meanger J, et al. Type 1 and 2 immunity following vaccination is influenced by nanoparticle size: formulation of a model vaccine for respiratory syncytial virus. Mol Pharm. 2007;4:73–84.

98. Douillard P, Stoitzner P, Tripp CH, Clair-Moninot V, Ait-Yahia S, McLellan AD, et al. Mouse lymphoid tissue contains distinct subsets of langerin/CD207 dendritic cells, only one of which represents epidermal-derived Langerhans cells. J Invest Dermatol. 2005;125:983–94.
99. McDermott R, Bausinger H, Fricker D, Spehner D, Proamer F, Lipsker D, et al. Reproduction of Langerin/CD207 traffic and Birbeck granule formation in a human cell line model. J Invest Dermatol. 2004;123:72–7.
100. Stoitzner P, Stossel H, Wankell M, Hofer S, Heufler C, Werner S, et al. Langerhans cells are strongly reduced in the skin of transgenic mice overexpressing follistatin in the epidermis. Eur J Cell Biol. 2005;84:733–41.
101. Stoitzner P, Tripp CH, Douillard P, Saeland S, Romani N. Migratory Langerhans cells in mouse lymph nodes in steady state and inflammation. J Invest Dermatol. 2005;125:116–25.
102. Hunger RE, Sieling PA, Ochoa MT, Sugaya M, Burdick AE, Rea TH, et al. Langerhans cells utilize CD1a and langerin to efficiently present nonpeptide antigens to T cells. J Clin Invest. 2004; 113:701–8.
103. Nishibu A, Ward BR, Jester JV, Ploegh HL, Boes M, Takashima A. Behavioral responses of epidermal Langerhans cells in situ to local pathological stimuli. J Invest Dermatol. 2006;126:787–96.
104. Udey MC. Langerhans cells on guard in the epidermis: poised to dSEARCH and ...? J Invest Dermatol. 2006;126:705–7.
105. Wollenberg A, Mommaas M, Oppel T, Schottdorf EM, Gunther S, Moderer M. Expression and function of the mannose receptor CD206 on epidermal dendritic cells in inflammatory skin diseases. J Invest Dermatol. 2002;118:327–34.
106. Wollenberg A, Wagner M, Gunther S, Towarowski A, Tuma E, Moderer M, et al. Plasmacytoid dendritic cells: a new cutaneous dendritic cell subset with distinct role in inflammatory skin diseases. J Invest Dermatol. 2002;119:1096–102.
107. Lambert PH, Laurent PE. Intradermal vaccine delivery: will new delivery systems transform vaccine administration? Vaccine. 2008;26:3197–208.
108. Kaushik S, Hord AH, Denson DD, McAllister DV, Smitra S, Allen MG, et al. Lack of pain associated with microfabricated microneedles. Anesth Analg. 2001;92:502–4.
109. McAllister DV, Allen MG, Prausnitz MR. Microfabricated microneedles for gene and drug delivery. Annu Rev Biomed Eng. 2000;2:289–313.
110. Matriano JA, Cormier M, Johnson J, Young WA, Buttery M, Nyam K, et al. Macroflux microprojection array patch technology: a new and efficient approach for intracutaneous immunization. Pharm Res. 2002;19:63–70.
111. Mikszta JA, Alarcon JB, Brittingham JM, Sutter DE, Pettis RJ, Harvey NG. Improved genetic immunization via micromechanical disruption of skin-barrier function and targeted epidermal delivery. Nat Med. 2002;8:415–9.
112. Mikszta JA, Dekker III JP, Harvey NG, Dean CH, Brittingham JM, Huang J, et al. Microneedle-based intradermal delivery of the anthrax recombinant protective antigen vaccine. Infect Immun. 2006;74:6806–10.
113. Alarcon JB, Hartley AW, Harvey NG, Mikszta JA. Preclinical evaluation of microneedle technology for intradermal delivery of influenza vaccines. Clin Vaccine Immunol. 2007;14:375–81.
114. Sullivan SP, Koutsonanos DG, Del Pilar Martin M, Lee JW, Zarnitsyn V, Choi SO, et al. Dissolving polymer microneedle patches for influenza vaccination. Nat Med. 2010;16:915–20.
115. Lechmann M, Murata K, Satoi J, Vergalla J, Baumert TF, Liang TJ. Hepatitis C virus-like particles induce virus-specific humoral and cellular immune responses in mice. Hepatology. 2001;34:417–23.
116. Murata K, Lechmann M, Qiao M, Gunji T, Alter HJ, Liang TJ. Immunization with hepatitis C virus-like particles protects mice from recombinant hepatitis C virus-vaccinia infection. Proc Natl Acad Sci U S A. 2003;100:6753–8.
117. Elmowalid GA, Qiao M, Jeong SH, Borg BB, Baumert TF, Sapp RK, et al. Immunization with hepatitis C virus-like particles results in control of hepatitis C virus infection in chimpanzees. Proc Natl Acad Sci U S A. 2007;104:8427–32.
118. Quan FS, Kim YC, Compans RW, Prausnitz MR, Kang SM. Dose sparing enabled by skin immunization with influenza virus-like particle vaccine using microneedles. J Control Release. 2010;147(3):326–32.
119. Quan FS, Kim YC, Vunnava A, Yoo DG, Song JM, Prausnitz MR, et al. Intradermal vaccination with influenza virus-like particles by using microneedles induces protection superior to that with intramuscular immunization. J Virol. 2010;84:7760–9.
120. Pulendran B. Variegation of the immune response with dendritic cells and pathogen recognition receptors. J Immunol. 2005;174:2457–65.
121. Pulendran B. Learning immunology from the yellow fever vaccine: innate immunity to systems vaccinology. Nat Rev. 2009;9:741–7.
122. Pulendran B, Li S, Nakaya HI. Systems vaccinology. Immunity. 2010;33:516–29.
123. Matsuo K, Ishii Y, Quan YS, Kamiyama F, Mukai Y, Yoshioka Y, et al. Transcutaneous vaccination using a hydrogel patch induces effective immune responses to tetanus and diphtheria toxoid in hairless rat. J Control Release. 2011;149:15–20.
124. Glenn GM, Flyer DC, Ellingsworth LR, Frech SA, Frerichs DM, Seid RC, et al. Transcutaneous immunization with heat-labile enterotoxin: development of a needle-free vaccine patch. Expert Rev Vaccines. 2007;6:809–19.
125. Glenn GM, Kenney RT. Mass vaccination: solutions in the skin. Curr Top Microbiol Immunol. 2006;304:247–68.

126. Glenn GM, Rao M, Matyas GR, Alving CR. Skin immunization made possible by cholera toxin. Nature. 1998;391:851.
127. Glenn GM, Taylor DN, Li X, Frankel S, Montemarano A, Alving CR. Transcutaneous immunization: a human vaccine delivery strategy using a patch. Nat Med. 2000;6:1403–6.
128. Shi Z, Zeng M, Yang G, Siegel F, Cain LJ, van Kampen KR, et al. Protection against tetanus by needle-free inoculation of adenovirus-vectored nasal and epicutaneous vaccines. J Virol. 2001;75:11474–82.
129. Eyles JE, Elvin SJ, Westwood A, Lebutt CS, Alpar HO, Somavarapu S, et al. Immunisation against plague by transcutaneous and intradermal application of subunit antigens. Vaccine. 2004;22:4365–73.
130. El-Ghorr AA, Williams RM, Heap C, Norval M. Transcutaneous immunisation with herpes simplex virus stimulates immunity in mice. FEMS Immunol Med Microbiol. 2000;29:255–61.
131. Belyakov IM, Hammond SA, Ahlers JD, Glenn GM, Berzofsky JA. Transcutaneous immunization induces mucosal CTLs and protective immunity by migration of primed skin dendritic cells. J Clin Invest. 2004;113:998–1007.
132. Combadiere B, Vogt A, Mahe B, Costagliola D, Hadam S, Bonduelle O, et al. Preferential amplification of CD8 effector-T cells after transcutaneous application of an inactivated influenza vaccine: a randomized phase I trial. PLoS One. 2010;5:e10818.
133. Kenney RT, Yu J, Guebre-Xabier M, Frech SA, Lambert A, Heller BA, et al. Induction of protective immunity against lethal anthrax challenge with a patch. J Infect Dis. 2004;190:774–82.
134. Takigawa M, Tokura Y, Hashizume H, Yagi H, Seo N. Percutaneous peptide immunization via corneum barrier-disrupted murine skin for experimental tumor immunoprophylaxis. Ann N Y Acad Sci. 2001;941:139–46.
135. Yagi H, Hashizume H, Horibe T, Yoshinari Y, Hata M, Ohshima A, et al. Induction of therapeutically relevant cytotoxic T lymphocytes in humans by percutaneous peptide immunization. Cancer Res. 2006;66:10136–44.
136. Frech SA, Dupont HL, Bourgeois AL, McKenzie R, Belkind-Gerson J, Figueroa JF, et al. Use of a patch containing heat-labile toxin from Escherichia coli against travellers' diarrhoea: a phase II, randomised, double-blind, placebo-controlled field trial. Lancet. 2008;371:2019–25.
137. Frech SA, Kenney RT, Spyr CA, Lazar H, Viret JF, Herzog C, et al. Improved immune responses to influenza vaccination in the elderly using an immunostimulant patch. Vaccine. 2005;23:946–50.
138. Etchart N, Hennino A, Friede M, Dahel K, Dupouy M, Goujon-Henry C, et al. Safety and efficacy of transcutaneous vaccination using a patch with the live-attenuated measles vaccine in humans. Vaccine. 2007;25:6891–9.
139. Atmar RL, Patel SM, Keitel WA. Intanza((R)): a new intradermal vaccine for seasonal influenza. Expert Rev Vaccines. 2010;9:1399–409.
140. Dupasquier M, Stoitzner P, van Oudenaren A, Romani N, Leenen PJ. Macrophages and dendritic cells constitute a major subpopulation of cells in the mouse dermis. J Invest Dermatol. 2004;123:876–9.
141. Iwamura H, Saito Y, Sato-Hashimoto M, Ohnishi H, Murata Y, Okazawa H, et al. Essential roles of SIRPalpha in homeostatic regulation of skin dendritic cells. Immunol Lett. 2011;135:100–7.
142. Kissenpfennig A, Henri S, Dubois B, Laplace-Builhe C, Perrin P, Romani N, et al. Dynamics and function of Langerhans cells in vivo: dermal dendritic cells colonize lymph node areas distinct from slower migrating Langerhans cells. Immunity. 2005;22:643–54.
143. Kissenpfennig A, Malissen B. Langerhans cells—revisiting the paradigm using genetically engineered mice. Trends Immunol. 2006;27:132–9.
144. Laurent PE, Bonnet S, Alchas P, Regolini P, Mikszta JA, Pettis R, et al. Evaluation of the clinical performance of a new intradermal vaccine administration technique and associated delivery system. Vaccine. 2007;25:8833–42.
145. Moser K, Kriwet K, Naik A, Kalia YN, Guy RH. Passive skin penetration enhancement and its quantification in vitro. Eur J Pharm Biopharm. 2001;52:103–12.
146. Naito S, Maeyama J, Mizukami T, Takahashi M, Hamaguchi I, Yamaguchi K. Transcutaneous immunization by merely prolonging the duration of antigen presence on the skin of mice induces a potent antigen-specific antibody response even in the absence of an adjuvant. Vaccine. 2007;25: 8762–70.
147. Ochoa MT, Loncaric A, Krutzik SR, Becker TC, Modlin RL. "Dermal dendritic cells" comprise two distinct populations: CD1+ dendritic cells and CD209+ macrophages. J Invest Dermatol. 2008;128:2225–31.
148. Suvarna SK, Cotton DW. Dermal dendrocytes and other factor XIIIa-positive cells. J Pathol. 1993;171:251–2.
149. Young F, Marra F. A systematic review of intradermal influenza vaccines. Vaccine. 2011;29:8788–801.
150. Zaba LC, Fuentes-Duculan J, Steinman RM, Krueger JG, Lowes MA. Normal human dermis contains distinct populations of CD11c+ BDCA-1+ dendritic cells and CD163+ FXIIIA+ macrophages. J Clin Invest. 2007;117:2517–25.
151. Zaba LC, Krueger JG, Lowes MA. Resident and "inflammatory" dendritic cells in human skin. J Invest Dermatol. 2009;129:302–8.

Suppressing the Skin Immune System

17

Adnan Nasir and Anthony Gaspari

17.1 Introduction

A number of factors can lead to exaggerated immunity or autoimmunity. Traditional strategies for suppressing immunity include drugs and devices to suppress the immune system. Drug inhibitors of B cells and T cells include corticosteroids and antimetabolites. Devices for the inhibition of immunity include ultraviolet light and photopheresis apparati. Nanotechnology drugs and devices aim to enhance the effectiveness of these modalities while reducing their toxicities.

Immunosuppressive agents such as cyclosporine A have demonstrated utility in the management of atopic dermatitis and psoriasis [1–4]. Drawbacks to systemic administration include side effects such as nephrotoxicity, hypertension, dyslipidemia, and vulnerability to infection. Cyclosporine A has been incorporated into nanoparticles for topical administration and may demonstrate a better risk/benefit profile [5].

Corticosteroids are a mainstay of therapy in inflammatory skin disease. Topical and systemic steroids are plaqued with side effects including osteoporosis, osteonecrosis, myopathy, cataract, glaucoma, psychosis, hyperglycemia, cardiovascular, infection, gastrointestinal, and HPA axis suppression. Cutaneous side effects consist of atrophy, striae, perioral dermatitis, hypertrichosis, ACD, rosacea, acne, addiction, and pigment change. Prednicarbate solid lipid nanoparticles tend to remain in the upper dermis and epidermis to a greater extent than vehicle control [6, 7]. Such solid lipid formations are associated with fewer dermal and systemic side effects. Stealth steroids can also be delivered systemically with relatively few side effects [8]. Nanoparticulate stealth steroids tend to home to sites of inflammation, where leaky capillaries form due to neoangiogenesis (Fig. 17.1). Through a mechanism of enhanced plasma permeability and retention, liposomal steroids tend to accumulate at sites of inflammation in a process called passive targeting. Passively targeted stealth steroids require lower doses and remain in the circulation for longer periods in models of arthritis. Passive targeting of stealth liposomal corticosteroids can be passively targeted to sites of inflammation without incurring systemic toxicity has been demonstrated in a number of studies [9–12]. Because enhanced plasma permeability is seen in inflammatory skin diseases such as psoriasis and in tumors, systemic nanoparticulate stealth steroids may be beneficial in the management of these diseases [9–11, 13, 14].

Angiogenesis is also seen at sites of inflammation. Inflammatory diseases such as psoriasis

A. Nasir, M.D., Ph.D. (✉)
Department of Dermatology, UNC School of Medicine, 3100 Thurston Bowles Building, Chapel Hill, NC 27544, USA
e-mail: anasir@live.com

A. Gaspari, M.D.
Department of Dermatology, University of Maryland Medical Center, Baltimore, MD 21201, USA

A. Nasir et al. (eds.), *Nanotechnology in Dermatology*,
DOI 10.1007/978-1-4614-5034-4_17, © Springer Science+Business Media, LLC 2013

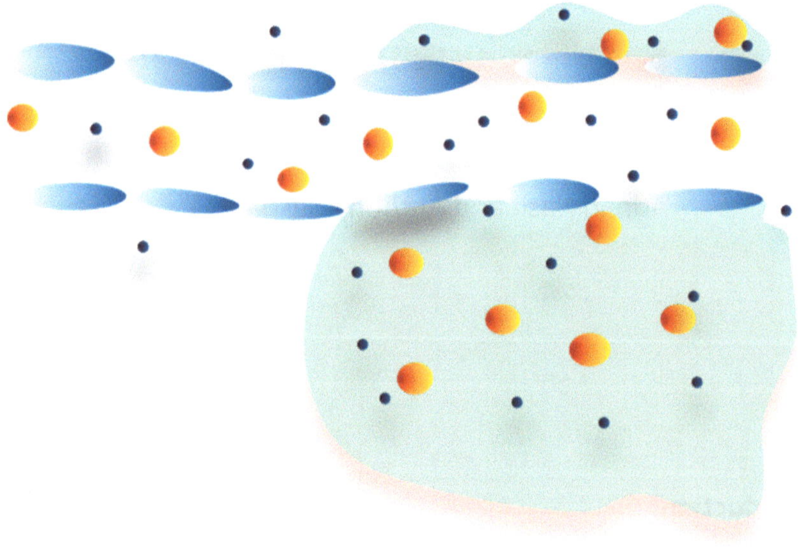

Fig. 17.1 Enhanced plasma permeation and retention (EPPR) is a feature of neovascularization. Capillaries typically have fenestrae which allow particles in the 10 nm size range to penetrate (*small blue dots*). Larger particles (≤150 nm) do not penetrate mature capillary walls. Neoangiogenic vessels—whether at sites of tumor or inflammation (depicted as an amorphous teal cloud around the capillary)—have larger fenestrae, which allow bigger nanoparticles (*larger orange dots*) to penetrate. Thus, EPPR allows for passive targeting of metastatic tumors and inflammatory foci

have increased expression of angiogenic factors such as angiopoietin. Angiogenesis is classically stimulated by vascular endothelial growth factor (VEGF) and inhibited by IFN-γ. Inhibition of angiogenesis has been used as a strategy for the management of some inflammatory and autoimmune reactions. Delivery of antiangiogenic factors using carbon nanotubes has been studied in animal models and in models of psoriasis, hypertrophic scar, and melanoma [9, 15–18]. Inhibition of pathologic angiogenesis with nanoparticulate antagomRs (inhibitors of miRNAs) has been shown to inhibit neovascularization [19].

One of the hallmarks of inflammation is calor, or increased temperature. Another form of passive targeting may be exploited by the use of thermosensitive polymers. These polymers can encapsulate nanoparticles and remain intact below a critical temperature threshold and release their contents above a critical temperature threshold. Such polymers may be useful in targeting inflammation severe enough to generate a critical temperature [20].

The genetic modification of the immune system has the potential for treating malignancy, genodermatosis, as well as infectious and inflammatory skin disorders [2]. Small inhibitor RNAs (siRNAs) are being used in a wide variety of model systems to treat skin disease [21–30].

Plasmapheresis has been used to remove circulating toxins, proteins, or cells which may contribute to skin disease. Magnetic nanoparticles coupled to targeting antibodies have been used in animal studies for the elimination of desired cells and proteins from the circulation (Fig. 17.2). The advantages of magnetic nanoparticles over traditional methods include avid binding kinetics, a renewable supply of receptors in case of saturation, and the ability to remove multiple agents simultaneously [31].

Magnetic nanoparticles and superparamagnetic iron oxide nanoparticles (SPION) coupled to drug can be used to target drug to tissue [32, 33]. Nanoparticles can be tuned to release drug in the presence of a magnetic field. Systemically administered immunosuppressive agents such as methotrexate could remain sequestered in magnetic nanoparticles [34] and localized and be simultaneously be released, for example, in psoriatic skin the presence of an externally applied magnetic field.

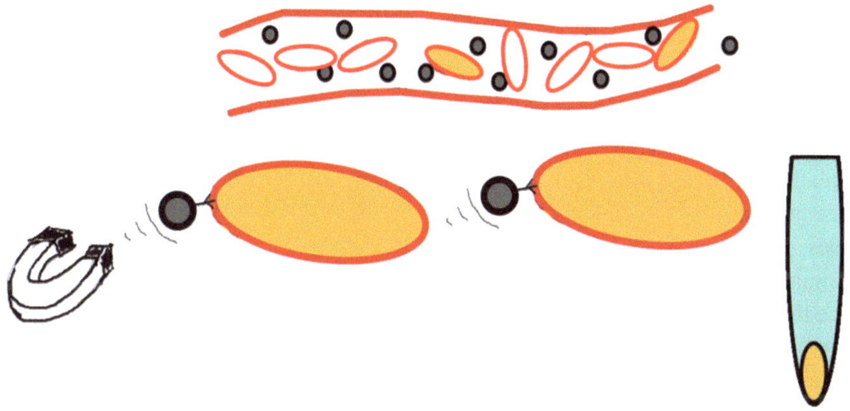

Fig. 17.2 Filtration of blood can be achieved using magnetic nanoparticles. Superparamagnetic nanoparticles coupled to a targeting molecule (such as an aptamer or antibody) can be used to tether targets as small as drugs and toxins, to macromolecules such as autoantibodies, to cells such as Sezary cells. Extracorporeally, the particles and their counterparts can be removed from the circulation using an externally applied magnetic field. They can be concentrated and eluted for further analysis. Saturation of binding sites is not an issue as it is with filtration because a replenishing supply of nanoparticles can be infused. Furthermore, if the binding is reversible, the particles themselves can be washed and recirculated in the patient

Phototherapy is typically administered by institutional, clinical, or home-based narrow band UVB units. Therapy is classically given over a body surface area which includes lesional and nonlesional skin. The disadvantages of these units are cost, convenience, and nonoptimal dosing of ultraviolet light. Fabrics impregnated with optical nanofibers may be able to monitor lesion size and shape through a collated camera feature and may simultaneously be able to deliver laser light in the narrow band UVB range in a pattern matching the patient's lesions. Such a wearable phototherapy device could be donned at bedtime for a steady low dose of therapy while the patient sleeps [35–41].

17.2 Immune Deviation

By shifting immunity from humoral to cell-mediated immunity of the Th1 variety, nanoencapsulation of antigens and oral delivery has been used to reduce allergy [42]. Diseases such as atopic dermatitis may benefit from such a shift. Alternatively, the reverse shift to the Th2 phenotype may be beneficial to patients with psoriasis [43].

Suppression of the immune response is a principal goal of allergy therapy. Nanoparticle preparations of vitamin D3, dexamethasone, and Der P1, Der P2, and ovalbumin have been used for sublingual immunotherapy in the treatment of dust mite allergy. Sublingual desensitization with OM-294-BA-MP lipid A derivative and positively charged nanoparticles reduced lung inflammation. CpG added to ragweed antigen enhances response to antigen and may be beneficial for hay fever.

17.3 Induction of Tolerance

Tolerance is important in preventing autoimmunity, transplant rejection, and in overcoming allergy. Tolerance can be induced by many routes of allergen exposure including sublingual, topical following Langerhans cell depletion, and cardiac. Regulatory T cells are important in the induction of tolerance. The use of Tregs in the induction of transplant tolerance be done using one of two methods. In the first, antigen-specific Tregs are expanded in vitro and infused in the patient as a form of autologous cell therapy. Alternatively, Tregs can be stimulated in vivo, for example, through activation of their constitutively expressed high affinity IL-2 receptor [44]. Nanoparticulate formulations of IL-2 for melanoma have been developed and may potentially be modified and adapted for this purpose.

Constitutive expression of co-stimulatory molecules such as B7-1 and B7-2 leads to exaggerated cutaneous immunity [45, 46]. Blockers of co-stimulation, such as belatacept, have been used to prevent transplant rejection [47]. Silica nanoparticles containing CTLA4-Ig have been used in animal models for the suppression of autoimmune thyroiditis [48, 49].

Dendritic cells play a role in the induction of Tregs through the production of IL-27, IL-10, and TGF-β [50]. The latter induces Fox-P3 expression in naïve T cells, which drives differentiation of the Treg phenotype. Many signals induce DC to produce TFG-β, including low-dose antigen, which binds DEC205 on dendritic cells. Other triggers include CpG DNA, dioxin, vitamin D3, RANKLE binding RANK, VIP, HGF, PGD2, corticosteroids, UVB, and haptens. Some bacterial peptides can cause DC to induce tolerance. PLGA nanoparticles can stimulate dendritic cells to induce Fox-P3 expression and Treg tolerance [51]. The cytokine IL-12 has also been shown to modify DC phenotype and enhance an immune response. Inhibition of IL-12, for example, by gene silencing or specific antibodies, has been shown to make DC tolerogenic in models of autoimmune disease and for the treatment of inflammatory diseases such as psoriasis [52–54]. Tolerance to allergens can be induced more rapidly, using less antigen, and with fewer doses by intralymphatic administration of antigen [55].

Fox-P3 expression is also controlled by miRNAs, which are highly expressed in T cells [56]. A wide variety of miRNAs which are involved in the regulation of immunity have been identified [57, 58]. Skin diseases such as psoriasis have associated variations in miRNA levels [59, 60]. Deficiency of a single miRNA, miR-146a has been shown to interfere with Treg activity and break down tolerance [61]. miR-146a has been associated with a number of autoimmune diseases as well as psoriasis [62]. The miRNA mir-142-3p is associated with systemic sclerosis and may be useful as a marker of disease severity and a target for therapy [63]. Nanoparticles targeting miRNA have been developed for cancer therapy [29]. And nanoparticle vectors comprised of lentiviruses are being used for the delivery of antagomRs, inhibitors of miRNAs [64]. Aptamers serving as competitive and allosteric inhibitors Dicer, which excises miRNAs from their double-stranded precursors, have also been developed [65]. miR-125b is an inhibitor of tumor necrosis factor (TNF)-α production and may be useful in the management of psoriasis and rheumatoid arthritis [66].

17.4 Fibrosis

A common sine qua non of inflammation is fibrosis. Fibrotic diseases are common and a number of instances significant associated morbidity and occasionally mortality have been described in the dermatologic literature [67]. Fibroblasts are involved in wound healing, scar formation, tissue remodeling, and in disease. Fibroblasts differentiate into myofibroblasts to cause wound contraction. Triggers of fibrosis can be noninflammatory (venous stasis in lipodermatosclerosis, proliferative in dermatofibroma, dermatofibrosarcomaprotuberans)orinflammatory. With the associated increased production of matrix metalloprotease, and release of profibrotic cytokines there is fibroblast proliferation and increased extracellular matrix synthesis, decreased extracellular matrix degradation, and resistance of fibroblasts to apoptosis. Inflammatory fibrosis is due to antifibrotic (TNF-α, IFN-γ, HGF) and profibrotic (TGF-β, IL-4, CCL2, CCL18, CTGF) cytokines. In scleroderma, there are antibodies to fibrillin-1 (an extracellular matrix protein), topoisomerase I (Scl-70), and autoreactive T cells. In nephrogenic systemic fibrosis, gadolinium can induce fibrosis. Animal models of fibrosis include Tsk-1 (which lacks inflammation and is due to an overproduction of fibrillin-1 due to gene duplication) and Tsk-2 (which has inflammatory cells; but is due to an unknown gene defect). Inhibition of IL-4 with antibodies or gene knockout improves fibrosis in Tsk mice.

Overall, transforming growth factor (TGF)-β might be a good target for the management of fibrosis. Scars and keloids are associated with excess activity of TGF-β. IL-10 has been shown to suppress keloid formation. Fetal wounds make less IL-6, and IL-8 and tend not to form keloids or hypertrophic scars [68–70]. Recombinant cytokines based on fetal expression patterns have been used to treat scars. These include Juvista (Avotermin) [71] TGF-β3, and Prevascar IL-10 (ilodecakin).

Other promising therapies include antiangiogenic factors, including VEGF inhibitors (e.g., bevacizumab) [18] and dexamethasone [68], phototherapy (photodynamic therapy (PDT), UVA-1 therapy, narrowband UVB therapy), transforming growth factor (TGF)-β3, tumor necrosis factor (TNF)-α inhibitors (etanercept), and recombinant human interleukin (rhIL-10), and possibly miRNA inhibitors [63]. CpG oligonucleotides also have ameliorated fibrosis in animal models [72]. The antimicrobial peptide cathelicidin (LL-37) may be useful in wound healing. Antimetabolites such as 5-FU have shown benefit in the treatment of keloids and nanoparticulate delivery systems may prove useful for therapy [73–75]. Self-assembling hydrogels functionalized with promoters of wound healing or inhibitors of fibrosis are one model of therapeutic exploration [76]. Alternatively, hydrogels can be impregnated with cells to aid in the wound healing process.

17.5 Biologics

Biologic agents are typically monoclonal antibodies designed to target particular receptors. In dermatology, they typically inhibit inflammation or autoimmunity. The design and engineering of these is precise, and they are in the nanometer size range and as such are a product of nanotechnology (Table 17.1). Nanoparticulate biologics can either be free or conjugated to nanoparticles, for example gold [77]. The nomenclature of biologics has to do with the origin (murine "omab"; hybrid murine/human "ximab"; mostly human "zumab"; and completely human "umab"), the target (cytokine receptors "cept"). Dermatologically important biologic nanoparticles fall into several categories: TNF inhibitors (etanercept, infliximab, adalimumab, certolizumab in PEG, golimumab), IL-5 inhibitors (mepolizumab), IL-12/23 inhibitors (ustekinumab, briakinumab), IFN-γ inhibitors (fontolizumab), CTLA4/B7 inhibitors (abatacept, belatacept), co-stimulation inhibitor (alefacept), and lymphocyte trafficking inhibitors (with equivocal results [78]). Biologics can be administered directly, or their delivery can be modified by nanoformulation [77, 79].

17.6 Passive Immunity

Passive immunization with antibodies can be used to augment or suppress immunity [80, 81]. Prolonged depot release of antibodies can be implemented with self-assembling peptide nanohydrogels [82]. Passive administration of decoy agents may also act to suppress immunity. This may be the mechanism of action of IVIg [83]. Thus, epitope masking is another mechanism of suppressing immunity. Certainly nanoparticles are small enough, can be designed with sufficient precision, to capture, mask, or alter the conformation of epitopes. This type of binding capacity has been seen with nanoparticle barrier creams

Table 17.1 "Biologics" are nanoengineered submicron particles

Definition of nanotechnology	"Biologic" drugs	Comment
Precise engineering	✓	Often synthesized in bioreactors
Nanoscopic dimensions	✓	Smaller than 100 nm
Reproducible	✓	Massive quantities produced
Targeted delivery	✓	High affinity for ligand
Size-dependent properties	✓	Bulk material and aggregates ineffective or have different properties, i.d., pentamers of IgM vs. monomers of IgG

"Biologics" are nanoparticles

against nickel. In a porcine model, nanoparticles containing calcium carbonate or calcium phosphate were able to capture nickel via ion exchange. The captured particles were easily washed away and nickel did not penetrate the epidermis in vitro. Per unit mass, the efficiency of ion exchange is 11-fold greater than ethylenediamine tetraacetic acid. The use of nanoparticules in this model for the sequestration of nickel has shown utility in the topical management of nickel-induced contact dermatitis [84, 85].

In summary, nanotechnology exploits a number of strategies for suppressing the immune system with specificity, high efficiency, and minimal toxicity. Systemic immunosuppressive agents can be delivered topically or localized to their desired area of operation either passively or using specific receptor/ligand pairs. Active drug is triggered to be released only at sites of inflammation. Triggering mechanisms can be physiologic, or physical, passive, or active. Biologic drugs already belong to the family of precisely engineered nanotherapeutics, but their functionality is being augmented with toxins or imaging beacons. Immune suppression can also involve passive immunity or epitope saturation.

References

1. Kim ST, et al. Topical administration of cyclosporin A in a solid lipid nanoparticle formulation. Pharmazie. 2009;64(8):510–4.
2. O'Rielly DD, Rahman P. Pharmacogenetics of psoriasis. Pharmacogenomics. 2011;12(1):87–101.
3. Schmitt J, Schmitt N, Meurer M. Cyclosporin in the treatment of patients with atopic eczema—a systematic review and meta-analysis. J Eur Acad Dermatol Venereol. 2007;21(5):606–19.
4. Steffan J, Favrot C, Mueller R. A systematic review and meta-analysis of the efficacy and safety of cyclosporin for the treatment of atopic dermatitis in dogs. Vet Dermatol. 2006;17(1):3–16.
5. Beauchesne PR, Chung NS, Wasan KM. Cyclosporine A: a review of current oral and intravenous delivery systems. Drug Dev Ind Pharm. 2007;33(3):211–20.
6. Baspinar Y, Keck CM, Borchert HH. Development of a positively charged prednicarbate nanoemulsion. Int J Pharm. 2010;383(1–2):201–8.
7. Juenger M, et al. Efficacy and safety of silver textile in the treatment of atopic dermatitis (AD). Curr Med Res Opin. 2006;22(4):739–50.
8. Horie RT, et al. Stealth-nanoparticle strategy for enhancing the efficacy of steroids in mice with noise-induced hearing loss. Nanomedicine (Lond). 2010;5(9):1331–40.
9. Banciu M, et al. Liposomal glucocorticoids as tumor-targeted anti-angiogenic nanomedicine in B16 melanoma-bearing mice. J Steroid Biochem Mol Biol. 2008;111(1–2):101–10.
10. Hofkens W, et al. Safety of glucocorticoids can be improved by lower yet still effective dosages of liposomal steroid formulations in murine antigen-induced arthritis: comparison of prednisolone with budesonide. Int J Pharm. 2011;416(2):493–8.
11. Kluza E, et al. Anti-tumor activity of liposomal glucocorticoids: the relevance of liposome-mediated drug delivery, intratumoral localization and systemic activity. J Control Release. 2011;151(1):10–7.
12. Momekova D, et al. Long-circulating, pH-sensitive liposomes sterically stabilized by copolymers bearing short blocks of lipid-mimetic units. Eur J Pharm Sci. 2007;32(4–5):308–17.
13. Crawshaw AA, Griffiths CE, Young HS. Investigational VEGF antagonists for psoriasis. Expert Opin Investig Drugs. 2012;21(1):33–43.
14. Creamer D, et al. Mediation of systemic vascular hyperpermeability in severe psoriasis by circulating vascular endothelial growth factor. Arch Dermatol. 2002;138(6):791–6.
15. Arbiser JL. Fumarate esters as angiogenesis inhibitors: key to action in psoriasis? J Invest Dermatol. 2011;131(6):1189–91.
16. Berrios RL, Arbiser JL. Novel antiangiogenic agents in dermatology. Arch Biochem Biophys. 2011;508(2):222–6.
17. Cheng J, et al. Nanotherapeutics in angiogenesis: synthesis and in vivo assessment of drug efficacy and biocompatibility in zebrafish embryos. Int J Nanomedicine. 2011;6:2007-21. Epub 2011 Sep 15. PubMed PMID: 21976976; PubMed Central PMCID: PMC3181060.
18. Diao JS, Xia WS, Guo SZ. Bevacizumab: a potential agent for prevention and treatment of hypertrophic scar. Burns. 2010;36(7):1136–7.
19. Anand S, et al. MicroRNA-132-mediated loss of p120RasGAP activates the endothelium to facilitate pathological angiogenesis. Nat Med. 2010;16(8):909–14.
20. Castonguay A, et al. Thermosensitive dendrimer formulation for drug delivery at physiologically relevant temperatures. Chem Commun (Camb). 2011;47(44):12146–8.
21. Alam MR, et al. Multivalent cyclic RGD conjugates for targeted delivery of small interfering RNA. Bioconjug Chem. 2011;22(8):1673–81.
22. Alshamsan A,et al. STAT3 Silencing in Dendritic Cells by siRNA Polyplexes Encapsulated in PLGA Nanoparticles for the Modulation of Anticancer Immune Response. Mol Pharm. 2010 Sep 14. [Epub ahead of print] PubMed PMID: 20804176.
23. Alshamsan A, et al. STAT3 knockdown in B16 melanoma by siRNA lipopolyplexes induces bystander immune response in vitro and in vivo. Transl Oncol. 2011;4(3):178–88.

24. Asai T, et al. Dicetyl phosphate-tetraethylenepentamine-based liposomes for systemic siRNA delivery. Bioconjug Chem. 2011;22(3):429–35.
25. Azuma M, Ritprajak P, Hashiguchi M. Topical application of siRNA targeting cutaneous dendritic cells in allergic skin disease. Methods Mol Biol. 2010; 623:373–81.
26. Cai H, et al. Inhibition of melanoma cell proliferation by targeting Wnt/beta-catenin pathway through Sox4 RNA interference. J Huazhong Univ Sci Technol Med Sci. 2011;31(4):565–9.
27. Chen L, et al. Knockdown of TSPAN1 by RNA silencing and antisense technique inhibits proliferation and infiltration of human skin squamous carcinoma cells. Tumori. 2010;96(2):289–95.
28. Chen Y, et al. Targeted nanoparticles deliver siRNA to melanoma. J Invest Dermatol. 2010;130(12):2790–8.
29. Chen Y, et al. Nanoparticles modified with tumor-targeting scFv deliver siRNA and miRNA for cancer therapy. Mol Ther. 2010;18(9):1650–6.
30. Leachman SA, et al. Therapeutic siRNAs for dominant genetic skin disorders including pachyonychia congenita. J Dermatol Sci. 2008;51(3):151–7.
31. Herrmann IK, et al. Device for continuous extracorporeal blood purification using target-specific metal nanomagnets. Nephrol Dial Transplant. 2011;26(9):2948–54.
32. Timko BP, Dvir T, Kohane DS. Remotely triggerable drug delivery systems. Adv Mater. 2010;22(44):4925–43.
33. Puri A, et al. Lipid-based nanoparticles as pharmaceutical drug carriers: from concepts to clinic. Crit Rev Ther Drug Carrier Syst. 2009;26(6):523–80.
34. Corem-Salkmon E, et al. Convection-enhanced delivery of methotrexate-loaded maghemite nanoparticles. Int J Nanomedicine. 2011;6:1595–602.
35. Sorin F, et al. Exploiting collective effects of multiple optoelectronic devices integrated in a single fiber. Nano Lett. 2009;9(7):2630–5.
36. Sorin F, et al. Resolving optical illumination distributions along an axially symmetric photodetecting fiber. Opt Express. 2010;18(23):24264–75.
37. Shapira O, et al. Surface-emitting fiber lasers. Opt Express. 2006;14(9):3929–35.
38. Ruff Z, et al. Polymer-composite fibers for transmitting high peak power pulses at 1.55 microns. Opt Express. 2010;18(15):15697–703.
39. Benoit G, et al. Dynamic all-optical tuning of transverse resonant cavity modes in photonic bandgap fibers. Opt Lett. 2005;30(13):1620–2.
40. Abouraddy AF, et al. Large-scale optical-field measurements with geometric fibre constructs. Nat Mater. 2006;5(7):532–6.
41. Nasir A. Nanodermatology: a bright glimpse just beyond the horizon—part I. Skin Therapy Lett. 2010;15(8):1–4.
42. Reisacher WR, Liotta D. The use of poly(D, L-lactic-co-glycolic) acid microspheres in the treatment of allergic disease. Curr Opin Otolaryngol Head Neck Surg. 2011;19(3):188–92.
43. Wippel-Slupetzky K, Stingl G. Future perspectives in the treatment of psoriasis. Curr Probl Dermatol. 2009;38:172–89.
44. Malek TR, et al. IL-2 family of cytokines in T regulatory cell development and homeostasis. J Clin Immunol. 2008;28(6):635–9.
45. Nasir A, et al. Exaggerated and persistent cutaneous delayed-type hypersensitivity in transgenic mice whose epidermal keratinocytes constitutively express B7-1 antigen. J Clin Invest. 1994;94(2):892–8.
46. Burns R, et al. Keratinocyte-derived, CD80-mediated costimulation is associated with hapten-specific IgE production during contact hypersensitivity to TH1 haptens. J Allergy Clin Immunol. 2005;115(2):383–90.
47. Wekerle T, Grinyo JM. Belatacept: from rational design to clinical application. Transpl Int. 2012;25(2):139–50.
48. Ascierto PA, Marincola FM, Ribas A. Anti-CTLA4 monoclonal antibodies: the past and the future in clinical application. J Transl Med. 2011;9:196.
49. Choi EW, et al. The effect of gene therapy using CTLA4Ig/silica-nanoparticles on canine experimental autoimmune thyroiditis. J Gene Med. 2008;10(7):795–804.
50. Kushwah R, Hu J. Role of dendritic cells in the induction of regulatory T cells. Cell Biosci. 2011;1(1):20.
51. Keijzer C, et al. PLGA, PLGA-TMC and TMC-TPP nanoparticles differentially modulate the outcome of nasal vaccination by inducing tolerance or enhancing humoral immunity. PLoS One. 2011;6(11):e26684.
52. Li R, et al. Gene silencing of IL-12 in dendritic cells inhibits autoimmune arthritis. J Transl Med. 2012;10:19.
53. Fitzgerald DC, et al. Intravenous tolerance effectively overcomes enhanced pro-inflammatory responses and experimental autoimmune encephalomyelitis severity in the absence of IL-12 receptor signaling. J Neuroimmunol. 2012;247(1–2):32–7.
54. Benson JM, et al. Therapeutic targeting of the IL-12/23 pathways: generation and characterization of ustekinumab. Nat Biotechnol. 2011;29(7):615–24.
55. Senti G, et al. Intralymphatic immunotherapy for cat allergy induces tolerance after only 3 injections. J Allergy Clin Immunol. 2012;129(5):1290–6.
56. Gao Y, et al. Molecular mechanisms underlying the regulation and functional plasticity of FOXP3(+) regulatory T cells. Genes Immun. 2012;13(1):1–13.
57. Jeker LT, et al. MicroRNA 10a marks regulatory T cells. PLoS One. 2012;7(5):e36684.
58. Ha TY. MicroRNAs in human diseases: from autoimmune diseases to skin, psychiatric and neurodegenerative diseases. Immune Netw. 2011;11(5):227–44.
59. Schneider MR. MicroRNAs as novel players in skin development, homeostasis and disease. Br J Dermatol. 2012;166(1):22–8.
60. Joyce CE, et al. Deep sequencing of small RNAs from human skin reveals major alterations in the psoriasis miRNAome. Hum Mol Genet. 2011;20(20): 4025–40.
61. Lu LF, et al. Function of miR-146a in controlling Treg cell-mediated regulation of Th1 responses. Cell. 2010;142(6):914–29.
62. Lofgren SE, et al. Genetic association of miRNA-146a with systemic lupus erythematosus in Europeans through decreased expression of the gene. Genes Immun. 2012;13(3):268–74.

63. Makino K, et al. Circulating miR-142-3p levels in patients with systemic sclerosis. Clin Exp Dermatol. 2012;37(1):34–9.
64. Surdziel E, Eder M, Scherr M. Lentivirus-mediated antagomir expression. Methods Mol Biol. 2010; 667:237–48.
65. Tyczewska A, et al. Selection of RNA oligonucleotides that can modulate human dicer activity in vitro. Nucleic Acid Ther. 2011;21(5):333–46.
66. Xu N, et al. MiR-125b, a microRNA downregulated in psoriasis, modulates keratinocyte proliferation by targeting FGFR2. J Invest Dermatol. 2011;131(7):1521–9.
67. Krieg T, Abraham D, Lafyatis R. Fibrosis in connective tissue disease: the role of the myofibroblast and fibroblast-epithelial cell interactions. Arthritis Res Ther. 2007;9 Suppl 2:S4.
68. Wu WS, et al. Dexamethasone induction of keloid regression through effective suppression of VEGF expression and keloid fibroblast proliferation. J Invest Dermatol. 2006;126(6):1264–71.
69. Que SK, Bergstrom KG. What's new in treatment of keloids? New applications for common therapies, new treatments to come. J Drugs Dermatol. 2011;10(5): 548–51.
70. Lee WJ, et al. Adenovirus-relaxin gene therapy for keloids: implication for reversing pathological fibrosis. Br J Dermatol. 2011;165(3):673–7.
71. Occleston NL, et al. Discovery and development of avotermin (recombinant human transforming growth factor beta 3): a new class of prophylactic therapeutic for the improvement of scarring. Wound Repair Regen. 2011;19 Suppl 1:s38–48.
72. Shen Y, et al. CpG oligodeoxynucleotides prevent the development of scleroderma-like syndrome in tight-skin mice by stimulating a Th1 immune response. J Invest Dermatol. 2005;124(6):1141–8.
73. Zhu L, et al. Chitosan-coated magnetic nanoparticles as carriers of 5-fluorouracil: preparation, characterization and cytotoxicity studies. Colloids Surf B Biointerfaces. 2009;68(1):1–6.
74. Zhang Y, et al. Folate-functionalized nanoparticles for controlled 5-fluorouracil delivery. J Colloid Interface Sci. 2011;354(1):202–9.
75. Yadav AK, et al. Development and characterization of hyaluronic acid decorated PLGA nanoparticles for delivery of 5-fluorouracil. Drug Deliv. 2010;17(8):561–72.
76. Branco MC, Schneider JP. Self-assembling materials for therapeutic delivery. Acta Biomater. 2009;5(3): 817–31.
77. Chamberland DL, et al. Photoacoustic tomography of joints aided by an etanercept-conjugated gold nanoparticle contrast agent-an ex vivo preliminary rat study. Nanotechnology. 2008;19(9):095101.
78. de Groot M, et al. Expression of the chemokine receptor CCR5 in psoriasis and results of a randomized placebo controlled trial with a CCR5 inhibitor. Arch Dermatol Res. 2007;299(7):305–13.
79. Julien DC, et al. Utilization of monoclonal antibody-targeted nanomaterials in the treatment of cancer. MAbs. 2011;3(5):467–78.
80. Cheung GY, Otto M. The potential use of toxin antibodies as a strategy for controlling acute Staphylococcus aureus infections. Expert Opin Ther Targets. 2012;16(6):601–12.
81. Ahmed AR, et al. Treatment of pemphigus vulgaris with rituximab and intravenous immune globulin. N Engl J Med. 2006;355(17):1772–9.
82. Koutsopoulos S, Zhang S. Two-layered injectable self-assembling peptide scaffold hydrogels for long-term sustained release of human antibodies. J Control Release. 2012;160(3):451–8.
83. Jordan SC, Toyoda M, Vo AA. Regulation of immunity and inflammation by intravenous immunoglobulin: relevance to solid organ transplantation. Expert Rev Clin Immunol. 2011;7(3):341–8.
84. Vemula PK, Anderson RR, Karp JM. Animal models for nickel allergy. Nat Nanotechnol. 2011;6(9):533.
85. Vemula PK, Anderson RR, Karp JM. Nanoparticles reduce nickel allergy by capturing metal ions. Nat Nanotechnol. 2011;6(5):291–5.

Nanotechnology in the Treatment of Infectious Diseases

18

Adam Friedman and Karin Blecher

18.1 Introduction

Nosocomial and community-acquired infections caused by multidrug-resistant (MDR) pathogens is rising at an alarming rate [1, 2]. Microbial resistance has developed as a result of the ease with which microorganisms can acquire and transfer antibiotic-resistant determinants as well as the inherent resistance in some species. In addition, the abuse of broad spectrum antibiotics has further influenced the development of antibiotic-resistant strains [3–6]. As a result, resistance to antibiotics and conventional therapies has become a public health threat resulting in increased patient morbidity and mortality, highlighting the need for novel approaches in the development of antimicrobial agents [7–9].

In response to this threat, nanotechnology has become a major area of interest due to its many unique characteristics. In particular, the physical and chemical properties of nanoparticles, including their high surface-to-volume ratio and small size, allows for the ability to surpass barriers and gain access to biological molecules including microorganisms [10]. In addition, the size, shape, and chemical characteristics of nanoparticles may be manipulated in order to facilitate such molecular interactions [10, 11]. The pursuit of developing nanotechnology for the treatment of infectious diseases, particularly skin disease, has diverged into two specific foci. The first utilizes materials that at the nanoscale have inherent antimicrobial properties, and the second incorporates known therapeutics into nano-vehicles in order to enhance delivery and improve efficacy. The following chapter provides a broad overview of the antimicrobial applications of nanotechnology-based drug delivery systems for the treatment of bacteria, viruses, and fungi.

18.2 Antimicrobial Nanomaterials

Nanomaterials consisting of biopolymers such as chitosan, or metals such as silver and metal oxides, may be promising agents for antimicrobial applications. These agents have the ability to bind to microbial cell walls resulting in increased membrane permeability and disruption through direct interactions, in addition to the ability to generate destructive free radicals.

18.2.1 Chitosan

Chitosan is a natural polysaccharide biopolymer derived from chitin, the principal structural component of the crustacean exoskeleton. The antimicrobial properties of chitosan result from its

A. Friedman, M.D., FAAD
Division of Dermatology, Department of Medicine,
Albert Einstein College of Medicine,
111 E 210th Street, Bronx, NY, 10467, USA

K. Blecher, M.D. (✉)
Department of Dermatology, Montifiore Medical Center,
111 East 210th Street, Bronx, NY, 10467, USA
e-mail: Karin.Belcher@gmail.com

A. Nasir et al. (eds.), *Nanotechnology in Dermatology*,
DOI 10.1007/978-1-4614-5034-4_18, © Springer Science+Business Media, LLC 2013

polycationic character in weakly acidic pH which favors interaction with negatively charged microbial cell walls and cytoplasmic membranes. These interactions result in decreased osmotic stability, membrane disruption, and eventual leakage of intracellular elements [12–14]. In addition, chitosan may enter the nuclei of bacteria and fungi and inhibit mRNA and protein synthesis by binding to microbial DNA [10, 13, 15]. When nano-scaled, chitosan has a higher surface to volume ratio, translating into higher surface charge density, increased affinity to bacteria and fungi, and greater antimicrobial activity [15]. Several studies have demonstrated the efficacy of chitosan nanoparticles against a variety of bacterial and fungal pathogens, including *Escherichia coli*, *Staphylococcus aureus* [12–15], and *Candida* [16].

One study reported the in vitro antifungal activity of low molecular weight chitosan (LMWC) against 105 clinical *Candida* isolates [16]. LMWC exhibited a superior antifungal activity, inhibiting roughly 90 % of the clinical isolates examined with 68.6 % complete inhibition. These *Candida* species included several fluconazole-resistant strains and less susceptible species such as *Candida glabrata*, suggesting chitosan may be a useful tool in the setting of antifungal drug resistance. Other studies have further supported the potential utility of chitosan as an antifungal agent [17–19]

In addition to its antifungal properties, compared to conventional antibiotics, chitosan nanoparticles were found to be more effective against *E. coli* and *S. aureus* isolates [15]. This included antibiotics such as doxycycline, chitosan alone, and acetic acid among others [14, 15]. In addition, chitosan's polycationic nature and high affinity to metal allow it to be used as a carrier system and platform stabilizer for a variety of other nanoparticle systems including metallic nanoparticles [20], nitric oxide-releasing nanoparticles (NO-NPs) [21], and drug-containing nanoparticles which allows for targeted delivery of various medications [15]. These chitosan platforms augment the antimicrobial properties of these nanoparticles. For example, the antimicrobial efficacy of silver-loaded membranes is enhanced with increasing chitosan contents of up to 70 %, resulting in larger zones of inhibition against both *S. aureus* and *E. coli* [13]. In addition, the mean inhibitory concentrations (MIC) of AgNps against *S. aureus* was significantly decreased upon addition of chitosan to the nanoparticle matrix (AgNps-9±4 vs. Chitosan-AgNps-1.25±0.75) [22]

As such, chitosan nanoparticles and platforms have demonstrated superior antimicrobial activities against a broad spectrum of pathogens. Although the use of chitosan has yet to be fully explored as a topical antimicrobial agent, its potential for the treatment of infectious disease and infectious skin conditions is vast.

18.2.2 Nano-Silver

Silver (Ag) has been used for centuries for the treatment of burns and wounds to prevent infection [23]. Although the mechanism behind its antimicrobial properties is not entirely known, it has been proposed that silver and silver ions (such as $AgNO_3$) penetrate bacterial cell walls and membranes by interacting with sulfur-containing proteins or thiol groups [24]. Once inside the cell, $AgNO_3$ targets and damages bacterial DNA and respiratory enzymes, leading to loss of the cell's replicating abilities and ultimately cell death [14, 23]. The small size and large surface area of silver nanoparticles (AgNps) make them better able to penetrate bacterial cell walls and membranes, and as a result, the antimicrobial effect is directly dependent on nanoparticle size and shape [10, 23, 25]. For example, smaller nanoparticles (<10 nm) that are triangular or truncated are more effective than larger particles that are round or rod-shaped [10, 24].

AgNps have been shown to be effective against a variety of pathogens, including viruses [23, 26, 27], fungi (*Candida albicans*) [28, 29], and many bacterial species. One study reported the anti-HIV activity of AgNPs, suggesting that these nanoparticles prevent viral replication at an early stage by inhibiting viral entry into host cells. This was accomplished through AgNP binding to gp120, thereby preventing CD4-dependent virion binding, fusion, and infectivity

[27]. Silver has shown efficacy against pathogenic viruses such as poliovirus, norovirus, and enteric adeno viruses [26]

In addition to antiviral capabilities, nano-silver has also demonstrated antifungal activity. For example, one study reported the antifungal effects of AgNps toward *C. albicans* which was found to be similar to the antifungal activity of amphotericin, with comparable MIC values [30]. The mode of action of AgNps was shown to be its ability to dissipate the membrane potential of *C. albicans*. Similarly to the mechanism of action of amphotericin B, the results indicated that AgNps also have the ability to form transmembrane pores causing leakage of cell constituents and eventually cell death. In addition to its antifungal capabilities, investigators demonstrated that AgNps have either no effect or a weakly cytolytic effect on human erythrocytes. This is compared to amphotericin B, which incurred a higher hemolytic activity suggesting that AgNps may be a safer treatment. These data have been further corroborated by other investigators [31–33].

Many studies have reported the antibiotic efficacy of AgNps against a broad spectrum of bacteria including *E. coli* [13, 14, 24, 25, 28, 29, 34–38], *S. aureus* [13, 28, 29, 34–36], *Bacillus subtilis* [25, 28], and *Salmonella typhi* [34]. In addition, AgNps have been found to augment the efficacy of other antimicrobial agents. In particular, the antibacterial activities of penicillin G, amoxicillin, erythromycin, clindamycin, and vancomycin increased against *S. aureus* and *E. coli* when mixed with AgNps, with erythromycin having the greatest synergy with the silver [35]. Microbes are also less likely to develop resistance against silver and AgNPs, as their broad range of targets would require multiple and simultaneous compensatory mutations [23, 24]. As a result, silver can be used to arrest bacterial resistance to antibiotics and enhance their efficacy [14].

In addition to its broad spectrum activity against viruses, fungi, and bacteria, the stronger interaction of AgNps with microbial surfaces might allow for the use of lower drug concentrations as compared to current conventional silver agents, limiting silver's known toxicity. Given these properties, there are many diverse applications for AgNps in the treatment and prevention of infectious diseases, including coatings on medical devices to prevent microbial colonization, wound dressings, and augmentation of antibiotics [17, 23].

18.2.3 Nano-Titanium Dioxide

When exposed to ultraviolet light, titanium dioxide (TiO_2) forms active oxygen species through a process called photocatalysis. These oxygen species, which include hydrogen peroxide and hydroxyl radicals, damage microbial cell membranes resulting in cell death [39, 40]. As a result, this antimicrobial property has been utilized in water and air purification, and recently has been investigated against pathogenic and opportunistic pathogens including bacteria and fungi [28]. For example, one study reported that TiO_2 nanoparticle-infused thin film composite (TFC) membranes reduced *E. coli* biofilm formation by disrupting the bacterial membrane and thus inhibiting bacterial attachment membrane surfaces [39, 40]. The antibiotic efficacy of TiO_2 can also be augmented when combined with other nanomaterials such as silver. For example, the efficacy of TiO_2–Ag nanoparticles (TiO_2–AgNPs) were investigated against gram-negative bacteria, gram-positive bacteria, and various fungi that are responsible for opportunistic infection and colonization of medical devices [28]. Although AgNPs were found to be more effective than TiO_2 and TiO_2–AgNPs against the bacterial strains tested (*E. coli, Acinetobacter baumanii, Pseudomonas aeruginosa, B. subtilis, Mycobacterium smegmatic, Mycobacterium bovis,* and *S. aureus*), TiO_2–AgNPs demonstrated better antifungal activity against *C. albicans* and *Aspergillus* when compared to AgNPs alone and conventional antifungals such as fluconazole [28]. Based on these results, it might be possible to combine metal nanoparticles to augment the antimicrobial activity of each alone to develop novel and more efficacious antimicrobial agents [10].

18.2.4 Nitric Oxide-Releasing Nanoparticles

Nitric oxide (NO) has many roles that encompass virtually every physiological system within the human body, including host defense [41]. When stimulated by microorganisms, macrophages upregulate the production and release of NO through the transcription of inducible nitric oxide synthase (iNOS) [42]. NO then exerts its antimicrobial effects through a variety of mechanisms including direct microbial DNA damage via peroxynitrite generation as well as interference of cellular respiration by inactivation of zinc metalloproteins. In addition, NO stimulates several innate antimicrobial pathways, enhancing the host's own immune response [41, 42]. Given these antimicrobial properties, there has been tremendous interest in the development of NO donors and delivery systems [42, 43]. One such delivery system capitalizes on the benefits of nanotechnology. Utilizing NO-NPs housing NO within a dry matrix, the system allows the release of gaseous NO-free radicals only after exposure to moisture [21, 44]. This delivery system is ideal for the topical treatment of cutaneous infections, as NO can be easily stored and applied to the skin, and provides sustained delivery to the affected areas over a prolonged period of time [10, 42]. These NO-NPs have been tested against a broad spectrum of pathogens including *S. aureus* [43, 45], *Acinetobacter* [46], *Enterococcus faecalis, E. coli, Streptococcus pyogenes, P. aeruginosa,* and *Klebsiella* [9, 47] with considerable efficacy. In a murine wound model, methicillin-resistant *S. aureus* (MRSA)-infected full thickness wounds treated with NO-NPs demonstrated accelerated wound closure and less bacterial burden when compared to wounds that were untreated or treated with control-nanoparticles (no nitric oxide release) [43]. Similar results were found when NO-NPs were either injected subcutaneous or applied topically into induced MRSA abscesses in mice [45]. Given the simplicity of their topical application, acceleration of wound healing, and excellent antimicrobial properties, the potential for NO-NPs in the treatment of infectious diseases is vast [10].

18.2.5 Soy Bean Oil-Derived Nanoemulsions

Nanoemulsions (NEs) are stable oil-in-water (O/W) solutions produced by mixing a water immiscible oil phase into an aqueous phase under high shear forces composed of nanometer-sized, positively charged droplets that have broad-spectrum activity against enveloped viruses, fungi, and bacteria [48–59]. In one study, a nanoemulsion composed of soybean oil, cetylpyridinium chloride (CPC), and Triton X-100 was investigated against *Streptococcus mutans*, a common causative agent of dental caries [48]. NE was shown to be bactericidal against both planktonic and biofilm cell membrane in a concentration-dependent manner. The mechanism is thought to be disruption of the bacterial cell membrane and cell wall as demonstrated by scanning electron microscopy [48]. Another nanoemulsion, NB-002, is currently being investigated as an antifungal agent. It is also composed of the cationic quaternary ammonium compound, CPC, oriented at its oil–water interface which stabilizes the nanoemulsion droplets and contributes to the anti-infective activity. The NB-002 nanoemulsion was shown to exert broad, uniformly fungicidal activity against numerous dermatophyte species including *Trichophyton rubrum, Trichophyton mentagrophytes*, and *Epidermophyton floccosum* as well as *C. albicans*. These dermatophytes are the primary causes of human fungal hair, skin, and nail disease [52, 60, 61]. The ability of NB-002 to serve as a topical treatment has also been evaluated ex vivo with human cadaver skin samples, demonstrating that NB-002 can enter through a transfollicular route to enter the epidermal and dermal layers with lateral diffusion occurring along tissue planes to sites distal from the application site [56, 62]. Clinical studies of NB-002 for the treatment of onychomycosis have provided supporting results in its use as an effective topical treatment [63].

18.3 Drug-Encapsulated Nanomaterials

Aside from microbial resistance, antibiotics are limited by their poor intracellular penetration and activity [64]. In addition, intracellular infections caused by facultative intracellular organisms, such as *Salmonella, Listeria*, and *Mycobacterium tuberculosis*, are difficult to eliminate as a result of several evolutionary mechanisms. These include: escape from the phagosome, inhibition of phagosome and lysosome fusion, resistance to lysosomal enzymes, and the ability to lay dormant, all of which lead to downregulation of many drug targets [10]. In an effort to enhance the delivery and efficacy of antibiotics, nanoparticles and liposomes have been investigated as potential drug carriers due to their ability to be endocytosed and released into phagocytic cells carrying intracellular pathogens [64, 65]. The specificity of delivery of nano-carriers to target tissues can be passive or active. In passive targeting, nano-carriers rely on the size of apertures in vessels at sites of inflammation and targeting is dependent on enhanced plasma permeability and retention. In active targeting, nano-carriers require receptor–ligand interactions. Activation of nano-carrier drug delivery by cellular activity, such as pH change or oxidative burst, can aid targeted therapy in these settings. In addition to passive and active targeting, extrinsic modes of targeting and activation, such as magnetic guidance and radio frequency-mediated drug release, may be used for effecting localized delivery of nano-drugs [10].

18.3.1 Antibiotic-Encapsulated Nanomaterials

Many studies have investigated the use of nanosized vehicles to deliver antibiotics and enhance microbial killing. These antibiotics include macrolides (azithromycin) [66], aminoglycosides [64], fluoroquinolones [64, 67], and β-lactams such as penicillin [68, 69], ampicillin [64, 70, 71], and cephalosporins [64, 69]. One study recently described the preparation and application of penicillin-bound polyarcylate nanoparticles against methicillin-susceptible (MSSA) and -resistant strains of *S. aureus* [68]. In this study, penicillin-bound NPs were found to have comparable antibacterial activity to penicillin G alone against MSSA, but much stronger activity against MRSA, suggesting that the activity of the penicillin-bound NPs were not affected by the β-lactamase enzyme produced by the bacterium. This likely represents the ability of the nano-vehicle to protect the antibiotic from enzymatic degradation by penicillinase, and was further confirmed by a control experiment. In this control, penicillin-bound NPs and penicillin G were screened against MSSA in the absence and presence of added β-lactamase. The addition of β-lactamase caused penicillin G to lose all of its antibiotic activity, but did not affect the activity of penicillin-bound NPs as demonstrated by zones of inhibition [68]. Thus, in addition to enhancing targeted drug delivery, antibiotic-encapsulated nanoparticles may overcome common mechanisms of antibiotic resistance, creating a new tool to combat infectious diseases.

18.3.2 Antifungal-Encapsulated Nanomaterials

In addition to targeting antibiotics, nanomaterials such as liposomes, have been employed to both enhance efficacy and improve delivery of antifungal agents. One example includes the liposomal encapsulation of amphotericin B. Amphotericin B is one of the most effective antifungal agents which causes direct fungicidal activity by binding to ergosterol in fungal cell membranes resulting in pore formation and ultimately cell death. However, the clinical use of this medication is limited due to severe, dose limiting side effects. Due to its lipophilic nature, amphotericin B is an ideal candidate for lipsomal encapsulation and was actually one of the first liposomal drugs to enter clinical trials in the late 1980s and continues to be a hot area of pursuit even today [72–87]. Several examples of products that have emerged from the initial studies include Abelcet®, Amphotec®, and Ambisome® [80, 81, 88]. However, the use of intravenous liposomal amphotericin B in the treatment of superficial dermatomycoses would be considered extreme, and no topical formulation is currently available.

Aside from amphotericin B, many other antifungals have been encapsulated into nanomaterials to enhance topical delivery and efficacy, including miconazole nitrate [89], econazole [90], nystantin [91–95], and fluconazole [96]. Targeted and focused delivery of fluconazole by conventional therapy is a major impediment in achieving its therapeutic efficacy against skin infections, such as cutaneous candidiasis. Liposomal technology has been employed to enhance topical delivery of this antifungal. In one study [96], fluconazole-loaded vesicular liposomes and niosomes were generated and incorporated into carbopol gel (1 %; w/w) for sustained, localized application against cutaneous candidiasis in immunosuppressed albino rats. The in vivo localization studies showed that liposomal gel produced 14.2-fold higher drug accumulation, compared with plain gel, while a 3.3-fold increase was appreciated in the case of an equivalent-dose application in the form of niosomal gel. The greatest clinical and microscopic improvement was seen with the liposomal gel, suggesting that this formulation may be an effective treatment of cutaneous candidal infections. Similar results were reported with liposomal econazole. In one controlled, double-blinded trial, the efficacy of liposomal econazole was evaluated as a drug carrier system [97]. Treatment either encompassed once-daily application for 14 days of econazole liposome gel 1 %, branded econazole cream 1 %, or a generic clotrimazole cream 1 %. Higher cure rates based on microscopy and culture were appreciated in the econazole liposome gel treatment group. In addition, tolerability was considered slightly better in the econazole liposome gel group, compared with the econazole cream and clotrimazole cream treatment groups [97]. As a result of enhanced delivery, efficacy, and tolerability, antifungal-encapsulated nanomaterials have exciting potential in the treatment of cutaneous fungal infections.

18.3.3 Antiviral-Encapsulated Nanomaterials

Antiviral agents have recently been incorporated into nanomaterials, to enhance its delivery and efficacy against HPV, a common cutaneous virus with multiple serotypes that are responsible for condyloma and even squamous cell carcinoma. Podophyllin (POD), an unstable plant extract with many active ingredients, can inhibit the growth of epithelial cells infected by HPV in the epidermis, and has been used as the first-line drug for the treatment of genital warts. However, standard topical preparations of POD, such as tincture and cream, can cause severe skin irritation as well as be systemically absorbed [98]. In an attempt to achieve higher accumulation in the epidermis and thus transdermal permeation, solid lipid nanoparticles loaded with POD (POD-SLNs) have been developed. The high accumulation of POD is beneficial to treat genital warts and reduce the systemic absorption and dermal irritation. In one study, POD-SLNs demonstrated 3.48 times increased accumulative amount of POD in porcine skin compared with 0.15 % tincture in vitro, and exhibited a strong localization of POD within the epidermis [99]. The penetration of POD-SLNs appeared to follow two pathways, along the stratum corneum and hair follicle route; both due to the small size and permeation enhancement of the SLN soybean lecithin [99].

In another study, podophyllotoxin solid lipid nanoparticles gel and routine podophyllotoxin gel preparation was applied, respectively, for the treatment of 97 volunteer patients with recurrent condyloma acuminatum [100]. In this randomized double-blinded study, clinical clearance of condyloma patients in the first treatment course with podophyllotoxin solid lipid nanoparticle gel reached 97.1 %, close to that with the routine preparation of 90.6 %; however, the nanoparticle preparation significantly reduced the recurrence rate and reported adverse effect [100]. This clinical study suggests that podophyllotoxin delivered via solid lipid nanoparticle gel has the potential to be used clinically to effectively treat and clear condyloma acuminatum and reduce its recurrence rate with only mild, tolerable adverse effect.

18.4 Nano-Vaccines

The immune system acts as the body's major defense against foreign pathogens, and comprises the innate and adaptive systems. The innate, or nonspecific, immune system responds to foreign pathogens once they have breached the body's physical barriers by recognizing microbial characteristics called pathogen-associated molecular patterns (PAMPs) [101]. PAMPs are recognized by pattern-recognition receptors (PRPs) that reside on a variety of host cells, initiating antigen uptake by antigen-presenting cells (APCs). APCs subsequently present the antigen to cells of the adaptive, or specific, immune system for the induction of specific memory T and B cells which lead to the downstream activation of CD8 and CD4 lymphocytes [20, 101]. This entire process is carefully coordinated by the interactions of various cytokines which often dictate the type of response that is generated for a given pathogen. For example, interferon-gamma (IFN-γ) promotes a T helper type-1 (Th1) response that mediates antibody-independent immune responses, while interleukin-4 (IL-4) and IL-5 result in a Th2 response, essential for antibody production [20, 102]. Thus the harmonized interactions of APCs, T cells, B cells, and inflammatory cytokines are imperative to an effective immune system.

Vaccines are largely responsible for the reduction in mortalities due to infectious diseases, as they initiate a powerful immune response that results in lasting and protective immunity. To be effective, a vaccine must induce both an innate and adaptive immune response in a manner that is safe and beneficial to the patient [20, 101]. There are various existing vaccines that display different degrees of immunogenicity and safety [20], and include live-attenuated, killed, or subunit vaccines. Although widely used, live-attenuated vaccines are unstable and have the potential to revert to a more virulent form within the host. Killed whole-virus and subunit vaccines do not carry this risk, but are less immunogenic than attenuated vaccines, requiring repeat dosing which drives cost and lowers vaccine availability [101]. In addition, most vaccinations require trained personnel for administration since an injection is needed. Due to the increased cost and difficulty with patient compliance associated with injectable vaccine, mucosal administration is becoming a favored route due to its noninvasive nature and ease of self-administration [20]. However, there are challenges in the development of mucosal vaccines, as the antigen has to pass several barriers before reaching an APC and must induce an immune response that is strong enough to elicit systemic immunity. For intranasal or inhalational routes, the vaccine components must be small enough to reach deep alveoli (<5 μm) and avoid clearance during exhalation (<500 nm) [20]. As a result of their size, foreign composition, and numerous carrying capabilities, nanoparticles have the ability to modulate various aspects of the immune response [103] and may improve the efficiency of mucosal and injected vaccines because of increased exposure time and uptake of antigens by APCs [104], improved immunogenicity of viral and bacterial components, and modulation of the cytokine response [105]. Numerous synthetic nanoparticle vaccines and immunostimulatory adjuvants have been created and include polymeric and non-polymeric-based particles. Epicutaneous delivery of vaccine may elicit a more robust immune response and may require less antigen as well [106–109].

18.4.1 Nanotechnology-Based Vaccines and Immunostimulatory Adjuvants

18.4.1.1 Synthetic Polymers

Polymeric-based nanoparticles can act as carriers for a wide range of materials including protein or DNA vaccines [20, 101]. The characteristics of the nanoparticle in vivo is largely dependent on the type of polymer used, and therefore may be manipulated to fit the requirements of a particular vaccine [20]. For example, to overcome compartmental differences of the digestive tract for oral vaccination, poly (ε-caprolactone) (PCL) polymers were used to create multicomponent particles. These particles are made up of nanoparticle-encapsulated DNA surrounded by a PCL microparticle, known as the nanoparticle-in-microsphere hybrid oral delivery system (NiMOS). The inner part of the NiMOS is composed of gelatin nanoparticles that are susceptible to proteolytic degradation, making them vulnerable to gastric contents. The outer PCL layer is resistant to proteases but is degraded by lipases. This layer therefore protects the inner particles during transit through the stomach until reaching the small intestine, where lipases degrade PCL, allowing for the release of the inner gelatin nanoparticle [20].

Although synthetic polymers facilitate vaccine delivery, they do not seem to be immunostimulatory on their own. Poly(lactide-co-glycolide) (PLGA) particles, for example, are efficiently taken up by macrophages and other APCs in vitro and in vivo. However, they do not lead to a significant pro-inflammatory cytokine response indicative of macrophage activation (IL-2, IL-6, IL-12, and TNF-α), rather having stimulatory profiles similar to that of saline [110]. However, when encapsulated with plasmid DNA, PLGA-based nanoparticles induced a stronger and more sustained serum IgG response after intranasal administration compared to naked DNA. It is thought that this response is due to the stability of the PLGA nanoparticles during delivery across nasal mucosa, allowing for internalization of the plasmid DNA by appropriate cells [111]. Thus, synthetic polymers cannot be used as immunostimulatory adjuvants alone, but they can enhance immunization via delivery of antigens across mucosal barriers.

Polymethyl methyl methacrylate (PMMA) nanoparticles have demonstrated adjuvant immunostimulatory properties. PMMA NP adjuvants induce a 100-fold increased antibody titer response to live HIV2 virus vaccine in murine models [112]. PMMA NP dendrimers have been shown in vivo to stimulate enhanced IgG and IgM anti-ovalbumin antibody production [113]. Carboxyfullerene nanoparticles augment immunity by stimulating neutrophils to destroy bacteria in a mouse model of *S. pyogenes*. Nanoparticles modified with toll-like receptor (TLR) agonists have been shown to enhance immune responsiveness. For example, PLGA NP modified with tetanus toxoid and MUC1 lipopeptide (a TLR ligand) induced enhanced T-cell responsiveness in vitro when compared to PLGA NP combined with tetanus toxoid or MUC1 lipopeptide alone [114].

Aptamers are polymeric oligonucleotides (either DNA or RNA) folded in a 3-dimensional configuration which are then selected for high affinity binding to a target. Targets for aptamers include proteins, peptides, and small molecules such as drugs, vitamins, and inorganic compounds. Aptamers can be selected from random nucleic acid libraries, and can be prepared to high degrees of purity, stability, and selectivity. Aptamers have been used to create compounds with antimicrobial activity, including the ability to inhibit HIV reverse transcriptase, vaccinia virus replication, and in vitro models of β-lactamase resistance in gram-positive and gram-negative bacteria [115–117]. Successful PEGylated delivery of an aptamer inhibitor of VEG-F has been used in clinical trials of macular degeneration [118].

18.4.1.2 Nanoemulsions

NEs encapsulate lipophilic or hydrophilic substances in a dispersed phase and are manufactured as water-in-oil (W/O) or O/W systems. These carrier systems can be used in the development of mucosal vaccines, as they can be endocytosed by cells on the mucosal surface (epithelial or M-cells) and subsequently delivered to APCs [20]. In addition, NEs are themselves immunostimulatory and can be used to boost the immune response of a mucosal vaccine. The current hepatitis B vaccine requires an intramuscular (IM) injection and consists of recombinant hepatitis B surface antigen (HBsAg) formulated with an aluminum salt (alum) adjuvant. This adjuvant elicits a Th2 immune response associated with an ineffective CD8 response to virally infected cells [119]. Moreover, adverse effects to the alum have also been reported and include nodules and erythema at the site of injection [120]. In contrast, recombinant HBsAg nanoemulsion (HBsAg-NE)-based vaccines induced a Th1 response and effective cellular immunity in animals, and did not produce local inflammation after injection [119]. When formulated into an intranasal vaccine, HBsAg-NE produced high levels of anti-HBsAg serum IgG antibodies that were comparable to IM HBsAg-NE and HBsAg-alum. In addition, intranasal HBsAg-NE induced a Th1 cytokine profile and significant levels of mucosal IgA antibodies [119]. NE-based vaccine has also been shown to be effective mucosal adjuvants for inactivated influenza virus [105] and vaccinia viruses [121] among other pathogens.

18.4.1.3 Immune-Stimulating Complexes

Immune-stimulating complexes (ISCOMs) are nanosized spherical micelles that act as carriers and immunostimulatory adjuvants due to their saponin-derived components such as Quil A, which originates from the bark of the *Quillaja* tree [20, 101, 102, 122]. As carriers, ISCOMs are efficiently taken up by APCs as a result of their particulate nature. In addition, even in the absence of antigen, they are able to activate and upregulate expression of MHC I and II on APCs, initiating an adaptive immune response [102]. This results from the ability of Quil A to induce proinflammatory cytokines such as IL-1, IL-6, IL-8, and IFN-γ. Furthermore, saponin-derived components are capable of producing a Th1-type response, but this may be modified by the type of antigen or adjuvant carried [20, 102]. For example, ISCOMs carrying *Leishmania* antigen were shown to induce high levels of IL-4, indicative of a Th2 polarized response [123]. Numerous antigens have been incorporated into ISCOMs and have been shown to induce protective immune responses, including influenza, hepatitis B, herpes virus, and bacteria such as *Helicobacter pylori* and *Corynebacterium* [102].

18.4.1.4 Cytidine-Phosphate-Guanosine (CpG) Motifs

Bacterial DNA sequences, specifically oligodeoxynucleotides containing unmethylated CpG motifs, have been found to be immunostimulatory and therefore have been used as vaccine adjuvants [20, 105, 124]. CpG motifs are recognized by APCs through activation of TLR-9. This activates secretion of IL-12 and expression of MHC and co-stimulatory molecules by monocytes and dendritic cells (DCs), subsequently inducing a Th1 immune response and generation of IgG antibodies [124]. When incorporated into nanoparticle formulations, CpG motifs induce a strong and sustained immune response. For example, a multiple nanoemulsion system (W/O/W) formulated with CpG and inactivated influenza virus (PELC/CpG) was found to induce a higher antigen-specific serum antibody response after only one dose when compared to two doses of a similar but non-adjuvanted vaccine. Furthermore, this PELC/CpG induced a higher antibody response than was obtained from vaccine adjuvanted with alum [105].

18.4.1.5 Chitosan

As discussed previously, chitosan can be used as a carrier system, and this property allows for improved delivery of vaccine antigens across mucous membranes [125]. It also acts as an immune adjuvant as a result of its ability to promote antigen uptake and cytokine production [20]. As a result, when incorporated with influenza surface glycoproteins HA and NA, chitosan nanoparticles induced a greater serum (IgG) and mucosal (IgA) antibody response than purified surface antigens alone after intranasal administration [125]. Similar results were found when *Streptococcus equi* protein antigen and cholera toxin B immune adjuvant were encapsulated into PCL nanoparticles modified with chitosan, spermine, or oleic acid adsorption enhancers, as chitosan-based particles induced a high serum IgG and secretory IgA response [126]. The high IgA response is thought to result from chitosan's strong interactions with sialic acid residues in mucin, which increases vaccine residence time and improves membrane transport by inducing the transient opening of mucosal tight junctions [20, 126]. Chitosan formulations have also shown promise as immunostimulatory adjuvants in intradermal vaccinations. *N*-trimethyl chitosan (TMC) nanoparticles were formulated with ovalbumin (OVA) or diphtheria toxoid (DT) and injected intradermally into mice. TMC nanoparticles were taken up by DCs in the skin and induced the expression of CD83, CD86, and MHC II, signifying DC maturation. Once internalized, the antigens (OVA and DT) were released from chitosan by lysosomal degradation, and subsequently processed and presented to T cells, inducing a Th2 antibody response. With both antigens, TMC induced higher IgG titers than either antigen alone, substantiating its role as an immunostimulatory adjuvant in intradermal vaccination [104].

18.4.2 Metallic Nanoparticles

When used as drug- or gene-delivery vehicles, nanoparticles must hide their contents during transport to their intended target in order to avoid immune system activation and cell toxicity. This property has been shown in metallic nanoparticles, substantiating their use as carrier systems. For instance, foreign nucleic acids are known to activate the innate immune response, complicating gene delivery. However, when conjugated with gold nanoparticles (AuNPs), the immune response is dampened. Compared to lipid-complexed DNA, the macrophage response to a polyvalent oligonucleotide-modified gold nanoparticle (DNA-AuNP) conjugate was 25-fold less, as evidenced by decreased levels of IFN-β despite higher amounts of DNA internalization [127]. It is thought that the DNA-Au NP conjugates allow a high oligonucleotide surface density, which reduces the ability of cellular DNA binding proteins to recognize nucleic acids on the nanoparticle. This was further validated by an increased innate immune response with decreasing DNA surface density [127]. The immune response is also highly dependent on the antigen that is being carried, particularly the peptide sequence of the conjugate. For example, AuNPs alone did not affect macrophage proliferation or cytokine response, while AuNP conjugates blocked macrophage proliferation and induced cytokine production depending on the specific peptide sequence of the conjugate. Even more interesting, although all AuNP conjugates induced TNF-α and IL-1β, only AuNP-CLPFFD-NH_2 induced a strong IL-6 response [128].

18.5 Conclusions

Nanotechnology allows for the creation of unique carrier systems that enhance molecular interactions, thus allowing nanoparticles to facilitate the body's response to foreign pathogens. This includes surpassing microbial resistance mechanisms and improving the innate and adaptive immune response. Incorporating conventional antimicrobials or other materials into nanoparticle platforms also allows for targeted drug delivery and minimizes drug resistance. In the case of antimicrobial agents, the large surface-to-volume ratio of nanoparticles increases drug penetration by disrupting the microbial cell wall or cytoplasmic membrane. When engineered to carry foreign antigens, nanoparticles may also act as immunostimulatory adjuvants as a result of increased uptake by APCs and subsequent Th1 or Th2 immune response.

Importantly, future studies will be needed to explore potential deleterious effects of these agents. Even today, there is growing concern regarding toxicity as a result of specific materials used in some particles (i.e., heavy metals) and their ability to penetrate vital organs, although the evidence regarding human toxicity is inconclusive to date [11]. Regardless, the future of nanotechnology in the antimicrobial arena is not only bright, but a requisite to successfully combat the medical crisis that has resulted from the continued emergence of resistant infectious organism.

References

1. Hoiby N, Bjarnsholt T, Givskov M, Molin S, Ciofu O. Antibiotic resistance of bacterial biofilms. Int J Antimicrob Agents. 2010;35(4):322–32.
2. Shenoy MS, Bhat GK, Kishore A, Hassan MK. Significance of MRSA strains in community associated skin and soft tissue infections. Indian J Med Microbiol. 2010;28(2):152–4.
3. Maragakis LL, Perl TM. How can we stem the rising tide of multidrug-resistant gram-negative bacilli? Infect Control Hosp Epidemiol. 2010;31(4):338–40.
4. Maragakis LL, Perencevich EN, Cosgrove SE. Clinical and economic burden of antimicrobial resistance. Expert Rev Anti Infect Ther. 2008;6(5):751–63.
5. Boucher HW. Challenges in anti-infective development in the era of bad bugs, no drugs: a regulatory perspective using the example of bloodstream infection as an indication. Clin Infect Dis. 2010;50 Suppl 1:S4–9.
6. Spellberg B, Guidos R, Gilbert D, et al. The epidemic of antibiotic-resistant infections: a call to action for the medical community from the Infectious Diseases Society of America. Clin Infect Dis. 2008;46(2):155–64.
7. Cosgrove SE. The relationship between antimicrobial resistance and patient outcomes: mortality, length of hospital stay, and health care costs. Clin Infect Dis. 2006;42 Suppl 2:S82–9.

8. Cosgrove SE, Carmeli Y. The impact of antimicrobial resistance on health and economic outcomes. Clin Infect Dis. 2003;36(11):1433–7.
9. Friedman A, Blecher K, Sanchez D, et al. Susceptibility of Gram-positive and -negative bacteria to novel nitric oxide-releasing nanoparticle technology. Virulence. 2011;2(3):217–21.
10. Blecher K, Nasir A, Friedman A. The growing role of nanotechnology in combating infectious disease. Virulence. 2011;2(5):395–401.
11. Kim BYS, Rutka JT, Chan WCW. Nanomedicine. N Engl J Med. 2010;363(25):2434–43.
12. Banergee M, Mallick S, Paul A, Chattopadhyay A, Ghosh S. Heightened reactive oxygen species generation in the antimicrobial activity of three component iodinated chitosan-silver nanoparticle composite. Langmuir. 2010;26(8):5901–8.
13. Ma Y, Zhou T, Zhao C. Preparation of chitosan-nylon-6 blended membranes containing silver ions as antibacterial materials. Carbohydr Res. 2008;343(2):230–7.
14. Sanpui P, Murugadoss A, Prasad PV, Ghosh SS, Chattopadhyay A. The antibacterial properties of a novel chitosan-Ag-nanoparticle composite. Int J Food Microbiol. 2008;124(2):142–6.
15. Qi L, Xu Z, Jiang X, Hu C, Zou X. Preparation and antibacterial activity of chitosan nanoparticles. Carbohydr Res. 2005;339:2693–700.
16. Alburquenque C, Bucarey SA, Neira-Carrillo A, Urzua B, Hermosilla G, Tapia CV. Antifungal activity of low molecular weight chitosan against clinical isolates of Candida spp. Med Mycol. 2010;48(8):1018–23.
17. Albasarah YY, Somavarapu S, Stapleton P, Taylor KMG. Chitosan-coated antifungal formulations for nebulisation. J Pharm Pharmacol. 2010;62(7):821–8.
18. Li RC, Guo ZY, Jiang PA. Synthesis, characterization, and antifungal activity of novel quaternary chitosan derivatives. Carbohydr Res. 2010;345(13): 1896–900.
19. Kulikov SN, Tiurin Iu A, Fassakhov RS, Varlamov VP. [Antibacterial and antimycotic activity of chitosan: mechanisms of action and role of the structure]. Zh Mikrobiol Epidemiol Immunobiol. 2009;5:91–7.
20. Chadwick S, Kriegel C, Amiji M. Nanotechnology solutions for mucosal immunization. Adv Drug Deliv Rev. 2010;62(4–5):394–407.
21. Friedman AJ, Han G, Navati MS, et al. Sustained release nitric oxide releasing nanoparticles: characterization of a novel delivery platform based on nitrite containing hydrogel/glass composites. Nitric Oxide. 2008;19(1):12–20.
22. Potara M, Jakab E, Damert A, Popescu O, Canpean V, Astilean S. Synergistic antibacterial activity of chitosan-silver nanocomposites on Staphylococcus aureus. Nanotechnology. 2011;22(13):135101.
23. Mahendra R, Yadav A, Gade A. Silver nanoparticles as a new generation of antimicrobials. Biotechnol Adv. 2009;27:76–83.
24. Pal S, Tak Y, Song JM. Does the antibacterial actiivty of silver nanoparticles depend on the shape of the nanoparticles? A study of the gram-negative bacterium *Escherechia coli*. Appl Environ Microbiol. 2007;27(6):1712–20.
25. Ruparelia JP, Chatterjee AK, Duttagupta SP, Mukherji S. Strain specificity in antimicrobial activity of silver and copper nanoparticles. Acta Biomater. 2008;4(3): 707–16.
26. Butkus MA, Labare MP, Starke JA, Moon K, Talbot M. Use of aqueous silver to enhance inactivation of coliphage MS-2 by UV disinfection. Appl Environ Microbiol. 2004;70(5):2848–53.
27. Lara HH, Ayala-Nunez NV, Ixtepan-Turrent L, Rodriguez-Padilla C. Mode of antiviral action of silver nanoparticles against HIV-1. J Nanobiotechnol. 2010;8:1.
28. Martinez-Gutierrez F, Olive PL, Banuelos A, et al. Synthesis, characterization, and evaluation of antimicrobial and cytotoxic effect of silver and titanium nanoparticles. Nanomedicine. 2010;6(5):681–8.
29. Ghosh S, Kaushik R, Nagalakshmi K, et al. Antimicrobial activity of highly stable silver nanoparticles embedded in agar-agar matrix as a thin film. Carbohydr Res. 2010;345(15):2220–7.
30. Kim KJ, Sung WS, Suh BK, et al. Antifungal activity and mode of action of silver nano-particles on Candida albicans. Biometals. 2009;22(2):235–42.
31. Paulo CS, Vidal M, Ferreira LS. Antifungal nanoparticles and surfaces. Biomacromolecules. 2010;11(10):2810–7.
32. Panacek A, Kolar M, Vecerova R, et al. Antifungal activity of silver nanoparticles against Candida spp. Biomaterials. 2009;30(31):6333–40.
33. Esteban-Tejeda L, Malpartida F, Esteban-Cubillo A, Pecharroman C, Moya JS. The antibacterial and antifungal activity of a soda-lime glass containing silver nanoparticles. Nanotechnology. 2009;20(8):085103.
34. Shrivastava S, Bera T, Roy A, Singh G, Ramachandararao P, Dash D. Characterization of enhanced antibacterial effects of novel silver nanoparticles. Nanotechnology. 2007;18:1–9.
35. Shahverdi AR, Fakhimi A, Shahverdi HR, Minaian S. Synthesis and effect of silver nanoparticles on the antibacterial activity of different antibiotics against Staphylococcus aureus and Escherichia coli. Nanomedicine. 2007;3(2):168–71.
36. Kim JS, Kuk E, Yu KN, et al. Antimicrobial effects of silver nanoparticles. Nanomedicine. 2007;3(1):95–101.
37. Wu Y, Jia W, An Q, Liu Y, Chen J, Li G. Multiaction antibacterial nanofibrous membranes fabricated by electrospinning: an excellent system for antibacterial applications. Nanotechnology. 2009;20(24):245101.
38. Banerjee M, Mallick S, Paul A, Chattopadhyay A, Ghosh SS. Heightened reactive oxygen species generation in the antimicrobial activity of a three component iodinated chitosan-silver nanoparticle composite. Langmuir. 2010;26(8):5901–8.
39. Kwak S, Kim SH, Kim SS. Hybrid organic/inorganic reverse osmosis (RO) membrane for bactericidal antifouling: 1. Preparation and characterization of TiO nanoparticle self-assembled aromatic polyamide thin-film-composite (TFC) membrane. Environ Sci Technol. 2001;35(11):2388–94.

40. Kim SH, Kwak S, Sohn B, Park TH. Design of TiO_2 nanoparticle self-assembled aromatic polyamide thin-film-composite (TFC) membrane as an approach to solve biofouling problem. J Membr Sci. 2003;211:157–65.
41. Han G, Zippin JH, Friedman A. From bench to bedside: the therapeutic potential of nitric oxide in dermatology. J Drugs Dermatol. 2009;8(6):586–96.
42. Englander L, Friedman A. Nitric oxide nanoparticle technology: a novel antimicrobial agent in the context of current treatment of skin and soft tissue infection. J Clin Aesthet Dermatol. 2010;3(6):45–50.
43. Martinez LR, Han G, Chacko M, et al. Antimicrobial and healing efficacy of sustained release nitric oxide nanoparticles against Staphylococcus aureus skin infection. J Invest Dermatol. 2009;129(10):2463–9.
44. Cabrales P, Han G, Roche C, Nacharaju P, Friedman AJ, Friedman JM. Sustained release nitric oxide from long-lived circulating nanoparticles. Free Radic Biol Med. 2010;49(4):530–8.
45. Han G, Martinez LR, Mihu MR, Friedman AJ, Friedman JM, Nosanchuk JD. Nitric oxide releasing nanoparticles are therapeutic for Staphylococcus aureus abscesses in a murine model of infection. PLoS One. 2009;4(11):e7804.
46. Mihu MR, Sandkovsky U, Han G, Friedman JM, Nosanchuk JD, Martinez LR. Nitrix oxide releasing nanoparticles are therapeutic for *Acinetobacter baumanni* wound infections. Virulence. 2010;1(2):62–7.
47. Friedman AJ, Blecher K, Schairer D, et al. Improved antimicrobial efficacy with nitric oxide releasing nanoparticle generated S-nitrosoglutathione. Nitric Oxide. 2011;25(4):381–6.
48. Karthikeyan R, Amaechi BT, Rawls HR, Lee VA. Antimicrobial activity of nanoemulsion on cariogenic Streptococcus mutans. Arch Oral Biol. 2011;56(5):437–45.
49. Ziani K, Chang YH, McLandsborough L, McClements DJ. Influence of surfactant charge on antimicrobial efficacy of surfactant-stabilized thyme oil nanoemulsions. J Agric Food Chem. 2011;59(11):6247–55.
50. Hemmila MR, Mattar A, Taddonio MA, et al. Topical nanoemulsion therapy reduces bacterial wound infection and inflammation after burn injury. Surgery. 2010;148(3):499–509.
51. Ciotti S, Eisma R, Ma L, Baker JR. In-vitro skin penetration of novel antimicrobial nanoemulsion formulations containing antifungal agents. J Invest Dermatol. 2009;129:S78.
52. Fothergill AW, McCarthy DI, Sutcliffe JA, Rinaldi MG. Antifungal activity of NB-002 a topical nanoemulsion, against rare fungal pathogens of onychomycosis. J Am Acad Dermatol. 2009;60(3):AB117.
53. Jones T, Ijzerman M, Flack M. A randomized, double-blind, vehicle-controlled trial of a novel topical antifungal nanoemulsion (NB-002) in subjects with distal subungual onychomycosis. J Am Acad Dermatol. 2009;60(3):AB102.
54. Lipuma JJ, Makidon PE, Foster BK, Keoleian JC, Rathinavelu S, Kailkin LM, Baker JR Jr. In vitro activities of a novel nanoemulsion against Burkholderia and other multi-drug resistant cystic fibrosis-associated bacterial species. Antimicrob Agents Chemother. 2009;53(1):249–55.
55. Pengon S, Limmatvapirat C, Limmatvapirat S. Preparation and evaluation of antimicrobial nanoemulsion containing herbal extracts. Drug Metab Rev. 2009;41:85.
56. Hamouda T, Flack M, Baker J. Development of a novel topically applied antifungal agent (NB-002) based on nanoemulsion technology. J Am Acad Dermatol. 2008;58(2):AB90.
57. Teixeira PC, Leite GM, Domingues RJ, Silva J, Gibbs PA, Ferreira JP. Antimicrobial effects of a microemulsion and a nanoemulsion on enteric and other pathogens and biofilms. Int J Food Microbiol. 2007;118(1):15–9.
58. Hamouda T, Myc A, Donovan B, Shih AY, Reuter JD, Baker JR. A novel surfactant nanoemulsion with a unique non-irritant topical antimicrobial activity against bacteria, enveloped viruses and fungi. Microbiol Res. 2001;156(1):1–7.
59. Hamouda T, Hayes MM, Cao ZY, et al. A novel surfactant nanoemulsion with broad-spectrum sporicidal activity against Bacillus species. J Infect Dis. 1999;180(6):1939–49.
60. Pannu J, McCarthy A, Martin A, et al. NB-002, a novel nanoemulsion with broad antifungal activity against dermatophytes, other filamentous fungi, and Candida albicans. Antimicrob Agents Chemother. 2009;53(8):3273–9.
61. Pannu J, Sutcliffe J, Ma LF, Ciotti S. Antifungal activity and mechanism of action of NB-002, a novel topical antifungal, against the major pathogens of onychomycosis. J Am Acad Dermatol. 2009;60(3):AB114.
62. Jones T, Flack M, Ijzerman M, Baker J. Safety, tolerance, and pharmacokinetics of topical nanoemulsion (NB-002) for the treatment of onychomycosis. J Am Acad Dermatol. 2008;58(2):AB83.
63. Ijzerman M, Baker J, Flack M, Robinson P. Efficacy of topical nanoemulsion (NB-002) for the treatment of distal subungual onychomycosis: a randomized, double-blind, vehicle-controlled trial. J Am Acad Dermatol. 2010;62(3):AB76.
64. Pinto-Alphandary H, Andremont A, Couvreur P. Targeted delivery of antibiotics using liposomes and nanoparticles: research and applications. Int J Antimicrob Agents. 2000;13(3):155–68.
65. Vieira DB, Carmona-Ribeiro AM. Cationic nanoparticles for delivery of amphotericin B: preparation, characterization and activity in vitro. J Nanobiotechnol. 2008;6:6.
66. Mohammadi G, Valizadeh H, Barzegar-Jalali M, et al. Development of azithromycin-PLGA nanoparticles: physicochemical characterization and antibacterial effect against Salmonella typhi. Colloids Surf B Biointerfaces. 2010;80(1):34–9.
67. Dillen K, Vandervoort J, Van den Mooter G, Ludwig A. Evaluation of ciprofloxacin-loaded Eudragit RS100 or RL100/PLGA nanoparticles. Int J Pharm. 2006;314(1):72–82.
68. Turos E, Reddy GS, Greenhalgh K, et al. Penicillin-bound polyacrylate nanoparticles: restoring the

activity of beta-lactam antibiotics against MRSA. Bioorg Med Chem Lett. 2007;17(12):3468–72.
69. Abeylath SC, Turos E, Dickey S, Lim DV. Glyconanobiotics: novel carbohydrated nanoparticle antibiotics for MRSA and Bacillus anthracis. Bioorg Med Chem. 2008;16(5):2412–8.
70. Fattal E, Rojas J, Youssef M, Couvreur P, Andremont A. Liposome-entrapped ampicillin in the treatment of experimental murine listeriosis and salmonellosis. Antimicrob Agents Chemother. 1991;35(4):770–2.
71. Fontana G, Pitarresi G, Tomarchio V, Carlisi B, San Biagio PL. Preparation, characterization and in vitro antimicrobial activity of ampicillin-loaded polyethylcyanoacrylate nanoparticles. Biomaterials. 1998;19(11–12):1009–17.
72. Shim YH, Kim YC, Lee HJ, et al. Amphotericin B aggregation inhibition with novel nanoparticles prepared with poly(epsilon-caprolactone)/poly(n, n-dimethylamino-2-ethyl methacrylate) diblock copolymer. J Microbiol Biotechnol. 2011;21(1):28–36.
73. Sheikh S, Ali SM, Ahmad MU, et al. Nanosomal amphotericin B is an efficacious alternative to Ambisome for fungal therapy. Int J Pharm. 2010;397(1–2):103–8.
74. Burgess BL, Cavigiolio G, Fannucchi MV, Illek B, Forte TM, Oda MN. A phospholipid-apolipoprotein A-I nanoparticle containing amphotericin B as a drug delivery platform with cell membrane protective properties. Int J Pharm. 2010;399(1–2):148–55.
75. Shao K, Huang RQ, Li JF, et al. Angiopep-2 modified PE-PEG based polymeric micelles for amphotericin B delivery targeted to the brain. J Control Release. 2010;147(1):118–26.
76. Jung SH, Lim DH, Lee JE, Jeong KS, Seong H, Shin BC. Amphotericin B-entrapping lipid nanoparticles and their in vitro and in vivo characteristics. Eur J Pharm Sci. 2009;37(3–4):313–20.
77. Amaral AC, Bocca AL, Ribeiro AM, et al. Amphotericin B in poly(lactic-co-glycolic acid) (PLGA) and dimercaptosuccinic acid (DMSA) nanoparticles against paracoccidioidomycosis. J Antimicrob Chemother. 2009;63(3):526–33.
78. Fukui H, Koike T, Saheki A, Sonoke S, Tomii Y, Seki J. Evaluation of the efficacy and toxicity of amphotericin B incorporated in lipid nano-sphere (LNS (R)). Int J Pharm. 2003;263(1–2):51–60.
79. Ritter J. Amphotericin B and its lipid formulations. Mycoses. 2002;45:34–8.
80. Bekersky I, Boswell GW, Hiles R, Fielding RM, Buell D, Walsh TJ. Safety, toxicokinetics and tissue distribution of long-term intravenous liposomal amphotericin B (AmBisome (R)): a 91-day study in rats. Pharm Res. 2000;17(12):1494–502.
81. Bekersky I, Boswell GW, Hiles R, Fielding RM, Buell D, Walsh TJ. Safety and toxicokinetics of intravenous liposomal amphotericin B (AmBisome (R)) in beagle dogs. Pharm Res. 1999;16(11):1694–701.
82. Johnson EM, Ojwang JO, Szekely A, Wallace TL, Warnock DW. Comparison of in vitro antifungal activities of free and liposome-encapsulated nystatin with those of four amphotericin B formulations. Antimicrob Agents Chemother. 1998;42(6):1412–6.
83. Gulati M, Bajad S, Singh S, Ferdous AJ, Singh M. Development of liposomal amphotericin B formulation. J Microencapsul. 1998;15(2):137–51.
84. Hiemenz JW, Walsh TJ. Lipid formulations of amphotericin B. J Liposome Res. 1998;8(4):443–67.
85. Hillery AM. Supramolecular lipidic drug delivery systems: from laboratory to clinic—a review of the recently introduced commercial liposomal and lipid-based formulations of amphotericin B. Adv Drug Deliv Rev. 1997;24(2–3):345–63.
86. Anstey NM, Stewart LM, Packard M, Graney WF, Bartlett JA. Open-label titration study of the safety of RMP-7 in patients with the acquired immune deficiency syndrome. Int J Antimicrob Agents. 1996;6(4):183–7.
87. Joly V, Farinotti R, Saintjulien L, Cheron M, Carbon C, Yeni P. In-vitro renal toxicity and in-vivo therapeutic efficacy in experimental murine cryptococcosis of amphotericin-B (fungizone) associated with intralipid. Antimicrob Agents Chemother. 1994;38(2):177–83.
88. Bekersky I, Fielding RM, Buell D, Lawrence I. Lipid-based amphotericin B formulations: from animals to man. Pharm Sci Technol Today. 1999;2(6):230–6.
89. Bhalekar MR, Pokharkar V, Madgulkar A, Patil N. Preparation and evaluation of miconazole nitrate-loaded solid lipid nanoparticles for topical delivery. AAPS PharmSciTech. 2009;10(1):289–96.
90. Naeff R. Feasibility of topical liposome drugs produced on an industrial scale. Adv Drug Deliv Rev. 1996;18(3):343–7.
91. Nystatin—liposomal. AR 121, Nyotran. Drugs R D. 1999;1(2):181–3.
92. Groll AH, Petraitis V, Petraitiene R, et al. Safety and efficacy of multilamellar liposomal nystatin against disseminated candidiasis in persistently neutropenic rabbits. Antimicrob Agents Chemother. 1999;43(10):2463–7.
93. Moribe K, Maruyama K. Pharmaceutical design of the liposomal antimicrobial agents for infectious disease. Curr Pharm Des. 2002;8(6):441–54.
94. Wallace TL, Paetznick V, Cossum PA, LopezBerestein G, Rex JH, Anaissie E. Activity of liposomal nystatin against disseminated Aspergillus fumigatus infection in neutropenic mice. Antimicrob Agents Chemother. 1997;41(10):2238–43.
95. Wasan KM, Ramaswamy M, Cassidy SM, Kazemi M, Strobel FW, Thies RL. Physical characteristics and lipoprotein distribution of liposomal nystatin in human plasma. Antimicrob Agents Chemother. 1997;41(9):1871–5.
96. Gupta M, Goyal AK, Paliwal SR, et al. Development and characterization of effective topical liposomal system for localized treatment of cutaneous candidiasis. J Liposome Res. 2010;20(4):341–50.
97. Korting HC, Klovekorn W, Klovekorn G, et al. Comparative efficacy and tolerability of econazole liposomal gel 1%, branded econazole conventional cream 1% and generic clotrimazole cream 1% in tinea pedis. Clin Drug Investig. 1997;14(4):286–93.

98. Wiesenthal A, Hunter L, Wang SG, Wickliffe J, Wilkerson M. Nanoparticles: small and mighty. Int J Dermatol. 2011;50(3):247–54.
99. Chen H, Chang X, Du D, et al. Podophyllotoxin-loaded solid lipid nanoparticles for epidermal targeting. J Control Release. 2006;110(2):296–306.
100. Xie FM, Zeng K, Chen ZL, et al. [Treatment of recurrent condyloma acuminatum with solid lipid nanoparticle gel containing podophyllotoxin: a randomized double-blinded, controlled clinical trial]. Nan Fang Yi Ke Da Xue Xue Bao. 2007;27(5):657–9.
101. Plummer, EM, Manchester, M. Viral nanoparticles and virus-like particles: platforms for contemporary vaccine design. Wiley Interdisciplinary Reviews: Nanomedicine and Nanobiotechnology. John Wiley 24 Sep 2010.
102. Sun HX, Xie Y, Ye YP. ISCOMs and ISCOMATRIX. Vaccine. 2009;27(33):4388–401.
103. Singh R, Lillard Jr JW. Nanoparticle-based targeted drug delivery. Exp Mol Pathol. 2009;86(3):215–23.
104. Bal SM, Slutter B, van Riet E, et al. Efficient induction of immune responses through intradermal vaccination with N-trimethyl chitosan containing antigen formulations. J Control Release. 2010;142(3):374–83.
105. Huang M-H. Emulsified nanoparticles containing inactivated influenza virus and CpG oligodeoxynucleotides critically influences the host immune response in mice. PLoS One. 2010;5(8):e12270.
106. Nasir A. Nanotechnology in vaccine development: a step forward. J Invest Dermatol. 2009;129(5):1055–9.
107. Liu L, Zhong Q, Tian T, Dubin K, Athale SK, Kupper TS. Epidermal injury and infection during poxvirus immunization is crucial for the generation of highly protective T cell-mediated immunity. Nat Med. 2010;16(2):224–7.
108. Combadiere B, Vogt A, Mahe B, et al. Preferential amplification of CD8 effector-T cells after transcutaneous application of an inactivated influenza vaccine: a randomized phase I trial. PLoS One. 2010;5(5):e10818.
109. Mahe B, Vogt A, Liard C, et al. Nanoparticle-based targeting of vaccine compounds to skin antigen-presenting cells by hair follicles and their transport in mice. J Invest Dermatol. 2009;129(5):1156–64.
110. Semete B, Booysen LI, Kalombo L, et al. In vivo uptake and acute immune response to orally administered chitosan and PEG coated PLGA nanoparticles. Toxicol Appl Pharmacol. 2010;249(2):158–65.
111. Csaba N, Sanchez A, Alonso MJ. PLGA:poloxamer and PLGA:poloxamine blend nanostructures as carriers for nasal gene delivery. J Control Release. 2006;113(2):164–72.
112. Rajananthanan P, Attard GS, Sheikh NA, Morrow WJ. Evaluation of novel aggregate structures as adjuvants: composition, toxicity studies and humoral responses. Vaccine. 1999;17(7–8):715–30.
113. Dykman LA, Sumaroka MV, Staroverov SA, Zaitseva IS, Bogatyrev VA. [Immunogenic properties of the colloidal gold]. Izv Akad Nauk Ser Biol. 2004;1:86–91.
114. Diwan M, Elamanchili P, Lane H, Gainer A, Samuel J. Biodegradable nanoparticle mediated antigen delivery to human cord blood derived dendritic cells for induction of primary T cell responses. J Drug Target. 2003;11(8–10):495–507.
115. Moore MD, Cookson J, Coventry VK, et al. Protection of HIV neutralizing aptamers against rectal and vaginal nucleases: implications for RNA-based therapeutics. J Biol Chem. 2011;286(4):2526–35.
116. Kim SK, Sims CL, Wozniak SE, Drude SH, Whitson D, Shaw RW. Antibiotic resistance in bacteria: novel metalloenzyme inhibitors. Chem Biol Drug Des. 2009;74(4):343–8.
117. Saccucci L, Crance JM, Colas P, Bickle M, Garin D, Iseni F. Inhibition of vaccinia virus replication by peptide aptamers. Antiviral Res. 2009;82(3):134–40.
118. Tolentino M. Systemic and ocular safety of intravitreal anti-VEGF therapies for ocular neovascular disease. Surv Ophthalmol. 2011;56(2):95–113.
119. Makidon PE, Bielinska AU, Nigavekar SS, et al. Pre-clinical evaluation of a novel nanoemulsion-based hepatitis B mucosal vaccine. PLoS One. 2008;3(8):e2954.
120. Muttil P, Prego C, Garcia-Contreras L, et al. Immunization of Guinea pigs with novel hepatitis B antigen as nanoparticle aggregate powders administered by the pulmonary route. AAPS J. 2010;12(3):330–7.
121. Bielinska AU. A novel, killed-virus nasal vaccinia virus vaccine. Clin Vaccine Immunol. 2008;14(2):348–58.
122. Helgeby A, Robson NC, Donachie AM, et al. The combined CTA1-DD/ISCOM adjuvant vector promotes priming of mucosal and systemic immunity to incorporated antigens by specific targeting of B cells. J Immunol. 2006;176(6):3697–706.
123. Maloy KJ, Donachie AM, Mowat AM. Induction of Th1 and Th2 CD4+ T cell responses by oral or parenteral immunization with ISCOMS. Eur J Immunol. 1995;25(10):2835–41.
124. Brunner C, Seiderer J, Schlamp A, et al. Enhanced dendritic cell maturation by TNF-alpha or cytidine-phosphate-guanosine DNA drives T cell activation in vitro and therapeutic anti-tumor immune responses in vivo. J Immunol. 2000;165(11):6278–86.
125. Bacon A, Makin J, Sizer PJ, et al. Carbohydrate biopolymers enhance antibody responses to mucosally delivered vaccine antigens. Infect Immun. 2000;68(10):5764–70.
126. Florindo HF, Pandit S, Lacerda L, Goncalves LMD, Alpar HO, Almeida AJ. The enhancement of the immune response against S. equi antigens through intranasal administration of poly-3-caprolactone-based nanoparticles. Biomaterials. 2009;30:879–91.
127. Massich MD, Giljohann DA, Seferos DS, Ludlow LE, Horvath CM, Mirkin CA. Regulating immune response using polyvalent nucleic acid-gold nanoparticle conjugates. Mol Pharm. 2009;6(6):1934–40.
128. Bastus NG, Sanchez-Tillo E, Pujals S, et al. Homogeneous conjugation of peptides onto gold nanoparticles enhances macrophage response. ACS Nano. 2009;3(6):1335–44.

19 Nanotechnology for the Histologic Diagnosis of Infectious Diseases: A Dermatopathologists Perspective

Dirk M. Elston

Nanotechnology holds tremendous promise for the diagnosis and treatment of infectious diseases. This chapter will focus on diagnostic advances that utilize nanotechnology. Immunohistochemical and polymerase chain reaction assays have become mainstream diagnostic techniques because of their sensitivity, rapid turn-around time, and low cost. They are now routinely being used to diagnose infectious diseases as well as to provide important clinical information on drug resistance and toxin production.

19.1 Bacterial Pathogens

Routine culture techniques have changed little in the past 50 years, and most bacteriology labs still employ agar plates and routine cultures for the diagnosis of common bacterial infections. Nanotechniques are beginning to revolutionize the field, as they can easily be automated and provide rapid and cost-effective assays for bacterial pathogens. They are currently being used for the diagnosis of bacterial infections as well as to identify virulence factors and genes that impart drug resistance.

In situ hybridization (ISH) is performed predominantly by fluorescent (FISH) or chromogenic (CISH) techniques, and both are in common use. Polymerase chain reaction (PCR) and direct fluorescent antibody (DFA) assays are also useful for some bacterial pathogens, although they are not used to the extent that they are for viral infections.

19.1.1 Nanotechnology Assays in Common Use for Bacterial Pathogens

Immunohistochemical (IHC) assays have become the gold standard for the detection of *Treponema pallidum* (*T. pallidum*). The sensitivity and ease of use of the IHC stain for *T. pallidum*, compared to traditional silver stains, have revolutionized the diagnosis of syphilis in histologic sections. Figure 19.1 demonstrates a Steiner stain in a patient with syphilis. Figure 19.2 demonstrates the ease with which tone can identify the spirochetes with the aid of an immunohistochemical stain. In one study, IHC detected the organisms in 17/35 cases, compared with 9/35 for Dieterle staining and 14/36 for PCR [1]. Other studies have demonstrated from 71 to 91% sensitivity in patients with secondary syphilis, compared with only 41% using a silver stain [2, 3].

PCR assays are commonly being used to assay for the presence of the *erm* gene that imparts inducible resistance to macrolides such as clindamycin among staphylococcal isolates as well as a range of other bacteria to include atypical mycobacteria [4, 5]. This assay is rapid and

D.M. Elston, MD (✉)
Ackerman Academy of Dermatopathology,
145 East 32nd Street, 10th Floor, New York, NY, 10016, USA
e-mail: delston@dermpathdiagnostics.com

A. Nasir et al. (eds.), *Nanotechnology in Dermatology*,
DOI 10.1007/978-1-4614-5034-4_19, © Springer Science+Business Media, LLC 2013

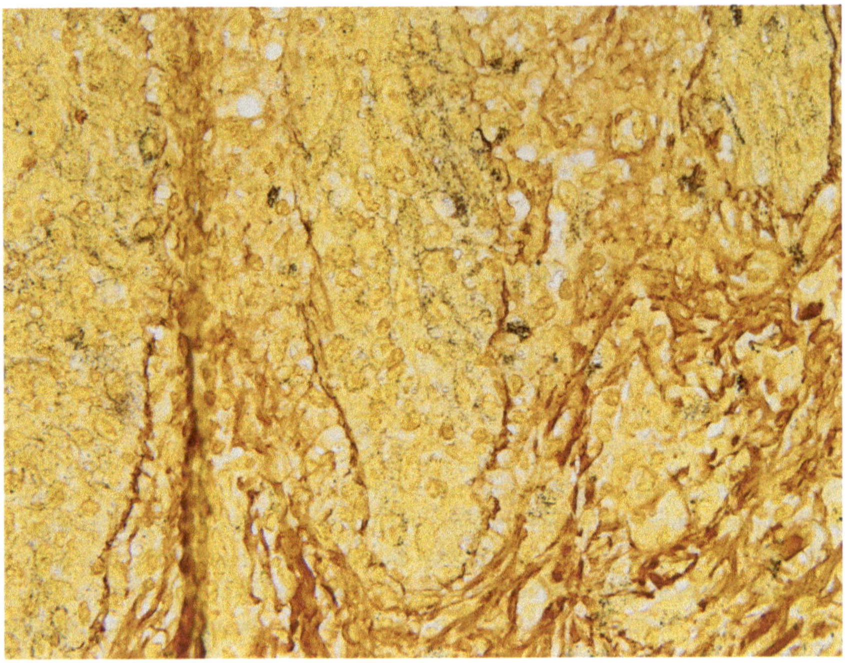

Fig. 19.1 A skin biopsy of syphilis stained with a Steiner stain (×200)

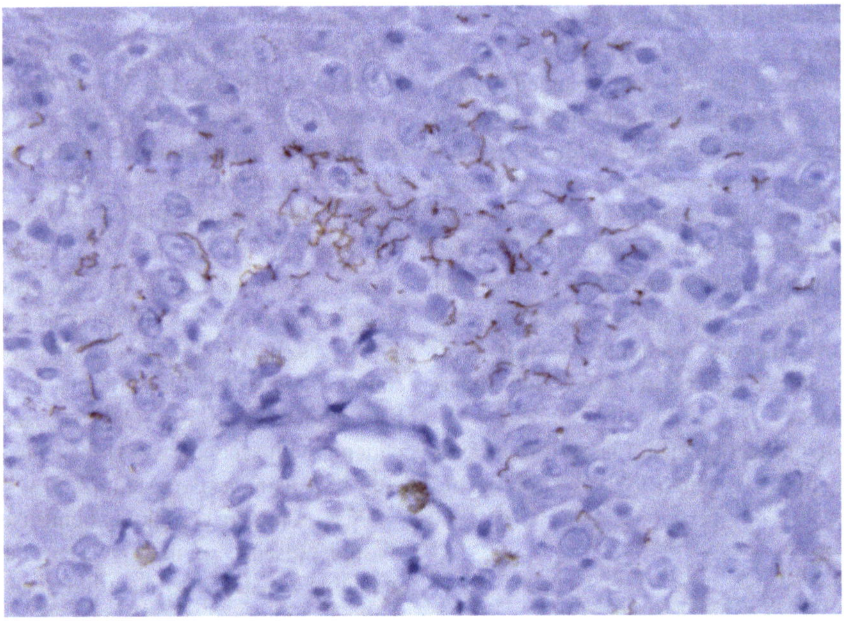

Fig. 19.2 *Treponema pallidum* spirochetes in syphilis are much easier to identify with the aid of an immunohistochemical stain (×400)

highly cost-effective when compared to traditional bacterial plating with measurement of zones of inhibition and D-testing, in which an erythromycin disk is placed next to a clindamycin disk to demonstrate the presence of inducible resistance. PCR assays are also commonly being used to detect the Panton-Valentine leukocidin virulence factor in community type strains of MRSA (CA-MRSA).

IHC is now being used for the detection of *Rickettsia*, although immunofluorescent techniques have been used more commonly in the past [6]. IHC antibodies are available to detect and differentiate spotted fever group and typhus group organisms. In one study, both immunofluorescence and immunoperoxidase staining showed a detection rate of 90% [7].

19.1.2 Emerging Nanotechnology Assays for Bacterial Disease

A colorimetric IHC assay employing a multistep indirect immunoalkaline phosphatase method with anti-*B. anthracis* cell wall (EAII-6G6-2-3) and *anti-B. anthracis* capsule (FDF-1B9) monoclonal antibodies have been developed, but are not yet in widespread use [8]. PCR assays also show considerable promise in this setting. During the bioterrorism scare of 2001, 117 skin biopsy samples were tested using IHC techniques by the Infectious Disease Pathology Activity (IDPA) of the Centers for Disease Control and Prevention (CDC). Of the 117 specimens, 8 tested positive for *B. anthracis* by IHC [9]. An advantage of the technique over routine culture is that the IHC assay can demonstrate nonviable bacillary fragments in formalin-fixed tissues even after 10 days of antibiotic treatment [10]. Skin biopsies should be performed both at the center and periphery of the eschar.

Bartonella henselae can be difficult to culture, and IHC assays are being developed.

A study comparing IHC with PCR and silver staining in patients with cat scratch lymphadenitis demonstrated the highest yield with Steiner silver stain 11 (46%), compared with PCR 9 (38%) and IHC 6 (25%), suggesting that more work needs to be done before these assays replace routine silver staining [11]. Even though sensitivity is low, specificity is quite high with both IHC and PCR methods. All three methods can be performed on formalin-fixed, paraffin-embedded tissue.

Brucella melitensis is a zoonotic pathogen that can cause sepsis in humans. An IHC assay demonstrated 30% sensitivity in detecting the organism [12]. Both IHC and DFA have been used to demonstrate the presence of *Francisella tularensis*, the etiologic agent of tularemia, in formalin-fixed tissues [13]. Two monoclonal antibodies are being used for enzyme immunoassays (EIA). The monoclonal T14 antibody demonstrated no cross-reactivity with *Y. pestis*, *Y. pseudotuberculosis*, *Y. enterocolitica*, *V. cholera*, *E. coli*, *S. typhimurium*, *Fr. novicida*, *Br. melitensis*, *Br. abortus*, *Br. suis*, *Br. ovis*, or *Br. neotomae*. A similar antibody, FB11, demonstrated no cross-reactivity with *Fr. novicida*, *Br. abortus*, *Br. suis*, *Br. melitensis*, *Br. ovis*, *Y. pestis*, *Y. enterocolitica*, *Y. pseudotuberculosis*, *E. coli*, or *V. cholerae*.

In the case of *Helicobacter pylori*, IHC has proved superior to routine histochemistry, in at least some studies. In one study, *H. pylori* was demonstrated by both techniques in 27 of 48 cases. In two cases, the immunostain was negative, but the bacteria were identified with a modified Giemsa stain. In five cases, the bacteria were only identified with the immunostain [14]. Another study showed a 66% detection rate with the antibody, comparing favorably to silver stains and PCR [15]. In yet other studies, PCR and ISH proved superior to IHC staining [16]. The IHC assay is dependent on both the fixation method and the staining system employed. Antigen retrieval techniques using high temperature and a 10 mM citrate buffer, at pH 6.0 have improved detection in formalin-fixed paraffin embedded tissue.

IHC testing for leptospirosis has demonstrated 78% sensitivity, although it has not been clearly superior to silver staining in some studies [17, 18]. Some antibody systems require antigen retrieval using trypsin. Antigen retrieval may also be performed on preheated slides by microwaving for 1.5 min at 630 W followed by 5 min at 180 W in Tris (pH 10) buffer. The product insert should be consulted for the antibody system used, and internal validation should always be performed.

In the setting of Lyme disease, IHC has demonstrated sensitivity as high as 96% with specificity of 99.4%, compared to 45.5% sensitivity and 100% specificity for PCR [19]. In the case of the plague organism, *Yersinia pestis*, bacilli can be identified in formalin-fixed tissues using IHC, DFA, and PCR assays, although culture remains the gold standard [20]. The same is true for *Pseudomonas aeruginosa* [21].

19.2 Tests for Mycobacteria

While cutaneous tuberculin testing, interferon gamma release assays, and culture remain the gold standards in the diagnosis of tuberculosis, other techniques are more widely used to detect atypical mycobacteria. Auramine-Rhodamine staining of routine tissue sections is quite sensitive, but requires a fluorescent microscope. An anti-bacille Calmette-Guerin (BCG) antibody immunostain has been developed that shows better staining for fungi than for mycobacteria [22, 23]. In early leprosy, false-positive staining was noted in 16% of patients, limiting the utility of the test [24]. Antibodies to detect *Mycobacterium avium* subsp. *Paratuberculosis* have demonstrated a wide range of sensitivity from 5% to as high as 93% [25, 26].

19.3 Nanotechnology Techniques to Detect Viruses

ISH, DFA, and PCR assays are widely used for the diagnosis of viral diseases. Both conventional and real-time PCR have been used, and multiplex PCR assays have been developed to detect mixed pathogens, including a panel to test for multiple sexually transmitted diseases with a single assay [27–29]. In contrast, immunohistochemistry (IHC) is currently used to detect only a limited number of viruses such as Human Herpes Virus 8 (HHV-8) in the setting of Kaposi's sarcoma (Fig. 19.3). With a monoclonal antibody to the HHV-8 latent nuclear antigen-1, IHC achieves

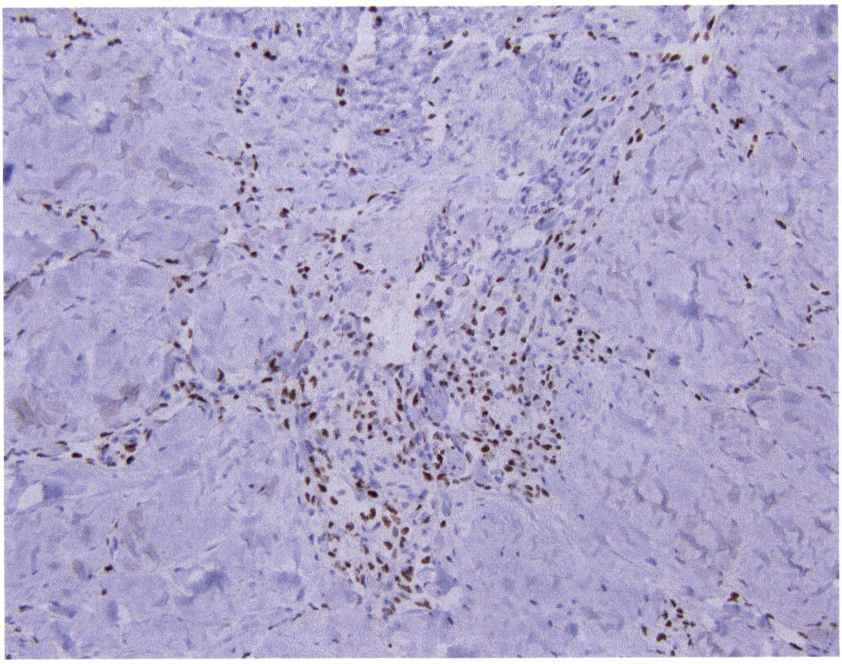

Fig. 19.3 Immunohistochemical staining for HHV-8 in patch stage Kaposi's sarcoma (×200)

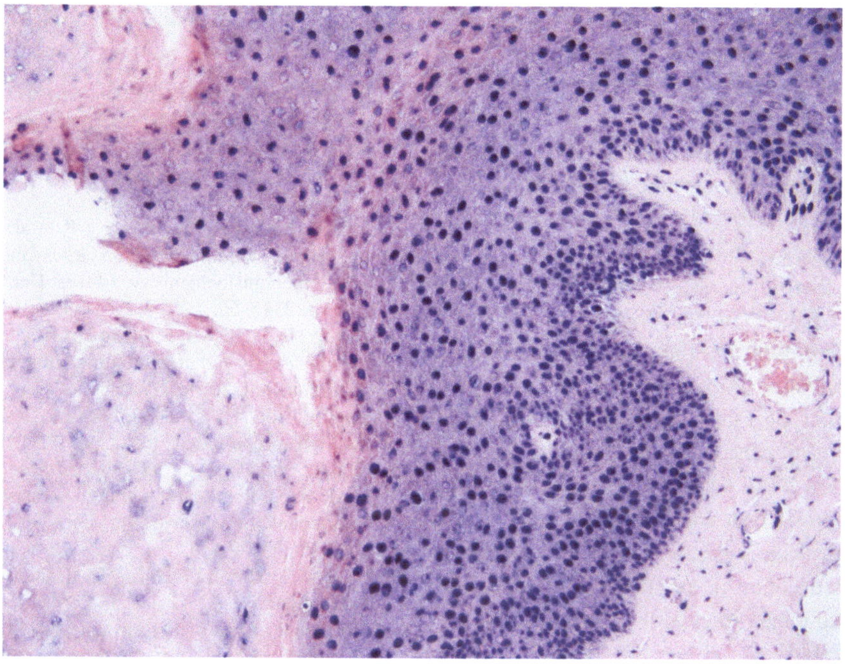

Fig. 19.4 Chromogenic in situ hybridization for high risk HPV virus in condyloma accuminatum (×200)

sensitivity and specificity as high as 100% in paraffin-embedded tissue [30].

ISH is the gold standard for detection of Epstein–Barr virus (EBV) in tissue. IHC has also been used, but not as widely [31]. In one study, IHC had a sensitivity of 44% and specificity of 93%, compared with 94% and 69%, respectively for ISH. PCR has been used more recently, with results similar to ISH, but ISH has the advantage of demonstrating localization of the virus to the nuclei of lesional tissue [32].

For the detection of cytomegalovirus (CMV), PCR is superior to either IHC or ISH if fresh tissue is available [33, 34]. The other assays may have some advantage in fixed tissue [35].

In one study of patients infected with hepatitis B, 83% and 67% of the samples were positive with immunohistochemistry and in situ RT-PCR, respectively [36]. In another study, ISH performed slightly better than the other two assays—19 of 20 detected with ISH, 18 of 20 with RT-PCR, and only 16 of 20 with IHC [37].

In the setting of Human papillomavirus (HPV) infection, ISH is used most commonly (Fig. 19.4), although both PCR and IHC have also been used [38]. HPV 6 and 11 are the predominant low risk virus types associated with condylomata, while HPV types 16, 18, 31, 33, 35, 39, 45, 51, 52, 56, 58, and 66 are more closely associated with cervical cancer. ISH is generally performed as a cocktail for "low risk" and "high risk" types of HPV to help guide management. IHC for p16 (INK4a), a marker of cell cycle dysregulation, is used as a surrogate marker for HPV in cervical biopsies, but this marker is not suitable as a surrogate marker for HPV infections in skin [39].

For the detection of Herpes Simplex (HSV) and the Varicella Zoster virus (VZV), DFA, is the most widely used assay, although ISH assays have also been developed for examination of formalin-fixed, paraffin-embedded tissue. DFA and ISH have very high sensitivity and specificity. PCR is more sensitive than IHC for detection of herpes simplex virus. In one study, the former detected the virus in approximately 90% of patients while the latter was positive in approximately 50% [40].

In response to the worldwide outbreaks of H1N1 swine influenza and H5N1avian influenza,

IHC tests have been developed to allow for detection of influenza A virus in tissue. The monoclonal antibodies are specific for the viral strain [41].

Parvovirus B19 causes fifth disease, aplastic crisis in children with chronic hemolytic anemia, fetal death related to hydrops fetalis, the purpuric gloves and socks syndrome, and an adult arthritis syndrome. Serologic assays are used most commonly, but IHC, PCR, and ISH are also used to confirm the presence of the virus in tissue [42].

A nuchal skin biopsy is often performed for the detection of rabies. Fluorescent and PCR techniques are most commonly used [43, 44]. The tests are fairly sensitive, but the antibodies used may cross react with European bat lyssavirus-1, Duvenhage virus, and Mokola virus [45]. For assays on brain tissue, the specimen should be stored no longer than 24 h in formalin prior to embedding [46].

19.4 Nanotechnology Tests for Fungal and Protozoan Pathogens

The anti-BCG antibody immunostain has been the most common IHC stain in use, although DAKO recently modified the antibody and it has lost reactivity with many fungal species. Other more specific antibodies are being developed, and PCR and ISH assays are also available. PCR has been shown to be more sensitive than IHC using the BCG immunostain for the detection of *Histoplasma capsulatum* [47]. The monoclonal antibody EB-A1 has been used to detect *Aspergillus* species, with good sensitivity. Cross-reactivity occurs with *Pseudallescheria boydii.* Polyclonal antibodies have shown a higher degree of cross reactivity with other fungi [48]. ISH has been used to detect *Blastomyces dermatitidis, Coccidioides immitis, and Cryptococcus neoformans* [49, 50].

IHC testing is quite good for the detection of *Pneumocystis carinii* [51, 52]. PCR appears superior to IHC for the detection of leishmaniasis [53].

References

1. Behrhof W, Springer E, Bräuninger W, Kirkpatrick CJ, Weber A. PCR testing for Treponema pallidum in paraffin-embedded skin biopsy specimens: test design and impact on the diagnosis of syphilis. J Clin Pathol. 2008;61(3):390–5.
2. Buffet M, Grange PA, Gerhardt P, et al. Diagnosing treponema pallidum in secondary syphilis by PCR and immunohistochemistry. J Invest Dermatol. 2007; 127(10):2345–50.
3. Hoang MP, High WA, Molberg KH. Secondary syphilis: a histologic and immunohistochemical evaluation. J Cutan Pathol. 2004;31(9):595–9.
4. Jung JH, Yoon EJ, Choi EC, Choi SS. Development of TaqMan probe-based real-time PCR method for erm(A), erm(B), and erm(C), rapid detection of macrolide-lincosamide-streptogramin B resistance genes, from clinical isolates. J Microbiol Biotechnol. 2009;19(11):1464–9.
5. Esteban J, Martín-de-Hijas NZ, García-Almeida D, Bodas-Sánchez A, Gadea I, Fernández-Roblas R. Prevalence of erm methylase genes in clinical isolates of non-pigmented, rapidly growing mycobacteria. Clin Microbiol Infect. 2009;15(10):919–23.
6. Bacchi CE, Gown AM, Bacchi MM. Detection of infectious disease agents in tissue by immunocytochemistry. Braz J Med Biol Res. 1994;27(12): 2803–20.
7. White WL, Patrick JD, Miller LR. Evaluation of immunoperoxidase techniques to detect Rickettsia rickettsii in fixed tissue sections. Am J Clin Pathol. 1994;101(6):747–52.
8. Tatti KM, Greer P, White E, Shieh WJ, Guarner J, Ferebee-Harris T, et al. Morphologic, immunologic, and molecular methods to detect bacillus anthracis in formalin-fixed tissues. Appl Immunohistochem Mol Morphol. 2006;14(2):234–43.
9. Shieh W-J, Guarner J, Paddock C, Greer P, Tatti K, Fischer M, et al., The Anthrax Bioterrorism Investigation Team. The critical role of pathology in the investigation of bioterrorism-related cutaneous anthrax. Am J Pathol. 2003;163:1901–10.
10. Guarner J, Jernigan J, Shieh W, Tatti K, Flannagan L, Stephens D, et al. Pathology and pathogenesis of bioterrorism-related inhalational anthrax. Am J Pathol. 2003;163:701–9.
11. Caponetti GC, Pantanowitz L, Marconi S, Havens JM, Lamps LW, Otis CN. Evaluation of immunohistochemistry in identifying Bartonella henselae in cat-scratch disease. Am J Clin Pathol. 2009;131(2):250–6.
12. Ilhan F, Yener Z. Immunohistochemical detection of Brucella melitensis antigens in cases of naturally occurring abortions in sheep. J Vet Diagn Invest. 2008;20(6):803–6.
13. Zeidner NS, Carter LG, Monteneiri JA, Petersen JM, Schriefer M, Gage KL, et al. An outbreak of Francisella tularensisincaptiveprairiedogs:animmunohistochemical analysis. J Vet Diagn Invest. 2004;16(2):150–2.

14. Wabing HR. Comparison of immunohistochemical and modified Giemsa stains for demonstration of *Helicobacter pylori* infection in an African population. Afr Health Sci. 2002;2(2):52–5.
15. Ashton-Key M, Diss TC, Isaacson PG. Detection of Helicobacter pylori in gastric biopsy and resection specimens. J Clin Pathol. 1996;49(2):107–11.
16. Ciesielska U, Dziegiel P, Jagoda E, Podhorska-Okołów M, Zabel M. The detection of Helicobacter pylori in paraffin sections using the PCR technique and various primers as compared to histological techniques. Folia Morphol (Warsz). 2004;63(2):229–31.
17. Wild CJ, Greenlee JJ, Bolin CA, Barnett JK, Haake DA, Cheville NF. An improved immunohistochemical diagnostic technique for canine leptospirosis using antileptospiral antibodies on renal tissue. J Vet Diagn Invest. 2002;14(1):20–4.
18. Saglam YS, Yener Z, Temur A, Yalcin E. Immunohistochemical detection of leptospiral antigens in cases of naturally occurring abortions in sheep. Small Rumin Res. 2008;74(1–3):119–22.
19. Lebech A, Clemmensen O, Hansen K. Comparison of in vitro culture, immunohistochemical staining, and PCR for detection of *Borrelia burgdorferi* in tissue from experimentally infected animals. J Clin Microbiol. 1995;33(9):2328–33.
20. Guarner J, Shieh WJ, Greer PW, Gabastou JM, Chu M, Hayes E, et al. Immunohistochemical detection of Yersinia pestis in formalin-fixed, paraffin-embedded tissue. Am J Clin Pathol. 2002;117(2):205–9.
21. Schmengler K, Goldmann T, Brade L, Sánchez Carballo PM, Albrecht S, Brade H, et al. Monoclonal antibody S60-4-14 reveals diagnostic potential in the identification of Pseudomonas aeruginosa in lung tissues of cystic fibrosis patients. Eur J Cell Biol. 2010;89(1):25–33.
22. Szeredi L, Glávits R, Tenk M, Jánosi S. Application of anti-BCG antibody for rapid immunohistochemical detection of bacteria, fungi and protozoa in formalin-fixed paraffin-embedded tissue samples. Acta Vet Hung. 2008;56(1):89–99.
23. Prinz BM, Michaelis S, Kettelhack N, Mueller B, Burg G, Kempf W. Subcutaneous infection with Mycobacterium abscessus in a renal transplant recipient. Dermatology. 2004;208(3):259–61.
24. Schettini AP, Ferreira LC, Milagros R, Schettini MC, Pennini SN, Rebello PB. Enhancement in the histological diagnosis of leprosy in patients with only sensory loss by demonstration of mycobacterial antigens using anti-BCG polyclonal antibodies. Int J Lepr Other Mycobact Dis. 2001;69(4):335–40.
25. Martinson SA, Hanna PE, Ikede BO, Lewis JP, Miller LM, Keefe GP, et al. Comparison of bacterial culture, histopathology, and immunohistochemistry for the diagnosis of Johne's disease in culled dairy cows. J Vet Diagn Invest. 2008;20:51–7.
26. Huntley JFJ, Whitlock RH, Bannantine JP, Stabel JR. Comparison of diagnostic detection methods for *Mycobacterium avium* subsp. *paratuberculosis* in North American Bison. Vet Pathol. 2005;42:42–51.
27. Muvunyi CM, Dhont N, Verhelst R, Crucitti T, Reijans M, Mulders B, Simons G, Temmerman M, Claeys G, Padalko E. Evaluation of a new multiplex polymerase chain reaction assay STDFinder for the simultaneous detection of 7 sexually transmitted disease pathogens. Diagn Microbiol Infect Dis. 2011;[Epub ahead of print] PubMed PMID:21798683.
28. Dhiman N, Wright PA, Espy MJ, Schneider SK, Smith TF, Pritt BS. Concurrent detection of herpes simplex and varicella-zoster viruses by polymerase chain reaction from the same anatomic location. Diagn Microbiol Infect Dis. 2011;70(4):538–40.
29. Ramamurthy M, Alexander M, Aaron S, Kannangai R, Ravi V, Sridharan G, et al. Comparison of a conventional polymerase chain reaction with real-time polymerase chain reaction for the detection of neurotropic viruses in cerebrospinal fluid samples. Indian J Med Microbiol. 2011;29(2):102–9.
30. Patel RM, Goldblum JR, Hsi ED. Immunohistochemical detection of human herpes virus-8 latent nuclear antigen-1 is useful in the diagnosis of Kaposi sarcoma. Mod Pathol. 2004;17(4):456–60.
31. Truong CD, Feng W, Li W, Khoury T, Li Q, Alrawi S, et al. Characteristics of Epstein-Barr virus-associated gastric cancer: a study of 235 cases at a comprehensive cancer center in U.S.A. J Exp Clin Cancer Res. 2009;28:14.
32. Suh N, Liapis H, Misdraji J, Brunt EM, Wang HL. Epstein-Barr virus hepatitis: diagnostic value of in situ hybridization, polymerase chain reaction, and immunohistochemistry on liver biopsy from immunocompetent patients. Am J Surg Pathol. 2007;31(9):1403–9.
33. Strickler J, Manivel J, Copenhaver C, Kubic V. Comparison of in situ hybridization and immunohistochemistry for detection of cytomegalovirus and herpes simplex virus. Hum Pathol. 1990;21:443–8.
34. Bajanowski T, Wiegand P, Brinkmann B. Comparison of different methods for CMV detection. Int J Legal Med. 1994;106:219–22.
35. Lu DY, Qian J, Easley KA, Waldrop SM, Cohen C. Automated in situ hybridization and immunohistochemistry for cytomegalovirus detection in paraffin-embedded tissue sections. Appl Immunohistochem Mol Morphol. 2009;17(2):158–64.
36. Benkoël L, Biagini P, Dodero F, De Lamballerie X, De Micco P, Chamlian A. Immunohistochemical detection of C-100 hepatitis C virus antigen in formaldehyde-fixed paraffin-embedded liver tissue. Correlation with serum, tissue and in situ RT-PCR results. Eur J Histochem. 2004;48(2):185–90.
37. Qian X, Guerrero RB, Plummer TB, Alves VF, Lloyd RV. Detection of hepatitis C virus RNA in formalin-fixed paraffin-embedded sections with digoxigenin-labeled cRNA probes. Diagn Mol Pathol. 2004;13(1):9–14.
38. Jeon JH, Shin DM, Cho SY, Song KY, Park NH, Kang HS, et al. Immunocytochemical detection of HPV16 E7 in cervical smear. Exp Mol Med. 2007;39(5):621–8.

39. Mulvany NJ, Allen DG, Wilson SM. Diagnostic utility of p16INK4a: a reappraisal of its use in cervical biopsies. Pathology. 2008;40(4):335–44.
40. Oda Y, Katsuda S, Okada Y, Kawahara EI, Ooi A, Kawashima A, et al. Detection of human cytomegalovirus, Epstein-Barr virus, and herpes simplex virus in diffuse interstitial pneumonia by polymerase chain reaction and immunohistochemistry. Am J Clin Pathol. 1994;102(4):495–502.
41. He F, Du Q, Ho Y, Kwang J. Immunohistochemical detection of Influenza virus infection in formalin-fixed tissues with anti-H5 monoclonal antibody recognizing FFWTILKP. J Virol Methods. 2009;155(1):25–33.
42. Escher F, Kuhl U, Sabi T, Suckau L, Lassner D, Poller W, et al. Immunohistological detection of Parvovirus B19 capsid proteins in endomyocardial biopsies from dilated cardiomyopathy patients. Med Sci Monit. 2008;14(6):CR333–8.
43. Ogawa T, Gamoh K, Aoki H, Kobayashi R, Etoh M, Senda M, et al. Validation and standardization of virus neutralizing test using indirect immunoperoxidase technique for the quantification of antibodies to rabies virus. Zoonoses Public Health. 2008;55(6):323–7.
44. Lembo T, Niezgoda M, Velasco-Villa A, Cleaveland S, Ernest E, Rupprecht CE. Evaluation of a direct, rapid immunohistochemical test for rabies diagnosis. Emerg Infect Dis. 2006;12(2):310–3.
45. Inoue S, Sato Y, Hasegawa H, Noguchi A, Yamada A, Kurata T, et al. Cross-reactive antigenicity of nucleoproteins of lyssaviruses recognized by a monospecific antirabies virus nucleoprotein antiserum on paraffin sections of formalin-fixed tissues. Pathol Int. 2003;53(8):525–33.
46. Wacharapluesadee S, Ruangvejvorachai P, Hemachudha T. A simple method for detection of rabies viral sequences in 16-year old archival brain specimens with one-week fixation in formalin. J Virol Methods. 2006;134(1–2):267–71.
47. Bialek R, Ernst F, Dietz K, Najvar LK, Knobloch J, Graybill JR, et al. Comparison of staining methods and a nested PCR assay to detect Histoplasma capsulatum in tissue sections. Am J Clin Pathol. 2002;117(4):597–603.
48. Fukuzawa M, Inaba H, Hayama M, Sakaguchi N, Sano K, Ito M, et al. Improved detection of medically important fungi by immunoperoxidase staining with polyclonal antibodies. Virchows Arch. 1995;427(4):407–14.
49. Hayden RT, Qian X, Roberts GD, Lloyd RV. In situ hybridization for the identification of yeast-like organisms in tissue sections. Diagn Mol Pathol. 2001;10(1):15–23.
50. Hayden RT, Qian X, Procop GW, et al. In situ hybridization for the identification of filamentous fungi in tissue sections. Diagn Mol Pathol. 2002;11(2):119–26.
51. Blumenfeld W, Kovacs JA. Use of a monoclonal antibody to detect Pneumocystis carinii in induced sputum and bronchoalveolar lavage fluid by immunoperoxidase staining. Arch Pathol Lab Med. 1988;112(12):1233–6.
52. Arastéh KN, Simon V, Musch R, Weiss RO, Przytarski K, Futh UM, et al. Sensitivity and specificity of indirect immunofluorescence and Grocott-technique in comparison with immunocytology (alkaline phosphatase anti alkaline phosphatase = APAAP) for the diagnosis of Pneumocystis carinii in broncho-alveolar lavage (BAL). Eur J Med Res. 1998;3(12):559–63.
53. Amato VS, Tuon FF, de Andrade HF, Jr BH, Pagliari C, Fernandes ER, et al. Immunohistochemistry and polymerase chain reaction on paraffin-embedded material improve the diagnosis of cutaneous leishmaniasis in the Amazon region. Int J Dermatol. 2009;48(10):1091–5.

Nanotechnology for the Diagnosis of Parasitic Infections

20

Fernando Sérgio Rodrigues Ribeiro Teles, Maria da Luz Martins, Maria Raquel Vieira, and Luís Joaquim Pina da Fonseca

20.1 Introduction

Infectious diseases can trigger a large range of clinical symptoms in humans and animals. Most of them are nonspecific signs, especially in their earlier stages, usually rendering differential diagnosis difficult. Many infections also result in cutaneous manifestations, which may present as isolated symptoms as a result of a more complex, or even serious, systemic disorder. In some cases, cutaneous signs may allow for the generation of an accurate diagnosis of disease by facile and direct examination. However, in many cases, the need for additional diagnostic tools and techniques are required. Repeated histopathological analysis of multiple skin samples in cases of equivocal diagnosis leads to significant cost and discomfort to the patient, despite the frequent reliance on clinical findings for diagnosis in dermatology [1]. As such, rapid and specific diagnosis of skin and soft tissues infections require improved diagnostic methods and tools. Ideally, these should be able to provide simple, cost-effective, rapid, specific and sensitive detection, identification and quantification of the etiologic agent. A rapid and accurate diagnosis of infection not only allows prompt initiation of therapy, but also, when resistant strains are detected, allows for a change of therapeutic regimen tailored to the patient's pathogen profile. In addition, quantitative methods of detection, able to accurately determine the effective burden of a microorganism in host tissues, may constitute an important means to predict disease progression and prognosis. Aesthetic concerns associated with current diagnostic methods must also be taken into account then managing infections. Some diagnostic techniques such as punch and excisional biopsies may result in considerable disfigurement

F.S.R.R. Teles (✉)
Laboratory of Mycology/Microbiology Unit, Instituto de Higiene e Medicina Tropical (IHMT), Universidade Nova de Lisboa (UNL),
Rua da Junqueira, 100, 1349-008, Lisboa, Portugal and Centre for Malaria and Tropical Diseases (CMDT), IHMT, UNL,
Lisboa, Portugal
e-mail: fteles@ihmt.unl.pt

M.d.L. Martins
Laboratory of Mycology/Microbiology Unit, Instituto de Higiene e Medicina Tropical (IHMT), Universidade Nova de Lisboa (UNL),
Rua da Junqueira, 100, 1349-008, Lisboa, Portugal and Departamento de Ciências da Vida, Centro de Recursos Microbiológicos (CREM), UNL,
Caparica, Portugal
e-mail: luz@ihmt.unl.pt

M.R. Vieira
Serviço de Dermatologia, Hospital Curry Cabral,
Rua da Beneficência, 8, 1069-166, Lisboa, Portugal
e-mail: mraquelvieira@gmail.com

L.J.P. da Fonseca
IBB—Instituto de Biotecnologia e Bioengenharia, DBE—Bioengineering Department, Instituto Superior Técnico,
Avenida Rovisco Pais, 1049-001, Lisboa, Portugal
e-mail: luis.fonseca@ist.utl.pt

A. Nasir et al. (eds.), *Nanotechnology in Dermatology*,
DOI 10.1007/978-1-4614-5034-4_20,

and scarring. New diagnostic approaches based on biosensor technology, especially involving nanotechnological structures, have been developed in all stages of the research cycle from proof-of-concept, to prototype, to early clinical trials. Apart from technical and commercial constraints in developed countries, shifting from lab-based molecular analysis to point-of-care testing faces many ethical concerns and obstacles. Although this is less so for the diagnosis of infectious than genetic diseases [2]. Nanotechnology represents a promising approach to develop and utilize novel and improved tests to diagnose dermatological infections.

20.2 Conventional and Molecular Diagnosis of Dermatological Infections

Cutaneous signs of infectious diseases are very frequent, yet rarely specific [3]. In dermatology, in addition to skin examination and possibly serology, the role of cutaneous biopsy is a useful adjunct. The combination can provide a strong correlation between clinical and histopathological patterns (as discussed in Chap. 19). Conventional laboratory diagnosis using traditional techniques such as swab culture, or tissue culture, or the results of special histologic stains can require a significant amount of time. Some methods may lack the desired sensitivity or specificity to establish a rapid diagnosis [4]. This can result in delay of proper therapy.

The ever-increasing number of available methods for amplification of biological targets and signals enables sensitive and specific pathogen detection in biological fluids. Polymerase chain-reaction (PCR) and related techniques for the amplification of nucleic-acids, including quantitative real-time PCR, have constituted the core of biomolecular diagnosis in dermatology [1, 5]. The advent of genome sequencing and PCR have enormously strengthened DNA databases, thus favoring the development of DNA microarrays (DNA biochips), inspired by conventional planar silicon-based microcircuitry, and their use for molecular diagnostics [6]. These technologies allow for high-throughput analysis, precision and reproducible design and, in dermatology, major applications have included expression profiling and tumor typing [7–9]. Biochip arrays of nucleic acids (and of proteins), however, are still too complex and expensive for in-the-field and point-of-care applications. Thus, innovative and improved approaches for the decentralized diagnosis of infectious diseases are urgently needed.

20.3 Nanomaterials for Diagnosis of Dermatological Diseases: An Overview

In dermatology, the main technological advances have been in the arena of drug delivery and, to a lesser extent, new diagnostics. In both cases, the main underlying approach has been the design and manipulation of nanoparticles (NPs) and other nanostructures as immobilization supports, labels, and signal amplifiers for bioprobes, to which they confer the required sensitivity and selectivity for accurate analyte detection in biorecognition systems [10]. Unlike with traditional labeling procedures (e.g., with fluorophores), the nanosized scale of NPs minimizes physical interference with the biorecognition events. NP assays are usually associated with colorimetric detection (by visual observation or spectrophotometry), which is simpler and more economical than fluorescence. Especially in the case of patients with cancer, NP-based biosensing is a crucial approach to optimizing disease management. As for drug delivery, in vivo imaging requires NP penetration into the skin. For NP penetrating the skin, biocompatibility, high-targeting efficiency, low short- and long-term toxicity, and favorable cost-effectiveness and in vivo kinetics are required [11]. NPs such as quantum dots, gold nanoparticles, and carbon nanotubes (CNTs) are superior alternatives to fluorescent tags, having higher stability, providing lower background noise [12], and being amenable for the fabrication of high-density nanoarrays for high-throughput testing. In addition, NP optical signal intensity does not decrease with miniaturization. The versatility of NPs optical and electrical methodologies.

As labels is derived from the variety of analytical techniques that can be used for their detection [8], including optical and electrochemical. NPs are also easily synthesized, can be chemically functionalized, and can form controlled self-assembled structures [12]. Inorganic metallic NPs include gold nanoparticles (Au-NPs). Biomolecule/Au-NP conjugates can be synthesized by physical adsorption or covalent coupling. The latter method creates stronger and more controlled linkages than the former, but is less amenable for reuse and requires prior modification with chemical functional groups.

In general, metallic NPs have been used in two different ways to tag biotargets: (a) in solid-phase assays (e.g., lateral-flow tests), where the bioprobe is anchored to a solid support, and (b) in aggregation assays, making use of the strong optical absorption of metallic NPs. The aggregation assay involves the use of two different NPs, each one carrying a moiety complementary to specific portions of the biotarget. For example, dispersed Au-NPs emit light in the red color spectrum, while aggregated Au-NPs emit blue light. Extensive specific double-binding events (Au-NP_1/target/Au-NP_2) create an interconnected three-dimensional network and lead to blue-shifting in a solution. This blue shift is easily detected by the naked eye. Simple blue shift assays are very promising for applications in low-income regions and the developing world [13].

Quantum dots (QDs), a type of fluorescent NP, are also important in nanodiagnostics. They differ from conventional organic fluorophores in that they are much brighter and more photostable. In addition, they have broad absorption spectra and can be tuned to fluoresce at discrete wavelengths, correlating with particle size and composition. Furthermore, their fluorescence emission wavelength is far distant from the absorption wavelength allowing for good signal/noise separation. Alternatively, a single light source can be used for the simultaneous excitation of QDs emitting at different wavelengths. This can permit multiplexed detection in complex biological media, after tagging each QD with a different bionanoprobe. A major in vivo application of QDs in dermatology is in real-time imaging of skin tumors, after QD penetration in cancer cells and targeting of tumor markers via antibody conjugation [8, 10]. Upon excitation by near-infrared (NIR) light for deep tissue penetration, QD-tagged tumors can be easily visualized up to a depth of 1–2 cm. At NIR wavelengths, tissue autofluorescence is negligible and therefore the signal-to-noise ratio (S/N) is very high [10]. One application of QDs is in the setting sentinel lymph-node identification and mapping, which has been used for melanomas and other invasive skin cancers, relying on infrared fluorescence detection [9, 14]. The small size and anionic surface charge of QDs facilitate rapid travel along the lymphatics, as well as penetration and anchoring to the nodes. Conjugation of QDs with suitable ligands further enables monitoring of metastases in the same lymph nodes [15]. Thus, QDs overcome the drawbacks of dye-based or radioactive labels, and may demonstrate superior performance for sentinel node location [9].

Magnetic NPs (MGNPs), including superparamagnetic iron oxide nanoparticles (SPION) constitute another important class of NPs. When a MGNP-bound bioprobe specifically binds its target, brief exposure to a strong magnetic field results in collective alignment of the individual magnetic forces, yielding an overall measurable signal. Advantages of MGNPs include inexpensive manufacturing, biocompatibility, physicochemical stability, and utility in analysis of turbid/opaque media. Moreover, they render high S/N ratios as a result of the nonmagnetic nature (and hence low background noise) of biological samples. In addition—or as an alternative—to magnetic detection, MGNPs can also be used in a biotarget pre-concentration step, which is especially relevant in cases of highly diluted biological media [12].

Unlike the above-mentioned types of inorganic NPs, organic NPs have been developed and used more recently for biorecognition purposes. Among these are polymeric NPs (PNPs), which combine the improved physical and chemical properties of NPs with the ease of organic polymer processing—they are less toxic and can be chemically modified and functionalized with greater ease than inorganic NPs [16]. They can be found in ultrathin films, nanotubes, nanowires, or nanocomposites doped or encapsulated with labels such as electroactive species, Au-NPs or CNTs, for voltaic [17] or photoluminescent [16] detection. Liposomes are also well-known organic NPs, consisting of cell-like small vesicles of amphiphilic molecules composed of a hydrophobic lipid bilayer that separates two aqueous environments. A common layout consists of filling the liposome core with dye molecules (e.g., rhodamine) for detection by fluorescence or other optical method, in probe-target sandwich formats [17]. Since many dye molecules (with high quantum yields) are present in each liposome NP, the scheme renders high signal amplification and hence improved sensitivity. Another approach relies on incorporating antigenic proteins onto the bilayer (proteoliposomes) followed by solid-phase immobilization and subsequent detection of specific antibodies within a sample solution [18].

Dendrimers compose another class of NPs; they are highly branched treelike polymeric structures (e.g., of proteins, DNA, and other polymers) with many more surface reactive groups than proteins of similar form and size [17]. Dendrimers thus provide enhanced sensitivity, as well as wide linear range and short response time due to analyte diffusion through the branched structure. Moreover, electrical charges from a biochemical reaction may accumulate inside the into the structure of porous dendrimer NP, leading to significant changes in electrical properties, which can be exploited for biosensing applications [18].

20.4 Nanotechnology and Advanced Platforms for Diagnosis of Infectious Diseases

One of the most anticipated applications of nanodiagnostics is point-of-care diagnosis. This has important benefits for the diagnosis of infectious diseases in low-income regions and in the developing world. Indeed, nanotechnologies are amenable to label-free detection and can circumvent some of the current limitations and impediments of microarray technology [8], namely the inherent complexity, cost, and need for surface scanning-based data acquisition [12]. Such diagnostic tests also circumvent the typical handicap due to short shelf-life of common biochemical reagents employed in conventional molecular diagnostics, especially when used in high-temperature remote regions in developing countries. By yielding almost immediate results, such tests facilitate patient adherence to therapeutics and, consequently, disease containment. The expected enhanced sensitivity should also enable measurements in complex, dilute, and opaque media, e.g., blood and urine (where pathogen loads are usually too low for detection without a prior sample amplification step), with minimal sample pretreatment. Single biotarget (whole cell or biomolecule) detection is ideal, which is of the utmost importance for early disease diagnosis. Ultrasensitive detection and highly specific identification of multiple pathogens in complex biological fluids can be performed through multiplexed DNA- or protein arrays. In this regard, optical detection is frequently preferred, since dyes of different colors can tag different probes.

Difficulties for spectral color distinction may arise, leading to signal interferences and poor resolution. Grafting bioprobes to surfaces of nanoarray formats may circumvent this limitation. A growing nanotechnology-based approach is the spectral barcode, namely the QD-based barcode assay and the bio-barcode-based amplification assay (BCA), both employing several types of NPs. The QD-based barcode consists of polystyrene microparticles (MPs) filled with QDs of several colors. The spectral pattern of each MP is a product of the composition and ratio of its contents, QDs of different colors. In a particular setup for DNA detection, each MP is bound to several single-stranded DNA (ssDNA) chains with a given sequence, while the complimentary target is tagged with a fluorophore. The specific biorecognition event thus renders the fluorescence signal of the bound target and the fluorescence intensity profile (signature) of the corresponding barcode. The BCA assay is based on the formation of a triple conjugate between two functionalized NPs (or an NP and an MP) bridged by the biotarget. In a typical assay, a metallic NP is coated with several ssDNA chains of identical sequence. These are designed to specifically recognize a terminal portion of the ssDNA target (e.g., 5′ terminus). These constitute the barcode signal. A second NP (or MP) is a MGNP bound to several ssDNA chains of a sequence partially complementary to the opposite terminus of the target chain (e.g., 3′ terminus). The sandwich conjugate is then captured by an applied magnetic field to concentrate the bound target, and unbound NPs are washed away. Upon heating, ssDNA bound to its complement on magnetic NP is denatured and released. This ssDNA can be amplified by PCR and sequenced. The barcode then hybridizes to a chip-spotted complementary sequence, and a new sandwich is formed upon binding of a metallic NP functionalized with barcode-complimentary DNA sequences. Thus, the procedure addresses amplification of the target by PCR and of the final signal as a result of the high number of target chains released from each triple conjugate [13]. Similar schemes have also been designed for specific detection of proteins, by covering NP surfaces with antibodies rather than ssDNA. These and other nanotechnology-based detection systems, are still too complex for decentralized and in-the-field diagnosis, which requires avoiding the use of complex, bulky, and expensive lab-based detection devices, requires avoiding the use of complex, bulky, and expensive lab-based detection devices.

Both established and newer detection principles are usually amenable for integration with nanoengineered structures onto portable and user-friendly diagnostic platforms with micromechanical structures and elements (e.g., disposable lateral-flow test-strips). For example, lab-on-a-chip platforms have emerged as a second-generation chip technology where all steps of the bioassay (including sample pretreatment) are compartmentalized in modules, with the entire assay automated and performed with minute amounts of sample and reagents. This allows for low cost, low sample contamination, high speed, disposability, small power demand and read-out analysis. For nucleic-acid testing, these devices may include a module for target amplification, although appropriate transduction methods, signal amplification, and processing are often envisaged as a way to achieve PCR-free detection. The layout usually relies on micro- and nanofluidics, essentially an adaptation of classical biochips where liquids flow within channels and chambers through the assay. A unique advantage of fluidic systems is that the biorecognition event occurs in a liquid, in vivo-like environment rather than directly onto a solid support [12]. The production of such diagnostic devices has been driven by well-equipped research laboratories, at the research (prototype) level, and by pharmaceutical and electronics industries, at the development (end product) level. Major challenges include the need for processes and materials that generate more cost-effective fabrication and a wider acceptance of this technology as a new diagnostic paradigm by physicians and patients. Table 20.1 lists some published nanotechnology-based diagnostic schemes (in untreated and pretreated clinical samples and in synthetic media) for dermatology-relevant human pathogens. These pathogens were essentially chosen among those listed in the literature [3].

20.5 Nanotechnology for Assessment of Drug Resistance and Disease Management

Determination and monitoring of pathogen-resistant profiles to drugs are crucial to assess the efficacy of therapeutic interventions and, ultimately, to anticipate disease prognosis. Usually this is done by microbial incubation under various drug concentrations and monitoring of culture turbidity as a measure of cell viability. Molecular biology methods for genomic and proteomic determinations have also been used as alternatives to the turbidity principle. Yet, all these methods are time-consuming (requiring no less than 24 h), expensive, and laborious, being therefore unsuitable for routine use in low-income regions, where drug resistance for several diseases is prevalent and burdensome. Unfortunately, many traditional molecular biology methods, by detecting molecular

Table 20.1 Referenced works about nanotechnology applications to the diagnosis of pathogens of dermatological diseases

Pathogen	Nanomaterial	Biotarget	Detection	Support	LOD	References
MRSA	Au-NP	*mecA* gene	Light scattering	Glass slide	33 zmol	[19]
S. epidermidis	Au-NP	16S rRNA gene	Microgravimetry	QCM array	1.3×10^3 CFU/mL	[20]
M. tuberculosis	Au-NP	RNA polymerase gene oligo	Colorimetry	Test solution	0.75 μg	[21]
S. aureus	MGNP	Cells	NMR	Microfluidic chip	10^{-12} M	[22]
HIV and HBV	QD	Viral particle epitopes	Fluorescence	Microfluidic chip	10^{-10}–10^{-12} M	[23]
HBV, HIV, smallpox	Au-NP	Synthetic oligos	Fluorescence	Microarray	~500 fM	[24]
Staphylococcus	MGNP	Bacterium cells	Fluorescence	Test solution	10 CFU/mL	[25]
L. interrogans	Au-NP	Cell epitope	Microscopic agglutination (colorimetry)	Test solution	10 cells/mL	[26]
B. anthracis	Polymeric NP	Bacterial spore dipicolinic acid	Ion-based luminescence	Test solution	10 pM	[16]
N. meningitidis	Au-NP	Cell protein 85 (OMP85)	Microgravimetry	QCM	300 ng/mL	[27]
L. amazonensis	Phospholipid liposomes	Anti-leishmania antibodies	Capacitance	Interdigitated electrodes	10^{-5} mg/mL	[18]
T. pallidum	QD	TP47 polyclonal antibody	Fluorescence	Lateral flow	2 ng/mL	[28]
T. gondi, rubella, CMV, HSV	QD	Specific antibodies	Fluorescence	Multiplexed microarrays	–	[29]
T. cruzi	Au-NP	Anti-*T. cruzi* IgG	Electrochemistry	Screen-printed carbon electrode	~3.1 ng/mL	[30]
S. typhi	Au-NP	0901 cell antigen	Colorimetry	Lateral flow	~1.1×10^5 CFU/mL	[31]
VZV	QD	4F9 and g62 IgG	Fluorescence	Test solution	–	[32]
HSV	Ag-NP	Synthetic oligo	DPV	Array chip	5 aM	[33]
DENV	Au-NP/PANI hybrid nanocomposite	Serum lectin	Impedance	Test solution	1:150 dilution	[34]
Candida sp.	Au-NP	Cell proteins	SEM and UVRS	Test solution	–	[35]

DPV differential pulse voltammetry; *LOD* limit of detection; *MRSA* methicillin-resistant *Staphylococcus aureus*; *QCM* quartz crystal microbalance; *SEM* scanning electron microscopy; *UVRS* ultraviolet resonance Raman spectroscopy

markers of infection rather than viable pathogen cells, are frequently unable to correlate disease states with metabolic profiles.

Recently, new nanotechnology-based schemes have been proposed to assess drug resistance. The principle of mechanical bending of a silicon nanocantilever array was employed to detect a simulated binding between mucopeptides (bacterial cell wall precursor analogues) immobilized onto the cantilevers and the vancomycin antibiotic molecules free in the test sample [36]. The bending effect indicates drug–target interaction, whose quantification was possible by measuring the extent of the deformation. The assay occurred at clinically relevant concentrations, and showed a final sensitivity of 10 nM. Given the amenability to miniaturization and multiplexing, cantilevers can be fabricated in array formats for high-throughput rapid concomitant testing of many drugs. However, the cantilever format cannot provide any information about the viability of the simulated interacting pathogen cell [6]. As such, a cyclic voltammetry technique was used to detect O_2 depletion in bacterial suspensions with different concentrations of several antibiotics, upon chemical reduction of added ferricyanide to ferrocyanide [37]. Response currents, measured after 10 min incubations, indicated the extent of bacterial respiratory activity, which decreased with increasing initial antibiotic concentrations in the reacting vessel. The two-electrode amperometric apparatus comprised an array of disposable screen-printed working electrodes. These were prepared by printing of a carbon-based ink onto flexible plastic-like polycarbonate supports, followed by chemical modification with chitosan (or poly-L-lysine), in order to minimize surface contamination by antibiotics and bacteria. Yet, the method requires relatively complex apparatus, complex reagents, and high microbial densities [6]. Therefore, other nanotechnological schemes which do not involve surface reactions have come into focus that utilize nutrient consumption rates in solution as a measure of microbial metabolism. In one assay, the surface plasmon resonance (SPR) of dextran-coated gold NPs was used to quantify carbohydrates in a bacterial suspension by the addition of concanavalin A (Con A) [38]. This assay relies on carbohydrate depletion from the medium as a result of bacterial growth (in the presence of non-inhibitory antibiotic concentrations or of an ineffective antibiotic), in which condition Con A is able to bridge the carbohydrate molecules to the functionalized NPs, thus forming small, dispersed clusters. Thus, many free NPs bind microbial surface carbohydrates, rendering low shifts in the NP's plasmonic band. Under inhibition of microbial growth, extensive Con A-mediated gold nanoclustering occurs, leading to large SPR shifts, resultant from the presence of high amounts of nonconsumed carbohydrates. The aggregation phenomenon and the resulting SPR shifts, which occur in less than 3 h, depend on several parameters, like NP size, pH, and ionic strength, but they are not relevant in this case since the proposed assay works only under strict physiological conditions (in which microbial cell integrity is maintained). A major handicap of this technique is the inability to test-in turbid/opaque media (usually the case of biological fluids) owing to excessive light interference with the NPs' plasmonic band. Therefore, a similar assay was proposed, but with proton detection by magnetic relaxation of water molecules [39]. The assay employed SPION, either coated with dextran plus Con A (competitive scheme), or silica, and directly bound to Con A (noncompetitive scheme). Active microbial metabolism decreases carbohydrate levels and increases magnetic relaxation time variations ($\Delta T2$), when compared to nonviable media.

Fungal culture is time-consuming and often requires prolonged incubation periods for species identification or may lack the requisite specificity and sensitivity necessary for diagnosis. A visual microarray has been developed to circumvent these challenges for the diagnosis of *Candida.* Immobilized probes recognizing different species of *Candida* (*glabrata*, *parapsilosis*, *krusei*, *lustaniae*, *guilliermondii*, *keyfr*) and *Cryptococcus neoformans* were placed on a microarray. The ribosomal DNA of clinical isolates is amplified by PCR and hybridized to the microarray. Hybridizing signals are amplified with gold nanoparticles and silver deposition. The results can be assessed visually or using a flatbed scanner [40].

Gold nanoparticles <10 nm also interact with the proteins of microorganisms such as *Escherichia coli* and *Candida* species. This interaction can be detected by UVRS (ultraviolet resonance Raman spectroscopy) in the scanning electron microscope [35].

Aptamers are oligonucleotides of ssDNA or ssRNA which fold into a unique three-dimensional conformation. Using a selection process, aptamers with high affinity for particular ligands can be enriched and amplified. Aptamers can be functionalized to other moieties, such as magnetic nanoparticles (MNP), and Yb/Er or Yb/Tm upconversion nanophosphor (UCNP) nanoparticle. UCNPs contain rare earth phosphors which are excited by near-infrared fluorescent light (NIRL). They are water soluble, easily bioconjugated, and exhibit narrow photoluminescence without autofluorescence. These types of dual label particles have been successfully used to detect and quantify two different pathogens in a mixture, such as *Salmonella typhi*, and *Staphylococcus aureus* [41]. Because of the depth of penetration of NIRL, UCNP's can be used for the in vivo imaging of cutaneous infections.

Parasitic infections such as Leishmaniasis are difficult to culture and difficult to speciate. The incidence of the migration of vectors across temperate zones in step with global climate change is increasing the population at risk. A major confounding diagnosis in Leishmaniasis in Chagas' disease is caused by *Trypanosoma cruzi*. Distinguishing the two requires invasive sampling for culture or histopathology. Nanostructured films containing immobilized antigens from *Leishmania amazonensis* and *T. cruzi* have been employed in a low-cost biosensor system [42]. Statistical analysis of electrical impedance allows for the distinction of antibody–antigen interactions in sera from patients who are positive for Chagas' disease, leishmaniasis, or both. The assay is rapid, and generalizable to other pathogens. Electrodeposited DNA-gold nanoparticles have been shown in aptasensors to detect *Leishmania infantum* in real time [43]. In assays requiring PCR, loop-mediated mechanisms allow for ambient temperature isothermal amplification of DNA, such as kinetoplast DNA [44], and have been shown to be highly sensitive method for detecting *T. cruzi* infection [45]. These techniques might be adapted for Leishmania and nematode infections [46, 47].

Scabies is occasionally difficult to diagnose by visual inspection. Identification of the organism from skin scrapings is rapid and simple, especially in cases of Norwegian scabies. However, in some instances, only a few organisms are scattered over a large body surface area, and localization or identification of the organism is difficult. Reflectance confocal microscopy and epiluminescence microscopy have been used [48]. A serologic assay measuring IgE antibody to scabies antigens has high sensitivity and specificity and could be adapted for nanoserology or direct visualization using NIRL in vivo or in complex and cloudy mixtures using arrays [49, 50].

A number of other parasites important to dermatology continue to elude diagnosis. Their disease manifestations can be quite atypical and dramatic in cases of immunocompromise due, for example, to chemotherapy or immune deficiency. In cases of real or imagined ectoparasitic infestations which are difficult to diagnose, such as *Cimex lectularius* infestation, patients incur significant cost and anxiety with each round of clinical evaluation [51]. A serologic array of common parasitic infections which is specific, sensitive, and rapid could increase clinician confidence, reduce patient uncertainty, and reduce costs through clear and accurate inclusionary or exclusionary results. Suitable components of such an array would include recognition of *Toxocara* spp., *Pediculosis* spp., *Sarcoptes* spp., *Tunga* spp., and *Demodex* spp. One other interesting development of nanotechnology is theranostics. Theranostic tools simultaneously diagnose and treat disease. An example would be an antibody coupled to an imaging probe and toxin. The antibody would identify its target, which could be imaged, and the toxin would destroy the target. Some ticks suppress the immune system by secreting a salivary protein which binds and inhibits CD4 lymphocytes [52]. A theranostic probe would overcome this inhibition, and simultaneously detect the presence of the salivary protein. Ticks can also induce paralysis, for example, through a neurotoxin such as holocyclotoxin. A theranostic probe of tick paralysis would effectively seek out the toxin, identify it, and reverse its effects simultaneously.

Filarial diseases have numerous manifestations. Early and accurate diagnosis of filarial disease can prevent potentially severe complications, such as elephantiasis and blindness, or differentiating elephantiasis caused by filaria or podoconiosis [53]. Nanotechnology is amenable to DNA-based diagnostic techniques, and such methodologies are being developed for the evaluation of filarial diseases [54–58]. Some filaria, such as onchocerciasis, requires an endosymbiont (bacteria of the *Wohlbachia* spp.) for survival and pathogenicity [58, 59]. Diagnostic and theranostic studies aimed at this bacterial genus may be useful in the diagnosis and treatment of onchocerciasis. Protein-based methodologies might rely on inhibitory mechanisms that some filaria use to stimulate T regs and generate non-complement fixing antibodies [60]. When antibodies for diagnostic and theranostic assays are not readily available, bioinformatic methodologies can be used to generate them [61]. This has been successfully demonstrated for parasitic nematodes.

Other parasites may harbor transmissible diseases. This is of particular concern among migrants, travelers, the indigent population, and in crowded residences such as military institutions, prisons, and college dormitories. *Pediculosis* is a vector for a number of disease-causing organisms, for example *Bartonella* spp., and this can be detected with molecular techniques which are easily adapted to the nanotechnologies described in this chapter [62, 63].

While performing as well as conventional methods for determination of antibiotic susceptibility, nanotechnology-based techniques are amenable for in-the-field use. Despite having been applied to the detection of other pathogens, they are amenable for diagnosis of pathogens of dermatological infections as well.

Tools and techniques employing nanotechnology have been developed for the diagnosis and management of nondermatologic diseases, such as parasitic and nonparasitic infections, inflammatory diseases, and tumors. Proof-of-concept adaptation of these methods to parasitic skin infections has been demonstrated in some animal models and has the potential to demonstrate utility in human skin disease.

References

1. Murphy MJ. Introduction to molecular diagnostic testing in dermatology and dermatopathology. In: Murphy MJ, editor. Molecular diagnostic in dermatology and dermatopathology. Connecticut, CT: Humana Press; 2011. p. 1–12.
2. Fortina P, Surrey S, Kricka LJ. Molecular diagnostics: hurdles for clinical implementation. Trends Mol Med. 2002;8:264–6.
3. Liles WC, Van Voorhis WC. Fever in travelers to tropical countries. In: Root RK, Waldvogel F, Corey L, Stamm WE, editors. Clinical infectious diseases. A practical approach. New York, NY: Oxford University Press; 1999. p. 859–74.
4. Bowden RA. Hematopoietic stem cell transplantation. In: Root RK, Waldvogel F, Corey L, Stamm WE, editors. Clinical infectious diseases. A practical approach. New York, NY: Oxford University Press; 1999. p. 829–38.
5. Sra KK, Torres G, Rady P, Hughes TK, Payne DA, Tyring SK. Molecular diagnosis of infectious diseases in dermatology. J Am Acad Dermatol. 2005;53: 749–65.
6. Kaittanis C, Santra S, Perez JM. Emerging nanotechnology-based strategies for the identification of microbial pathogenesis. Adv Drug Deliv Rev. 2010;62:408–23.
7. Wessagowit V, South AP. Dermatological applications of DNA array technology. Clin Exp Dermatol. 2002;27:485–92.
8. Jain KK. Nanotechnology in clinical laboratory diagnostics. Clin Chim Acta. 2005;358:37–54.
9. Nasir A. The future of nanotechnology in dermatology. US Dermatol. 2008;3:9–13.
10. Hia J, Nasir A. Photonanodermatology: the interface of photobiology, dermatology and nanotechnology. Photodermatol Photoimmunol Photomed. 2011;27:2–9.
11. Cai W, Gao T, Hong H, Sun J. Applications of gold nanoparticles in cancer nanotechnology. Nanotechnol Sci Appl. 2008;1:17–32.
12. Teles FRR, Fonseca LP. Trends in DNA biosensors. Talanta. 2008;77:606–23.
13. Hauck TS, Giri S, Gao Y, Chan WCW. Nanotechnology diagnostics for infectious diseases prevalent in developing countries. Adv Drug Deliv Rev. 2010;62:438–48.
14. Kim S, Lim YT, Soltesz EG, et al. Near-infrared fluorescent type II quantum dots for sentinel lymph node mapping. Nat Biotechnol. 2004;22:93–7.

15. Ballou B, Ernst LA, Andreko S, et al. Sentinel lymph node imaging using quantum dots in mouse tumor models. Bioconjug Chem. 2007;18:389–96.
16. Oh W-K, Jeong YS, Song J, Jang J. Fluorescent europium-modified polymer nanoparticles for rapid and sensitive anthrax sensors. Biosens Bioelectron. 2011;29:172–7.
17. Taylor EN, Webster TJ. The use of superparamagnetic nanoparticles for prosthetic biofilm prevention. Int J Nanomedicine. 2009;4:145–52. Epub 2009 Sep 10. PubMed PMID: 19774113; PubMed Central PMCID: PMC2747349.
18. Perinoto A, Maki RM, Colhone MC, et al. Biosensors for efficient diagnosis of leishmaniasis: innovations in bioanalytics for a neglected disease. Anal Chem. 2010;82:9763–8.
19. Storhoff JJ, Lucas AD, Garimella V, Bao YP, Müller UR. Homogeneous detection of unamplified genomic DNA sequences based on colorimetric scatter of gold nanoparticle probes. Nat Biotechnol. 2004;22:883–8887.
20. Xia H, Wang F, Huang Q, et al. Detection of *Staphylococcus epidermidis* by a quartz crystal microbalance nucleic acid biosensor array using Au nanoparticle signal amplification. Sensors. 2008;8:6453–70.
21. Baptista PV, Koziol-Montewka M, Paluch-Oles J, et al. Gold-nanoparticle-probe-based assay for rapid and direct detection of *Mycobacterium tuberculosis* DNA in clinical samples. Clin Chem. 2006;52:1433–4.
22. Lee H, Sun E, Ham D, Weissleder R. Chip-NMR biosensor for detection and molecular analysis of cells. Nat Med. 2008;14:869–74.
23. Klostranec JM, Xiang Q, Farcas GA, et al. Convergence of quantum dot barcodes with microfluidics and signal processing for multiplexed high-throughput infectious disease diagnostics. Nano Lett. 2007;7:2812–8.
24. Stoeva SI, Lee J-S, Thaxton CS, Mirkin CA. Multiplexed DNA detection with barcoded nanoparticle probes. Angew Chem Int Ed. 2006;45:3303–6.
25. Gao J, Li L, Ho P-L, et al. Combining fluorescent probes and biofunctional magnetic nanoparticles for rapid detection of bacteria in human blood. Adv Mater. 2006;18:3145–8.
26. Chirathaworn C, Chantaramalai T, Sereemaspun A, Kongthong N, Suwancharoen D. Detection of *Leptospira* in urine using anti-*Leptospira*-coated gold nanoparticles. Comp Immunol Microbiol Infect Dis. 2011;34:31–4.
27. Reddy SB, Mainwaring DE, Al Kobaisi M, Zeephongsekul P, Fecondo JV. Acoustic wave immunosensing of a meningococcal antigen using gold nanoparticle-enhanced mass sensitivity. Biosens Bioelectron. 2012;31:382–7.
28. Yang H, Li D, He R, et al. A novel quantum dots-based point of care test for syphilis. Nanoscale Res Lett. 2010;5:875–81.
29. Yang H, Guo Q, He R, et al. A quick and parallel analytical method based on quantum dots labeling for ToRCH-related antibodies. Nanoscale Res Lett. 2009;4:1469–74.
30. Pereira SV, Bertolino FA, Fernández-Baldo MA, et al. A microfluidic device based on a screen-printed carbon electrode with electrodeposited gold nanoparticles for the detection of IgG anti-*Trypanosoma cruzi* antibodies. Analyst. 2011;136:4745–51.
31. Preechakasedkit P, Pinwattana K, Dungchai W, et al. Development of a one-step immunochromatographic strip test using gold nanoparticles for the rapid detection of *Salmonella typhi* in human serum. Biosens Bioelectron. 2012;31:562–6.
32. Tang X, He J, Partin J, Vafai A. Comparative analysis of direct fluorescence, zenon labeling, and quantum dot nanocrystal technology in immunofluorescence staining. J Immunoassay Immunochem. 2010;31:250–7.
33. Li H, Sun Z, Zhong W, Hao N, Xu D, Chen H-Y. Ultrasensitive electrochemical detection for DNA arrays based on silver nanoparticle aggregates. Anal Chem. 2010;82:5477–83.
34. Andrade CAS, Oliveira MDL, de Melo CP, et al. Diagnosis of dengue infection using a modified gold electrode with hybrid organic–inorganic nanocomposite and *Bauhinia monandra* lectin. J Colloid Interface Sci. 2011;362:517–23.
35. Naja G, Hrapovic S, Male K, Bouvrette P, Luong JHT. Rapid detection of microorganisms with nanoparticles and electron microscopy. Microsc Res Tech. 2008;71:742–8.
36. Ndieyira JW, Watari M, Barrera AD, et al. Nanomechanical detection of antibiotic-mucopeptide binding in a model for superbug drug resistance. Nat Nanotechnol. 2008;3:691–6.
37. Mann TS, Mikkelsen SR. Antibiotic susceptibility testing at a screen-printed carbon electrode array. Anal Chem. 2008;80:843–8.
38. Nath S, Kaittanis C, Tinkham A, Perez JM. Dextran-coated gold nanoparticles for the assessment of antimicrobial susceptibility. Anal Chem. 2008;80:1033–8.
39. Kaittanis C, Nath S, Perez JM. Rapid nanoparticle-mediated monitoring of bacterial metabolic activity and assessment of antimicrobial susceptibility in blood with magnetic relaxation. PLoS One. 2008;3:e3253.
40. Lu W, Gu D, Chen X, Xiong R, Liu P, Yang N, et al. Application of an oligonucleotide microarray-based nano-amplification technique for the detection of fungal pathogens. Clin Chem Lab Med. 2010;48(10):1507–14.
41. Duan N, Wu S, Zhu C, Ma X, Wang Z, Yu Y, et al. Dual-color upconversion fluorescence and aptamer-functionalized magnetic nanoparticles-based bioassay for the simultaneous detection of Salmonella typhimurium and Staphylococcus aureus. Anal Chim Acta. 2012;723:1–6.
42. Perinoto ÂC, Maki RM, Colhone MC, Santos FR, Migliaccio V, Daghastanli KR, et al. Biosensors for efficient diagnosis of leishmaniasis: innovations in

bioanalytics for a neglected disease. Anal Chem. 2010;82(23):9763–8.
43. Moreno M, González VM, Rincón E, Domingo A, Domínguez E. Aptasensor based on the selective electrodeposition of protein-linked gold nanoparticles on screen-printed electrodes. Analyst. 2011;136(9):1810–5.
44. Nasir A, Cook GA, Donelson JE. Sequences of two kinetoplast minicircle DNAs of Trypanosoma (Nannomonas) congolense. Mol Biochem Parasitol. 1987;24(3):295–300.
45. Kuboki N, Inoue N, Sakurai T, Di Cello F, Grab DJ, Suzuki H, et al. Loop-mediated isothermal amplification for detection of African trypanosomes. J Clin Microbiol. 2003;41(12):5517–24.
46. Reithinger R, Dujardin JC. Molecular diagnosis of leishmaniasis: current status and future applications. J Clin Microbiol. 2007;45(1):21–5.
47. Takagi H, Itoh M, Kasai S, Yahathugoda TC, Weerasooriya MV, Kimura E. Development of loop-mediated isothermal amplification method for detecting Wuchereria bancrofti DNA in human blood and vector mosquitoes. Parasitol Int. 2011;60(4):493–7.
48. Cinotti E, Perrot JL, Labeille B, Vercherin P, Chol C, Besson E, Cambazard F. Reflectance confocal microscopy for quantification of Sarcoptes scabiei in Norwegian scabies. J Eur Acad Dermatol Venereol. 2012 May 23. doi:1111/j.1468-3083.2012.04555.x. PubMed PMID: 22621304.
49. Jayaraj R, Hales B, Viberg L, Pizzuto S, Holt D, Rolland JM, et al. A diagnostic test for scabies: IgE specificity for a recombinant allergen of Sarcoptes scabiei. Diagn Microbiol Infect Dis. 2011;71(4):403–7.
50. Gaudin JC, Rabesona H, Choiset Y, Yeretssian G, Chobert JM, Sakanyan V, et al. Assessment of the immunoglobulin E-mediated immune response to milk-specific proteins in allergic patients using microarrays. Clin Exp Allergy. 2008;38(4):686–93.
51. Szalanski AL, Tripodi AD, Austin JW. Multiplex polymerase chain reaction diagnostics of bed bug (Hemiptera: Cimicidae). J Med Entomol. 2011;48(4):937–40.
52. Juncadella IJ, Anguita J. The immunosuppressive tick salivary protein, Salp15. Adv Exp Med Biol. 2009;666:121–31.
53. Davey G. Podoconiosis, non-filarial elephantiasis, and lymphology. Lymphology. 2010;43(4):168–77.
54. Fink DL, Kamgno J, Nutman TB. Rapid molecular assays for specific detection and quantitation of Loa loa microfilaremia. PLoS Negl Trop Dis. 2011;5(8):e1299. doi:10.1371/journal.pntd.0001299.
55. Fink DL, Fahle GA, Fischer S, Fedorko DF, Nutman TB. Toward molecular parasitologic diagnosis: enhanced diagnostic sensitivity for filarial infections in mobile populations. J Clin Microbiol. 2011;49(1): 42–7.
56. Mehlotra RK, Gray LR, Blood-Zikursh MJ, Kloos Z, Henry-Halldin CN, Tisch DJ, et al. Molecular-based assay for simultaneous detection of four Plasmodium spp. and Wuchereria bancrofti infections. Am J Trop Med Hyg. 2010;82(6):1030–3.
57. Thanchomnang T, Intapan PM, Lulitanond V, Sangmaneedet S, Chungpivat S, Taweethavonsawat P, et al. Rapid detection of Dirofilaria immitis in mosquito vectors and dogs using a real-time fluorescence resonance energy transfer PCR and melting curve analysis. Vet Parasitol. 2010;168(3–4):255–60.
58. Nuchprayoon S. DNA-based diagnosis of lymphatic filariasis. Southeast Asian J Trop Med Public Health. 2009;40(5):904–13.
59. Tamarozzi F, Halliday A, Gentil K, Hoerauf A, Pearlman E, Taylor MJ. Onchocerciasis: the role of Wolbachia bacterial endosymbionts in parasite biology, disease pathogenesis, and treatment. Clin Microbiol Rev. 2011;24(3):459–68. doi:10.1128/CMR.00057-10.
60. Adjobimey T, Hoerauf A. Induction of immunoglobulin G4 in human filariasis: an indicator of immunoregulation. Ann Trop Med Parasitol. 2010;104(6):455–64.
61. Kron MA, Cichanowicz S, Hendrick A, Liu A, Leykam J, Kuhn LA. Using structural analysis to generate parasite-selective monoclonal antibodies. Protein Sci. 2008;17(6):983–9.
62. Sasaki T, Kobayashi M, Agui N. Detection of Bartonella quintana from body lice (Anoplura: Pediculidae) infesting homeless people in Tokyo by molecular technique. J Med Entomol. 2002;39(3):427–9.
63. Roux V, Raoult D. Body lice as tools for diagnosis and surveillance of reemerging diseases. J Clin Microbiol. 1999;37(3):596–9.

Application of Nanomedicine in Wound Healing

21

Xuelai Liu and Kenneth K.Y. Wong

In recent years nanotechnology has been emerging as a rapidly growing and challenging research field, and many nanomaterials have gained access into daily life, including biomedical applications. In this chapter, we review the use of various nanomatericals in nanomedicine in the field of skin wound healing.

21.1 Introduction to Nanomaterials in Medicine

Emergence of new materials and technology and their broad application is invariably identified as the symbol of scientific progress in any era. With recent advancement and development of nanotechnology, it has been made possible to engineer many kinds of materials into various shapes and sizes at the nanometer level. The sizes of these nanoparticles usually vary from 1 nm to 100 nm, which is much smaller than the generally targeted cells in the organism (Fig. 21.1), and will exhibit remarkably differential and amazing biological activities due to their unusual physical and chemical properties due to enhanced surface area to volume ratio for a given mass [1, 2]. This means that nano-sized particles are more likely to have increased cell surface interactions and to gain access into internal cell environments, resulting in significant biomedical effect or efficacy. In recent years, burgeoning interest in the medical applications of nanotechnology has led to the emergence of a new scientific field—Nanomedicine [3–8]. Nanomedicine, as the most important cross-link subject between nanoscience and biomedicine, is defined as the process of disease prevention, diagnosis, and therapeutic using nanomaterials or nanotechnology [9, 10].

The great potentials of nanotechnology in medicine will contribute to the improvement in health and make remarkable achievements in new-generation biological materials, artificial internal organ, interventional therapy, drug carrier systems, blood purification, and bio-macromolecule separation. These will all add up to better clinical care for our patients. So far the benefit to skin wound healing attributed to nanomedicine has mainly been attributed to the use of nanomaterials. In this chapter we will focus on the current knowledge and applications of nanomaterials, and their effects on wound healing.

21.2 Skin Wound and Process of Healing

In the clinical setting, the types of wounds encountered and their classification are vastly different. They can be divided according to the anatomical location in the body: mucosal wound,

X. Liu (✉)
Department of Surgery, The University of Hong Kong, Queen Mary Hospital, Pokfulam Road, Hong Kong SAR, China
e-mail: liuxlai@hku.hk

K.K.Y. Wong, kkywong@hku.hk
Department of Surgery, The University of Hong Kong, Sassoon Road, Hong Kong SAR, China

A. Nasir et al. (eds.), *Nanotechnology in Dermatology*,
DOI 10.1007/978-1-4614-5034-4_21, © Springer Science+Business Media, LLC 2013

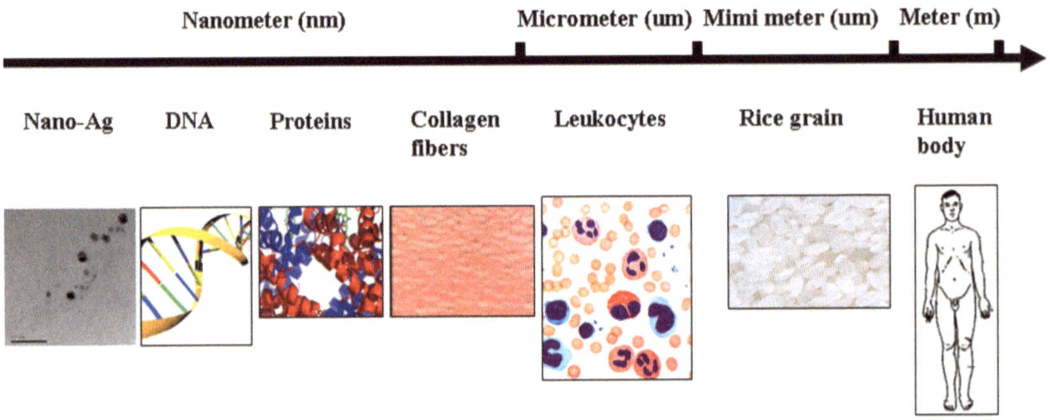

Fig. 21.1 Schematic diagram showing comparisons of various materials in terms of size

skin wound, muscle wound, bone wound, etc.; according to the mechanism of injury: abrasion, laceration, penetrating wound, etc. Furthermore, the healing processes after wounding are also highly complex. Nonetheless, the commonly seen wounds in daily life are skin wounds, as the skin is the superficial barrier of the body and is the largest organ of the body. Its normal healing represents classic characteristics of wounds repair and regeneration.

In terms of anatomical structure, the skin consists of two mutually dependent layers, the epidermis and dermis. Both of them, as one integrated entity, are residents on the fatty subcutaneous layer. The epidermis is made up of four to five layers of keratinocytes and derived primarily from surface ectoderm, but is colonized by pigment-containing melanocytes of neural crest origin. On the other hand, the dermis is a thicker layer containing connective tissue, which is derived primarily from mesoderm and contains a large amount of fibroblasts, collagen, elastic fibers, and blood vessels, as well as sensory structures. The skin plays important roles in physiological functions involving anti-invasion of foreign pathogens, prevention of dehydration, and regulation of human body temperature, as well as absorption. As a result, its normal healing after wounding is critical for both morphological and functional recovery. In general, four essential processes: activation of coagulation, inflammation, proliferation, and tissue remodeling, are involved (Fig. 21.2). During the early stage of wound, the damage in skin will mean loss of integrity of blood vessels in the dermal layer. The exposure of platelets to extra cellular matrix (ECM) will stimulate immediate coagulation and fibrin clot formation [11]. This event occurs rapidly and usually lasts about 5–10 min. This process is essential in activating the subsequent inflammation reaction, as platelet aggregation will stimulate mass cells to release pro-inflammatory factors (i.e., tumor necrosis factor-α/TNF-α and platelet-derived growth factor/PDGF, etc.), which in turn induce leucocytes migration and stimulate angiogenesis [11–13]. However, these same inflammatory cytokines can also damage normal tissues around the wound site. Thus, the double-edged sword effect of inflammation can contribute to wound healing on one hand, while on the other hand, may delay wound healing with over persistence of the inflammatory reaction. With the formation of granulation tissue on the skin wounds, the proliferation stage also starts as the mast cells activate the fibroblasts around wound site, and drive them to start proliferating and migrating into granulation tissues [14, 15]. At the same time, some differentiated myofibroblasts in granulation tissue, arisen from proliferating fibroblasts, express the alpha-smooth muscle actins (α-SMA) and generate the contractile force to help wound contraction during the healing process [16]. At this stage, the generated granulation in the wound bed provides a platform for re-epithelialization. Further production of cytokines by proliferated fibroblasts will further active kerati-

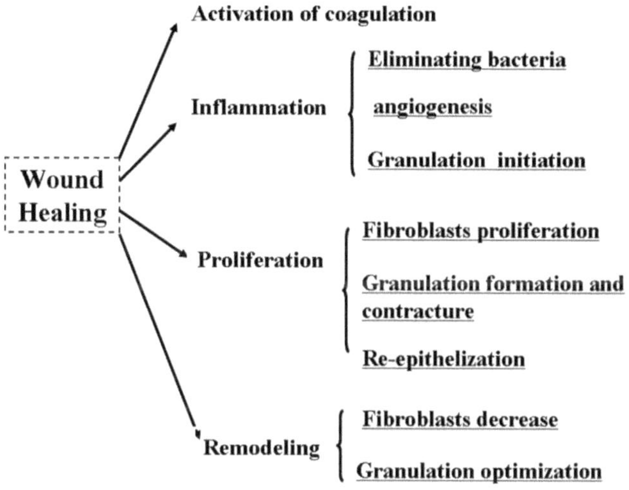

Fig. 21.2 A simplified diagram of various events taking place during wound healing

nocytes proliferation during re-epithelization toward the wound centre. In the proliferation stage, it is the contraction of the wound, from fibroblast proliferation and migration, as well as re-epithelization through keratinocytes, which minimize the overall wound area. Once the minimization of wound occurs, the remodeling stage is initiated. The main characteristics for this stage are that fibroblasts in granulation tissue will decrease in numbers and the regenerated ECM (mostly collagen) further optimizes and achieves more compact and organized alignment. For ECM, collagen starts to be laid down as early as in the inflammatory stage, although in small amount. Ideal collagen deposition also plays an essential role in wound healing. A large amount of collagen formation due to over-proliferation of fibroblasts in proliferation stage will not only have negative effect on further angiogenesis and granulation contracture and resulting in the delayed healing but also reduces overall mechanical function due to disorganized deposition.

21.3 Metallic Nanoparticles

So far, many metals and metallic compounds have been engineered into nano-scale level and proved to have biomedical properties. Of all, silver nanoparticles (AgNPs) have been most widely used and are regarded as a promising candidate metal for help tissue repair and regeneration during wound healing. Historically, silver has been utilized widely since ancient times for war wounds. In the past decades, the main formats of silver used for the burn wound treatment were nitrate or sulfadiazine compounds [17–22], as they were effective antimicrobial agents whose mechanism of action involved blocking respiratory chain system in microorganism [23, 24]. Regarding the understandings of AgNPs, initial research mostly focused on their antibacterial effects and their superiority when compared with the other formats of silver reagents [25–28]. Nonetheless, recent data revealed AgNPs' potential anti-inflammatory effects on various wounds [29–33], and better efficacy in the promotion of wound healing [34, 35].

The mechanisms of enhanced antibacterial action by AgNPs include: (1) larger overall contact surface of AgNPs with bacteria due to larger surface area to volume ratio; (2) easy penetration of AgNps into bacteria; (3) degradation of bacterial cell membrane due to interaction of AgNPs with membrane proteins containing sulfur; (4) interfering with DNA replication fidelity through binding to DNA; as well as (5) inducing oxidative stress reaction. So far AgNPs have been proved to be effective in killing *E. coli*, *Microcystis aeruginosa*, *Shewanella oneidensis*. In vitro studies also showed that AgNPs had the fastest killing kinetics against five bacteria tested [36–40]. With the early knowledge of AgNPs' better efficacy, development of a commercially available nanocrystalline silver-coated dressing, Acticoat®, came into fruition for burn wound treatment and care. This product has already been utilized in clinics extensively. In this regard, we also created a burn wound model in mice, then topically applied AgNPs solution, and observed acceleration of wound healing, even when compared with traditional antibiotics. This finding suggested that the mechanism of wound healing promotion by AgNPs was not solely due to antibacterial action and other potential factors or mechanisms were involved [41]. Furthermore, Sibbald et al. also investigated the use of Acticoat® dressing to evaluate the effect of AgNPs in clinical patients with chronic venous leg ulcers. It was found that AgNPs-coated dressing had beneficial effects on protecting the wound sites from bacterial contamination. Others further explored the possibility of bacteria resistance to AgNPs through a randomized clinical study. It was concluded that the frequency of burn wound sepsis and secondary bacteremia were much less in patients treated with AgNPs than those treated with the 0.5 % silver nitrate solution. Also the sustained release of AgNPs played an important role in the reduction of likelihood in bacteria resistance [42, 43].

In addition to antibacterial property, AgNPs-mediated anti-inflammatory effect is another essential factor in promotion of wound healing. In this regard, Tian et al. also investigated the anti-inflammatory effects of AgNPs during burn wound healing. Quantitative real-time RT-PCR results showed that the expression of IL-6 and TGF-β1 in healing tissue stayed at a lower level in AgNPs-treated wounds, while levels of IL-10, VEGF, and IFN-γ stayed higher, in comparison to control at all time points during healing. These suggested AgNPs could modulate inflammatory cytokines expression in wound healing, and this provided another explanation why wound healing was enhanced [41]. In this regard, the findings of reduced inflammation by AgNPs was supported by Sibbald, who found reduction of neutrophilic infiltration during wound healing, although the exact mechanism was still unknown [43]. Wright et al. also found that AgNPs induced promotion in wound healing through the modulation of various cytokines after tropical application of AgNPs on a contaminated wound model in pigs, and the enhancement of wound healing was associated with reduced local matrix metalloproteinase (MMP) levels [34].

A surprising biological effect of AgNPs was seen in another experimental model of skin wound healing. In an excisional wound, we found that AgNPs could inhibit the proliferation of fibroblasts in skin and drove their differentiation toward myofibroblasts in both in vitro and in vivo experiments [44]. It would appear that the alpha smooth-muscle actins (α-SMA) produced play an important role in wound contraction, which would greatly minimize the open wound area by pulling neighboring tissues toward the wound center. This differentiation and following reduced wound area could also provide feeder layer and resident space for keratinocytes in the re-epitheliazation processes. Furthermore, we also found that AgNPs had the potential to promote the proliferation and migration of keratinocytes in skin epidermal layer, indicating the re-epithelization process could be prompted by AgNPs, through which the wound repair was contributed. In any tissue repair, the need to provide an adequate supply of oxygenated blood is vital. Hernández found that AgNPs had the potential to drive endothelial regeneration in blood vessels, through an enhanced angiogenesis process. Similarly, during skin wound healing, angiogenesis induced by AgNPs via an upregulation of vascular endothelial growth factor (VEGF) would enhance capillary regeneration [45].

Based on the accelerated skin wound healing mediated by AgNPs, Kwan et al. investigated and compared the healed skin function by investigating their mechanical properties. Through quantitative and qualitative tests, they found that AgNPs could modulate the collagen deposition and alignment and achieve better mechanical tension in repaired skin [46]. Moreover, their study suggested AgNPs had the potential to enhance the cosmetic appearance after healing.

21.4 Peptide Nanostructures

Molecular engineering of various self-assembling peptide scaffolds has emerged as an active area in the past few decades. Some peptide nanostructures have provided innovative therapies in medicine, and are useful to understand cell biology in 3D environment [47–51]. Nanotechnology can provide new strategies for regenerative medicine. These novel nanomaterials can be easily acquired through supramolecular chemistry and self-assembly strategies. The assembled scaffolds closely mimic the native ECM, with the scaffold's fibers and pores being much smaller than cells. The scaffolds can be further modified and functionalized to enhance their interaction with other cells and tissues [52–55].

One example of this is peptide hydrogel, which has been shown to have great cyto-compatibility in a variety of mammalian cells, including fibroblasts, mesenchymal stem cells, hepatocytes, and articular chondrocytes [56–58]. Cell culture media stimulate the peptides to fold into fibrils and subsequently assemble into hydrogels. Cultured cells are homogeneously distributed within peptide hydrogels, and some may even be encapsulated cells without affecting their viability [59]. In addition, some newly developed peptide hydrogels also showed great potential in cell culture [60–62]. These peptides hydrogels with above ability not only support cell attachment [63–66] but also promote the differentiation of liver progenitor cells into hepatocyte spheroids, which will contribute greatly to the liver tissue repair and regeneration after injury, as well as embryonic stem cells into osteogenic cells [67, 68]. Report has indicated that this differentiation could be enhanced by the presence of hydroxyapatite (HA) particles into the nanofiber scaffold. In this regard, injectable delivery of encapsulated neonatal cardiomyocytes within the RADA 16-II hydrogels has also been evaluated. Result showed that cells could survive after delivery, and the peptide hydrogels provide better microenvironments for endogenous endothelial cells [69].

Recently, some peptide amphiphile (PA) systems have been tailor-made by display of bioactive epitopes, aiming to optimize tissue-engineered scaffolds for a specific cell type. For example, the Arg-Gly-Asp-Ser (RGDS) cell adhesion epitope has been incorporated into a PA molecular design. This RGDS presenting PA network provided an enhanced support for bone marrow mononuclear cells and enamel organ epithelial cells when compared with the PA network without the RGDS epitope [70–72]. Another modification has been done, in which PA is incorporated into cell adhesion sequence and an enzyme-cleavable site. The generated hydrogels provided a cell-responsive matrix, which is specifically favorable to dental stem cells [73]. Furthermore, researchers have explored a new assembly strategy by mixing high molecular weight hyaluronic acid, to produce a self-sealing closed sac. This sac filled with liquid could encapsulate human mesenchymal stem cells (MSCs) and be used to deliver cells into an isolated environment for regenerative medicine [74]. In recent year, new self-assembling peptides conjugated with aromatic groups on the N-terminal end of peptides, could alter pH value, and even incorporate sensitive domains, were found to support chondrocyte survival and proliferation. These can be used to promote the chondrocyte regeneration after bone injury and following functional recovery [68, 75–78].

The self-assembling peptide hydrogel biomaterials revealed the advantages of regenerative medicine and their benefits when compared to the traditional biomaterials. First of all, these assembled nanostructures offer a reliable 3D microenvironment for cell growth and differentiation in wound healing. Secondly, most of these nanomaterials are made from natural amino acids, and possess better inherent biocompatibility and biodegradability. Thirdly, the multiple peptide-based signaling sequences or other molecular units can be incorporated into the nanostructure to tailor-make for specific targets. Finally, these nanomaterials appear not to be immunogenic and are thus free from tissue rejection once applied in wound treatment and care [79, 80].

21.5 Polymeric Nanostructures

Polymers are organic molecules with long chain in nano-scale level, which is composed of covalently bound repeating units called monomers linked together to form a backbone. Usually, the chain length is defined by the molecular weight (MW), which is a product of the number of repeat units and the molecular weight of the individual monomer. The polymers are biomaterials commonly utilized in practice, involving surgical tools, catheters, device coatings, implantable devices, and vascular grafts [81]. The common natural polymers include DNA, hyaluronic acid, and collagen, while synthetic polymers mainly include polyethylene (PE), polystyrene (PS), polypropylene (PP), polyvinyl chloride (PVC), polymethylmethacrylate (PMMA), polytetrafluoroethylene (PTFE), polyamides (nylon), and polyurethanes, as well as polysiloxanes (silicone) [82–85]. Some advantages of polymer applications include the ease of manufacturing, lower cost, and versatility. So far the polymeric nanostructures have been widely used in scaffolds, biological carriers, vascular grafts, surgical sutures, but more new materials are being developed.

Polymers are usually divided into two categories: thermoplastics and thermosets. Thermosets cannot be reshaped once moulded. A rise in temperature will result in chemical decomposition and cross-link. An important example of this nanostructure is vulcanized rubber, which is utilized widely in orthopedic prostheses. On the other hand, thermoplastic polymers are solid materials which can be reshaped once the temperature rises above glass transition temperature. Both nylon and polyethylene are examples of thermoplastics, which are used for the manufacture of catheters/dialysis membranes and implant coatings, respectively [86]. Polyurethane, as the important polymer nanostructure, plays an important role in wound healing. In experiments, it was found that polyurethane could stimulate angiogenesis during wound healing and could indirectly enhance cell proliferation during re-epithelization around wound site [87–89].

Additionally, polymeric hydrogel is another useful category for wound healing. The hydrogel scaffold has the potential to facilitate not only inflammatory cell infiltration in the early stage of healing but also the infiltration of angiogenic cells into regenerative tissue in subsequent phases of wound healing. Furthermore, additional growth factors, cytokines, or cells can enable neovascularization, and the improvement of microenvironment around wound [90–95]. Therefore, these have been utilized in both infectious and normal wounds treatment. One novel finding recently showed that the cross-linked polymer networks could swell in aqueous environment [96–98]. This unique feature has the potential to further contribute the wound healing, as the ECM modulation is an important link in normal healing.

21.6 The Future

Nanomedicine is an emerging but important field in medicine. It is breaking through potential obstacles in practice, leading the scientific and technological field into a new world. The current focus of research on the creation of new nanotechnology and the development of new nanomaterials certainly need to engage more in the exploration of physiochemical, biomedical characteristics, and understanding of mechanism, as well as potential benefits in practical application. Furthermore, these nanomaterials with better biomedical properties should be manufactured into clinical products and serve the people.

In the future, the research and development in nanotechnology and nanomaterials will lead to the development of more miniature devices, resulting in micro-trauma, swift response and accurate, real-time, remote sensing, dynamic state, and intelligence, and cause a great revolution in diagnosis and treatment, including the tissue repair and regeneration in wound healing.

References

1. Aiken III JD, Finke RGJ. J Mol Catal A Chem. 1999;145:1–44.
2. Albrecht MA, Evans CW, Colin L. Green chemistry and the health implications of nanoparticles. Green Chem. 2006;8:417–32.

3. Nanomedicine: grounds for optimism, and a call for papers. Lancet. 2003;362:673
4. Freitas Jr RA. Nanodentistry. J Am Dent Assoc. 2000;131:1559–66.
5. Freitas Jr RA. The future of nanofabrication and molecular scale devices in nanomedicine. Stud Health Technol Inform. 2002;80:45–59.
6. Bogunia-Kubik K, Sugisaka M. From molecular biology to nanotechnology and nanomedicine. Biosystems. 2002;65:123–38.
7. Haberzettl CA. Nanomedicine: destination or journey? Nanotechnology. 2002;13:R9–13.
8. Emerich DF, Thanos CG. Nanotechnology and medicine. Expert Opin Biol Ther. 2003;3:655–63.
9. Robert A, Freitas Jr RA. What is nanomedicine? Nanomedicine. 2005;1:2–9.
10. Jain KK. Nanodiagnostics: application of nanotechnology in molecular iagnostics. Expert Rev Mol Diagn. 2003;3:153–61.
11. Lau K, Paus R, Tiede S, et al. Exploring the role of stem cells in cutaneous wound healing. Exp Dermatol. 2009;18:921–33.
12. Malaviya R, Ikeda T, Ross E, et al. Mast cell modulation of neutrophil influx and bacterial clearance at sites of infection through TNF-alpha. Nature. 1996;381:77–80.
13. Szpaderska A, Egozi E, Gamelli R, et al. The effect of thrombocytopenia on dermal wound healing. J Invest Dermatol. 2003;120:1130–7.
14. Weber A, Knop J, Maurer M. Pattern analysis of human cutaneous mast cell populations by total body surface mapping. Br J Dermatol. 2003;148:625–31.
15. Maurer M, Theoharides T, Granstein RD, et al. What is the physiological function of mast cells? Exp Dermatol. 2003;12:886–910.
16. Jo Ellen F, QiJing L, Lina W, et al. The cxc chemokine cCAF stimulates differentiation of fibroblasts into myofibroblasts and accelerates wound closure. J Cell Biol. 2002;156(1):161–72.
17. Stickler DJ. Biomaterials to prevent nosocomial infections: is silver the gold standard? Curr Opin Infect Dis. 2000;13:389–93.
18. Warriner R, Burrell R. Infection and the chronic wound: a focus on silver. Adv Skin Wound Care. 2005;18 Suppl 1:1–2.
19. Hussain S, Ferguson C. Silver sulphadiazine cream in burns. Emerg Med J. 2006;23:929–32.
20. Stephen-Haynes J, Toner L. Assessment and management of wound infection: the role of silver. Br J Community Nurs. 2007;12:S6–12.
21. Muangman P, Muangman S, Opasanon S, et al. Benefit of hydrocolloid SSD dressing in the outpatient management of partial thickness burns. J Med Assoc Thai. 2009;92:1300–5.
22. Uygur F, Oncul O, Evinc R, et al. Effects of three different topical antibacterial dressings on Acinetobacter baumannii-contaminated full-thickness burns in rats. Burns. 2009;35:270–3.
23. Modak SM, Fox CL. Biochem Pharmacol. 1973;22:2391–404.
24. Bragg PD, Rainnie DJ. The effect of silver ions on the repiratory chain of Escherichia Coli. Can J Microbiol. 1974;20:883–9.
25. Feng QL, Wu J, Chen GQ, et al. A mechanistic study of the antibacterial effect of silver ions on Escherichia coli and Staphylococcus aureus. J Biomed Mater Res. 2000;52:662–8.
26. Sondi I, Salopek-Sondi B. Silver nanoparticles as antimicrobial agent: a case study on E. coli as a model for Gram-negative bacteria. J Colloid Interface Sci. 2004;275:177–82.
27. Morones JR, Elechiguerra JL, Camacho A, et al. The bactericidal effect of silver nanoparticles. Nanotechnology. 2005;16:2346–53.
28. Song HY, Ko KK, Oh LH, et al. Fabrication of silver nanoparticles and their antimicrobial mechanisms. Eur Cells Mater. 2006;11 Suppl 1:58.
29. Shin SH, Ye MK, Kim HS, et al. The effects of nano-silver on the proliferation and cytokine expression by peripheral blood mononuclear cells. Int Immunopharmacol. 2007;7(13):1813–8.
30. Bhol KC, Schechter PJ. Effects of nanocrystalline silver (NPI 32101) in a rat model of ulcerative colitis. Dig Dis Sci. 2007;52:2732–42.
31. Nadworny PL, Wang J, Tredget EE, et al. Anti-inflammatory activity of nanocrystalline silver in a porcine contact dermatitis model. Nanomedicine. 2008;4:241–51.
32. Wong KK, Cheung SO, Huang L, et al. Further evidence of the anti-inflammatory effects of silver nanoparticles. ChemMedChem. 2009;4:1129–35.
33. Foldbjerg R, Olesen P, Hougaard M, et al. PVP-coated silver nanoparticles and silver ions induce reactive oxygen species, apoptosis and necrosis in THP-1 monocytes. Toxicol Lett. 2009;190:156–62.
34. Wright JB, Lam K, Buret AG, et al. Early healing events in a porcine model of contaminated wounds: effects of nanocrystalline silver on matrix metalloproteinases, cell apoptosis, and healing. Wound Repair Regen. 2002;10:141–51.
35. Warriner R, Burrell R. Infection and the chronic wound: a focus on silver. Adv Skin Wound Care. 2005;18 Suppl 1:2–12.
36. Yang W, Shen C, Ji Q, et al. Food storage material silver nanoparticles interfere with DNA replication fidelity and bind with DNA. Nanotechnology. 2009;20:085102.
37. Park MH, Kim KH, Lee HH, et al. Selective inhibitory potential of silver nanoparticles on the harmful cyanobacterium Microcystis aeruginosa. Biotechnol Lett. 2010;32:423–8.
38. Wang H, Law N, Pearson G, et al. Impact of silver(I) on the metabolism of Shewanella oneidensis. J Bacteriol. 2010;192:1143–50.
39. Dror-Ehre A, Mamane H, Belenkova T, Markovich G, Adin A. Silver nanoparticle-E. coli colloidal interaction in water and effect on E. coli survival. J Colloid Interface Sci. 2009 Nov 15;339(2):521–6. Epub 2009 Jul 28. PubMed PMID: 19726047.

40. Yin HQ, Langford R, Burrell RE. Comparative evaluation of the antimicrobial activity of ACTICOAT antimicrobial barrier dressing. J Burn Care Rehabil. 1999;20:195–200.
41. Tian J, Wong KKY, Ho CM, et al. Topical Delivery of Silver Nanoparticles Promotes Wound Healing. ChemMedChem. 2007;2:129–36.
42. Tredget EE, Shankowsky HA, Groeneveld A, et al. Randomized study evaluating the efficacy and safety of acticoat silver-coated dressing for the treatment of burn wounds. J Burn Care Rehabil. 1998;19(6):531–7.
43. Sibbald RG, Contreras-Ruiz J, Coutts P, et al. Bacteriology, inflammation, and healing: a study of nanocrystalline silver dressings in chronic venous leg ulcers. Adv Skin Wound Care. 2007;20:549–58.
44. Liu XL, Lee PY, Ho CM, et al. Silver nanoparticles mediate differential responses in keratinocytes and fibroblasts during skin wound healing. Chem Med Chem. 2010;5:468–75.
45. Rosas-Hernández H, Jiménez-Badillo S, Martínez-Cuevas PP, et al. Effects of 45-nm silver nanoparticles on coronary endothelial cells and isolated rat aortic rings. Toxicol Lett. 2009;191:305–13.
46. Kwan KH, Liu XL, To MK, et al. Modulation of collagen alignment by silver nanoparticles results in better mechanical properties in wound healing. Nanomedicine. 2011;7:497–504.
47. Hsieh PC, Davis ME, Gannon J, et al. Controlled delivery of PDGF-BB for myocardial protection using injectable self-assembling peptide nanofibers. J Clin Invest. 2006;116:237–48.
48. Tysseling-Mattiace VM, Sahni V, Niece KL, et al. Self-assembling nanofibers inhibit glial scar formation and promote axon elongation after spinal cord injury. J Neurosci. 2008;28:3814–23.
49. Hamada K, Hirose M, Yamashita T, et al. Spatial distribution of mineralized bone matrix produced by marrow mesenchymal stem cells in self-assembling peptide hydrogel scaffold. J Biomed Mater Res A. 2008;84:128–36.
50. Schneider A, Garlick JA, Egles C. Self-assembling peptide nanofiber scaffolds accelerate wound healing. PLoS One. 2008;3:e1410.
51. Koutsopoulos S, Unsworth LD, Nagai Y, et al. Controlled release of functional proteins through designer self-assembling peptide nanofiber hydrogel scaffold. Proc Natl Acad Sci U S A. 2009;106:4623–8.
52. Zhang S. Beyond the Petri dish. Nat Biotechnol. 2004;22:151–2.
53. Zhang S. Fabrication of novel biomaterials through molecular self-assembly. Nat Biotechnol. 2003;21: 1171–8.
54. Zhang S. Designer self-assembling Peptide nanofiber scaffolds for study of 3-d cell biology and beyond. Adv Cancer Res. 2008;99:335–62.
55. Gelain F, Lomander A, Vescovi AL, et al. Systematic studies of a self-assembling peptide nanofiber scaffold with other scaffolds. J Nanosci Nanotechnol. 2007;7:424–34.
56. Kretsinger JK, Haines LA, Ozbas B, et al. Cytocompatibility of self-assembled beta-hairpin peptide hydrogel surfaces. Biomaterials. 2005;26: 5177–86.
57. Pochan DJ, Schneider JP, Kretsinger J, et al. Thermally reversible hydrogels via intramolecular folding and consequent self-assembly of a de novo designed peptide. J Am Chem Soc. 2003;125: 11802–3.
58. Schneider JP, Pochan DJ, Ozbas B, et al. Responsive hydrogels from the intramolecular folding and self-assembly of a designed peptide. J Am Chem Soc. 2002;124:15030–7.
59. Haines-Butterick L, Rajagopal K, Branco M, et al. Controlling hydrogelation kinetics by peptide design for three-dimensional encapsulation and injectable delivery of cells. Proc Natl Acad Sci U S A. 2007; 104:7791–6.
60. Semino CE, Kasahara J, Hayashi Y, et al. Entrapment of migrating hippocampal neural cells in three-dimensional peptide nanofiber scaffold. Tissue Eng. 2004;10:643–55.
61. Wang S, Nagrath D, Chen PC, et al. Three-dimensional primary hepatocyte culture in synthetic self-assembling peptide hydrogel. Tissue Eng Part A. 2008;14: 227–36.
62. Kisiday J, Jin M, Kurz B, et al. Self-assembling peptide hydrogel fosters chondrocyte extracellular matrix production and cell division: implications for cartilage tissue repair. Proc Natl Acad Sci U S A. 2002; 99:9996–10001.
63. Genove E, Shen C, Zhang S, et al. The effect of functionalized self-assembling peptide scaffolds on human aortic endothelial cell function. Biomaterials. 2005;26: 3341–51.
64. Narmoneva DA, Oni O, Sieminski AL, et al. Self-assembling short oligopeptides and the promotion of angiogenesis. Biomaterials. 2005;26:4837–46.
65. Sieminski AL, Was AS, Kim G, et al. The stiffness of three-dimensional ionic self-assembling peptide gels affects the extent of capillary-like network formation. Cell Biochem Biophys. 2007;49:73–83.
66. Semino CE, Merok JR, Crane GG, et al. Functional differentiation of hepatocyte-like spheroid structures from putative liver progenitor cells in three-dimensional peptide scaffolds. Differentiation. 2003;71: 262–70.
67. Garreta E, Genove E, Borros S, et al. Osteogenic differentiation of mouse embryonic stem cells and mouse embryonic fibroblasts in a three-dimensional self-assembling peptide scaffold. Tissue Eng. 2006;12: 2215–27.
68. Davis ME, Motion JP, Narmoneva DA, et al. Injectable self-assembling peptide nanofibers create intramyocardial microenvironments for endothelial cells. Circulation. 2005;111:442–50.
69. Webber MJ, Tongers J, Renault MA, et al. Development of bioactive peptide amphiphiles for therapeutic cell delivery. Acta Biomater. 2010;6:3–11.

70. Webber MJ, Kessler JA, Stupp SI. Emerging peptide nanomedicine to regenerate tissues and organs. J Intern Med. 2010;267:71–88.
71. Huang Z, Sargeant TD, Hulvat JF, et al. Bioactive nanofibers instruct cells to proliferate and differentiate during enamel regeneration. J Bone Miner Res. 2008; 23:1995–2006.
72. Galler KM, Cavender A, Yuwono V, et al. Self-assembling peptide amphiphile nanofibers as a scaffold for dental stem cells. Tissue Eng Part A. 2008;14:2051–8.
73. Capito RM, Azevedo HS, Velichko YS, et al. Self-assembly of large and small molecules into hierarchically ordered sacs and membranes. Science. 2008;319:1812–6.
74. Jayawarna V, Smith A, Gough JE, et al. Three-dimensional cell culture of chondrocytes on modified di-phenylalanine scaffolds. Biochem Soc Trans. 2007;35:535–7.
75. Yang Z, Gu H, Zhang Y, et al. Small molecule hydrogels based on a class of antiinflammatory agents. Chem Commun (Camb). 2004;21(2):208–9.
76. Toledano S, Williams RJ, Jayawarna V, et al. Enzyme-triggered self-assembly of peptide hydrogels via reversed hydrolysis. J Am Chem Soc. 2006; 128:1070–1.
77. Thornton K, Smith AM, Merry CL, et al. Controlling stiffness in nanostructured hydrogels produced by enzymatic dephosphorylation. Biochem Soc Trans. 2009;37:660–4.
78. Smith AM, Williams RJ, Tang C, et al. Fmoc-Diphenylalanine self assembles to a hydrogel via a novel architecture based on π–π interlocked beta-sheets. Adv Mater. 2008;20:37–41.
79. Davis ME, Hsieh PCH, Takahashi T, et al. Local myocardial insulin-like growth factor 1 (IGF-1) delivery with biotinylated peptide nanofibers improves cell therapy for myocardial infarction. Proc Natl Acad Sci U S A. 2006;103:8155–60.
80. Binyamin G, Shafi BM, Mery CM. Biomaterials: a primer for surgeons. Semin Pediatr Surg. 2006;15:276–83.
81. Schackelford JF. Introduction to materials science for engineers. 3rd ed. New York, NY: Macmillan Publishing Company; 1992.
82. Park JB, et al. In: Park JB, Bronzino JD, editors. Biomaterials: principles and applications. Danvers, MA: CRC Press; 2003.
83. Ngo YH, Li D, Simon GP, et al. Paper surfaces functionalized by nanoparticles. Adv Colloid Interface Sci. 2011;163:23–38.
84. Moreira-Teixeira LS, Georgi N, Leijten J, et al. Cartilage tissue engineering. Endocr Dev. 2011;21: 102–15.
85. Bhat SV. Biomaterials second edition. Harrow, U.K.: Alpha Science International; 2005.
86. Matsumoto Y, Arai K, Momose H, et al. Development of a wound dressing composed of a hyaluronic acid sponge containing arginine. J Biomater Sci Polym Ed. 2009;20:993–1004.
87. Scherer SS, Pietramaggiori G, Mathews JC, et al. The mechanism of action of the vacuum-assisted closure device. PRSJ journal. 2008;122:786–97.
88. Heit YI, Dastouri P, Helm D, et al. Foam pore size is a critical interface parameter of suction-based wound healing devices. Plast Reconstr Surg. 2012;129(3):589–97.
89. Younan GJ, Heit YI, Dastouri P, et al. Mast cells are required in the proliferation and remodeling phases of microdeformational wound therapy. Plast Reconstr Surg. 2011;128:649e–58.
90. Edwards J. Hydrogels and their potential uses in burn wound management. Br J Nurs. 2010; 19:S12, S14-16
91. Anumolu SS, Menjoge AR, Deshmukh M, et al. Doxycycline hydrogels with reversible disulfide crosslinks for dermal wound healing of mustard injuries. Biomaterials. 2011;32:1204–17.
92. Seidlits SK, Drinnan CT, Petersen RR, et al. Fibronectin-hyaluronic acid composite hydrogels for three-dimensional endothelial cell culture. Acta Biomater. 2011;7:2401–9.
93. Jiang B, Larson JC, Drapala PW, et al. Investigation of lysine acrylate containing poly(N-isopropylacrylamide) hydrogels as wound dressings in normal and infected wounds. J Biomed Mater Res B Appl Biomater. 2012;100(3):668–76. doi:10.1002/jbm.b.31991.
94. Hill PS, Apel PJ, Barnwell J, et al. Repair of peripheral nerve defects in rabbits using keratin hydrogel scaffolds. Tissue Eng Part A. 2011;17:1499–505.
95. Sun G, Zhang X, Shen YI, et al. Dextran hydrogel scaffolds enhance angiogenic responses and promote complete skin regeneration during burn wound healing. Proc Natl Acad Sci U S A. 2011;108:20976–81.
96. Anderson JM. Biological responses to materials. Annu Rev Mater Res. 2001;31:81–110.
97. Langer R, Peppas NA. Advances in biomaterials, drug delivery and bionanotechnology. AIChE J. 2003; 49:2990–3006.
98. Ratner BD, Bryant SJ. Biomaterials: where we have been and where we are going. Annu Rev Biomed Eng. 2004;6:41–75.

Nanotoxicology

22

Alison C. Lowe, Lindsey A. Hunter-Ellul, and Michael G. Wilkerson

The "science of engineered nanodevices and nanostructures that deals with their effects in living organisms" [1].

22.1 Introduction

The development of biomedical and cosmetic products containing nanoparticles is on the rise, and there is a growing concern about their use and potential toxicity. Nano-sized TiO_2 alone has an estimated annual global production of 12,500 metric tons with annual US production being 3,630 metric tons [2]. Entire populations are at risk for exposure to nanoparticles due to the wide availability of these products [3], and many questions remain unanswered regarding the short- and long-term effects of nanoparticles on cytotoxicity, systemic toxicity, and carcinogenicity [4].

Topically applied nanomaterials have been shown to cause inflammation, cytotoxicity, and genotoxicity in various studies [5]. The ability of nanomaterials to penetrate beyond the epidermal and/or dermal layers is important to consider, along with their subsequent bioaccumulation and systemic effects. Data often conflict when comparing in vitro and in vivo studies, and the use of animal models brings the generalizability of study results into question as well. For example, rodent skin is more penetrable than porcine skin, which is more penetrable than human skin [6, 7].

Studies thus far suggest that the risk for acute nanoparticle-related systemic toxicity is low. However, acute effects, including inflammatory reactions and pulmonary tissue injury, have been demonstrated following inhalational exposure to certain nanoparticles such as nano-sized TiO_2 [8, 9]. In vitro studies have demonstrated several cellular effects of nanoparticles, including oxidative stress–induced alterations in gene expression and changes in cell signaling pathways such as calcium and cytokine pathways [10–14]. The chronic consequences of these nanoparticle-induced effects are currently unclear (Table 22.1).

Local and systemic adverse effects consist of primarily inflammatory reactions. Other observed pathological effects include the generation of reactive oxygen species (ROS) and subsequent oxidative stress, and the disruption of proteins, DNA, mitochondria, and membrane structures. These inflammatory processes have resulted in fibrosis, granulomatous reactions, atherogenesis, and even increased expression of acute phase proteins, like C-reactive protein and

A.C. Lowe (✉)
School of Medicine, University of Texas,
301 University Boulevard, Galveston, TX 77555, USA
e-mail: acpugsle@utmb.edu

L.A. Hunter-Ellul
Department of Dermatology, University of Texas,
301 University Boulevard, Galveston, TX 77555, USA
e-mail: lihunter@utmb.edu

M.G. Wilkerson
Department of Dermatology, University of Texas,
301 University Boulevard, 4.112 McCullough Building,
Galveston, TX 77555, USA
e-mail: mgwilker@utmb.edu

A. Nasir et al. (eds.), *Nanotechnology in Dermatology*,
DOI 10.1007/978-1-4614-5034-4_22, © Springer Science+Business Media, LLC 2013

Table 22.1 Potential nanoparticle-induced toxic effects [8–15]

Potential adverse effects (chronic)
Inflammation
Oxidative stress
Cytotoxicity
Alteration of gene expression
Changes to cell signaling pathways
Foreign body granulomatous reaction
Carcinogenicity

fibrinogen [15]. It is clear that additional in vivo studies using human subjects must be undertaken in order to adequately predict the short- and long-term effects of nanomaterials. This chapter will focus on potential nanotoxicologic mechanisms and the effects elicited in present day research.

22.2 Nanoparticle Properties

Nano-sized materials come in an array of shapes, sizes, and surface coatings (i.e., pegylation). These properties confer potential therapeutic and toxic effects including specific routes of penetration and internalization, cell targeting and selectivity, as well as delayed clearance and biopersistence. For example, incidental inhalational exposure is potentially greater for nanoparticles vs. nanofibers, as the shape of nanoparticles renders them more easily aerosolized [16]. The toxicity of certain nanomaterials is additionally influenced by their distribution and agglomeration tendencies, catalytic activity, solubility, contaminants acquired during manufacture, exposure time, and route of exposure [4].

All potential routes of exposure must be considered when evaluating potential nanoparticle toxic effects. For instance, topically applied nanomaterials carry the obvious risk of dermal exposure, but the risks for concurrent GI and respiratory exposure must also be considered, particularly with chronically applied nanomaterials. GI exposure is likely to occur due to hand–oral transmission, and the risk for respiratory exposure stems from the aerosolization of nanomaterials [8].

22.3 Size

Nanomaterials cannot be regarded as smaller versions of their respective bulk substances. Nano-sized substances possess unique physical, chemical, and biological properties, resulting in distinctive behaviors and effects [3, 17]. Compared to larger sized particles of equal mass, nanomaterials have a greater surface area, providing for increased reactivity and added biological interactions. This is compounded by their size, which permits enhanced mobility not only through the human body but also through the environment as well [1, 18, 19].

22.4 Shape

Nanomaterials may be shaped as spheres, fibers, tubes, rings, or planes. These various shapes of nanoparticles confer distinctive potential adverse effects, particularly with inhalational exposure [1]. The degree of pulmonary toxicity is primarily determined by particle biopersistence, shape, and length [20]. When compared to exposure to nanospheres, there is a greater risk of pulmonary fibrotic change following exposure to nanofibers, including both natural (i.e., asbestos, a naturally occurring silicate) and man-made (i.e., carbon nanotubes) nanofibers [21, 22].

The needle-like fiber shape of carbon nanotubes has been likened to that of asbestos, and the potential development of mesothelioma or a similar disease process following chronic exposure to these nanoparticles warrants discussion [23, 24]. Specific fiber characteristics contribute to the potential to induce mesothelioma. Fiber length is significant, as longer asbestos and carbon nanotube fibers elicit changes in the mesothelium when compared to shorter fibers. These changes include granuloma formation as well as exudation of protein and polymorphonuclear cells [25]. A “frustrated phagocytosis” process occurs, serving as the body’s normal defense against foreign bodies that are too large to be cleared by macrophages alone [26, 27]. Fiber number is another important consideration, as greater quantities lead to a more intense or continued inflammatory reaction, which has the potential to result in fibrosis, genotoxicity, and/or carcinogenicity [28, 29]. The hallmark of asbestos-induced mesothelioma is chronic inflammation and the prolonged latency period between exposure and presentation of symptoms. Asbestos exposure classically occurs 25–30 years prior to neoplastic development, which suggests that biopersistence of carbon nanotubes may also be a key factor in potential carcinogenicity [23].

Though carbon nanotubes appear to act like asbestos fibers, the ability to cause mesothelioma remains in question [25]. It is unlikely that the level of exposure experienced by the general consumer will reach high enough levels to induce mesothelioma. Workers, however, face a much greater risk for carbon nanotube–induced mesothelioma due to their exponentially greater levels of exposure [25]. To further complicate matters, as with asbestos and mesothelioma, any potentially devastating effects may not be apparent until decades following exposure.

22.5 Surface Modifications

Surface coatings can make nanoparticles more biocompatible, significantly altering toxicological effects. As an example, neutral and pegylated (i.e., polyethylene glycol-coated) quantum dots have been shown to be nontoxic to human epidermal keratinocytes. Conversely, quantum dots with amine surface coatings exhibit cytotoxicity, while those with carboxylic surface coatings elicit both inflammatory and cytotoxic reactions [30]. Moreover, the addition of a coating to nanoparticles reduces the risk of both particle destabilization and agglomeration, each of which can enhance toxicity [14, 16, 17].

22.6 Contaminants

Contaminants found in engineered nanoparticles can contribute to toxicity. Contaminants may be introduced during virtually any step of the manufacturing process, and include dispersive agents like surfactants, side products, organics, endotoxins, etc. [4, 8]. It has been documented that inhaled carbon nanotubes elicit acute pulmonary inflammatory effects. Further investigation, however, suggests that the pulmonary toxicity may be attributable, at least in part, to impurities from the manufacturing process [31]. Additionally, though fullerenes have been declared toxic to bacteria and fish [1, 32], much of the potential for toxicity lies with the associated remnants of solubilization, including tetrahydrofuran (THF). Fullerene compounds that were processed using THF had LC_{50} values ten times lower than compounds processed by sonication (i.e., ultrasound-assisted dispersion), suggesting that THF significantly influences toxicity [32, 33] (Table 22.2).

22.7 Accumulation

The ability and tendency for nanoparticles to accumulate within cells and tissues are noteworthy, as numerous well-known disease processes stem from the accumulation of various materials. When the lungs become overwhelmed by exposure to silica, asbestos, coal dust, or smoke, disease processes such as fibrosis, COPD, or pneumoconiosis can develop [28, 34]. Hepatic cirrhosis may follow the accumulation of copper or iron within the liver and the successive development of Wilson's disease or hemochromatosis, respectively [35, 36]. Accumulation of amyloid in brain tissue can lead do neurodegenerative changes such as those seen in Alzheimer's or Parkinson's disease [37]. The cutaneous lesions of nephrogenic systemic fibrosis are associated with the accumulation and deposition of gadolinium following administration of intravenous gadolinium-based contrast agents in patients with renal impairment [38].

22.8 Agglomeration

Nanoparticle agglomeration reduces overall particle surface area, leading to fewer surface interactions and possibly less toxicity.

Table 22.2 The various and unique characteristics of nanoparticles and the resulting contribution to potential toxicity [1, 3, 4, 16–19, 21, 22, 30, 32, 33]

Specific nanoparticle characteristics and their effect on toxicology	
Size	Increased surface area leads to increased reactivity and more biological interactions
	Greater penetration with an inverse relationship between penetration depth and particle size
Shape	Nanofibers induce a greater risk of pulmonary fibrotic change than do nanospheres
Surface coating	Neutral and pegylated quantum dots shown to be nontoxic
	Amine-coated nanoparticles exhibit cytotoxicity
	Carboxylic nanoparticles demonstrate both cytotoxicity and inflammatory reactions
Contaminants	LC_{50} of compounds processed with tetrahydrofuran (THF) is ten times lower than compounds processed with sonication only
Charge	Due to interactions with negatively charged cell membranes, cationic nanomaterials demonstrate increased cellular uptake when compared with anionic nanomaterials
The unique physical, chemical, and biological properties of nanoparticles also influence toxicity by altering	
Degree of exposure	
Route of exposure	
Route of penetration	
Internalization	
Cell targeting	
Catalytic activity	
Agglomeration	
Solubility	
Biopersistence	

However, some studies indicate that agglomeration (i.e., clumping) can lead to particle destabilization and disintegration, potentially releasing smaller or more toxic components within the organism or environment [4]. Agglomeration, which may increase with time, has been demonstrated by many man-made nanomaterials [39]. One example is the TiO_2 nanoparticle agglomerates that have been observed in intracellular organelles. Additionally, agglomeration itself may contribute to cellular uptake of nanoparticles [40], which occurs via invagination of cell membranes, forming vesicles around nanoparticles [41]. This appears to be a size-dependent process, as cells are unable to internalize larger agglomerates (Table 22.3).

Table 22.3 Well-known disease processes attributed to material accumulation in cells and/or tissues [28, 34–38]

Diseases secondary to material accumulation	
Silica, asbestos, coal dust, smoke	Fibrosis
	Mesothelioma
	Pneumoconiosis
	Chronic obstructive pulmonary disease
Iron	Hemochromatosis
Copper	Wilson's disease
Amyloid	Alzheimer's disease
Gadolinium	Nephrogenic systemic fibrosis

22.9 Exposure

Nanoparticles are currently utilized as ingredients in medical products, cosmetics, fuel cells, tires, electronics, and textiles. The potential for exposure exists throughout the nanoparticle life cycle, during manufacture, use, or following disposal [1]. Exposure to the two broad sources of nanoparticles, anthropogenic and engineered nanomaterials, may occur via inhalation, ingestion, transdermal uptake, or injection. The primary routes of exposure in humans include the GI and respiratory tracts. Incidental exposure is primarily due to inhalation of airborne nanoparticles from combustion-derived anthropogenic sources, such as engines and power plants. Naturally occurring nano-sized fullerenes occur after combustion processes such as volcanoes or forest fires [1].

The chronic toxic effects of nanoparticles will likely be seen initially and with greater severity in the workers involved with nanomaterial manufacture. These workers possess an exponentially larger risk for exposure compared to the general population [1]. Moreover, the threat posed to these workers is considerably higher because they are exposed to raw, uncompounded, unmodified nanoparticles. As previously mentioned, the modification processes typically render nanoparticles more stable and reduce their toxic potential [4].

22.10 Skin

A variety of products have the potential for dermal exposure and penetration including the more evident ones such as photoprotection, barrier creams, cosmetics, antiseptics, transdermal drug delivery systems, and stain-resistant clothing. A somewhat less obvious source of dermal exposure stems from airborne particle accumulation. Airborne nanoparticles, such as carbon nanotubes and carbon black, have been shown to agglomerate and settle onto surfaces where they have the potential to make dermal contact [8].

In vitro studies of topically applied nanoparticles have routinely demonstrated significant cytotoxic and proinflammatory effects [41–43]. For example, in vitro studies of keratinocytes exposed to high doses of carbon nanotubes demonstrated internalization, oxidative stress, and proinflammatory effects with free radical formation and depletion of cellular antioxidants. However, in vivo studies with topically applied nanomaterials have demonstrated only mild irritation without evidence of photo-irritation or sensitization [44, 45]. Such disparities may be explained by the fact that in vivo intact skin barriers appear to limit dermal penetration. Compared to topical application, a more severe foreign body granuloma-like reaction is elicited by subcutaneous implantation of carbon nanotubes. As this granulomatous reaction better reflects the effects of topically applied nanoparticles described by many in vitro studies, these findings support the idea of decreased in vivo penetration of topically applied nanoparticles [46, 47]. Additionally, many in vitro studies have used high doses of nanoparticles that are not representative of concentrations associated with human exposure [1].

Silver nanoparticles are widely used in wound and burn dressings, as well as for disinfecting water [4]. Despite the beneficial bactericidal properties, overwhelming or chronic exposure to silver can potentially result in argyria, an irreversible bluish change in skin color [47, 48]. Furthermore, the bactericidal effect may alter the balance of gut microflora species, with potential consequences similar to those of antibiotic-associated diarrhea and pseudomembranous colitis [49].

Dermal exposure may also occur with the nanoparticles used in textile modification, as with fibers used in stain- and wrinkle-resistant clothing [3]. However, textiles undergo extensive processing from raw material to finished product. This processing significantly diminishes the likelihood that any potentially harmful materials found in the original fibers would pose a significant health threat to consumers [50]. Again, workers are the primary individuals most at risk for toxicity, as they face exposure to the raw materials.

The toxic dermal effects of nanoparticles are of significant concern. There is an immense potential for dermal exposure, putting one at risk of systemic effects associated with skin penetration, given the absorptive capability and abundance of blood/tissue macrophages, lymphatics, Langerhans cells, and sensory nerve endings [1].

Table 22.4 Potential modes of dermal exposure to nanomaterials [3, 8]

Dermal exposure
Photoprotection
Barrier creams
Cosmetics
Antiseptics
Transdermal drug delivery
Stain-resistant clothing
Airborne particles[a]
Cutaneous uptake of systemically/internally applied nanoparticles[b]

[a]Secondary to particle agglomeration and settling onto surfaces
[b]Nephrogenic systemic fibrosis following intravenous exposure to gadolinium

Presently, no one knows what short- or long-term side effects are associated with exposure to chronic low levels of nanoparticles, which warrants further investigation (Table 22.4).

22.11 Barrier Functions

The skin's stratum corneum layer serves as a barrier. Topical substances may penetrate through this barrier intercellularly or by diffusion through lipid bilayers, and the degree of penetration is dependent upon size and physiochemical properties. Hair follicles have been shown to act as an alternate infiltrative pathway for nanomaterials, thus increasing the degree of penetration and absorption, which is further enhanced by the depot capability of hair follicles [51–58]. Certain particulate formulations demonstrate the ability to target specific cell populations, and some preferentially target hair follicles over the stratum corneum [53]. For example, following short-term exposure, TiO_2 nanoparticles cannot penetrate past the upper stratum corneum of intact skin, but instead may accumulate in hair follicle infundibula [6, 59, 60]. Potential follicular targets include Langerhans cells, matrix cells, melanocytes, and the sebaceous gland [61, 62]. The reservoir function of the follicular infundibulum can be up to ten times that of the stratum corneum, providing for the pharmacological goal of sustained drug release [63, 64] as well as increased duration of exposure.

Negatively charged polystyrene spheres exhibit similar behavior to topical TiO_2 nanoparticles [65]. Conversely, topically applied quantum dots express limited dermal penetration, with a small fraction of the applied dose accumulating in the dermis. Smaller, spherical particles result in a greater degree of penetration compared to larger, elliptical particles [66]. Limited dermal penetration has also been demonstrated by small (<10 nm) iron and maghemite (an iron oxide) nanoparticles [67].

22.12 Nanoparticle Uptake

The particle uptake ability of the dermis has been well demonstrated. There has even been significant soil particle load evident in inguinal lymph nodes of people who are routinely barefoot [68, 69]. Quantum dot studies have validated that nanoparticles localize to lymph nodes following dermal uptake, likely via dermal macrophages and Langerhans cells [70–72]. This localization ability is important from a toxicological standpoint because further translocation from lymph nodes to the blood, and then systemwide, is possible [73].

Within hours of topical exposure, quantum dots of various sizes, shapes, and surface coatings can penetrate through an intact stratum corneum with subsequent epidermal and dermal localization [66]. Cellular uptake of quantum dots by human epidermal keratinocytes appears to be independent of surface coating, with particle localization to the cytoplasm, cytoplasmic organelles (i.e., vacuoles), or even intranuclearly [30].

Uptake by olfactory and tracheobronchial sensory nerves has been demonstrated, and skin exposure to herpes virus with subsequent retrograde axonal and dendritic transport is well established [74, 75]. Therefore, nanoparticle uptake by the abundant dermal sensory nerves must be considered. It is likely that nanoparticle charge and size may be governing factors for dermal neuronal transport [76–79].

22.13 Internalization and Transport

Understanding the various particle–tissue interactions is important for predicting potential toxic effects, in addition to having important therapeutic implications. One example of these tissue interactions is the adsorption of tissue proteins on the particle surface, which may effect changes in particle surface characteristics, including charge [4]. Other biokinetic behaviors of nanoparticles include cellular endocytosis, transcytosis, translocation to circulatory and neurologic systems, and systemic distribution [1].

Intracellular uptake is affected by surface charge. Positively charged particles interact with the negatively charged cell membrane, contributing to increased cellular uptake of cationic particles when compared to anionic size-matched particles [80]. While cellular uptake of nanoparticles may be therapeutically desired, interactions with the cell membrane can disrupt its integrity, contributing to overall toxicity [81].

Various mechanisms for internalization of nanomaterials have been demonstrated, including uptake via clathrin-coated pits or caveolae. Caveolae are plentiful in pulmonary capillaries and type 1 pneumocytes but are sparse in capillaries of the brain [82]. When the CNS is the desired target, surface modifications can facilitate penetration of the blood–brain barrier. One example is an apolipoprotein coating, which allows for LDL-receptor-mediated endocytosis of nanoparticles [83–85].

Nanoparticles undergo transcytosis across respiratory epithelium into the interstitium. Studies suggest that albumin and lecithin, both known components of alveolar lining fluid, facilitate endocytosis of nanoparticles [86]. Within the interstitium, nanoparticles may cause local effects or gain access to the circulatory system directly or via lymphatics, providing for potential systemic access. Migration to the interstitium was apparent within 24 h after inhalation, and higher doses (i.e., those with a greater surface area) of nanoparticles resulted in decreased pulmonary epithelial inflammation when compared to lower doses. This was explained by the fact that after 24 h at least 50 % of the particles reach the pulmonary interstitium, resulting in a shift of the inflammatory response out of the alveoli and away from the pulmonary epithelium [87]. After leaving the alveoli, inhaled nanoparticles may continue on to cause systemic toxicity. One potential mechanism for extrapulmonary spread is uptake by nerves in respiratory epithelium with subsequent spread to the CNS via axonal translocation [1].

The argument can be made that at such high doses, the observed interstitialization of nanoparticles is not representative of realistic exposure conditions. However, most of these studies have been performed on rats, and research has shown greater rates of interstitialization in larger vs. smaller mammals [87, 88].

22.14 Local Effects

There is an inverse relationship between penetration depth and particle size. When substances are loaded onto nanoparticles as carriers, the result is increased penetration depth and permeation rate of the substance [80]. Both augmented penetration and selective cell targeting contribute to the potential dermal toxicity of topically applied nanoparticles, as both mechanisms provide for increased particle number and localized particle accumulation [4].

Metal oxide nanoparticles, such as zinc oxide and titanium dioxide, are the most common nanomaterials found in topically applied consumer products (i.e., sunscreen and cosmetics) [5]. These nanoparticles can serve as photon blockers, and TiO_2 particles are a major component of sunscreens. However, studies have demonstrated that exposure to UV light results in a particle-mediated increase in ROS and carcinogenicity [89, 90]. These effects of photogenotoxicity and photocytotoxicity were not reproduced by numerous regulatory studies, however [43]. Additionally, coating of nano-scale TiO_2 stabilizes the particles, precluding photocatalysis [91]. The photocatalytic property of TiO_2 nanoparticles is actually desirable in some circumstances, as when utilized as a disinfecting agent.

UV exposure promotes ROS formation and peroxidation of the lipid membrane of prokaryotes, including bacteria [92].

Dermal exposure to nano-sized TiO_2 results in oxidative stress and free radical formation, as well as increased keratinization, dermal atrophy, lipid peroxidation, and collagen depletion [93]. Penetration past the stratum corneum appears to be particle size and exposure time-dependent. A study using nano-sized TiO_2 demonstrated penetration limited to the stratum corneum after 24 h, into the epidermal prickle cell layer and basal cell layer after 30 days, and through the skin to various organs (i.e., liver, lung, and heart) after 60 days of exposure [93].

The different crystal forms of TiO_2 nanoparticles induce different levels of ROS production following exposure to UV irradiation. Either the rutile form or a combination of rutile plus anatase is the most commonly used forms in sunscreen preparations [5, 94]. In Chinese hamster ovary (CHO) cells, both the level of hydroxyl radical formation and subsequent cytotoxicity are dependent upon the presence and dose of the anatase crystal form. Following UV exposure, rutile and amorphous forms generate fewer OH radicals and do not significantly affect CHO cell viability [5, 95]. Surface modifications such as coating reduce both the level of ROS produced and the threat of carcinogenicity [96, 97].

The integrity of nanoparticle surface modifications is not yet known, and surface endurance will likely influence the chronic consequences of nanoparticle exposure [4]. Surface modifications may be metabolized by cells, or disrupted by oxygen and UV light during use or following disposal. The potential results of surface weathering include destabilization and leaching of core materials into the body and/or the environment [1].

22.15 Inflammation

Nanoparticles are immunomodulators, exhibiting inflammatory, pro-oxidant, and antioxidant properties. Again, the modification of size, surface charge, and surface functionalization groups directs the body's response to nanoparticles and provides the potential for tighter control of nanoparticle effects as well as increased stability [4]. For example, nano-sized particles elicit a stronger cellular immune response compared to microparticles, which elicit a stronger humoral response [4, 98]. Additionally, pegylation prevents opsonization, therefore evading recognition by the reticuloendothelial system. Coating of nanoparticles reduces the risk of both particle destabilization and agglomeration [99, 100].

Surface charge is an important modulator of cytotoxic and inflammatory effects, likely via influencing cellular localization or due to variations in tissue receptor sensitivity to particle charge [101, 102]. Carboxylic acid–coated quantum dots induce the release of the proinflammatory cytokines IL-1β, IL-6, and IL-8. Pegylated quantum dots do not result in increased cytokine release from keratinocytes, and PEG-amine-coated quantum dots induce a minor increase in IL-6 and IL-8. Additionally, cell viability has been shown to be reduced following exposure to carboxylic acid–coated and PEG-amine-coated quantum dots, but not PEG-coated quantum dots, and the cytotoxic effects appear to be modulated by exposure time as well [103]. Cytotoxicity could be due to ROS production or leaching of core materials, with the latter being less likely due to capping of the quantum dots used in the study [30]. Inflammatory reactions may be restricted to the site of exposure, or may also include systemic immunogenic effects [103].

Various nanoparticles selectively target mitochondria, and there is evidence of mitochondrial distribution upon endocytosis. Certain nanoparticles influence mitochondrial ROS formation, which may overwhelm antioxidant actions, resulting in an oxidative stress response [1]. Some studies have shown oxidative stress–related inflammatory reactions and cytotoxicity to be dose-dependent [4]. The exact mechanism has not yet been elicited, but ROS may be produced by (a) photo excitation resulting in free electrons, (b) metabolism of nanoparticles to reactive intermediates, or (c) inflammatory responses involving macrophage release of free radicals [4].

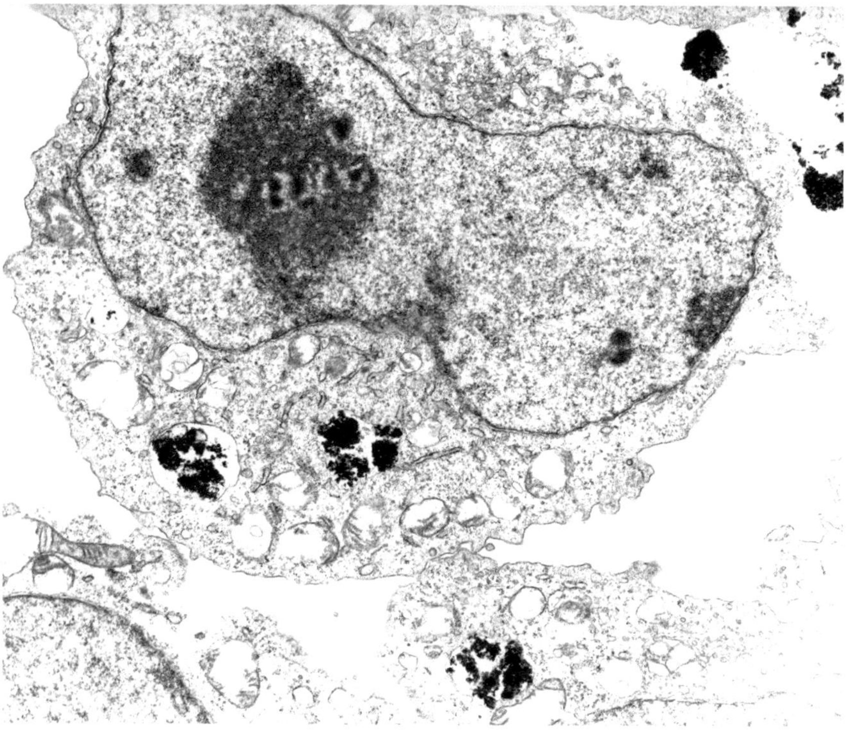

Fig. 22.1 Transmission electron microscope image demonstrating cellular uptake of agglomerated nanoparticles in CHO cells following exposure to nano-TiO_2 [108]. Agglomeration itself may influence cellular uptake, and also contributes to nanoparticle toxic effects, particle destabilization, and airborne half-life [4, 14, 16, 17, 40, 109]. *CHO* Chinese Hamster ovary

The toxic effects of airborne nanoparticles appear to be a consequence of oxidative stress [104], and the degree of inflammation is associated with particle surface area and surface charge [1, 105, 106]. The large surface area of nano-sized particles allows for more interactions with tissue sensory receptors, such as those found in the plasma membranes of brain microglia or respiratory epithelial cells. As an example, negatively charged airborne nanoparticles activate the TRPV1 vanilloid receptor found on these cells, which stimulates proinflammatory cytokine release [101, 102].

Mitochondrial injury following exposure to cationic polystyrene nanoparticles is attributed to generation of ROS and subsequent structural organelle damage and increased calcium uptake [107]. Nano-sized TiO_2 has been shown to increase ROS even in an acellular environment, suggesting that internalization is not a prerequisite for nano-sized TiO_2-induction of oxidative stress [108]. Notably, the in vivo mechanism is likely to be different than the in vitro mechanism due to the unrealistically high doses used in most studies, which may overwhelm endogenous defense mechanisms (Fig. 22.1).

22.16 Systemic Effects

In order to elicit systemic consequences, nanomaterials must first obtain access to the blood either directly or via the lymphatic system. The efficacy of both the barrier function and clearance mechanisms at the site of exposure impacts the actual systemic dose [8]. Upon reaching the circulatory system, the reticuloendothelial system (including the liver and spleen) is a primary site of nanoparticle deposition secondary to uptake by liver Kupffer cells and macrophages within spleen sinusoids. Nanoparticle interaction with self-proteins of the reticuloendothelial system holds the potential to elicit autoimmune reactions via changes in nanoparticle antigenicity [110]. When other somatic sites are being therapeutically targeted, reticuloendothelial system localization can be virtually eliminated by pegylation and other surface coatings that reduce opsonization [111, 112].

It must be taken into consideration that the strong epidermal localization exhibited by certain topically applied nanoparticles likely reduces systemic translocation and thereby decreases systemic toxicity [5]. However, some nanoparticles, such as quantum dots, have shown uptake by skin cells even when not targeted for skin localization. This nonselective uptake is an important consideration for nanoparticles used internally, as they could potentially result in cytotoxic and immunogenic effects in the skin [30]. One such example is the characteristic cutaneous lesion of nephrogenic systemic fibrosis following intravenous exposure to gadolinium-containing agents in patients with impaired renal function [38]. Intracellular uptake of nanoparticles may be a mechanism by which nanoparticles localize to the avascular epidermis and thus evade clearance by macrophages [113]. Though the systemic toxicity may be reduced by such behaviors, increased epidermal concentration of nanoparticles, and a prolonged period of exposure with the possibility of agglomeration or protective surface modification break-down, could result in further local adverse effects.

22.17 Respiratory

Inhalation of airborne nanoparticles has potentially detrimental effects not only in the respiratory tract but also systemically as well. Local consequences of inhalation may include inflammation, exacerbation of chronic lung disease (i.e., asthma), genotoxicity, and carcinogenicity [16]. Studies have shown diffuse nanoparticle deposition throughout healthy respiratory tracts, with greater deposition in lungs affected by asthma or chronic obstructive pulmonary disease [1].

As with dermal exposure, specific particle characteristics determine the degree of pulmonary toxicity. Deposition of inhaled particles is size-dependent. For example, study results demonstrate that 90 % of 1-nm particles deposit in the nasopharynx, 10 % in the tracheobronchial tree, and virtually no deposition in the alveoli. Conversely, roughly 30 % of 5-nm particles deposit in each of the three locations, while approximately 50 % of 20-nm particles deposit in the alveoli [1].

The inflammatory reaction to inhaled nanoparticles is more severe following exposure to ultrafine particles as compared to those with the same chemical composition and mass concentrations, but of a larger size [114, 115]. When describing dose–response relationships of inhaled nanoparticles, accuracy is enhanced by conveying the "dose" as a function of surface area instead of mass [10, 87, 116–120]. For example, pulmonary exposure to 20 nm TiO_2 elicits a greater inflammatory neutrophilic response than 250 nm TiO_2 particles of equal mass, likely due to variations in particle surface area [4].

Particle accumulation has a similar effect, with the larger agglomerated particles being less toxic than smaller, individual particles with a larger surface area. However, the potential destabilization and disintegration effects of particle agglomeration must also be considered [4]. This is complicated by the likelihood that during the time it takes for particles to accumulate, concurrent surface changes render the particles themselves less toxic. The airborne half-life of agglomerated particles is longer than that of free particles, potentially increasing the risk of inhalation exposure [109]. Moreover, as their size increases further, agglomerated particles tend to settle out of the air, reducing inhalation risk and increasing the risk of dermal (object to skin) and GI (object/hand to mouth) exposure [8].

22.18 GI Tract

The GI tract as a route of exposure is relevant in occupational, consumer, and environmental scenarios. A primary method of GI exposure is via the mucociliary escalator when inhaled nanoparticles are cleared from the respiratory tract and then swallowed [8]. Exposure may also occur after topically applied or aggregated airborne particles settle onto surfaces, including skin, and are subsequently ingested via hand-to-mouth or object-to-mouth transmission [8]. Ingestion of contaminated food and water is also pertinent. Estimations suggest that each day the average person consumes 10^{12} submicron-sized particles as food additives, like nano-sized TiO_2 and aluminosilicates [121].

As with other systems, uptake and distribution of nanoparticles within the GI tract is dependent upon size and various surface characteristics, with increased absorption rates of smaller, hydrophobic, and neutral particles [122]. Uptake following oral exposure occurs primarily by Peyer's patches, with subsequent movement into mesenteric lymphatics, then systemically to blood, liver, spleen, bone marrow, and kidneys [123].

22.19 Cardiovascular System

There is epidemiological evidence of cardiovascular effects following exposure to ambient airborne nanoparticles [124, 125]. Translocation of nanoparticles across respiratory epithelium and the subsequent adverse cardiovascular effects are directed by size, surface functionalization, and charge. For example, cationic particles (i.e., gold and polystyrene particles) may cause hemolysis and blood clotting [126]. In vitro studies of pulmonary exposure to carbon nanotubes have demonstrated platelet aggregation effects, and in vivo studies showed increased rates of thrombosis [127]. The potential health risks of these nanoparticle effects have not been elicited. However, taken altogether, they could in theory place individuals at higher risk for vascular diseases such as coronary artery disease or cerebrovascular accidents (Table 22.5).

22.20 Neurologic System

Concern regarding nanotoxic effects in the CNS stems from the hypothesis that nanoparticle exposure can cause neurodegenerative disease. Naturally occurring nano-sized biogenic magnetite can be found in bacteria, protozoa, and animals (even human brains), and has been associated with neurodegenerative processes, including Alzheimer's disease [1]. A number of studies have demonstrated the ability of nanoparticles to cross the blood–brain barrier. Adverse neuronal effects have been demonstrated by animal studies following exposure to ambient airborne nanoparticle pollutants [128]. These effects include increased levels of TNF-α, decreases in dopaminergic neurons, and inflammatory and neurodegenerative changes in olfactory, cortical, and subcortical structures [129–131].

In vitro exposure to nano-size TiO_2 has been shown to stimulate the oxidative burst reaction in brain microglia, and to interfere with one or more of the mitochondrial electron transport chain complexes, with resulting formation of ROS [104]. These effects were not shown to be toxic to the microglia, likely due to functioning endogenous antioxidant mechanisms such as the reduction of superoxide to hydrogen peroxide (H_2O_2) by superoxide dismutase [104, 132, 133]. However, the H_2O_2 and any remaining superoxide can potentially diffuse out of the brain microglial cells [134, 135], resulting in oxidative stress and neurodegenerative changes similar to those seen in Alzheimer's disease, amyotropic lateral sclerosis, and Parkinson's disease [132, 136, 137].

When largemouth bass were exposed to uncoated, water-soluble colloidal fullerenes, nonfatal lipid peroxidation was seen in the brain, as was glutathione depletion in the gill. Hypothesized mechanisms include (a) hematogenous or axonal fullerene translocation to the brain with damage caused by the subsequent redox activity, (b) oxyradicals from microglial cells, and (c) reactive fullerene metabolites [1]. Further investigation, however, suggests that the oxidative toxicity was primarily due to residual THF, which was used to solubilize the fullerene nanoparticles [138, 139].

Proposed mechanisms of translocation following inhalation exposure stem from knowledge of the movement of the meningitis (i.e., enteroviruses) and herpes viruses through olfactory and trigeminal neurons, respectively [74, 75]. These mechanisms are (a) transynaptic

Table 22.5 System-specific exposure and the potential toxicologic consequences [1, 16, 41–43, 93, 104, 123, 126, 127]

Exposure and potential toxic effects[a]	
Dermal effects	Oxidative stress
	Free radical formation
	Lipid peroxidation
	Collagen depletion
	Dermal atrophy
	Increased keratinization
	Glutathione depletion
	Genotoxicity
	Cytotoxicity
	Systemic distribution
Respiratory effects	Inflammation
	Exacerbation of chronic lung disease
	Genotoxicity and carcinogenicity
	Systemic distribution
Gastrointestinal effects	Exposure followed by uptake via Peyer's patches then spread to mesenteric lymphatics
	Systemic distribution
Cardiovascular effects	Hemolysis
	Platelet aggregation
	Thrombosis
Neurological effects	Generation of reactive oxygen species
	Neurodegenerative changes

[a]As elicited by present day research

transport through sensory nerve endings of olfactory and tracheobronchial epithelium, which has been demonstrated by murine studies, and (b) receptor-mediated endocytosis [84, 140]. Initial uptake may then be followed by retrograde transport of nanoparticles through axons and dendrites to access deeper brain structures including the frontal cortex, striatum, and cerebellum [141–143]. Of note, it has been suggested by some that at least some of this deposition is a result of circulatory transport [144].

22.21 Carcinogenicity

The tendency for some topically applied nanoparticles to preferentially accumulate in the hair follicle, particularly the bulge region and hair matrix, raises concern for carcinogenicity due to the considerable representation of stem cells in these locations [61, 62]. Experiments with mice have shown that topical gene delivery can induce gene expression, a desired effect regarding conditions such as alopecia, but with potentially dangerous carcinogenic effects as well [145].

Human epidermal cells exposed to zinc oxide nanoparticles demonstrate dose-dependent genotoxic effects with significant levels of DNA damage. Additional effects following ZnO exposure include enhanced lipid peroxidation, glutathione depletion, and decreased catalase and superoxide dismutase activity [146].

PEG-coated and carboxylic acid–coated quantum dots exhibit nuclear localization. Other quantum dots exhibit localization within the cytoplasm, which may also provide access to the nucleus [30]. It is currently unknown whether intranuclear localization of quantum dots is associated with genotoxic effects, but quantum dots coated with biotin have been shown to nick supercoiled DNA in vitro [147]. It is reasonable to hypothesize that these localization characteristics of quantum dots may apply to other engineered nanoparticles as well [30].

The results of murine studies have earned TiO_2 a group 2B classification (possible human carcinogen) by the International Agency for Research on Cancer, though the carcinogenic risk of TiO_2 or nano-sized TiO_2 in humans has not been exhibited as of yet [148–151]. Nano-sized TiO_2 increases lipid peroxidation, and production of ROS such as superoxide, hydrogen peroxide, hydroxyl radicals, and nitric oxide [107, 152–154]. Discordance amongst studies exists, surrounding the capability of nano-sized TiO_2 to induce DNA damage and genotoxicity [107, 152, 155]. These inconsistencies may be explained by the use of different cell lines, concentrations of nanomaterials, and/or exposure times. Different cell lines have varying thresholds for DNA repair and upregulation of antioxidant or other adaptive mechanisms. They likely become overwhelmed at differing concentrations and exposure times, which would account for the variations in observed DNA damage and genotoxicity [108].

Murine studies show an increased incidence of lung tumors following chronic inhalation of TiO_2 nanoparticles as well as tumors subsequent to lower exposure concentrations of nano-sized TiO_2 vs. micron-sized TiO_2 [156, 157]. However, epidemiological studies of workers involved with the manufacture of TiO_2 have demonstrated no increased risk of lung cancer and no risk of any other life-threatening complications. This inconsistency is most likely due to the previously mentioned lung overload that is characteristic of many animal studies [156].

22.22 Exacerbating Factors

As noted previously, due to the wide variety of nanoparticle applications, the entire population is at risk for exposure, though the risk posed to some populations may be considerably greater. These populations include those with a disrupted skin barrier and those with significantly increased exposure. Healthy, intact skin serves as an effective barrier to certain nano-sized materials. Damaged skin, however, is associated with increased nanoparticle absorption due to malfunction of the skin's protective barrier function [8]. Barrier malfunction is seen with conditions such as irritant, contact, or atopic dermatitis; insults including abrasions and UV-induced damage; or physical stress such as flexion (i.e., wrist movements) [15, 158, 159]. In fact, when enhanced permeation is desirable, the skin barrier may be intentionally disrupted, resulting in translocation to viable skin and interaction with appropriate target cell populations [160, 161]. The potential for enhanced toxicity in these patients is especially pertinent due to development of nanoparticle barrier creams targeted toward improving the disrupted skin barrier [4].

Studies have demonstrated that nanoparticles cannot penetrate beyond the upper stratum corneum of intact skin, and therefore cannot access viable epidermis [6, 59, 60]. UV can weaken the barrier function of the skin, possibly by altering expression of tight junction proteins, which has been demonstrated after exposure to UVB [162, 163]. In fact, following exposure to UV light, quantum dot nanoparticles have been shown to penetrate viable epidermis and dermis where the resultant photochemical reactions could alter cellular DNA and DNA functions [96, 97, 164]. Surface modifications, such as coating, enhance particle stability, halting the photochemical reactions, at least in the short term [91] (Table 22.6).

Perhaps the greatest threat of nanotoxicity is to the workers involved in the research and manufacture of nanoparticles due to the clearly increased incidence of exposure. Workplace measurements have demonstrated only low concentrations of ambient nanoparticles. However, actual particle number can be very high even at very low concentrations with 10 $\mu g/m^3$ of 20-nm particles amounting to greater than 1×10^6 particles/cm^3. As the risk has not yet been definitively determined, it is recommended that those at greatest risk for

Table 22.6 Populations at increased risk for nanoparticle exposure and subsequent chronic effects [8, 15, 158, 159, 162, 163]

Sub-populations at increased risk	
Persons associated with nanoparticle production	Workers involved in research, manufacture, and distribution of nanomaterials
	Due to (1) increased exposure risk and (2) exposure to raw, unmodified nanomaterials
	Effects may not manifest for decades following exposure
Compromised skin barrier	Irritant, contact, or atopic dermatitis
	Skin abrasions
	UV light–induced damage[a]
	Physical stress (i.e., flexed skin)
Inadequacy or exhaustion of defense mechanisms	Overwhelming exposure resulting in exhaustion of antioxidant defense mechanisms
	Cells/tissues deficient in endogenous defense mechanisms

[a]UVB exposed skin exhibits altered expression of tight junction proteins

exposure maintain caution by the use of proper containment and ventilation techniques, as well as adequate personal protective equipment [8].

22.23 Adaptation

Studies suggest that in the absence of exacerbating factors, such as UV radiation, nano-sized TiO_2, carbon black, and carboxylated polystyrene do not result in a significant amount of oxidative stress whereas polystyrene and incidental nanoparticles (i.e., those found in pollution) do elicit oxidative stress reactions [107]. After translocation to deep within the body, the likelihood of photoactivation is minimal [108]. However, nano-sized TiO_2 has been shown to produce detectable levels of superoxide independent of photoactivation [152, 165]. The lack of cytotoxic oxidative stress may be explained by the variety of cellular mechanisms that cells possess for detoxification of superoxide and other ROS [104].

Upregulation of detoxification mechanisms, such as production of superoxide dismutase and peroxidase, is likely to be at least partially responsible for the lower levels of ROS seen in CHO cells chronically exposed (60 days) to nano-sized TiO_2 as compared to cells acutely exposed (2 days). Other adaptive mechanisms including extrusion or uptake inhibition are likely contributory as well [108]. This adaptive ability does not eliminate concerns, however. Potential adverse effects may occur when cellular adaptive mechanisms become exhausted, or when cells deficient in such mechanisms encounter additional insult [108].

22.24 Ecotoxicologic Effects

Some nanomaterials are manufactured by the ton. With these, and even other nanoparticles produced on a smaller scale, some amount of material is likely to reach the environment during the manufacture, distribution, and/or usage of the final product. For example, nanoparticles found in sunscreens and cosmetics are released into the environment when washed off of the wearer. Nanoparticles in these and many other products such as tires and electronics may also be leached into the environment following disposal [1, 166].

Due to silver nanoparticle utilization as a water disinfectant, the ecotoxicological effects are of great concern. Studies indicate that silver nanoparticles cause significantly greater toxicity in certain fish and algae species than do silver ions of equal mass [167, 168]. The tendency of silver nanoparticles to bioaccumulate in aquatic organisms is a likely contributor to toxicity [169]. Adverse and potentially teratogenic effects of silver nanoparticles in fish include oxidative stress, cardiac arrythmias, and spinal cord deformities [170–173].

Biodegradable polymer-based particles, including poly-lactic acid (PLA) and poly glycolic-*co*-lactic acid (PGLA), are currently being studied. The use of biodegradable nanomaterials is thought to increase safety and may help alleviate ecotoxicologic concerns [174].

22.25 Conclusion

The current data regarding the toxicology of nanoparticles is incomplete and, at times, conflicting. Potential toxic effects are variable, but are dictated by specific particle physiochemical characteristics including particle size, shape, and surface coatings. Toxicity is influenced by particle distribution, agglomeration, catalytic activity, solubility, contaminants, exposure time, and route of exposure. The acute effects following inhalation, ingestion, and dermal exposure predominantly consist of inflammatory reactions. Tissue damage appears to be mediated by the formation of ROS and subsequent oxidative stress.

The cytotoxic, genotoxic, and carcinogenic potential of nanoparticles is less apparent. In vitro studies have demonstrated these effects following high doses and prolonged exposure to various nanoparticles, but the data from many in vivo studies do not support these findings as of yet. The disparity is mostly attributable to the unrealistic exposure conditions utilized in many in vitro studies. It is possible to make future in vitro studies more realistic and thus more generalizable to humans; however, this may require extended lengths of exposure time since disease processes involving particle exposure often manifest decades later (i.e., asbestos and mesothelioma). Long-term effects will most likely be observed in workers currently involved in the manufacturing and processing of raw nanomaterials.

Leaching of nanoparticles during manufacture, distribution, and following disposal has the potential for causing significant ecotoxicologic effects. These effects will need to be closely monitored and evaluated as production and use of nanomaterials continues to rise. It is conceivable that due to their extremely small size, nanoparticles may require innovative storage and disposal containers and protocols in order to minimize environmental contamination.

The potential biomedical, industrial, and consumer applications of nanomaterials makes nanotechnology an exciting and ever-growing field of study, which will have a significant impact on the general public and the field of medicine. As with most medical interventions and many consumer products, the benefits of nanoparticles are not without risks, and the decision regarding nanoparticle use should be guided by a comprehensive cost-benefit analysis. Such decisions will be made possible when the acute and chronic toxic effects of nanoparticles are more fully elucidated.

References

1. Oberdörster G, Oberdörster E, Oberdörster J. Nanotoxicology: an emerging discipline evolving from studies of ultrafine particles. Environ Health Perspect. 2005;113:823–39.
2. Eastern Research Group. State of the science literature review: nano titanium dioxide environmental matters. Washington: United States Environmental Protection Agency; 2010. p. 5–7.
3. Thomas T, Thomas K, Sadrieh N, Savage N, Adair P, Bronaugh R. Research strategies for safety evaluation of nanomaterials, part VII: evaluating consumer exposure to nanoscale materials. Toxicol Sci. 2006;91:14–9.
4. Papakostas D, Rancan F, Sterry W, Blume-Peytavi U, Vogt A. Nanoparticles in dermatology. Arch Dermatol Res. 2011;303:533–50.
5. Wiesenthal A, Hunter L, Wang S, Wickliffe J, Wilkerson MG. Nanoparticles: small and mighty. Int J Dermatol. 2011;50:247–54.
6. Newman MD, Stotland M, Ellis JI. The safety of nanosized particles in titanium dioxide- and zinc oxide-based sunscreens. J Am Acad Dermatol. 2009;61:685–92.
7. Magnusson BM, Walters KA, Roberts MS. Veterinary drug delivery: potential for skin penetration enhancement. Adv Drug Deliv Rev. 2001;50:205–27.
8. Stern ST, McNeil SE. Nanotechnology safety concerns revisited. Toxicol Sci. 2008;101:4–21.
9. Warheit DB, Webb TR, Sayes CM, Colvin VL, Reed KL. Pulmonary instillation studies with nanoscale TiO2 rods and dots in rats: toxicity is not dependent upon particle size and surface area. Toxicol Sci. 2006;91:227–36.

10. Brown DM, Stone V, Findlay P, MacNee W, Donaldson K. Increased inflammation and intracellular calcium caused by ultrafine carbon black is independent of transition metals or other soluble components. Occup Environ Med. 2000;57:685–91.
11. Brown DM, Wilson MR, MacNee W, Stone V, Donaldson K. Size-dependent proinflammatory effects of ultrafine polystyrene particles: a role for surface area and oxidative stress in the enhanced activity of ultrafines. Toxicol Appl Pharmacol. 2001;175:191–9.
12. Li N, Sioutas C, Cho A, Schmitz D, Misra C, Sempf J, et al. Ultrafine particulate pollutants induce oxidative stress and mitochondrial damage. Environ Health Perspect. 2003;111:455–60.
13. Donaldson K, Brown D, Clouter A, et al. The pulmonary toxicology of ultrafine particles. J Aerosol Med. 2002;15:213–20.
14. Donaldson K, Stone V. Current hypotheses on the mechanisms of toxicity of ultrafine particles. Ann Ist Super Sanita. 2003;39:405–10.
15. Choksi AN, Poonawalla T, Wilkerson MG. Nanoparticles: a closer look at their dermal effects. J Drugs Dermatol. 2010;9(5):475–81.
16. Borm PJ, Robbings D, Haubold S, et al. The potential risks of nanomaterials: a review carried out for ECETOC. Part Fibre Toxicol. 2006;3:11.
17. Hallock MF, Greenley P, DiBerardinis L, Kallin D. Potential risks of nanomaterials and how to safely handle materials of uncertain toxicity. J Chem Health Safety. 2009;16:16–23.
18. Kimbrell G. Nanotechnology and nanomaterials in consumer products: regulatory challenges and necessary amendments. FDA Public Meeting on Nanotechnology. Available at http://Nano_FDA_Public%20Meeting_10_06.pdf. Accessed 20 Dec 2011.
19. Amato I. Making the right stuff. Sci News. 1989;136: 108–10.
20. Donaldson K, Tran CL. An introduction to the short-term toxicology of respirable industrial fibers. Mutat Res. 2004;553:5–9.
21. Greim H, Borm P, Schins R, et al. Toxicity of fibers and particles—report of the workshop held in Munich, Germany, 26–27 October 2000. Inhal Toxicol. 2001;13:737–54.
22. IARC (International Agency for Research on Cancer). Manmade vitreous fibres. IARC Monogr Eval Carcinog Risks Hum. 2002;81:1–418.
23. Donaldson K, Aitken R, Tran L, et al. Carbon nanotubes: a review of their properties in relation to pulmonary toxicology and workplace safety. Toxicol Sci. 2006;92:5–22.
24. Royal Society and Royal Academy of Engineering. Nanoscience and nanotechnologies: opportunities and uncertainties. London: Royal Society; 2004. p. 35–49.
25. Poland CA, Duffin R, Kinloch I, et al. Carbon nanotubes introduced into the abdominal cavity of mice show asbestos-like pathogenicity in a pilot study. Nat Nanotechnol. 2008;3(7):423–8.
26. Kamp DW, Graceffa P, Pryor WA, Weitzman SA. The role of free radicals in asbestos-induced diseases. Free Radic Biol Med. 1992;12:293–315.
27. Ye J, Shi X, Jones W. Critical role of glass fiber length in TNF-alpha production and transcription factor activation in macrophages. Am J Physiol. 1999;276:426–34.
28. Mossman BT, Churg A. Mechanisms in the pathogenesis of asbestosis and silicosis. Am J Respir Crit Care Med. 1998;157:1666–80.
29. IARC (International Agency for Research on Cancer). Mechanisms of mineral fibre carcinogenesis. IARC Monogr Eval Carcinog Risks Hum. 1999;140:11–34.
30. Ryman-Rasmussen JP, Riviere JE, Monteiro-Riviere NA. Surface coatings determine cytotoxicity and irritation potential of quantum dot nanoparticles in epidermal keratinocytes. J Invest Dermatol. 2007; 1:143–53.
31. Warheit DB, Laurence GR, Reed KL, Roach DH, Reynolds GAM, Webb TR. Comparative pulmonary toxicity assessment of single-walled carbon nanotubes in rats. Toxicol Sci. 2004;77:117–25.
32. Lyon DY, Fortner JD, Sayes CM, Colvin VL, Hughe JB. Bacterial cell association and antimicrobial activity of a C60 water suspension. Environ Toxicol Chem. 2005;24:2757–62.
33. Lovern SB, Klaper R. Daphnia magna mortality when exposed to titanium dioxide and fullerene (C60) nanoparticles. Environ Toxicol Chem. 2006;5:1132–7.
34. de Vuyst P, Camus P. The past and present of pneumocondioses. Curr Opin Pulm Med. 2000;6:151–6.
35. Stremmel W, Meyerrose KW, Niederau C, et al. Wilson disease: clinical presentation, treatment, and survival. Ann Intern Med. 1991;115:720–6.
36. Niederau C, Fischer R, Sonnenberg A, et al. Survival and causes of death in cirrhotic and in noncirrhotic patients with primary hemochromatosis. N Engl J Med. 1985;313:1256–62.
37. Blessed G, Tomlinson BE, Roth M. The association between quantitative measures of dementia and of senile change in the cerebral grey matter of elderly subjects. Br J Psychiatry. 1968;114:797–811.
38. Wiginton CD, Kelly B, Oto A, et al. Gadolinium-based contrast exposure, nephrogenic systemic fibrosis, and gadolinium detection in tissue. Am J Roentgenol. 2008;190:1060–8.
39. Gorelik J, Shevchuk A, Ramalho M, et al. Scanning surface confocal microscopy for simultaneous topographical and fluorescence aging: application to single virus-like particle entry into a cell. Proc Natl Acad Sci U S A. 2002;99:16018–23.
40. Skebo JE, Brabinski CM, Schrand AM, Schlager JJ, Hussain SM. Assessment of metal nanoparticle agglomeration, uptake, and interaction using high-illuminating system. Int J Toxicol. 2007;26: 135–41.
41. Manna SK, Sarkar S, Barr J, et al. Single-walled carbon nanotube induces oxidative stress and activates

nuclear transcription factor-kB in human keratinocytes. Nano Lett. 2005;5:1676–84.
42. Sayes CM, Gobin AM, Ausman KD, Mendez J, West JL, Colvin VL. Influence of length on cytotoxicity of multi-walled carbon nanotubes against human acute monocytic leukemia cell line THP-1 in vitro and subcutaneous tissue of rats in vivo. Mol Biosyst. 2005;1:176–82.
43. Shvedova AA, Castranova V, Kisin ER, et al. Exposure to carbon nanotube material: assessment of nanotube cytotoxicity using human keratinocyte cells. J Toxicol Environ Health A. 2003;66:1909–26.
44. SCCNFP. Opinion of the scientific committee on cosmetic products and non-food products intended for consumers concerning titanium dioxide. European Commission, Brussels, Belgium. Available at http://ec.europa.eu/health/ph_risk/committees/sccp/sccp_opinions_en.htm. Accessed 24 Oct 2000
45. SCCNFP. Opinion of the scientific committee on cosmetic products and non-food products intended for consumers concerning zinc oxide. European Commission, Brussels, Belgium. Available at http://ec.europa.eu/health/ph_risk/committees/sccp/sccp_opinions_en.htm. Accessed 24 June 2003.
46. Sato Y, Yokoyama A, Shibata K, et al. Influence of length on cytotoxicity of multi-walled carbon nanotubes against human acute monocytic leukemia cell line THP-1 in vitro and subcutaneous tissue of rats in vivo. Mol Biosyst. 2005;1:176–82.
47. Rosenman KD, Moss A, Kon S. Argyria: clinical implications of exposure to silver nitrate and silver oxide. J Occup Environ Med. 1979;21:430–5.
48. Hill WR. Argyria: the pharmacology of silver. South Med J. 1941;34:340.
49. Sawosz E, Binek M, Grodzik M, et al. Influence of hydrocolloidal silver nanoparticles on gastrointestinal microflora and morphology of enterocytes of quails. Arch Anim Nutr. 2007;61:444–51.
50. Wakelyn PJ. Cotton yarn manufacturing. In: Ivester AL, Neefus JD, editors. ILO encyclopedia of occupational health and safety. 4th ed. Geneva, Switzerland: International Labour Office; 1994. p. 89.9–89.11.
51. Blume-Peytavi U, Massoudy L, Patzelt A, et al. Follicular and percutaneous penetration pathways of topically applied minoxidil foam. Eur J Pharm Biopharm. 2010;76:450–3.
52. Knorr F, Lademann J, Patzelt A, Sterry W, Blume-Peytavi U, Vogt A. Follicular transport route-research progress and future perspectives. Eur J Pharm Biopharm. 2009;71:173–80.
53. Lademann J, Otberg N, Richter H, et al. Investigation of follicular penetration of topically applied substances. Skin Pharmacol Physiol. 2001;21:274–82.
54. Meidan VM. Methods for quantifying intrafollicular drug delivery: a critical appraisal. Expert Opin Drug Deliv. 2010;7:1095–108.
55. Michel M, L'Heureux N, Pouliot R, Xu W, Auger FA, Germain L. Characterization of a new tissue-engineered human skin equivalent with hair. In Vitro Cell Dev Biol Anim. 1999;35:318–26.
56. Otberg N, Richter H, Schaefer H, Blume-Peytavi U, Sterry W, Lademann J. Variations of hair follicle size and distribution in different body sites. J Invest Dermatol. 2004;122:14–9.
57. Schaefer H, Lademann J. The role of follicular penetration. A differential view. Skin Pharmacol Appl Skin Physiol. 2001;14:23–7.
58. Tenjarla S. Microemulsions: an overview and pharmaceutical applications. Crit Rev Ther Drug Carrier Syst. 1999;16:461–521.
59. Cross SE, Innes B, Roberts MS, Tsuzuki T, Robertson TA, McCormick P. Human skin penetration of sunscreen nanoparticles: invitro assessment of a novel micronized zinc oxide formulation. Skin Pharmacol Physiol. 2007;20:148–54.
60. Lademann J, Weigmann H, Rickmeyer C, Barthelmes H, Schaefer H, Mueller G, et al. Penetration of titanium dioxide microparticles in a sunscreen formulation into the horny layer and the follicular orifice. Skin Pharmacol Appl Skin Physiol. 1999;12:247–56.
61. Meidan VM, Bonner MC, Michniak BB. Transfollicular drug delivery—is it a reality? Int J Pharm. 2005;306:1–14.
62. Vogt A, Mandt N, Lademann J, Schaefer H, Blume-Peytavi U. Follicular targeting—a promising tool in selective dermatotherapy. J Investig Dermatol Symp Proc. 2005;10:252–5.
63. Lademann J, Otberg N, Jacobi U, Hoffman RM, Blume-Peytavi U. Follicular penetration and targeting. J Investig Dermatol Symp Proc. 2005;10:301–3.
64. Schaefer-Korting M, Mehnert W, Korting HC. Lipid nanoparticles for improved topical application of drugs for skin diseases. Adv Drug Deliv Rev. 2007; 59:427–43.
65. Alvarez-Roman R, Naik A, Kalia YN, Guy RH, Fessi H. Skin penetration and distribution of polymeric nanoparticles. J Control Release. 2004;99: 53–62.
66. Ryman-Rasmussen JP, Riviere JE, Monteiro-Riviere NA. Penetration of intact skin by quantum dots with diverse physiochemical properties. Toxicol Sci. 2006;91:159–65.
67. Baroli B, Ennas MG, Loffredo F, Isola M, Pinna R, Lopez-Quintela MA. Penetration of metallic nanoparticles in human full-thickness skin. J Invest Dermatol. 2007;127:1701–12.
68. Corachan M, Tura JM, Campo E, Soley M, Traveria A. Poedoconiosis in Aequatorial Guinea. Report of two cases from different geological environments. Trop Geogr Med. 1988;40:359–64.
69. Blundell G, Henderson WJ, Price EW. Soil particles in the tissues of the foot in endemic elephantiasis of the lower legs. Ann Trop Med Parasitol. 1989; 83(4):381–5.
70. Kim S, Lim YS, Soltesz EG, et al. Near infrared fluorescent type II quantum dots for sentinel lymph node mapping. Nat Biotechnol. 2004;22:93–7.
71. Ohl L, Mohaupt M, Czeloth N, Hintzen G, Kiafard Z, Zwirner J, et al. CCR7 governs skin dendritic cell

migration under inflammatory and steady-state conditions. Immunity. 2004;21:279–88.
72. Sato K, Imai Y, Irimura RT. Contribution of dermal macrophage trafficking in the sensitization phase of contact hypersensitivity. J Immunol. 1998;161: 6835–44.
73. Oberdörster G, Morrow PE, Spurny K. Size dependent lymphatic short term clearance of amosite fibers in the lung. Ann Occup Hyg. 1988;32:149–56.
74. Kennedy P, Chaudhuri A. Herpes simplex encephalitis. J Neurol Neurosurg Psychiatry. 2002;73:237–8.
75. Terasaki S, Kameyama T, Yamamoto S. A case of zoster in the 2nd and 3rd branches of the trigeminal nerve associated with simultaneous herpes labialis infection—a case report. Kurume Med J. 1997;44:61–6.
76. Arvidson B. A review of axonal transport of metals. Toxicology. 1994;88:1–14.
77. Malmgren L, Olsson Y, Olsson T, Kristensson K. Uptake and retrograde axonal transport of various exogenous macromolecules in normal and crushed hypoglossal nerves. Brain Res. 1978;153:477–93.
78. Olsson T, Kristensson K. Neuronal uptake of iron: somatopetal axonal transport and fate of cationized and native ferretin, and iron-dextran after intramuscular injections. Neuropathol Appl Neurobiol. 1981; 7:87–95.
79. Oldfors A, Fardeau M. The permeability of the basal lamina at the neuromuscular junction. An ultrastructural study of rat skeletal muscle using particulate tracers. Neuropathol Appl Neurobiol. 1983;9: 419–32.
80. Jung S, Otberg N, Thiede G, et al. Innovative liposomes as a transfollicular drug delivery system: penetration into porcine hair follicles. J Invest Dermatol. 2006;126:1728–32.
81. Mecke A, Uppuluri S, Sassanella TM, Lee DK, Ramamoorthy A, Baker Jr JR, et al. Direct observation of lipid bilayer disruption by poly(amidoamine) dendrimers. Chem Phys Lipids. 2004;132:3–14.
82. Gumbleton M. Caveolae as potential macromolecule trafficking compartments within alveolar epithelium. Adv Drug Deliv Rev. 2001;49:281–300.
83. Kreuter J. Nanoparticulate systems for brain delivery of drugs. Adv Drug Deliv Rev. 2001;47:65–81.
84. Kreuter J. Influence of the surface properties on nanoparticle-mediated transport of drugs to the brain. J Nanosci Nanotechnol. 2004;4:484–8.
85. Kreuter J, Shamenkov D, Petrov V, et al. Apolipoprotein-mediated transport of nanoparticle-bound drugs across the blood–brain barrier. J Drug Target. 2002;10:317–25.
86. Kato T, Yashiiro T, Murata Y, et al. Evidence that exogenous substances can be phagocytized by alveolar epithelial cells and transported into blood capillaries. Cell Tissue Res. 2003;311:47–51.
87. Oberdörster G, Ferin J, Gelein R, Soderholm SC, Finkelstein J. Role of the alveolar macrophage in lung injury: studies with ultrafine particles. Environ Health Perspect. 1992;97:193–9.
88. Nikula KJ, Avila KJ, Griffith WC, Mauderly JL. Lung tissue responses and sites of particle retention differ between rats and cynomolgus monkeys exposed chronically to diesel exhaust and coal dust. Fundam Appl Toxicol. 1997;37:37–53.
89. Gonzalez S, Fernandez-Lorente M, Gilaberte-Calzada Y. The latest on skin photoprotection. Clin Dermatol. 2008;26:614–26.
90. Nakagawa Y, Wakuri S, Sakamoto K, Tanaka N. The photogenotoxicity of titanium dioxide particles. Mutat Res. 1997;394:125–32.
91. Allen NS, Edge M, Sandoval G, Verran J, Stratton J, Maltby J. Photocatalytic coatings for environmental applications. Photochem Photobiol. 2005;81:279–90.
92. Tsuang YH, Sun JS, Huang YC, Lu CH, Chang WH, Wang CC. Studies of photokilling of bacteria using titanium dioxide nanoparticles. Artif Organs. 2008;32:167–74.
93. Wu J, Liu W, Xue C, et al. Toxicity and penetration of TiO_2 nanoparticles in hairless mice and porcine skin after subchronic dermal exposure. Toxicol Lett. 2009;191:1–8.
94. Dussert AS, Gooris E. Characterization of the mineral content of a physical sunscreen emulsion and its distribution onto human stratum corneum. Int J Cosmet Sci. 1997;19:119–29.
95. Uchino T, Tokunaga H, Ando M, et al. Quantitative determination of OH radical generation and its cytotoxicity induced by TiO_2-UVA treatment. Toxicol In Vitro. 2002;16:629–35.
96. Van RI. Beyond skin feel: innovative methods for developing complex sensory profiles with silicones. J Cosmet Dermatol. 2006;5:61–7.
97. Wamer WG, Yin JJ, Wei RR. Oxidative damage to nucleic acids photosensitized by titanium dioxide. Free Radic Biol Med. 1997;23:851–8.
98. Kanchan V, Panda AK. Interactions of antigen-loaded polylactide particles with macrophages and their correlation with the immune response. Biomaterials. 2007;28:5344–57.
99. Mohamed F, van der Walle CF. PLGA microcapsules with novel dimpled surfaces for pulmonary delivery of DNA. Int J Pharm. 2006;311:97–107.
100. O'Reilly RK, Hawker CJ, Wooley KL. Cross-linked block copolymer micelles: functional nanostructures of great potential and versatility. Chem Soc Rev. 2006;35:1068–83.
101. Veronesi B, Oortgiesen M. Neurogenic inflammation and particulate matter (PM) air pollutants. Neurotoxicology. 2001;22:795–810.
102. Pooler M, Makwana O, Carter J, Beck-Speier I, Kreyling W, Veronesi B. Electrostatic charge on nanoparticles activates CNS macrophages (microglia). The Toxicologist CD- An Official Journal of the Society of Toxicology. 2005,;84..
103. Ullrich SE. Dermal application of JP-8 jet fuel induces immune suppression. Toxicol Sci. 1999;52:61–7.
104. Long T, Saleh N, Tilton R, et al. Titanium dioxide (P25) produced reactive oxygen species in immortalized brain microglia (BV2): implications for

nanoparticle neurotoxicity. Environ Sci Technol. 2006;40:4346–52.
105. Renwick LC, Brown D, Clouter A, Donaldson K. Increased inflammation and altered macrophage chemotactic responses caused by two ultrafine particle types. Occup Environ Med. 2004;61:422–47.
106. Veronesi B, de Haar C, Lee L, Oortgiesen M. The surface charge of visible particulate matter predicts biological activation in human bronchial epithelial cells (BEAS-2B). Toxicol Appl Pharmacol. 2002;178:144–54.
107. Xia T, Kovochich M, Brant J, et al. Comparison of the abilities of ambient and manufactured nanoparticles to induce cellular toxicity according to an oxidative stress paradigm. Nano Lett. 2006;6:1794–807.
108. Wang S, Hunter L, Arslan Z, Wilkerson M, Wickliffe J. Chronic exposure to nanosized, anatase titanium dioxide is not cyto- or genotoxic to Chinese hamster ovary cells. Environ Mol Mutagen. 2011;52:614–22.
109. Preining O. The physical nature of very, very small particles and its impact on their behavior. J Aerosol Sci. 1998;29:481–95.
110. Nel A, Xia T, Li N. Toxic potential of materials at the nanolevel. Science. 2006;311:622–7.
111. Ackerman MA, Chan WCW, Laakkonen P, Bhatia SN, Ruoslahti E. Nanocrystal targeting in vivo. Proc Natl Acad Sci U S A. 2002;99:12617–21.
112. Ogawara K, Furumoto K, Takakura Y, Hashida M, Higaki K, Kimura T. Surface hydrophobicity of particles is not necessarily the most important determinant in their in vivo disposition after intravenous administration in rats. J Control Release. 2001;77:191–8.
113. Monteiro-Riviere NA, Inman AO, Wang YY, Nemanich RJ. Surfactant effects on carbon nanotube interactions with human keratinocytes. Nanomedicine. 2005;1:293–9.
114. Donaldson K, Stone V, Clouter A, Renwick L, MacNee W. Ultrafine particles. Occup Environ Med. 2001;58:211–6.
115. Oberdörster G, Finkelstein JN, Johnston C, Gelein R, Cox C, Baggs R, et al. Acute pulmonary effects of ultrafine particles in rats and mice. Res Rep Health Eff Inst. 2000;96:5–74.
116. Donaldson K, Li XY, MacNee W. Ultrafine (nanometre) particle mediated lung injury. J Aerosol Sci. 1998;29:553–60.
117. Driscoll KE. Role of inflammation in the development of rat lung tumors in response to chronic particle exposure. Inhal Toxicol. 1996;8:139–53.
118. Oberdörster G, Yu CP. The carcinogenic potential of inhaled diesel exhaust: a particle effect? J Aerosol Sci. 1990;21:S397–401.
119. Tran CL, Buchanan D, Cullen RT, Searl A, Jones AD, Donaldson K. Inhalation of poorly soluble particles. II. Influence of particle surface area on inflammation and clearance. Inhal Toxicol. 2000;12:1113–26.
120. Tran CL, Jones AD, Cullen RT, Donaldson K. Influence of particle characteristics on the clearance of low toxicity dusts from lungs. J Aerosol Sci. 1998;29:S1269–70.
121. Lomer MCE, Thompson RPH, Powel JJ. Fine and ultrafine particles of the diet: influence on the mucosal immune response and association with Chron's disease. Proc Nutr Soc. 2002;61:123–30.
122. Hussain N, Jaitley V, Florence AT. Recent advances in the understanding of uptake of microparticulates across the gastrointestinal lymphatics. Adv Drug Deliv Rev. 2001;50:107–42.
123. Jani P, Halbert GW, Langridge J, Florence AT. Nanoparticle uptake by the rat gastrointestinal mucosa: quantitation and particle size dependency. J Pharm Pharmacol. 1990;42:821–6.
124. Pekkanen J, Peters A, Hoek G, Tiittanen P, Brunekreef B, de Hartog J, et al. Particulate air pollution and risk of ST-segment depression during repeated submaximal exercise tests among subjects with coronary heart disease. The Exposure and Risk assessment for Fine and Ultrafine Particles in Ambient Air (ULTRA) study. Circulation. 2002;106:933–8.
125. Wichmann H-E, Spix C, Tuch T, Wolke G, Peters A, Heinrich J, et al. Daily mortality and fine and ultrafine particles in Erfurt, Germany. Part I: role of particle number and particle mass. Res Rep Health Eff Inst. 2000;98:5–86.
126. Kainthan RK, Gnanamani M, Ganguli M, et al. Blood compatibility of novel water soluble hyperbranched polyglycerol-based multivalent cationic polymers and their interaction with DNA. Biomaterials. 2006;27:5377–90.
127. Radomski A, Jurasz P, Alonso-Escolano D, Drews M, Morandi M, Malinski T, et al. Nanoparticle-induced platelet aggregation and vascular thrombosis. Br J Pharmacol. 2005;146:882–93.
128. Peters A, Veronesi B, Calderon-Garciduenas L, Gehr P, Chen LC, Geiser M, et al. Translocation and potential neurological effects of fine and ultrafine particles: a critical update. Part Fibre Toxicol. 2006;3:1–13.
129. Campbell A, Oldham M, Becaria A, et al. Particulate matter in polluted air may increase biomarkers of inflammation in mouse brain. Neurotoxicology. 2005;26:133–40.
130. Calderon-Garciduenas L, Azzarelli B, Acuna H, et al. Air pollution and brain damage. Toxicol Pathol. 2002;30(3):373–89.
131. Hussain SM, Javorina AK, Schrand AM, Duhart HM, Ali SF, Schlager JJ. The interaction of manganese nanoparticles with PC-12 cells induces dopamine depletion. Toxicol Sci. 2006;92:456–63.
132. Block ML, Hong JS. Microglia and inflammation-mediated neurodegeneration: multiple triggers with a common mechanism. Prog Neurobiol. 2005;76:77–98.
133. Emerit J, Edeas M, Bricaire F. neurodegenerative diseases and oxidative stress. Biomed Pharmacother. 2004;58:39–46.
134. De Georgi F, Lartique L, Bauer MK, et al. The permeability transition pore signals apoptosis by directing Bax translocation and multimerization. FASEB J. 2002;16:607–9.
135. Fernandes MA, Santos MS, Vicente JA, et al. Effects of 1,4-dihydropyridine derivatives (cere-

brocrast, gammapyrone, glutapyrone, and diethone) on mitochondrial bioenergetics and oxidative stress: a comparative study. Mitochondrion. 2003; 3:47–59.
136. Colton CA, Chernyshev ON, Gilbert DL, Vitek MP. Microglial contribution to oxidative stress in Alzheimer's disease. Ann N Y Acad Sci. 2000;889:292–307.
137. Liu B, Hong JS. Role of microglia in inflammation-mediated neurodegenerative diseases: mechanisms and strategies for therapeutic intervention. J Pharmacol Exp Ther. 2003;304:1–7.
138. Andrievsky G, Klochkov V, Derevyanchenko L. Is the C60 fullerene molecule toxic. Fuller Nanotub Carbon Nanostruct. 2005;13:363–76.
139. Zhu S, Oberdörster E, Haasch ML. Toxicity of an engineered nanoparticle (fullerene, C60) in two aquatic species, Daphnia and fathead minnow. Mar Environ Res. 2006;62:S5–9.
140. Müller RH, Keck CM. Drug delivery to the brain-realization by novel drug carriers. J Nanosci Nanotechnol. 2004;4:471–83.
141. Bodain D, Howe HA. Experimental studies on intraneural spread of poliomyelitis virus. Bull Johns Hopkins Hosp. 1941;69:248–67.
142. Bodain D, Howe HA. The rate of progression of poliomyelitis virus in nerves. Bull Johns Hopkins Hosp. 1941;69:79–85.
143. Howe HA, Bodain D. Portals of entry of poliomyelitis virus in the chimpanzee. Proc Soc Exp Biol Med. 1940;43:718–21.
144. Elder A, Gelein R, Silva V, et al. Translocation of inhaled ultrafine manganese oxide particles to the central nervous system. Environ Health Perspect. 2006;114:1172–8.
145. Nakamura M, Jo J, Tabata Y, Ishikawa O. Controlled delivery of T-box21 small interfering RNA ameliorates autoimmune alopecia (alopecia areata) in a C3H/HeJ mouse model. Am J Pathol. 2008;172:650–8.
146. Sharma V, Shukla RK, Saxena N, et al. DNA damaging potential of zinc oxide nanoparticles in human epidermal cells. Toxicol Lett. 2009;185:21–218.
147. Green M, Howman E. Semiconductor quantum dots and free radical induced DNA nicking. Chem Commun. 2005;7:121–3.
148. Borm PJ, Schins RP, Albrecht C. Inhaled particles and lung cancer: Part B. Paradigms and risk assessment. Int J Cancer. 2004;110:3–14.
149. Baan R, Straif K, Grosse Y, Secretan B, El Ghissassi F, Cogliano V. Carcinogenicity of carbon black, titanium dioxide, and talc. Lancet Oncol. 2006;7:295–6.
150. Dankovic D, Kuempel E, Wheeler M. An approach to risk assessment for TiO2. Inhal Toxicol. 2007;19:205–12.
151. IARC (International Agency for Research on Cancer). Carbon black, titanium dioxide, talc. IARC Monogr Eval Carcinog Risks Hum. 2010;93:1–466.
152. Gurr J-R, Wang ASS, Chen C-H, Jan K-Y. Ultrafine titanium dioxide particles in the absence of photoactivation can induce oxidative damage to human bronchial epithelial cells. Toxicology. 2005;213:66–73.
153. Kang JL, Moon C, Lee HS, et al. Comparison of the biological activity between ultrafine and fine titanium dioxide particles in RAW 264.7 cells associated with oxidative stress. J Toxicol Environ Health A. 2008;71:478–85.
154. Kang SJ, Kim BM, Lee YJ, Chung HW. Titanium dioxide nanoparticles trigger p53-mediated damage response in peripheral blood lymphocytes. Environ Mol Mutagen. 2008;49:399–405.
155. Falck G, Lindberg H, Suhonen S, et al. Genotoxic effects of nanosized and fine TiO2. Hum Exp Toxicol. 2009;28:339–52.
156. Lee KP, Trochimowicz HJ, Reinhardt CF. Pulmonary response of rats exposed to titanium dioxide (TiO2) by inhalation for two years. Toxicol Appl Pharmacol. 1985;79:179–92.
157. Heinrich U, Fuhst R, Rittinghausen S, et al. Chronic inhalation exposure of Wistar rats and two different strains of mice to diesel exhaust, carbon black, and titanium dioxide. Inhal Toxicol. 1995;7:533–56.
158. Rouse JG, Yang J, Ryman-Rasmussen JP, Barron AR, Monteiro-Riviere NA. Effects of mechanical flexion on the penetration of fullerene amino acid-derivatized peptide nanoparticles through skin. Nano Lett. 2007;7:155–60.
159. Tinkle SS, Antonini JM, Rich BA, Roberts JR, Salmen R, DePree K, et al. Skin as a route of exposure and sensitization in chronic beryllium disease. Environ Health Perspect. 2003;111:1202–8.
160. Vogt A, Combadiere B, Hadam S, et al. 40 nm, but not 750 or 1,500 nm, nanoparticles enter epidermal CD1a+cells after transcutaneous application on human skin. J Invest Dermatol. 2006;126: 1316–22.
161. Vogt A, Mahe B, Costaglioloa D, et al. Transcutaneous anti-influenza vaccination promotes both CD4 and CD8 T cell immune responses in humans. J Immunol. 2008;180:1482–9.
162. Jiang SJ, Chen JY, Lu ZF, Yao J, Che DF, Zhou XJ. Biophysical and morphological changes in the stratum corneum lipids induced by UVB irradiation. J Dermatol Sci. 2006;44:29–36.
163. Yamamoto T, Kurasawa M, Hattori T, Maeda T, Nakano H, Sasaki H. Relationship between expression of tight junction-related molecules and perturbed epidermal barrier function in UVB-irradiated hairless mice. Arch Dermatol Res. 2008;300:61–8.
164. Mortensen LJ, Oberdörster G, Pentland AP, Delouise LA. In vivo skin penetration of quantum dot nanoparticles in the murine model: the effect of UVR. Nano Lett. 2008;8:2779–87.
165. Bhattacharys K, Davoren M, Boertz J, Schins RP, Hoffmann E, Dopp E. Titanium dioxide nanoparticles induce oxidative stress and DNA-adduct formation but not DNA-breakage in human lung cells. Part Fibre Toxicol. 2009;6:17.
166. Daughton C, Ternes T. Pharmaceuticals and personal care products in the environment: agents of subtle change? Environ Health Perspect. 1999; 107:907–38.

167. Fabrega J, Luoma SN, Tyler CR, Galloway TS, Lead JR. Silver nanoparticles: behaviour and effects in the aquatic environment. Environ Int. 2011;37:517–31.
168. Navarro E, Piccapietra F, Wagner B, et al. Toxicity of silver nanoparticles to *Chlamydomonas reinhardtii*. Environ Sci Technol. 2008;42:8959–64.
169. Louma SN, Rainbow PS. Metal contamination in aquatic environments: science and lateral management. Cambridge: Cambridge University Press; 2008.
170. Asharani PV, Wu YL, Gong ZY, Valiyaveettil S. Toxicity of silver nanoparticles in zebrafish models. Nanotechnology. 2008;19:255102.
171. Yeo MK, Pak SW. Exposing zebrafish to silver nanoparticles during caudal fin regeneration disrupts caudal fin growth and p53 signaling. Mol Cell Toxicol. 2008;4:311–7.
172. Bilberg K, Malte H, Wang T, Baatrup E. Silver nanoparticles and silver nitrate cause respiratory stress in Eurasian perch (*Perca fluviatilis*). Aquat Toxicol. 2010;96:159–65.
173. Scown T, Santos E, Johnston B, et al. Effects of aqueous exposure to silver nanoparticles of different sizes in rainbow trout. Toxicol Sci. 2010;115:521–34.
174. Mahe B, Vogt A, Liard C, Duffy D, Abadie V, Bonduelle O, et al. Nanoparticle-based targeting of vaccine compounds to skin antigen-presenting cells by hair follicles and their transport in mice. J Invest Dermatol. 2009;129:1156–64.

Toxicogenomic Evaluation of Nanomaterials

23

John M. Veranth, Sancy A. Leachman, and Philip J. Moos

23.1 Introduction

The ongoing search for specialty substances with unique physical and chemical properties has motivated the development of numerous types of metal oxide nanomaterials, but concerns remain regarding biological effects of particles that are comparable in size to ultrafine air pollution ($d < 100$ nm). Much of the nanoparticle toxicological research has been on inhalation and intact-skin dermal exposures. Lung epithelial cells exposed to oxide nanoparticles show diverse responses, most notably upregulation of pro-inflammatory signaling pathways. This chapter addresses potential etiologies, mechanisms, and methodologies for investigating potential unforeseen toxicities of nanoparticles that are likely to occur in skin as nanoproducts become more and more available.

Cutaneous exposure to manufactured oxide nanomaterials can occur in both occupational settings and from deliberate application of sunscreens, lotions, and cosmetics containing nano-sized materials. This chapter focuses on metal oxide nanoparticles that are currently produced as specialty chemicals. The substances SiO_2, Al_2O_3, TiO_2, and ZnO are all used in skin care and cosmetic formulations as well as other consumer products. Carbon black from hydrocarbon sources is not approved for foods or cosmetics, but carbon black is widely used in pigments, inks, and photocopy toners, and as a filler in rubber products. Carbon nanotubes are being regulated as unique nanomaterials. In addition to consumer uses, all these materials are widely used in industrial processes where occupational exposure can occur. Specialty chemical suppliers are actively marketing new grades of these materials with ever smaller primary particle size as they pursue the nanomaterials market.

ZnO and TiO_2 are of particular interest since they are used in many consumer products; they are used in sunscreens (generally < 20 %), cosmetics, and paints. Nanoformulations of ZnO and TiO_2 can function as physical barriers to UV radiation and therefore are effective sunscreens that aid in the reduction of sunburn, photoaging, and resultant skin cancer. However, the visible light scattering efficiency of ZnO and TiO_2 increases rapidly as the particle size increases from molecular size to diameters comparable to the wavelength of light. This high scattering efficiency results in the white appearance of sunscreens that contain larger particle sizes ZnO and TiO_2 for physical protection from UV radiation. Therefore,

J.M. Veranth (✉) • P. J. Moos (✉)
Department of Pharmacology and Toxicology, University of Utah, L.S. Skaggs Pharmacy,
Rm. 201, 30 S 2000 East, Salt Lake City, UT, 84112, USA
e-mail: john.veranth@m.cc.utah.edu; philip.moos@utah.edu

S.A. Leachman
Department of Dermatology, University of Utah, Huntsman Cancer Institute,
2000 Circle of Hope, Salt Lake City, UT, 84112, USA
e-mail: sancy.leachman@hci.utah.edu

A. Nasir et al. (eds.), *Nanotechnology in Dermatology*,
DOI 10.1007/978-1-4614-5034-4_23, © Springer Science+Business Media, LLC 2013

the decrease in the particle sizes to nano-sized particulate (<100 nm) decreases the white light scatter but can still protect against UV radiation.

Consumer web sites are expressing concern regarding the use of nanoparticles in cosmetics and topical skin products. The FDA has considered specific regulations for nano-formulations of some materials, but current regulations do not make any distinctions between ZnO and TiO_2 nanoparticles and supermicron-sized particles of the same chemical substances [21 CFR parts 172 and 700]. Concern for the unintended consequences of nanotechnology is the focus of consumer groups, and popular culture has fictionalized the danger in futuristic stories like Michael Crichton's "Prey." However, the danger of occupational exposure has been observed in the *nanotoxicity of polyacrylate nanomaterials resulting in pulmonary fibrosis and death under conditions of exposure without sufficient safeguards for the employees* [1, 2].

Laboratory studies with intact skin models suggest that the stratum corneum is an effective barrier and only a minute fraction of tested nanomaterials penetrate deeply [3]. However, there is also the potential for nanoparticle exposure and toxicity through damaged skin and other tissue epithelial barriers such as the oral, conjunctival, and nasal mucosa.

23.1.1 Synopsis of Chapter

Whole genome microarray analysis of changes in RNA expression following particle exposure is valuable as a tool to characterize the changes in major signaling, response, and damage repair pathways, and often provides insights not available from measuring limited endpoints, such as the secretion of selected cytokines. Analyzing changes in the relative expression of thousands of genes, coupled with powerful software for data analysis, can be a useful approach for screening unanticipated effects and for developing new hypotheses.

The basic techniques of gene array studies will be presented along with various experimental designs. Next, a number of recent experimental results will be presented to illustrate current research. Finally, techniques for confirming gene array results and extending the results to more realistic biological models will be discussed.

Work to date has not reported unusual toxicity from nano-sized powders of substances that are normally considered benign. However, a prudent precautionary approach will require additional ongoing research.

23.2 Current Speculative Theories of Nanoparticle Toxicology

Numerous hypotheses have been proposed for potential toxicological responses induced by nanoparticle exposure including both chemical-specific effects of particle core composition or surface chemistry, and non-specific effects associated with the small size and high specific surface area of nano-sized particles [4]. All nanoparticles have high surface area per mass (specific surface area). The geometry can greatly affect particle properties due to the large fraction of the total atoms situated on the surface for interaction. At the nanometer-scale, certain material properties become uniquely scale-dependent, especially in particles below 20 nm diameter. Inter-particle interactions and surface attraction (e.g., capillary forces), optical properties, melting points, conductivity, ionization potential, electron affinity, magnetism, and, significantly, surface energy and reactivity all change as the particles approach molecular dimensions. Because nanomaterials are being formulated to achieve unique size-dependent properties, it may be possible to couple size-dependent biological effects with the molecular function of the agent.

Many chemicals, both inorganic species and biological macromolecules, can be physically adsorbed on the surface of solids. The high specific surface area of nanomaterials increases the ability to act as a carrier for adsorbed materials. Adsorbed substances can lead to nonspecific effects that are dependent primarily on the available surface area and less on the chemical composition of the core substance. For example, endotoxin on the surface of industrial-grade nanopowders may cause pro-inflammatory signaling that is mistaken for a response to the particle core substance.

The general expectation is that transport by both passive (diffusion, dissolution, entrainment in cytoplasmic flow) and active processes (endocytosis, phagocytosis) is enhanced when low-solubility substances are in the form of smaller particles [4]. This is the basis of nanomaterials being developed for medical applications such as drug delivery and diagnostic imaging. Experimental evidence supports the trend of increasing transport with decreasing size, but quantification of rate versus size is difficult because other factors, such as surface chemistry, are also important. Size-dependent transport rates motivates the hypothesis that some observed metal oxide toxicity is due to overload of the elemental metal because of enhanced uptake of the particles followed by dissolution in the low pH environment within endosomes [5].

Mechanistic hypotheses that have been widely reported in the nanoparticle literature include particle induced pro-inflammatory signaling responses, oxidative signaling responses, and changes in metal metabolism pathways. Nanoparticles, although small, are much larger than the typical molecular agonists that activate cell signaling pathways. The specific proteins that serve as the proximal receptor that initiates cell responses to low solubility particles are not well understood and this is another area of current research.

An example of a speculative size-dependent toxicity mechanism is the report of nanoparticles "mechanically" interfering with gene transcription. Human epithelial cells were reported to take up and translocate nano-sized silica particles to the nucleus where the particles induced formation of aberrant clusters that inhibited replication, gene transcription, and cell proliferation [6]. Citation searches show ongoing related work on nanoparticle–protein interactions, but as yet no convincing evidence that this mechanism leads to in vivo toxicity. Thus, unexpected and unusual experimental results deserve research follow-up, but in vitro experiments are primarily directed to understanding biochemical processes and cell culture results do not necessarily indicate an immediate clinical concern.

23.3 Overview of Gene Array Studies

A major limitation of toxicology experiments that focus on a specific hypothesis is that only a limited number of biological endpoints are typically measured and it is possible to miss real changes that were not anticipated when developing the experimental design. Thus whole genome approaches have the value of screening for unanticipated changes. This is especially important in the early stages of research on the health and safety of novel chemical substances.

There have been whole-genome array studies of nanoparticle effects, but very few have specifically addressed skin-derived cells. Sen et al. reviewed gene expression profiling studies of the effects of complex environmental particle mixtures including diesel emissions, urban particulate matter, and metal fumes [7]. Soot nanoparticles, rich in polycyclic aromatic hydrocarbons, induce oxidative stress and inflammation pathways in mouse lungs [8]. Nano-TiO_2 activates inflammatory response and cell adhesion, but not oxidative stress genes in human keratinocytes [9]. In mouse lungs nano-TiO_2 exacerbates LPS-induced inflammation, but does not alter the gene expression pattern [10]. A comparison of 10- and 500-nm SiO_2 particles using RAW264.7 mouse macrophages showed high correlation in gene expression changes between the sizes, suggesting common modes of action [11]. Ultrafine NiO, but not C60 fullerene particles, activated genes for chemokines, oxidative stress, and matrix metalloproteinase 12 in rat lungs [12]. An inhalation comparison of multiwalled carbon nanotubes and α-quartz in rats revealed many interesting patterns such as that expression of genes likely playing a role in fibrosis was qualitatively similar for the two materials [13]. To date, few studies have been performed examining the global gene expression responses to nanomaterials [14–16], but these studies have provided new insight into toxicological mechanisms of nanomaterials.

23.4 Understanding Genomic Profiling Literature

This is a brief overview to help a clinician to read and understand papers that refer to whole genome array results. The main caveat is that a change in gene regulation only changes the concentration of specific mRNA sequences which the experiment measures at a specific time point after the treatment. Other techniques must be used to determine if patterns indicated by statistical analysis of the gene array result in actual changes in the concentration of the relevant proteins for the biological hypothesis being studied. Thus, genomic profiling is most important as an exploratory hypothesis-generating tool.

Gene profiling studies can use either tissue samples or isolated cells. Studies using skin biopsy samples have the most direct relevance, but results are confounded by the difficulty of isolating a single cell type from a tissue. Cell culture using skin-derived cell lines has advantage of being single cell type under reproducible in vitro conditions. Disadvantages of cell culture are that immortalized cells are by definition not normal, and isolated cells lack the normal intracellular, neurological, and hormonal signaling found in an intact organism. Cell culture methodology is best suited for understanding fundamental biological processes.

A typical in vitro experiment consists of growing the cells in culture media, then adding the test substance, such as nanoparticles or a positive control drug, to the cell culture. After a period of time (e.g., 4 h) the cells are harvested for genomic analysis. Important considerations include growing the cells under consistent and reproducible conditions since our own experience (and other reports [17]) indicates that small differences in growth conditions can lead to many transcriptional changes when measuring global changes by cDNA microarrays.

Total cellular RNA is recovered, typically using commercial RNA purification kits. If the total amount of RNA is low, the RNA is then amplified, again typically using commercially available reagent kits. Variability in microarrays can come from multiple sources. Biological variability as well as "noise" can be introduced into the experiment from samples of different origin or in the processing of the RNA. Therefore, care is given to evaluate RNA quality and reproducibility prior to microarray analysis. We use a Nanodrop spectrophotometer to measure RNA concentration and a BioRad Experion automated electrophoresis station to monitor quality. RNA quality is critical for any 'omic analyses. RNA quality is most commonly evaluated and quantified by measuring multiple features of the electropherogram, and these data are reduced to a single RNA integrity number (RIN, from the Agilent BioAnalyzer) or RNA quality index (RQI, from the BioRad Experion). Figure 23.1 shows a typical high-quality electropherogram result as well as the "gel-view" of 12 samples. RNA should be of the highest quality, as a degraded sample will still provide expression data; however, the data from a degraded sample will not necessarily be easily distinguished at the end of the experiment and could lead to misinterpretation of the experiment.

Whole genome transcript profiling can be accomplished with various techniques from fluorescent-based methods like microarray (and quantitative PCR for more limited gene set analysis) to pyro-sequencing techniques (RNA-seq). Assuming high-quality RNA as the input, these techniques allow the relative assessment of gene expression levels; the RNA sequencing methods allow for even more information as it can identify single nucleotide polymorphisms, splice variants, etc.

For the toxicogenomic data presented in the following, we utilized Agilent human whole genome microarrays to competitively hybridize untreated samples with those treated with the nanomaterials. The RNA is reverse-transcribed, labeled during amplification, hybridized to the oligo array, the relative fluorescence of the fluorescent probes is quantified, and the data are processed (local background is subtracted, irregular spots flagged, global linear regression (lowess) normalization is performed, and this ratio is log transformed) prior to analysis. There

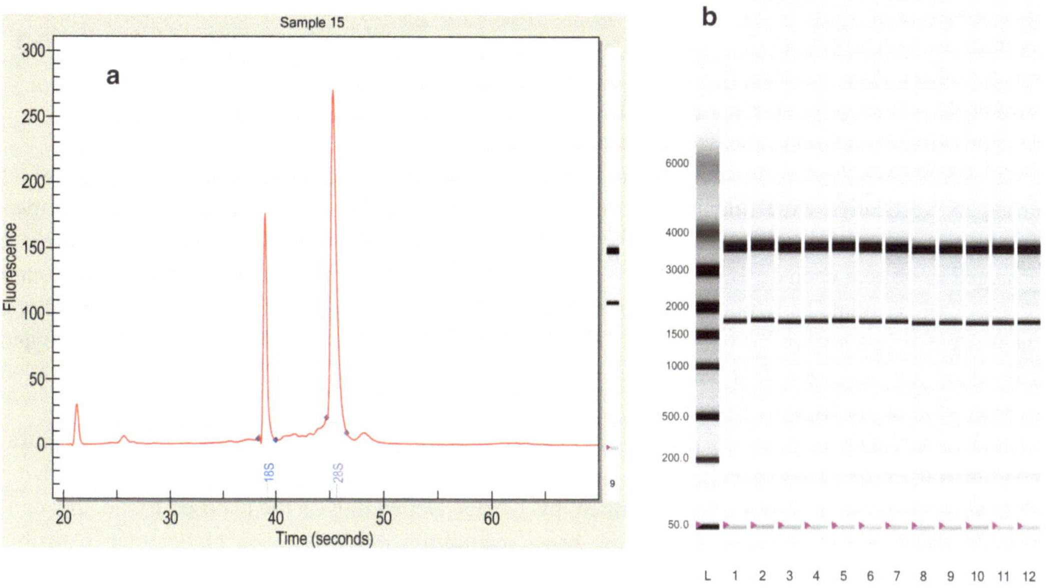

Fig. 23.1 (**a**) Example of an electropherogram of total RNA purified from BEAS-2B cells after treatment with nanosilica PM for 4 h. The insert at the right shows the conversion of the electropherogram to a "gel-view." (**b**) The quality and reproducibility of the RNA purification are displayed in the 12 samples shown in this "gel-view"

is no single method for the evaluation of expression data, but the supervised statistical approaches tend to utilize iterative analyses to estimate the false discovery rate (a method that controls the false positives in multiple comparison testing) since the family-wise error rate is a poor predictor of significance when the number of hypotheses tested (proportional to the number of genes) is large compared to the number of experiments.

23.5 Example of a Recent Study

The most useful type of hypothesis-generating result is the statistically significant observation of coincident changes in families of genes related to a specific metabolic pathway. For example, the upregulation of metal transport genes in the absence of changes in cell cycle and inflammation-related pathways might be interpreted as a metal oxide nanoparticle-induced intracellular metal overload that is being ameliorated by normal homostatic processes [4].

The application of toxicogenomic methods to a problem in dermatology will be illustrated by a recent set of experiments using nano-sized ZnO and TiO_2 applied to skin-derived cell lines. We used HaCaT and SK Mel-28 cells as representatives of skin-derived cells that are easily manipulated in cell culture. The skin-derived cell lines were treated with 1or 5 μg/cm^2 of ZnO or 5 or 10 μg/cm^2 of TiO_2, and 100 μM $ZnCl_2$ was used for the soluble Zn treatment. Based on preliminary experiments, these dose levels were selected to achieve a robust response but avoid cytotoxicity.

The initial hypothesis of this study was that the nano-sized particles of ZnO and TiO_2 (as well as other metal oxide nanoparticles) would induce pro-inflammatory signaling across a range of cell types [4]. The gene array data did not support this hypothesis, but rather suggested that the main effect of ZnO particles was upregulation of genes related to metal transport and metabolism. In contrast, an earlier study that exposed lung-derived cells to an inhalable desert dust and nano-sized SiO_2 particles showed a robust pattern of upregulation of cytokines (Fig. 23.2). It is important to note the scale of gene expression represented in the heatmaps and recall that the data are usually log transformed so that the direction (induction or suppression) of expression is apparent. The data are also generally hierarchically clustered with the genes arranged vertically, experiments horizontally, and the genes with similar expression levels arranged to be near each other. These genes may or may not be related in function. The assessment of the pathways or ontology that is influenced by the treatment is best assessed through annotation analysis to determine the statistical enrichment of genes in particular pathways or function.

Several important points are illustrated by these data. As seen in Fig. 23.2, a few pro-inflammatory genes are upregulated at the higher dose level even if we preselect genes related to inflammation. If, instead of a gene array experiment, the work had relied on quantitative PCR of only a few genes or had measured changes in only a few specific proteins, the data might have erroneously supported an inflammation hypothesis while missing the bigger picture that most inflammation genes are not altered by the particle treatment. Therefore, our data in this experiment specifically suggest that should ZnO nanomaterials in a dermal formulation penetrate the stratum corneum, we would not expect a pronounced inflammatory response in the live cell layers of the skin, but instead, we would expect the ZnO to dissolve and the free Zn ions would elicit a metal response with associated metal toxicity. The wealth of data provided by the gene array allows immediate searching for alternative hypotheses and disruption of other pathways.

Conversely, many important enzymes and signaling proteins are subject to post-transcriptional regulation and changes in messenger RNA are not necessarily correlated with changes in the protein concentration. Thus, gene array results are primarily hypothesis-generating, and can point out targeted strategies for further investigation using confirmatory assays at the protein level.

23.6 Dose Relevance for the Tissue and Importance of Comparing Detailed Particle Characteristics

There are several important points to consider when evaluating the clinical meaning of nanoparticle toxicology experiments. These include in vitro dose relevance, importance of the outer skin layer as a nanoparticle barrier, the metabolic differences between isolated cells in culture and an intact tissue, and the importance of detailed particle characteristics.

Most in vitro studies use particle concentrations equivalent to 1–100 μg/cm^2 on growing cells. Depending on the exposed tissue, this concentration range is much higher than plausible particle deposition in the lung, but is within the range that might occur in the digestive tract after ingestion of a nanoparticle containing consumer product. Application of a sunscreen or cosmetic to damaged skin would result in a similar or higher particle concentration. Thus, typical in vitro nanoparticle concentrations found in the toxicology literature are realistic for skin exposures, but numerous studies have reported that intact skin provides an effective barrier to nanoparticles such as quantum dots [18] and fullerenes [19]. Even sunburned skin shows limited transdermal transfer of TiO_2 and ZnO nanoparticles [3].

As noted earlier, experiments with cultured cells facilitate the study of gene transcription changes by allowing collection of RNA from a single cell type. Isolated cells, especially immortalized cell lines, are continuously dividing and therefore do not have “normal” metabolism. Further, the isolated cells lack the inputs from cytokines, hormones, and other physiological signals that regulate tissue responses in an intact organism. In vitro gene transcription studies can generate hypotheses, but in vivo follow-up research is essential to determine if the results are clinically important.

Nanoparticles of the same nominal chemical substance can have greatly different surface chemistry, which affects both chemical reactivity and

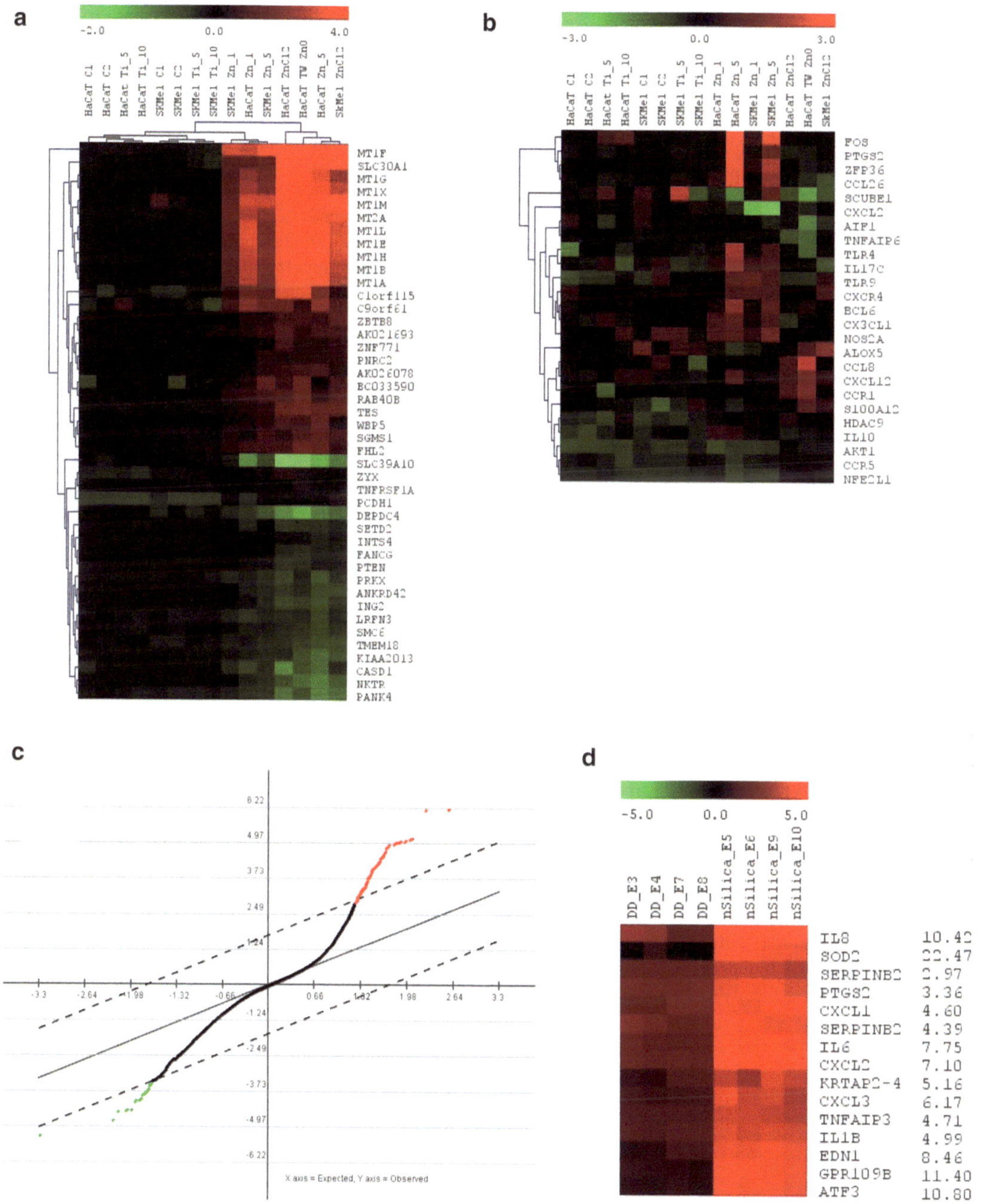

Fig. 23.2 (**a**) Hierarchical clustering following significance analysis of microarray (SAM) that identifies gene expression associated with metal ion binding, metabolism, and transport. These data demonstrate that the ZnO nanoparticles have a pronounced transcriptional response, but it is a metal response rather than a pro-inflammatory response. (**b**) Hierarchical clustering of inflammation-related genes that best discriminate the ZnO treatment from the TiO_2 and controls. While there are some alterations in gene expression, few genes show consistent responses across the ZnO treatments. (**c**) SAM plot for comparing cells treated with desert dust or nano-silica indicating 170 differentially expressed genes with a false discovery rate of 0.5% (highlighted in red for genes induced compared to the control expression level, and green for genes depressed compared to the control). Four independent experiments were performed and used in the analysis. (**d**) A select cluster of genes displaying the nanosilica-induced expression of several inflammatory genes (IL-8, PTGS2 (Cox-2), IL-6, IL-1β, TNFαIP3, CXCL1, CXCL3) is displayed in the "heatmap." Expression ratios (log_2) of the treated samples compared to non-treated cells are displayed. The scale is shown at the top, the genes are indicated by their gene symbols, and the "fold difference" between the treatments is indicated at the right

References

1. Song Y, Li X, Du X. Exposure to nanoparticles is related to pleural effusion, pulmonary fibrosis and granuloma. Eur Respir J. 2009;34(3):559–67.
2. Song Y, Tang S. Nanoexposure, unusual diseases, and new health and safety concerns. ScientificWorldJournal. 2011;11:1821–8.
3. Monteiro-Riviere NA, Wiench K, Landsiedel R, Schulte S, Inman AO, Riviere JE. Safety evaluation of sunscreen formulations containing titanium dioxide and zinc oxide nanoparticles in UVB sunburned skin: an in vitro and in vivo study. Toxicol Sci. 2011;123(1):264–80.
4. Moos PJ, Olszewski K, Honeggar M, et al. Responses of human cells to ZnO nanoparticles: a gene transcription study. Metallomics. 2011;3(11):1199–211.
5. dos Santos T, Varela J, Lynch I, Salvati A, Dawson KA. Quantitative assessment of the comparative nanoparticle-uptake efficiency of a range of cell lines. Small. 2011;7(23):3341–9.
6. Chen M, von Mikecz A. Formation of nucleoplasmic protein aggregates impairs nuclear function in response to SiO2 nanoparticles. Exp Cell Res. 2005;305(1):51–62.
7. Sen B, Mahadevan B, DeMarini DM. Transcriptional responses to complex mixtures: a review. Mutat Res. 2007;636(1–3):144–77.
8. Rouse RL, Murphy G, Boudreaux MJ, Paulsen DB, Penn AL. Soot nanoparticles promote biotransformation, oxidative stress, and inflammation in murine lungs. Am J Respir Cell Mol Biol. 2008;39(2):198–207.
9. Fujita K, Horie M, Kato H, et al. Effects of ultrafine TiO2 particles on gene expression profile in human keratinocytes without illumination: involvement of extracellular matrix and cell adhesion. Toxicol Lett. 2009;191(2–3):109–17.
10. Inoue K, Takano H, Ohnuki M, et al. Size effects of nanomaterials on lung inflammation and coagulatory disturbance. Int J Immunopathol Pharmacol. 2008; 21(1):197–206.
11. Waters KM, Masiello LM, Zangar RC, et al. Macrophage responses to silica nanoparticles are highly conserved across particle sizes. Toxicol Sci. 2009;107(2):553–69.
12. Fujita K, Morimoto Y, Ogami A, et al. Gene expression profiles in rat lung after inhalation exposure to C60 fullerene particles. Toxicology. 2009; 258(1):47–55.
13. Ellinger-Ziegelbauer H, Pauluhn J. Pulmonary toxicity of multi-walled carbon nanotubes (Baytubes) relative to alpha-quartz following a single 6 h inhalation exposure of rats and a 3 months post-exposure period. Toxicology. 2009;266(1–3):16–29.
14. Zollanvari A, Cunningham MJ, Braga-Neto U and Dougherty ER. Analysis and modeling of time-course gene-expression profiles from nanomaterial-exposed primary human epidermal keratinocytes. *BMC Bioinformatics*. 2009;10 Suppl 11:S10
15. Yang Y, Qu Y, Lu X. Global gene expression analysis of the effects of gold nanoparticles on human dermal fibroblasts. J Biomed Nanotechnol. 2010;6(3): 234–46.
16. Ma J, Lu X, Huang Y. Genomic analysis of cytotoxicity response to nanosilver in human dermal fibroblasts. J Biomed Nanotechnol. 2011;7(2):263–75.
17. Hughes TR, Marton MJ, Jones AR, et al. Functional discovery via a compendium of expression profiles. Cell. 2000;102(1):109–26.
18. Prow TW, Monteiro-Riviere NA, Inman AO, et al. Quantum dot penetration into viable human skin. Nanotoxicology. 2012;6(2):173–85.
19. Xia XR, Monteiro-Riviere NA, Riviere JE. Skin penetration and kinetics of pristine fullerenes (C60) topically exposed in industrial organic solvents. Toxicol Appl Pharmacol. 2010;242(1):29–37.
20. Warheit DB, Webb TR, Reed KL, Frerichs S, Sayes CM. Pulmonary toxicity study in rats with three forms of ultrafine-TiO2 particles: differential responses related to surface properties. Toxicology. 2007; 230(1):90–104.

The Language of Nanotechnology and Its Role in Intellectual Property, Regulatory Settings, and Consumer Perception

24

Jeffrey K. Mills

24.1 Introduction

One of the first lessons to which new law students are exposed is the critical role that language and word choice play in the legal field. Whether it is the interpretation of a contract between two parties or the meaning of a specific clause in a state law, every word is scrutinized and construed to determine the scope of what has been written. Even the U.S. Constitution is constantly being interpreted to determine whether new laws enacted by Congress and signed by the President are, in fact, Constitutional. Historical literalists, in fact, look to the *literal* meaning of the words that were first written in 1787 to determine the scope of this founding document [1].

It is interesting then to examine the words that we use to "talk" about nanotechnology and some of the unique problems that result in this technical field because of language. Specifically, this chapter will examine the role of language in three settings: intellectual property, regulatory issues, and consumer perception.

Jeffrey K. Mills, Ph.D., Esq., is a Partner at Fanelli Haag & Kilger PLLC, in Washington, DC. The opinions expressed in this chapter are those of the author and should not be attributed to other members of Fanelli Haag & Kilger PLLC or clients of the firm.

J.K. Mills (✉)
Fanelli Haag & Kilger PLLC,
1909 K Street, NW, Suite 1120, Washington, DC, 20006, USA
e-mail: jmills@fanellihaag.com

24.2 Language in Nanotechnology

The following example is provided in order to illustrate the "problem" facing the language of nanotechnology. If the picture shown in Fig. 24.1 was provided to a group of dermatologists, and they was asked to associate a single term with the skin condition illustrated, a vast majority would likely indicate that it is an example of "dry skin." In a similar example, if the same group of clinicians were asked to define the term "abscess," most would probably describe "a cavity containing pus and surrounded by inflamed tissue, formed as a result of suppuration in a localized infection," [2] or a similar definition.

In contrast, if a group of researchers, regardless of field, was asked to provide a term to describe the structures shown in Fig. 24.2, it is very conceivable that a number of different words could be applied. For example, nanoparticles, nanospheres, nanocolloids, nanocarriers, or some other nano-term could all be suggested. If the same group was then asked to define the term "nanoparticle," they may each provide varying, different, definitions. The first variation may simply be in the size scale of a nanoparticle. While less than 100 nm is a fairly common size range [3], particles of up to 150 nm or even 200 nm [4] often exhibit the same characteristics. An additional variation in definitions may relate to the number of dimensions that are within this size range; must it be all three, or perhaps just two, or even a single dimension, could be sufficient to

A. Nasir et al. (eds.), *Nanotechnology in Dermatology*,
DOI 10.1007/978-1-4614-5034-4_24, © Springer Science+Business Media, LLC 2013

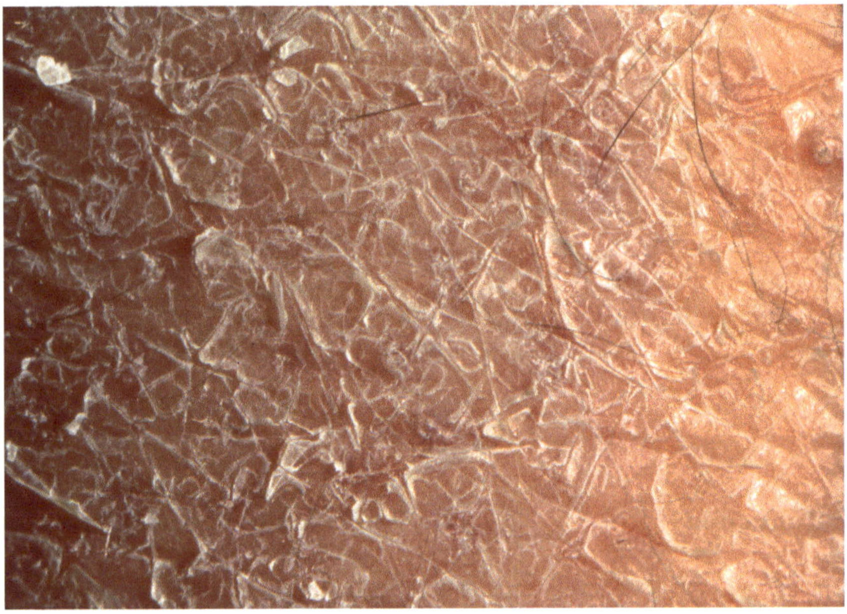

Fig. 24.1 Photograph of dry skin. *Source*: Otto H. Mills, jr, private teaching collection

characterize something as a nanoparticle. The results of this example, hopefully, demonstrate the unique challenges that are facing the field of nanotechnology, simply from the standpoint of language that is used to describe what is being developed and studied.

24.3 Why is Nanotechnology Unique?

The initial question, then, becomes why is nanotechnology unique. While many could argue that researchers have been studying aspects of nanotechnology since the beginning of modern science [5], it is only as recent as 2001 that it has come to the forefront of discussion as a field of research [6]. With this heightened focus coming within just the past 11 years, particularly from areas such as government regulation and public interest, development of standards and definitions for use in the field is still very much in its infancy [7]. Standards setting groups, such as British Standards Institute (BSI), ASTM International, International Organization for Standardization (ISO), and American National Standards Institute (ANSI), have all come out with suggested terminology for use in the field of nanotechnology. In many cases the terms have similar definitions. However, there is still variability that, in general, does not exist in other fields.

Perhaps the biggest reason for the unique situation facing nanotechnology is that in no other field are so many distinct and diverse players involved in the development of the science. A simple review of recent journal and book titles and their related target audiences reveals interested researchers in medicine and biology (*Nanomedicine: Nanotechnology, Biology, and Medicine*), electronics and optics (*Journal of Nanoelectronics and Optoelectronics*), chemistry (*Nanochemistry*), and physics and materials engineering (*Journal of Nanomaterials*).

For each of these distinct technical fields, researchers are often using the same terms to describe very different structures. In other cases, the same exact structure may be used by researchers in different areas. For example, quantum dots are employed in the field of medicine for imaging and diagnostics [8] and in the field of electronics in light-emitting diodes [9]; carbon nanotubes in the development of batteries [10] and gas sensors [11]. A search for the word "nanoparticle" reveals research in cancer therapy [12], solar cells [13], and DNA detection [14].

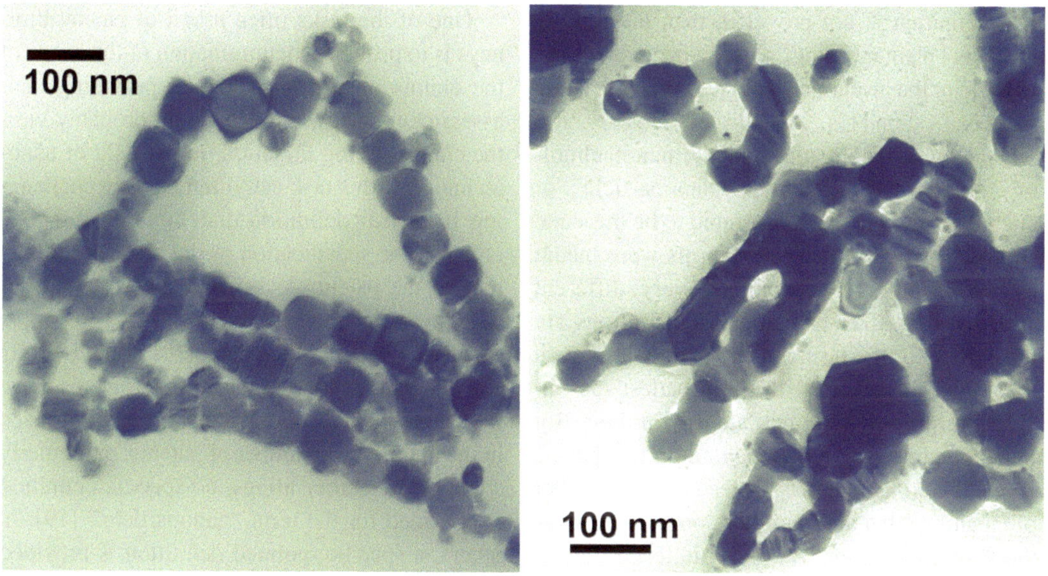

Fig. 24.2 Cubes of cobalt. *Source*: National Institute of Standards and Technology, AV Number 09PHY014

Thus, perhaps unlike any other field, nanotechnology is being impacted by researchers in nearly every discipline of science. In each case, these researchers have their own desires, ideas, and agendas, as to how terms in the field should be defined and used. As discussed in the following, the diverse, multidisciplinary impact on nanotechnology has created some unique challenges in areas such as patent law, regulatory issues, and public perception, simply based on the language that is used to talk about the field.

24.4 Language in Patent Law

When reading a patent, arguably the most important section is the claims that conclude the document. The claims define the scope of the patent, in other words, how broad or narrow an area the patent covers. One of the easiest ways to view the claims of a patent is as a fence surrounding a parcel of property. Much like a fence, the claims of a patent provide a means to keep others off of the property. If anyone crosses the boundaries of the fence, i.e., infringes a claim, the owner of the patent can stop the encroachment, and also potentially recover some money for any damages that may have occurred. However, a patent does not actually let a patent owner make or do anything—it only provides a right to exclude others. It is still necessary to consider if there may be other patents that prevent the patent owner from practicing. As with real property, simply because an owner may be able to put up a fence bordering the land, zoning restrictions may prevent building a skyscraper in the middle of a residential area.

The scope of the claims of a patent is determined by the terms that are used to describe the invention. While there is a requirement that the claims of a patent be specific and not vague or indefinite [15], an inventor is allowed to define the terms used in the claims however they see fit, and thus are able to be their own lexicographer [16].

During prosecution of a patent before the U.S. Patent & Trademark Office (USPTO), the Patent Examiner who is reviewing a patent application is initially charged with determining the scope of what is in fact covered by the claims of a patent, as well as whether or not the application meets a number of patentability requirements. During examination, Patent Examiners must give the claims "their broadest reasonable interpretation consistent with the specification" [17]. For terms utilized in the claims, this interpretation is read in light of any definitions provided in the specification.

If a definition is not provided, then the term is given its "plain meaning," as someone working in the field of the invention would interpret it, reading the specification [17].

For patent applications in the nanotechnology sector, terms such as "nanoparticle" or "nanocrystal" may be interpreted to be the same structure, when in fact these terms were meant by the inventor to encompass very different structures, or sizes of matter. If definitions are not provided, the terms in the claims are given their plain meaning, with the broadest reasonable interpretation by the Patent Examiner. But in nanotechnology, terms may not in fact have a single meaning, and may instead have a number of meanings that an Examiner could consider as the "plain meaning." As multiple terms can be used to describe the same structure, or the same term can have different meanings, it is critical to define terminology used in a patent application to make clear the intent of what is encompassed, and what is excluded, by the term. Otherwise, an inventor runs the risk of his or her claims being read so broadly that they encompass structures in the art that they are seeking to specifically exclude. While some may argue that vague or ambiguous terms can provide broader claim scope, the danger of not actually encompassing what is desired weighs in favor of using definitions whenever possible and prudent.

After a patent has issued, the scope of the claims is determined by a judge, if and when that patent is ever asserted against a potential infringer in litigation. Under current law, the scope of terms in a claim is to be determined in view of "intrinsic evidence" [18]. This means that a judge is to first look to the patent specification to determine if a definition was provided. If no definition is provided, the next step is to look to the prosecution history, i.e., the back and forth arguments between the patent examiner and the patent applicant/inventor, in order to determine if the term was discussed and therefore defined in the various exchanges. If there is still no reasonable definition that can be determined, only then are outside sources, for example, dictionaries or other scientific literature, to be consulted.

One of the tasks often asked of patent attorneys is to provide a determination of the scope of the claims of a granted patent, providing their assessment of how a judge would probably view the claims during litigation. In the field of nanotechnology, this task can often be very challenging if a clear definition does not present itself either in the specification or the prosecution history of the patent. As a result, one is often left with a very inconclusive analysis.

A brief review of the databases of the USPTO reveals that approximately 1,400 patents have issued in which the term "nanoparticle" was used in the claims. In a number of patents, particularly some of the earlier filings, no specific definition is provided for the term "nanoparticle." [19]. In others, a specific, pointed definition is provided by the inventor, "[t]he term 'nanoparticle' refers to an inorganic particle having a particle diameter or particle size of not greater than 200 nm" [20]. However, as may be expected, the definitions that are provided often vary dramatically in size, composition, and structure.

A number of concerns then are facing the nanotechnology community, when considering intellectual property. By not having specific, set definitions for terms that are used by numerous different practitioners across a multidisciplinary set of industries and sciences, a great amount of ambiguity has arisen. This comes to light during prosecution of a patent, when claims are rejected over art that clearly is not relevant or was specifically meant to be excluded, as the scope of a term is fluid and changing. With issued patents, concern exists over very broad patents covering structures beyond the scope of what was initially described, simply because of a poorly defined term. In addition, confusion over the field of use or other elements of a patent license can arise because definitions used in the license, and the related patents, may be changing or vague [21]. Finally, the issue of unclear claim scope provides frustration to those trying to determine whether or not it is safe to continue working in a particular area for fear of being sued.

Overall, the best strategy is to be as particular and consistent with the terminology that is selected for use in patents, and include definitions that ultimately limit any ambiguity that can result. In a legal area such as intellectual property where precision in language is critical, those practicing in the nanotechnology field must be aware of these issues and take steps to reduce their impact.

24.5 Language in Regulatory Settings

In its initial report dated July 25, 2007, the Nanotechnology Task Force of the U.S. Food and Drug Administration (FDA) specifically stated that the Task Force had "not adopted a precise definition for 'nanoscale materials,' 'nanotechnology,' or related terms to define the scope of its work" [22]. The Task Force determined that it would be "most productive to take a broadly inclusive approach in identifying potentially relevant studies, data, and other information" [22]. Nearly 4 years later, in its Guidance Document dated June 2011, the FDA again stated it still has not "established regulatory definitions of 'nanotechnology' 'nanoscale' or related terms" [23].

Coordinated with the publication of this Guidance Document was the release of a memorandum issued jointly by the Office of Science and Technology Policy (OSTP), the Office of Management and Budget (OMB), and the U.S. Trade Representative (USTR). In the memorandum, general principles were outlined for the overall regulatory approach that is to be taken with regard to developments in nanotechnology. The document indicated "[t]o further understanding, enhance risk management, improve risk and hazard communication, and promote coordination, simplification, and harmonization in regulatory actions, a working group established by OSTP, OMB and USTR will coordinate an approach and choice of terminology relevant to regulation and oversight of nanomaterials" [24]. In its Guidance Document, the FDA suggested that, while specific terms will not be assigned to FDA-regulated products, when considering whether a product contains nanomaterials or otherwise involves the application of nanotechnology, the FDA will ask:

1. Whether an engineered material or end product has at least one dimension in the nanoscale range (approximately 1–100 nm); or
2. Whether an engineered material or end product exhibits properties or phenomena, including physical or chemical properties or biological effects, that are attributable to its dimension(s), even if those dimensions fall outside the nanoscale range, up to one micrometer. [23]

The Guidance Document made a concerted effort to illustrate that its concern was with materials that were deliberately manipulated and controlled to produce specific sizes or properties, not simply those that occur naturally or in incidental or background levels.

While some critics question whether the FDA is properly addressing the regulation of products comprising nano-ingredients [25], the FDA "believes that nanoscale materials will present regulatory challenges that are similar to those posed by other new technologies FDA has dealt with in the past" [22]. Currently, there is no specific requirement that a product containing nanoscale materials, including cosmetic products, must be labeled as containing such materials [22]. A similar stance has been taken by the Therapeutic Goods Administration of Australia, "the TGA does not require any specific warnings about nanoparticles to be placed on labels of sunscreens," [26] and by Health Canada [27]. However, U.S. legislation introduced in June 2011 would allow the FDA to require that "minerals and other particulate ingredients be labeled as 'nano-scale' on a cosmetic ingredient label or list if not less than 1 % of the ingredient particles in the cosmetic are 100 nm or smaller in not less than 1 dimension" [28]. The bill is currently referred to the Subcommittee on Workforce Protections.

The European Union (EU) has taken a more concrete stance with regard to terminology, for example, defining "nanomaterial" as "an insoluble or biopersistant and intentionally manufactured material with one or more external dimensions, or an internal structure, on the scale from 1 to 100 nm" [29]. However, the EU acknowledges that "[i]n view of the various definitions of nanomaterials published by different bodies and the constant technical and scientific developments in the field of nanotechnologies, the Commission shall adjust and adapt [the definition of nanomaterial] to technical and scientific progress and to definitions subsequently agreed at international level" [29]. With regard to labeling of cosmetics, the EU currently requires, "[a]ll ingredients present in the form of nanomaterials [to] be clearly indicated in the list of ingredients. The names of such ingredients shall be followed by the word 'nano' in brackets" [29].

Thus, 11 years after the development of the National Nanotechnology Initiative, the issue of terminology and how to discuss nanotechnology and nanomaterials is still very much in flux. There has, however, been a clear effort to not limit nanotechnology simply to the size of an article, but also to examine the properties of a material to determine if they may have relevance to nanotechnology. In addition, as noted earlier, there is no specific requirement from the FDA to indicate on the label of a product that it contains nanoscale materials. This decision is left to the manufacturer.

However, the Federal Food, Drug & Cosmetic Act clearly requires that any labeling that is used be truthful and not misleading. Labeling must include material information, including any risk information. If the labeling is false or misleading, the product is "misbranded" and cannot be marketed [22]. The FDA is directly responsible for case-by-case determination of whether product labeling contains adequate information to support the safe and effective use of a product. For products not subject to premarket authorization, such as cosmetics, the labeling must also include material information and not include false or misleading information.

The FDA Nanotechnology Task Force concluded:

> If disclosure of information on the use of nanoscale materials is not material (and, therefore, is not required to be included in the product labeling), the manufacturer may still be able to include such information, as long as the information is not false or misleading. However, because claims regarding the use of nanoscale materials might be misleading and, therefore, misbrand a product, the Task Force would recommend encouraging manufacturers to consult with the agency concerning such labeling to avoid misbranding the product [22].

The Nanotechnology Task Force correctly identified a primary concern to those making products in the field—the risk of misbranding a nanotechnology product. However, as the FDA has yet to develop a formal definition for nanotechnology and other underlying terms in the field, and instead provides guidance that nanoscale products include both those that are less than 100 nm and those that exhibit properties or phenomena that are attributable to its dimension [23], the agency has done little to address the issues surrounding language in regulatory settings.

Calling a material a nanoparticle, a nanocrystal, a nanofiber, or some nano-based term on the label of a product may not be terribly concerning. The fact that definitions for each of these terms, including any distinguishing characteristics that they may have, are still in development is probably an indication that specific term choice is just that–a matter of choice. So long as the inclusion of a nano-term in the labeling of a product is not "false or misleading," then it will most likely comply with the current requirements of the FDA. Under the released Industry Guidelines, it appears that so long as a material exhibits phenomena attributable to its dimension (even a single dimension), labeling something as a nanomaterial would seem to be acceptable.

A question may arise though if claims are made about a product and attributed to the inclusion of a nanoscale material. For example, if a product is claimed to have "increased penetration due to the presence of nanocrystals," care should be taken to make sure that the penetration effect is in fact attributable to crystals that are less than 100 nm or have unique properties attributable to their dimension, and not simply because of a penetration enhancer or other additive in the formulation. As suggested by the Task Force, product manufacturers should be careful to not include information on labels regarding nanoscale products that are false and misleading, and thus consult closely with the FDA. Until more concrete definitions and standards are developed, it may be best to be cautious when applying nano-terms to product labels, and instead focus on the composition of the product.

24.6 Language in Consumer Perception

Currently, on the order of 1,000 or more products are identified as having some type of nanoscale material or otherwise involve nanotechnology [30]. These products include dermatologic products such as various sunscreens containing zinc oxide and/or titanium dioxide of less than 100 nm in dimension, and REVITALIFT®, described by L'Oreal as containing Nanosomes™ of Pro-Retinal A. The choice of a manufacturer to disclose the presence of nanomaterials in a product is often driven by a cost-benefit analysis, weighing the increased consumer interest that a product may receive because it has "nano" in the product name or description, with the concern that may be generated by concern over safety.

The message delivered to the public concerning nanotechnology in early stages was generally dominated by proponents touting the economic and societal benefits of technologies utilizing nanoscale materials [31]. However, there is an increasing concern that "[t]he U.S. public's perception of nanotechnology is up for grabs. It could divide along the lines of nuclear power, global warming and other contentious environmental issues absent a major public education and engagement effort by industry, government, civic groups and scientists" [32]. Critics have stressed the need for a unified approach from government, business, and educators when talking to the public about nanotechnology [32], as it is an informed public that tends to show higher levels of overall support than those who are less aware of nanotechnology [33].

Inconsistencies in terminology, while frustrating to the researcher and technically trained observer, can often be overcome by more closely examining the structure and properties of an article to establish similarities and distinguish differences. However, the public rarely has access to this information and rather must rely on the language that is used to describe the articles. What becomes clear then is that in order to provide a unified, public discussion, consistent vocabulary must be utilized by those delivering the message. Simply the choice of words provides a significant impact on how the public may view a product. If inconsistent terminology is used, there is a very real possibility of alienating the public to nanotechnology as a whole, simply because they become frustrated with the lack of uniformity and predictability.

To the extent possible, it is critical that researchers and product manufacturers look to industry standards and utilize them, and be consistent in their own uses. Considering that there currently is no requirement to label a product, or not label a product, as containing a nanomaterial, the nanotechnology industry must stress consistent, accurate use of language. Indicating a product contains a nanomaterial, when in fact it has particles millimeters in scale, causes confusion and distrust in a rapidly evolving technology that needs the support of the public. Absent this diligence, nanotechnology runs the risk of becoming a polarizing technology that may not achieve all that it promises to deliver.

24.7 Conclusion

Nanotechnology has been touted as an enabling technology that can revolutionize industries ranging from electronics to medicine to materials science. It is critical that uniform terms and definitions be developed to aid in "talking" about nanotechnology. Consistencies in the language of nanotechnology will provide stability and predictability in areas such as intellectual property, regulatory issues, and consumer perception, allowing this field to reach its full potential.

References

1. Barron JA, Dienes CT, McCormack M, Redish MH, editors. Constitutional law: principles and policy. 5th ed. Charlottesville, VA: Michie Law Publishers; 1996. p. 31–3.
2. Anderson KN, Anderson LE, Glanze WD, editors. Mosby's medical, nursing, and allied health dictionary. 4th ed. St. Louis, MO: Mosby; 1994. p. 6.
3. Nanoparticles-Vocabulary. British Standards Institute. PAS 71:2011. 2001. Available at: http://shop.bsig-

roup.com/en/Browse-by-Subject/Nanotechnology/?t=r. Accessed 27 Jan 2012
4. Rechberger W, Hohenau A, Leitner A, Krenn JR, Lamprecht B, Aussenegg FR. Optical properties of two interacting gold nanoparticles. Optics Communications. 2003;220:137–41.
5. Berger M. Nanotechnology in cosmetics—2000 Years Ago. 2006. Available at: http://www.newswiretoday.com/news/8233/. Accessed 27 Jan 2012
6. National nanotechnology initiative strategic plan, February 2011. Available at: http://www.nano.gov/node/581. Accessed 25 Jan 2012
7. Violet F. Analysis of technical standards in the field of nanotechnology. Nanotechnology Law & Business. 2010;7:299–307.
8. Michalet X, Pinaud FF, Bentolila LA, et al. Quantum dots for live cells, in vivo imaging, and diagnostics. Science. 2005;307:538–44.
9. Qingjiang S, Wang YA, Song LL, et al. Bright, multicoloured light-emitting diodes based on quantum dots. Nature Photonics. 2007;1:717–22.
10. Chena WX, Leea JY, Liuc Z. The nanocomposites of carbon nanotube with Sb and $SnSb_{0.5}$ as Li-ion battery anodes. Carbon. 2003;41:959–66.
11. Ong KZG, Zeng K, Grimes CA. IEEE Sens J. 2002;2:82–8.
12. Brannon-Peppas L, Blanchette JO. Nanoparticle and targeted systems for cancer therapy. Adv Drug Deliv Rev. 2004;56:1649–69.
13. Anderson IE, Breeze AJ, Olson JD, Yang L, Sahoo Y, Carter SA. All-inorganic spin-cast nanoparticle solar cells with nonselective electrodes. Appl Phys Lett. 2009;94:63101–4.
14. Park S-J, Taton TA, Mirkin CA. Array-based electrical detection of DNA with nanoparticle probes. Science. 2002;295:1503–6.
15. 35 United States Code (U.S.C.) § 112, 2nd paragraph. 2005
16. United States Patent & Trademark Office, Manual of Patent Examining Procedure (M.P.E.P.) § 2111.02. 8th Ed. July 2010
17. M.P.E.P. § 2111. 8th Ed. July 2010
18. Phillips v. AWH Corp., 415 F.3d. 1303, 1313 (Fed. Cir. 2005)
19. See for example, Desieno MA, Stetsko G, United States Patent No. 5,573,783. 12 Nov 1996
20. Anderson KS, Gaddam BN, Pocius AV, Yarwood JM. United States Patent No. 7,973,096. 5 July 2011
21. Rasha E, Gamota D, Jamison D, Miller J, Hermann K. Standards in nanotechnology. Nanotechnology Law & Business. 2004;1(2):185–92.
22. Nanotechnology. A report of the U.S. food and drug administration nanotechnology task force, 25 July 2007. Available at: http://www.fda.gov/ScienceResearch/SpecialTopics/Nanotechnology/NanotechnologyTaskForceReport2007/default.htm. Accessed 25 Jan 2012
23. Considering when an FDA-regulated product involves the application of nanotechnology. Guidance for industry, June 2011. Available at: http://www.fda.gov/ScienceResearch/SpecialTopics/Nanotechnology/ucm257926.htm. Accessed 25 Jan 2012
24. Policy principles for the U.S. decision-making concerning regulation and oversight of applications of nanotechnology and nanomaterials. 9 June 2011. Available at: http://www.whitehouse.gov/blog/2011/06/09/responsible-realization-nanotechnologys-full-potential. Accessed 25 Jan 2012
25. Bawa R. Regulating nanomedicine—can the FDA handle it? Curr Drug Deliv. 2011;8:227–34.
26. TGA fact sheet: sunscreens. 9 Feb 2010. Available at: http://www.tga.gov.au/safety/alerts-medicine-sunscreens-051202.htm. Accessed 25 Jan 2012
27. Policy statement on health Canada's working definition for nanomaterial. 6 Oct 2011. Available at: http://www.hc-sc.gc.ca/sr-sr/pubs/nano/pol-eng.php. Accessed 27 Jan 2012
28. Safe Cosmetics Act of 2011, H.R. 2359. 24 June 2011
29. Regulation (EC) No 1223/2009 of the European Parliament and of the Council of 30 November 2009 on Cosmetic Products. 30 Nov 2009. Available at: http://eur-lex.europa.eu/JOHtml.do?uri=OJ:L:2009:342:SOM:EN:HTML. Accessed 25 Jan 2012
30. The project on emerging nanotechnologies. 2012. Available at: http://www.nanotechproject.org/inventories/consumer/. Accessed 27 Jan 2012
31. Scheufele DA, Nanotechnology and public opinion. 24 Jan 2011. Available at: http://www.nanowerk.com/spotlight/spotid=19819.php. Accessed 25 Jan 2012
32. U.S. public's perception of nanotechnology is up for grabs. 14 Mar 2007. Available at: http://www.azonano.com/news.aspx?newsID=3840. Accessed 25 Jan 2012
33. Scheufele DA, Lewenstein BV. The public and nanotechnology: how citizens make sense of emerging technologies. J Nanopart Res. 2005;7:659–67.

Understanding Public Opinion of Nanotechnology

25

Andrew R. Binder

25.1 Understanding Public Opinion of Nanotechnology

A common refrain among scientists, policymakers, academics, and even consumer advocacy groups is that the public is simply too uninformed about, disinterested in, or otherwise afraid of groundbreaking emerging sciences like nanotechnology. Scapegoats for these problems abound, with blame placed alternately on a fickle media culture that emphasizes sensation over substance, the decline of science sections in newspapers, the use of nanotechnology as a narrative device in science fiction stories, or else simply the decline and stagnation of scientific literacy in the USA. But to what extent do these explanations hold up to empirical scrutiny? To what extent do they help us gain a better understanding of opinion formation about nanotechnology as opposed to distracting us from other important questions of public opinion? This chapter tackles these questions by evaluating empirical data on public opinion of nanotechnology from the past decade.

The chapter focuses on three areas. First, I provide a brief report on the most recent nationwide data on opinion of nanotechnology. This section also covers several problems associated with measuring public opinion about novel and unfamiliar technologies, and ways in which researchers can overcome these problems. The second section outlines prominent theoretical perspectives on how best to understand public opinion of nanotechnology. Finally, I conclude with similarities and differences of opinion in various subgroups and what they mean for both policymaking and the use of nanotechnology in consumer products.

25.1.1 Recent Trends in Awareness, Knowledge, and Opinion

Notable studies on public opinion of nanotechnology point to a set of generalized conclusions about the shape and state of opinion in the USA. These can best be broken down into levels of awareness, knowledge, and assessments of the risks vs. benefits of this emerging technology in the past decade. High interest in science among citizens equates to positive evaluations of nanotechnology, and the early emphasis on nanotechnology in science fiction and popular culture (e.g., in Michael Crichton's 2002 novel *Prey*) did not result in substantial levels of fear

---

Andrew R. Binder (Ph.D., University of Wisconsin-Madison, 2010) is an Assistant Professor in the Department of Communication and Associate Director of the Public Communication of Science and Technology (PCOST) project olina State University. His research focuses on public opinion, the interplay between science and politics, and risk communication.

A.R. Binder, PhD (✉)
Department of Communication, North Carolina State University,103 Winston Hall, Campus Box 8104, Raleigh, NC 27695-8104, USA
e-mail: arbinder@ncsu.edu

A. Nasir et al. (eds.), *Nanotechnology in Dermatology*,
DOI 10.1007/978-1-4614-5034-4_25,

or evaluations of nanotechnology as "pseudoscience" by the public [1]. While the public exhibited low levels of nano-specific knowledge in the early 2000s, their feelings of optimism, high levels of trust in scientists, and lack of worry seemed to translate into positive opinions [2]. Yet, according to a more recent meta-analysis of extant research [3], there remains a high degree of malleability of judgments regarding risks (and benefits) of this emerging technology.

Many of these insights from early studies of public opinion bear out in the most recent national-level survey trends available. These data come from systematic assessments published most recently in the *Science & Engineering Indicators, 2012* [4] and measured in the General Social Survey (GSS) in 2006, 2008, and 2010 [5]. In terms of awareness or familiarity, there is a clear longitudinal trend indicating that the number of people who have heard "nothing at all" about nanotechnology decreased from 2006 to 2010 (see Fig. 25.1), and the only category that seems to have increased substantially was those reporting to have heard "some" about it.

In each year of the GSS respondents were also asked to indicate if the following two statements were true, false, or if they did not know: (1) "Nanotechnology involves manipulating extremely small units of matter, such as individual atoms, in order to produce better materials." and (2) "The properties of nanoscale materials often differ fundamentally and unexpectedly from the properties of the same materials at larger scales." While both of these statements are factually true, only around one-quarter of the public answered the first question correctly; that percentage dropped to 16 % for the second question. Since 2006, there has been no significant change in levels of knowledge among citizens (see Fig. 25.2).

In contrast, there have been changes in terms of the overall benefits (vs. risks) of nanotechnology. Specifically, there has been statistically significant shift from optimistic outlook (i.e., benefits outweigh risks) to a more ambivalent outlook, with an increasing proportion of the public indicating that they cannot give a specific judgment on the matter. These trends are depicted in Fig. 25.3.

But to what degree do these data—the only comprehensive trend data available to date—help us to develop a more nuanced understanding of public opinion? The answer to this question is twofold. First, there are a number of pitfalls associated with measuring public opinion using these four questions and subsequently drawing broad-based conclusions. Stakeholders—including scientists, policymakers, medical researchers, and journalists—should take these disadvantages into account to avoid over-interpreting what appears to be widespread public optimism. The details on these pitfalls lie outside the scope of this chapter, but a recent in-depth exploration of them by Binder and colleagues reveals three important points [6]: (1) Responding to survey questions is a very complex process; (2) Asking global evaluation questions (e.g., "Do the risks of nanotechnology outweigh the benefits?") can result in responses that do not accurately reflect underlying attitudes; and (3) The optimal approach for survey research is to ask a number of context-specific questions, distill those underlying attitudes, and report the resulting balance of opinion in a responsible way to audiences.

Second, in recent years there have been more comprehensive explanatory models of public opinion formation that take into account not only the three recommendations outlined above but also the roles of demographic and communication variables. These models paint a much more nuanced picture that accounts for various "publics" that hide beneath the surface of aggregate data and trends such as those used in the *Science & Engineering Indicators*. The remainder of this chapter will consider each of these factors in detail before turning toward a set of more concrete conclusions about the future of research on public opinion of nanotechnology.

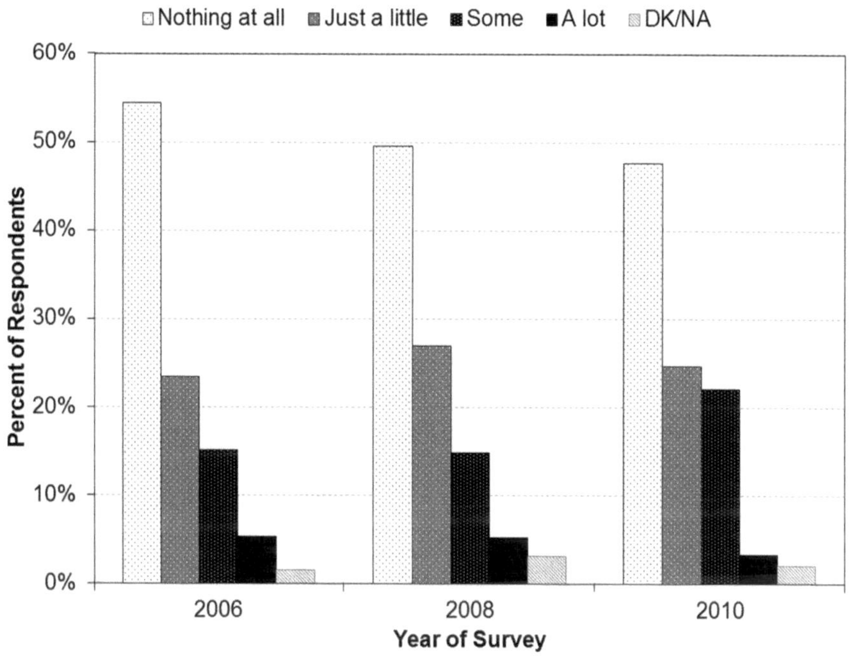

Fig. 25.1 Responses to the question "How much have you heard about nanotechnology?" *Note*: "DK/NA" indicates a response of "don't know" or no answer

25.1.2 Explaining Opinion Formation Toward Nanotechnology

The foregoing review of the state of public opinion research emphasizes two key questions. First, to what extent does *awareness* of nanotechnology tend to translate into explicit *opinions*? Second, how does *knowledge* play a role in opinion formation? In order to address these questions researchers from a variety of disciplines have come to rely on concepts of risk perception and related social dimensions of how people understand emerging technologies.

An assumption of the first theoretical approach is that individuals' perception of risk is an empirical phenomenon distinct from more traditional quantitative assessments of exposure to various societal hazards. The work of Paul Slovic within the psychometric paradigm [7] allowed for a more comprehensive understanding of how laypersons view societal risks. In summary, the psychometric approach revealed perceptions along two dimensions: dread risk (the extent to which the risk is feared) and unknown risk (the extent to which the hazard is known to science). According to Slovic's original 1987 analysis [8], examples classified as high on both these dimensions are some of the more controversial hazards, such as nuclear reactor accidents and DDT. Pharmaceutical products (e.g., valium, oral contraceptives) tend to be viewed as lower on the dread factor, yet also having consequences unknown to science. According to recent data, nanoparticles are viewed similarly to this second type of hazards [9].

A second insight from this literature holds that public risk perceptions do not correspond to quantitative risk assessments due to complex social relationships. The social amplification of risk framework [10] attempts to put many of these variables together in a comprehensive conceptual diagram. Applying the framework to a variety of emerging technologies, empirical case studies have built a broad base of knowledge on how those technologies come to be perceived as having substantial negative consequences for society, or else ignored or downplayed almost entirely, based upon social and cultural factors.

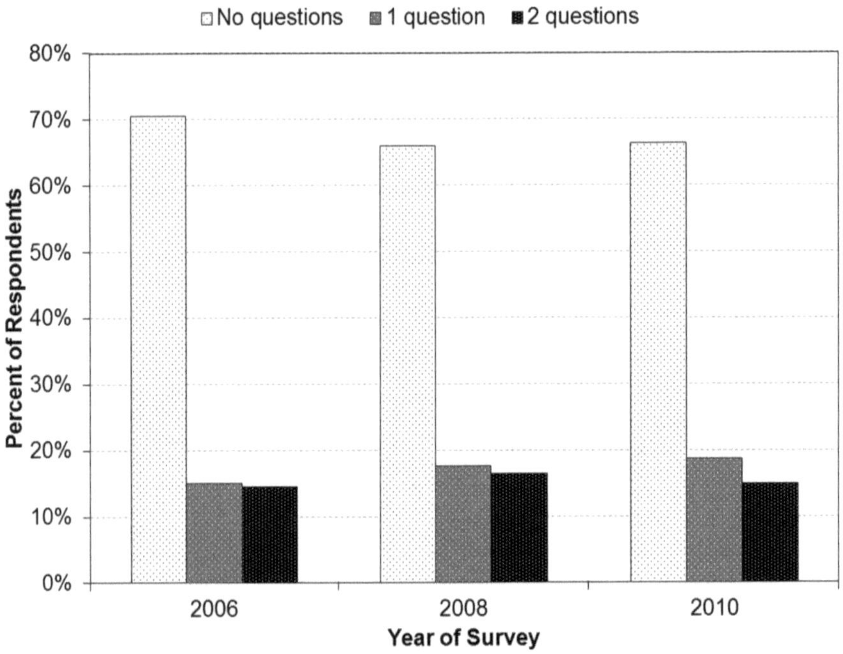

Fig. 25.2 Percentage of GSS respondents who correctly answered factual knowledge questions about nanotechnology

Based on these predominant theoretical frameworks, research on public opinion of nanotechnology has come to focus on two broad areas: (1) the development of informed opinion through processes of deliberation outlined by normative democratic theorists, and (2) the implications of perceived risk and communication for processes of opinion formation based on social and cognitive psychology. The first group of scholars has investigated what public opinion of nanotechnology would look like under ideal democratic circumstances, as pioneered in the field of political science [11]. These circumstances include full access to relevant information, the ability and motivation to participate in public discussions, and the expression of their subsequent opinion. Gauging informed public opinion has been a considerable focus of federally funded research from the National Science Foundation and other agencies. A good example of this approach was the National Citizens' Technology Forum (NCTF), which brought together small groups of ordinary citizens to discuss issues associated with nanotechnology. The deliberation process is designed to create informed opinions, and so it should come as no surprise that this is precisely what occurred: whereas a majority (55 %) of participants reported no opinion on the risk vs. benefits relationship for nanotechnology *before* deliberation, only a small fraction of participants (3 %) did so afterwards [12]. More strikingly, the same results revealed that the share of participants reporting that benefits outweigh risks doubled (from 23 to 46 %) and participants reporting that risks outweigh benefits increased nearly fivefold (from 6 to 28 %). While deliberation appears to create informed opinions, and those informed opinions seem to lean in a positive direction, more specific questions about using nanotechnology for a variety of human enhancements also revealed substantial variability in these informed opinions. These findings mirror those from other deliberative exercises [13]. In addition to pure shifts in opinion, deliberative exercises such as consensus conferences can improve self-efficacy and confidence in one's own abilities to confront complexities of emerging science and technology [12, 14].

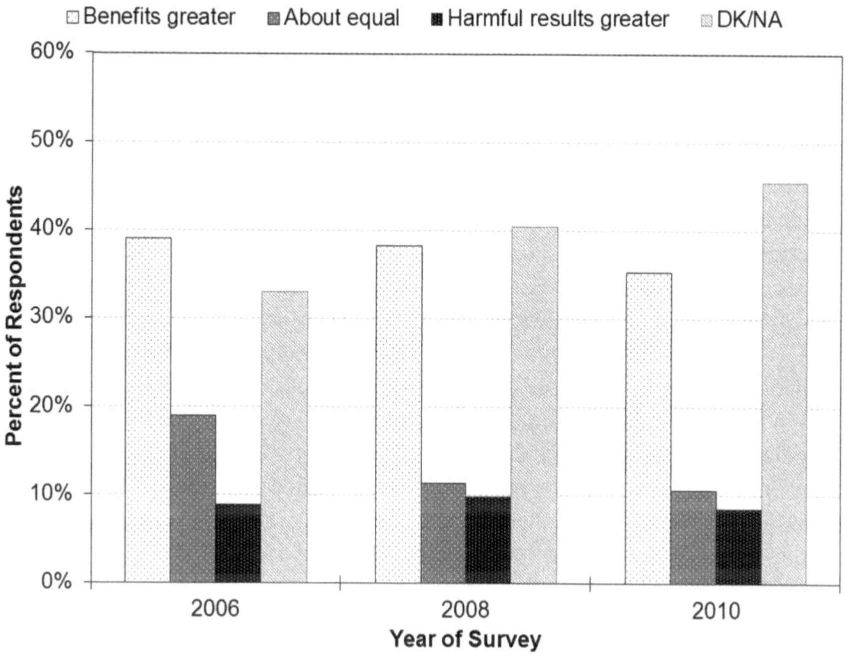

Fig. 25.3 Responses to the question "Do you think the benefits of nanotechnology will outweigh the harmful results or the harmful results will outweigh the benefits?" *Note*: "DK/NA" indicates a response of "don't know" or no answer. Response options "About equal" and "DK/NA" were unprompted in by the survey interviewer

In spite of these largely optimistic results, there are shortcomings to these types of exercises for a variety of reasons. First, their utility for informing policymakers and stakeholders is not self-evident [12]. Second, and more importantly, deliberative exercises are ideally aimed at citizens who are wholly uninformed and hold no opinion about nanotechnology. However, this imagined, ideal citizen participant in deliberative exercises does not match up with actual participants, who tend to be more highly educated, earning a higher income, and more politically liberal than most US citizens [15]. And while there may be a variety of means for sidestepping these apparent shortcomings of participant recruitment [16], the deliberative democracy paradigm nevertheless suffers from a variety of other drawbacks, including trade-offs in the ecological and external validity of their design and the sheer cost of sponsoring them.

In contrast to investigations rooted in the deliberative polling model, the second group of scholars has investigated what public opinion of nanotechnology looks like as it unfolds in real-world contexts and circumstances. This approach has tackled our two key questions (To what extent does awareness translate to opinion? What role does knowledge play?) by looking at citizen decision-making and information they encounter in everyday life. Research from social and cognitive psychology—as well as the fields of risk and science communication—has emphasized the concepts of low-information rationality and what that means for how people understand information from the dominant source on issues like nanotechnology: the mass media. In making assessments and forming opinions, individuals rely on heuristics and cognitive shortcuts when they encounter new information, such as the availability heuristic and the affect heuristic [17]. The former, for example, makes individuals more likely to rely on what first comes to mind when they hear about nanotechnology for the first time. The latter will be more likely to play a role depending on the emotional reaction that people experience during their first and subsequent encounters with information about nanotechnology.

Heuristics, typically used in psychological models of information processing, also have implications at more abstract levels of analysis. Following from research in sociology on "frames," or dominant interpretations of social action [18], scholars have outlined a number of implications for science. Chief among these is the idea that values and life experiences can shape people's views of the science-society nexus [19]. Media framing plays a role as well, as a reflection of public discourse about science. Frames in media can emphasize very different interpretations of an emerging science, and ultimately interact with citizens' own perspectives. Indeed, one person's understanding of nanotechnology based upon a "path to progress" may compete in the public sphere with another person's notion of nanotechnology as a "Pandora's Box" waiting to unleash negative impacts on society at large [20]. As a result, competing definitions used by policymakers and journalists, who have substantial control over how issues are portrayed in the media, can directly influence how audiences respond to these messages [21].

The evidence from this media-effects perspective has led to a more coherent framework of how audiences respond to media messages. Message frames emphasizing risks and benefits of nanotechnology can have an impact on public evaluations more so than frames based on general ideas about science [22]. Frames tend to have this influence on public attitudes regardless of how complex these messages are. For example, messages that are framed within a specific context have an impact dependent upon the dominant frames in the minds of various audience members, and this relationship holds regardless of whether or not that information is provided in conjunction with scientific evidence (e.g., "scientific studies show that…"). As a result, scholars have concluded that adding facts to frames do not necessarily have a differential (assumed, in many cases, to be positive among scientists) on audience perceptions or opinions [23]. These frames with which audiences approach science messages can also vary in accordance with visual images that they encounter. And since visual images of science and technology tend to be portrayed in polarized ways (either as "wondrous" or "demonical"), exposure to them can result in ambiguity embedded in individual interpretations of nanotechnology [24].

In terms of message attributes, emphasizing benefits do not necessarily translate into shifts on risk perception because other attributes—including details on the social context (e.g., the size and type of company using nanotechnology in their industry)—also impact perceptions [25]. Evaluations of risks and benefits also depend, to some degree, on preexisting cognitive associations that people have with nanotechnology in different domains of its applications. When risk perceptions are low, cognitive associations related to the medical field and machines/computers correspond to elevated ratings of the usefulness of nanotechnology [26].

As these studies have shown, specific mass media messages or classes of messages can have a contingent impact on public perception of nanotechnology. On balance, citizens tend to interpret nanotechnology positively, but this tendency could change depending upon (1) the distribution of available interpretations within the population, (2) the dominant application context, and (3) the types of people conveying those messages (e.g., university vs. industry scientists). Based on these same considerations, other research has focused on describing what media coverage looks like more broadly.

Since mass media content is, very often, a reaction to events in the outside world, longitudinal trends can be important. The emphasis within business-related news, for example, has shifted back and forth between benefits outweighing risks (2003–2004, 2008–2009) and risks outweighing benefits (2005–2007) within the past decade, with the peak volume of coverage occurring in 2004–2006 [27]. Corroborative evidence has demonstrated an overall three-to-one ratio of articles emphasizing that benefits outweigh risks compared to risks outweighing benefits, although half of articles do not clearly suggest this one way or another [28].

In that same time period, there has been an increasing overall focus on business and health aspects of nanotechnology. "Risk" as an attribute of this emerging technology appeared rather late in the first 20 years of news mentioning nanotechnology in the USA, with several thematic elements largely ignored in print news (e.g., regulation, environmental implications, and uncertainty) [29]. "Progress" was a prominent theme in early media coverage, with little discussion of responsibility for any societal impacts [30], whereas more recently themes of "runaway technology" and health risks were mentioned most often [31]. Articles emphasizing ethical, legal, and social implications (ELSI) of nanotechnology tended to be longer, appear in main news sections, and were relatively less likely to appear in US newspapers compared to non-US newspapers [28].

Based upon what researchers have concluded about the nature of nanotechnology media coverage—that depictions of nanotechnology are likely to be infrequent but predominantly positive—persistent public ambivalence toward nanotechnology may be puzzling. On the one hand, if media coverage appears mostly positive, then public attitudes should also be positive rather than increasingly ambivalent. On the other hand, perhaps due to the prominent themes in media coverage, citizens' benefit perceptions have been given greater weight in their decisions to support or oppose nanotechnology generally, controlling for the extent to which they were exposed to this media coverage [32]. The extent to which individuals are exposed to information on nanotechnology appears to play a crucial role in opinion formation.

In spite of differing assumptions, the two schools of thought outlined here (deliberative, informed opinion vs. heuristic, less informed opinion) both emphasize how very different subgroups of a population can have profound impacts on societal dynamics of public opinion. Regardless of whether these are fully informed ideal "scientific" citizens or individuals with little knowledge of or interest in science, the decisions they reach are partially due to levels of expertise, access to information, and values or world views. The final section of the chapter considers these subgroups of the overall population.

25.1.3 Considering Different Types of "Publics"

Most recent research efforts in the area of public opinion of nanotechnology have focused on (1) what scholars mean by the "public" in "public opinion," and (2) how relaxing the assumption that "the public" conforms to a single, homogeneous entity can build a better understanding of public opinion. One intuitive way of dividing "the public" into subgroups is to consider experts—scientists, researchers, and medical practitioners—as members of the public and to compare their views with those of lay citizens. Contrary to the knowledge-deficit hypothesis—that differing views on technology and risk are due to experts' higher levels of science literacy and knowledge—empirical data has demonstrated that scientists, like average citizens, rely on cognitive shortcuts in order to make judgments about nanotechnology [33]. As mentioned above, lay citizens tend to rely on perceived benefits when forming an opinion about nanotechnology [32]; in contrast, scientists rely much more on the perceived risks, and the subsequent need for regulation to oversee those risks, when reporting an opinion. At the same time, on balance nanoscientists view more benefits than risks of nanotechnology compared to the public, but they also see environmental and health issues as most in need of further attention [34]. This relative difference—that lay people view more potential risks than experts do—can be linked to their trust in governmental agencies [35]. Experts view public health and environmental issues most urgent in need of attention from policymakers and regulation [36], whereas policy considerations do not usually spring immediately to mind for lay citizens.

The expert–lay distinction is perhaps the most intuitive way to think of different types of publics, but there is also a wealth of research providing evidence regarding how different groups within the "lay" category view nanotechnology. Indeed, it is this area of research that has gone furthest in explaining the seeming tension between positive/ambivalent global evaluations of nanotechnology (seen in data from the GSS) and the type of public discourse emphasizing positive impacts in media coverage. One key contribution relates to social structure, and especially to socioeconomic status. Because of where nanotechnology tends to be discussed the most—in newspapers, and in particular lengthy articles in front page or business sections of newspapers—varying levels of socioeconomic status among citizens have a direct impact on their level of exposure to this information. Those individuals with a higher level of income and education are more likely (1) to be exposed initially to this information and (2) to learn from that exposure in the long term [37]. The resulting knowledge gap—the more affluent gain more in terms of informational resources, whereas the less affluent do not—is a pressing issue for outreach, engagement, and information dissemination exercises.

In addition to social structural considerations, considerable attention has been paid to the role of values in shaping public attitudes. Even for those most knowledgeable about nanotechnology (and who therefore tend to be slightly more favorable toward it), this relationship does not hold among individuals who are highly religious [38, 39]. Perhaps even more than issues such as stem cell research and agricultural biotechnology, nanotechnology may be seen as interfering with the natural order of things or playing God—especially with potential human-enhancement applications—among these highly religious groups of people. Knowledge of nanotechnology therefore does little to mitigate the resulting cognitive dissonance between deeply held beliefs about the nature of the world and the implications of those beliefs for nano-applications. It remains to be seen if this viewpoint will have a long-term impact on a politicization of nanotechnology and subsequent policies. Studies investigating "cultural cognition" of nanotechnology suggest that this might be the case in the future, as public evaluations are dependent upon the role that people see government playing in society [40].

As these broad areas of research demonstrate, the problem of understanding public opinion of nanotechnology is perhaps best approached by relaxing embedded assumptions regarding what "the public" means and instead subscribing to the idea of multiple, heterogeneous publics found within the broader population. The consequences of differing levels of expertise, socioeconomic status, and values or world views go a long way in explaining the differential impact of risk or benefit perceptions, access to and learning from mass media, and deeply held beliefs about the world and society and how those have a direct effect on attitudes and opinions of nanotechnology and other emerging scientific innovations.

25.1.4 Conclusions: Citizens, Consumer Products, and Nanotechnology

This chapter began by introducing substantial variability in recent public opinion in the general public. By outlining the key factors that play a role in shaping these dimensions of public opinion, this chapter has attempted to provide a more sound and coherent framework for understanding these dynamics and their implications for academics, policymakers, and practitioners. Several strands of ongoing and future research are already emerging based upon such implications. First, tackling questions of access to information and participation in deliberation exercises are of the utmost importance, and those who sponsor or lead such events would do well to heed the recommendations provided in a detailed typology of engagement mechanisms [41], which has insights for improving two-way communication flows between event sponsors and participants in a wide variety of contexts. A recent study on "upstream engagement" is one example of an approach rooted in increasing widespread public interest in learning more about nanotechnology through everyday information-seeking activities [42].

Second, as nanotechnology applications and products appear in consumer markets, the foregoing overview of public opinion can be useful in anticipating how different types of consumers may react to those products. Of course, this is not intended to suggest that research can or should be used to control media discourse, downplay potential public–relations mistakes, or distract the public from real and consequential hazards, should they arise. Rather, this conclusion is aimed at encouraging stakeholders to gain a better understanding of how individuals come to form opinions about these potential products and anticipate the boundaries of acceptance in making choices about which products to pursue for introduction in the marketplace. In this area, too, researchers have already provided partial answers to these types of questions. For example, the perceived risk of nanoparticles, relative to other potential societal and application hazards, is relatively low—the level of risk falls well below the risks associated with cigarette smoking, nuclear waste, stress, motor vehicle accidents, and sun tanning [9]. These data do not show, however, how the perception of risk changes once nanoparticles are introduced to consumer products like sunscreens. For perceptions of current (or possible) consumer products like these, the best corollary in peer-reviewed research concerns the nexus between food and nanotechnology [43], with the initial conclusion being that consumers want to know when and if they will be exposed to nanotechnology in order to ward off both real and imagined risks associated with it. Based in part on a solid understanding of public opinion formation and its consequences, the value of asking and answering these types of consumer-oriented questions early in the product-development process cannot be overstated.

References

1. Bainbridge WS. Public attitudes toward nanotechnology. J Nanopart Res. 2002;4(6):561–70.
2. Cobb MD, Macoubrie J. Public perceptions about nanotechnology: risks, benefits and trust. J Nanopart Res. 2004;6(4):395–405.
3. Satterfield T, Kandlikar M, Beaudrie CEH, Conti J, Harthorn BH. Anticipating the perceived risk of nanotechnologies. Nat Nanotechnol. 2009;4(11):752–8.
4. National Science Board. Science and Engineering Indicators 2012. Arlington, VA: National Science Foundation (NSB 12-01); 2012.
5. Smith TW, Marsden P, Hout M, Kim J. General social surveys, 1972–2010 [machine-readable data file]. Chicago: National Opinion Research Center. Storrs, CT: The Roper Center for Public Opinion Research, University of Connecticut; 2011.
6. Binder AR, Cacciatore MA, Scheufele DA, Shaw BR, Corley EA. Measuring risk/benefit perceptions of emerging technologies and their potential impact on communication of public opinion toward science. Public Understanding of Science. doi: 10.1177/0963662510390159.
7. Slovic P. Perception of risk. Sterlin, VA: Earthscan; 2000.
8. Slovic P. Perception of risk. Science. 1987;236:280.
9. Berube DM, Cummings CL, Frith JH, Binder AR, Oldendick R. Comparing nanoparticle risk perceptions to other known EHS risks. J Nanopart Res. 2011;13(8):3089–99.
10. Kasperson RE, Renn O, Slovic P, et al. The social amplification of risk: a conceptual framework. Risk Anal. 1988;8(2):177–87.
11. Fishkin JS. Democracy and deliberation. New Haven, CT: Yale University Press; 1991.
12. Cobb MD. Creating informed public opinion: citizen deliberation about nanotechnologies for human enhancements. J Nanopart Res. 2011;13(4):1533–48.
13. Pidgeon N, Harthorn BH, Bryant K, Rogers-Hayden T. Deliberating the risks of nanotechnologies for energy and health applications in the United States and United Kingdom. Nat Nanotechnol. 2009;4(2):95–8.
14. Powell M, Kleinman DL. Building citizen capacities for participation in nanotechnology decision-making: the democratic virtues of the consensus conference model. Public Underst Sci. 2008;17(3):329–48.
15. Powell M, Colin M, Kleinman DL, Delborne J, Anderson A. Imagining ordinary citizens? Conceptualized and actual participants for deliberations on emerging technologies. Sci Cult. 2011;20(1):37–70.
16. Hamlett PW, Cobb MD. Potential solutions to public deliberation problems: structured deliberations and polarization cascades. Policy Stud J. 2006;34(4): |629–48.
17. Kahneman D. Maps of bounded rationality: a perspective on intuitive judgment and choice. In: Frängsmyr T, editor. Les Prix Nobel: The Nobel Prizes 2002. Stockholm, Sweden: Nobel Foundation; 2003. p. 449–89.
18. Goffman E. Frame analysis: an essay on the organization of experience. New York, NY: Harper & Row; 1974.
19. Nisbet MC, Scheufele DA. What's next for science communication? Promising directions and lingering distractions. Am J Bot. 2009;96(10):1767–78.
20. Nisbet MC, Scheufele DA. The future of public engagement. Scientist. 2007;21(10):38–44.

21. Allan S, Anderson A, Petersen A. Framing risk: nanotechnologies in the news. J Risk Res. 2010;13(1):29–44.
22. Cobb MD. Framing effects on public opinion about nanotechnology. Sci Commun. 2005;27(2):221–39.
23. Druckman JN, Bolsen T. Framing, motivated reasoning, and opinions about emergent technologies. J Commun. 2011;61(4):659–88.
24. Landau J, Groscurth CR, Wright L, Condit CM. Visualizing nanotechnology: the impact of visual images on lay American audience associations with nanotechnology. Public Underst Sci. 2009;18(3):325–37.
25. Schutz H, Wiedemann PM. Framing effects on risk perception of nanotechnology. Public Underst Sci. 2008;17(3):369–79.
26. Cacciatore MA, Scheufele DA, Corley EA. From enabling technology to applications: the evolution of risk perceptions about nanotechnology. Public Underst Sci. 2011;20(3):385–404.
27. Dudo A, Choi DH, Scheufele DA. Food nanotechnology in the news. Coverage patterns and thematic emphases during the last decade. Appetite. 2011;56(1):78–89.
28. Stephens LF. News narratives about nano S&T in major US and non-US newspapers. Sci Commun. 2005;27(2):175–99.
29. Dudo A, Dunwoody S, Scheufele DA. The emergence of nano news: tracking thematic trends and changes in U.S. Newspaper coverage of nanotechnology. Journal Mass Commun Q. 2011;88(1):55–75.
30. Weaver DA, Lively E, Bimber B. Searching for a frame news media tell the story of technological progress, risk, and regulation. Sci Commun. 2009;31(2):139–66.
31. Friedman SM, Egolf BP. A longitudinal study of newspaper and wire service coverage of nanotechnology risks. Risk Anal. 2011;31(11):1701–17.
32. Scheufele DA, Lewenstein BV. The public and nanotechnology: how citizens make sense of emerging technologies. J Nanopart Res. 2005;7(6):659–67.
33. Corley EA, Scheufele DA, Hu Q. Of risks and regulations: how leading US nanoscientists form policy stances about nanotechnology. J Nanopart Res. 2009;11(7):1573–85.
34. Scheufele DA, Corley EA, Dunwoody S, Shih TJ, Hillback ED, Guston DH. Scientists worry about some risks more than the public. Nat Nanotechnol. 2007;2(12):732–4.
35. Siegrist M, Keller C, Kastenholz H, Frey S, Wiek A. Laypeople's and experts' perception of nanotechnology hazards. Risk Anal. 2007;27(1):59–69.
36. Besley JC, Kramer VL, Priest SH. Expert opinion on nanotechnology: risks, benefits, and regulation. J Nanopart Res. 2008;10(4):549–58.
37. Corley EA, Scheufele DA. Outreach going wrong? Scientist. 2010;24(1):22.
38. Brossard D, Scheufele DA, Kim E, Lewenstein BV. Religiosity as a perceptual filter: examining processes of opinion formation about nanotechnology. Public Underst Sci. 2009;18(5):546–58.
39. Scheufele DA, Corley EA, Shih TJ, Dalrymple KE, Ho SS. Religious beliefs and public attitudes toward nanotechnology in Europe and the United States. Nat Nanotechnol. 2009;4(2):91–4.
40. Kahan DM, Braman D, Slovic P, Gastil J, Cohen G. Cultural cognition of the risks and benefits of nanotechnology. Nat Nanotechnol. 2009;4(2):87–90.
41. Rowe G, Frewer LJ. A typology of public engagement mechanisms. Sci Technol Human Values. 2005;30(2):251–90.
42. Xenos MA, Becker AB, Anderson AA, Brossard D, Scheufele DA. Stimulating upstream engagement: an experimental study of nanotechnology information seeking. Soc Sci Q. 2011;92(5):1191–214.
43. Siegrist M, Keller C. Labeling of nanotechnology consumer products can influence risk and benefit perceptions. Risk Anal. 2011;31(11):1762–9.

Index

A. Nasir et al. (eds.), *Nanotechnology in Dermatology*,
DOI 10.1007/978-1-4614-5034-4,

O

P

MIX
Papier aus verantwortungsvollen Quellen
Paper from responsible sources
FSC® C105338

If you have any concerns about our products,
you can contact us on
ProductSafety@springernature.com

In case Publisher is established outside the EU,
the EU authorized representative is:
Springer Nature Customer Service Center GmbH
Europaplatz 3, 69115 Heidelberg, Germany

Printed by Libri Plureos GmbH
in Hamburg, Germany